Neutrino Astrophysics

Neutrino Astrophysics

John N. Bahcall

Institute for Advanced Study, Princeton, New Jersey

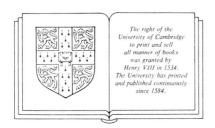

The right of the
University of Cambridge
to print and sell
all manner of books
was granted by
Henry VIII in 1534.
The University has printed
and published continuously
since 1584.

Cambridge University Press

Cambridge
New York   New Rochelle   Melbourne   Sydney

Published by the Press Syndicate of the University of Cambridge
The Pitt Building, Trumpington Street, Cambridge CB2 1RP
32 East 57th Street, New York, NY 10022, USA
10 Stamford Road, Oakleigh, Melbourne 3166, Australia

First Published 1989

Printed in the United States of America

Library of Congress Cataloging-in-Publication Data

Bahcall, John N.
   Neutrino astrophysics / John N. Bahcall
      p.   cm.
   ISBN 0-521-35113-8.   ISBN 0-521-37975-X (pbk.)
   1. Neutrino astrophysics. I. Title.
QB464.2.B34 1989   88-39165
523.01'97215–dc19   CIP

To Neta, Safi, Dan, Orli,
and my mother, Mildred

Contents

Tables

Figures

Preface

The contents

This book describes achievements in neutrino astrophysics and summarizes the major unsolved problems. The purpose is to share the fun of figuring out some of nature's puzzles.

Different aspects of neutrino astrophysics are presented in semi-autonomous chapters, each of which describes the main ideas and results of a separate topic without requiring detailed knowledge of what is in other chapters. In order to facilitate random sampling of individual chapters, a few of the key results are presented in more than one place. Derivations are not given, even when new theoretical results are presented, since this would have increased the length of the book to an unusable size. I have not hesitated, however, to express my own views of what is important or promising.

Some parts of the book are accessible to undergraduate students in physics and astronomy and some parts require, for their full appreciation, graduate study in one of the physical sciences. The essential aspects of all the chapters should be intelligible to a physics student who has had an introductory course in nuclear physics. Key technical terms are indicated by boldface type immediately prior to their definition. For a few of the topics, such as stellar evolution or nuclear fusion reactions in stars, the treatment is pedagogical and introductory. For other subjects, such as weak interaction cross sections or experimental techniques, only state-of-the-art results are presented. Neutrino oscillations are treated in detail because of the beauty and possible importance of recent theoretical developments.

Readers should shop around among the different chapters to find out what is most interesting for themselves.

Students and experienced researchers will find problems described in this book to which their expertise can be applied. Hopefully, some readers with special skills will find better ways of solving some of the problems.

Chapter 1 is entitled "Overview" and presents an informal summary of where we stand and where we may be heading in solar neutrino research. The style and content of this chapter are appropriate for a general colloquium for students and research workers in the physical sciences. The closing section of this chapter contains informal answers to questions that I am often asked about solar neutrinos.

Chapters 2 to 9 summarize the basis for making predictions for solar neutrino experiments. The topics covered include the theory of stellar evolution and solar models, nuclear fusion reactions, neutrino cross sections, estimates of the total uncertainties in the calculated neutrino fluxes, and suggested explanations that go beyond the standard model of electroweak interactions.

Chapters 10 to 14 describe the solar neutrino experiments that are in progress or are being developed actively.

Chapter 15 summarizes the theory of stellar collapse and describes the use of neutrino telescopes to study the formation of neutron stars. The bright supernova in the Large Magellanic Cloud, SN1987A, provides the observational focus for this discussion.

Chapter 16 describes the prospects for progress in the next decade and the main unsolved problems.

The Appendix is a reprint of an informal article on the development of the solar neutrino problem that was originally written by R. Davis, Jr. and myself for the W.A. Fowler Festschrift.

In order to make it easier for readers to find their way in the research literature, the reference section near the end of the book contains a list of some of the most accessible research papers in which further details are presented. The index, the last section of the book, is extensive and is designed to help students and researchers to locate discussions of specific topics, which often appear in several different contexts and chapters. The bibliographical notes, which are included at the end of each chapter, are provided for those who like to read original papers in order to find out how we got to where we are now. There is a special thrill in reading a paper in which a new result or idea was first presented.

To obtain a more detailed picture of what is included in this book, read the summary paragraphs at the beginning of each chapter. Taken together, the summary paragraphs present the content of the book in a succinct form, a sort of book within a book. The overview presented in Chapter 1 and the summaries are sufficient to be used by themselves as an introduction to neutrino astrophysics in undergraduate physics or astronomy courses.

The fun of neutrino astrophysics consists in relating fundamental questions in one field to detailed expertise in another. Hopefully, you will experience some of this pleasure by skipping between what you know well and what you do not know at all.

Look first at those chapters in which you have an *a priori* interest or expertise. Perhaps you already know something which can advance the subject. Nuclear physicists may wish to concentrate initially on Chapter 3 (nuclear fusion reactions) and Chapter 7 (uncertainties in predicted solar neutrino fluxes). Nuclear and particle theorists may wish to peruse Chapter 8 (neutrino cross sections). There are many facets of weak interactions that can be tested most directly by using the large separation, 10^8 km, that exists between source and target in solar neutrino experiments. Physicists who are interested in the weak interactions will want to look at Chapter 9 (particle physics explanations). Experimentalists with experience in nuclear or particle physics will want to look at Chapters 13 and 14 (which discuss applications of modern electronic detectors to solar neutrinos), while chemists will have a special interest in Chapters 10 to 12, which describe detectors in which chemistry plays a major role. Astronomers will find relevant the discussions in Chapter 2 (stellar evolution), Chapters 4 and 5 (standard and nonstandard solar models), and Chapter 15 (stellar collapses). Researchers in any of the related fields (physics, astronomy, chemistry, or geophysics) may find useful the summary in Chapter 16 of problems that need solving.

The solar neutrino problem is the result of work by hundreds of individuals who have made essential contributions to different aspects of the subject. Many investigations were necessary to establish that there is a problem, to define its implications, and to identify the most likely directions of future progress. The people who did this work know who they are and can take pride in what they have accomplished. To give here a realistic description of the individual contributions of the most essential works in each of these

areas would require a book in itself. Instead, I have made no attempt to assign credit properly since that would lengthen greatly
this book without adding much of immediate use to the student or
active researcher.

There are three necessary exceptions to this rule. The subject of
solar neutrinos would not exist without Ray Davis. For two decades,
his ^{37}Cl experiment has plagued, inspired, and confounded scientists
in different fields. I am indebted to Ray for a quarter of a century
of stimulation and joint effort. Willy Fowler, who first introduced
me to nuclear astrophysics, has been a constant source of wisdom
and guidance, as well as fun. The sound experimental basis for
nuclear astrophysics, essential to the understanding of solar neutrinos, is largely the product of Willy's insight and enthusiasm. Roger
Ulrich has been an esteemed collaborator on solar models for two
decades. He has always combined the highest standards of numerical precision with a dedication to incorporating the best possible
physical description, whatever the cost in human and computer-
time.

Many friends have read and corrected chapters in this book. The
science and presentation have been much improved by their generous
efforts. The text of the book was prepared in TEX by B. Schuver with
great skill, dedication, and aesthetic judgment, as well as constant
cheerfulness, all of which made the writing a pleasure.

The title

Why is the title of this book *Neutrino Astrophysics* instead of *Solar Neutrinos*? The title is more of an expression of hope than a
description of the book's contents; most of the detailed discussion
relates to solar neutrinos. However, many of the techniques of solar
neutrino astronomy are applicable to observing neutrinos from more
distant sources. To date, the only astronomical source beside the
Sun to be observed in neutrinos is supernova SN1987A, from which
about 20 neutrinos were detected. About 10^3 neutrinos have been
detected from the Sun over the past quarter century. The extension of solar neutrino astronomy to more distant sources is difficult,
but observations of SN1987A provided an historic beginning. As the
experimental techniques described in this book are developed and
made more sensitive, and as new ideas (theoretical and experimental) are explored, the observational horizon of neutrino astrophysics

may grow and the successor to this book may take on a different character, perhaps in a time as short as one or two decades.

Neutrino cooling of ordinary stars, other than the Sun, is not discussed in this book because there is no known way to detect this low intensity neutrino background. The flux of neutrinos from the cooling of stars is faint for the same reason that the sky is dark at night – all other stars are much farther away than the Sun. Similarly, neutrinos from the Big Bang are not considered here. They are numerous but of too low an energy to permit detection with existing techniques. Neutrinos from stellar cooling and from the Big Bang are important to physics and astronomy. Fortunately, they have been discussed in a number of excellent reviews.

Extremely high-energy neutrinos from known X-ray sources in the Galaxy or from luminous extragalactic sources are also not discussed since the production mechanisms are uncertain.

Neutrinos emitted in stellar collapses can be detected and are the subject of many detailed calculations. Chapter 15 discusses neutrinos from stellar collapse.

Why now?

The subject of solar neutrinos is undergoing a revolution, from a one-experiment to a multi-experiment field. Related theoretical work, both in astrophysics and in particle physics, is becoming more widespread. When things are changing so rapidly, the reader may well ask: Is this the right time for a book on neutrino astrophysics?

I think the answer is "yes." This book is intended to make the expansion of research more meaningful by providing a unified discussion of the field.

The subject of solar neutrinos has many seemingly independent aspects, both theoretical (involving nuclear, atomic and particle physics, as well as stellar evolution) and experimental (involving chemistry, nuclear physics, geochemistry, and astronomy). In this book, results from all of these disciplines are combined. Although there have been many conference discussions (including several conference proceedings dedicated to solar neutrinos) and thousands of papers written about solar neutrinos, no treatment relating the different aspects of the subject has previously appeared.

The theme

The theme of this book is the interplay of theory and observations of solar and supernova neutrinos. The intertwining of models and measurements is built into the way we express ourselves about neutrino astrophysics. For example, probably only a very few readers could say before reading Chapter 10 how many neutrino captures per day are observed in the ^{37}Cl experiment, while many more could state the approximate number of SNUs (the Solar Neutrino Unit) that are predicted and detected. Exemplifying this connection between theory and observation, a SNU (pronounced like "snew") was introduced as a convenient way of characterizing both experimental rates and theoretical predictions. A SNU is the product of solar neutrino fluxes (measured or calculated) and calculated neutrino interaction cross sections.[†]

Historically, the "solar neutrino problem" arose as the difference (in SNUs) between the observed and predicted capture rate in the ^{37}Cl experiment. The experimental results are exciting because they confirm or reject expectations from accepted theoretical models. For example, there would not be a solar neutrino problem if the theoretical results could lie anywhere between 1 and 10 SNU, bracketing the experimental value of 2 SNU. On the other hand, much of the interest in accurate models of the solar interior has been stimulated by the possibility of direct tests by solar neutrino experiments.

The relation between theory and observation is different in some areas of astronomy and physics; discoveries having fundamental implications have been made without a pre-existing conceptual model. In modern astronomy, pulsars, quasars, and X-ray sources are examples of serendipitous discoveries of great theoretical importance. In physics, the discoveries of the muon, of the tau particle, and of high-temperature conductivity show how experimental results can change our view of the relevant physics without pre-existing theoretical calculations. In neutrino astrophysics, theory and experiment are linked, for better or for worse.

The theoretical sections of this book focus on testable predictions and the experimental sections concentrate on the potential of different measurements for confronting theoretical models.

Princeton, New Jersey   John N. Bahcall
December 1988

[†] *Historical footnote.* The unit, SNU, pronounced "snew," was introduced as a pun which fortunately escaped the editor of Physical Review Letters [see footnote 10 of Bahcall, J.N. (1969) *Phys. Rev. Lett.* **23**, 251]. Numerically, a SNU is 10^{-36} events per target atom per second. This unit is most useful for characterizing results relating to detectors, such as the radiochemical ^{37}Cl and ^{71}Ga detectors, which have fixed energy thresholds.

1

Overview

Summary

How does the Sun shine? Does the neutrino have a mass? Can solar neutrinos be used to test the theory of stellar evolution? To explore the unification of strong, weak, and electromagnetic forces? These are some of the questions that motivate the study of neutrino astronomy.

A **neutrino** is a weakly interacting particle that travels at essentially the speed of light and has an intrinsic angular momentum of $1/2$ unit ($\hbar/2$). Neutrinos are produced on Earth by natural radioactivity, by nuclear reactors, and by high-energy accelerators. In the Sun, neutrinos are produced by weak interactions that occur during nuclear fusion. There are three known types of neutrinos, each associated with a massive **lepton** that experiences weak, electromagnetic, and gravitational forces, but not strong interactions. The known leptons are electrons, muons, and taus (in increasing order of their rest masses).

Neutrino astronomy is interesting for the same reason it is difficult. Because neutrinos only interact weakly with matter, they can reach us from otherwise inaccessible regions where photons, the traditional messengers of astronomy, are trapped. Hence, with neutrinos we can look inside stars and examine directly energetic physical processes that occur only in stellar interiors. We can study the interior of the Sun or the core of a collapsing star as it produces a supernova.

Large detectors, typically hundreds or thousands of tons of material, are required to observe astronomical neutrinos. These detectors must be placed deep underground to avoid confusing the rare astronomical neutrino events with the background interactions caused by cosmic rays and their secondary particles, which are relatively common near the surface of the Earth.

For two decades, the only operating solar neutrino experiment yielded results in conflict with the most accurate theoretical calculations. This conflict between theory and observation, which has recently been confirmed by a new experiment, is known as the **solar neutrino problem**.

More is known about the Sun than about any other star and the calculations of neutrino emission from the solar interior can be done with relatively high precision. Hence, the solar neutrino discrepancy has puzzled (and worried) astronomers who want to use neutrino observations to try to understand better how the Sun and other stars shine. The solar neutrino problem could be a clue to something new under the Sun.

Neutrinos from the Sun and from supernovae provide particle beams for probing the weak interactions on energy or time scales that cannot be achieved with traditional laboratory experiments. Since neutrinos from the Sun and from supernovae travel astronomical distances before they reach the Earth, experiments performed with these particle beams are sensitive to weak interaction phenomena that require long path lengths in order for slow weak interaction effects to have time to occur. The effects of tiny neutrino masses ($\gtrsim 10^{-6}$ eV), unmeasurable in the laboratory, can be studied with solar neutrinos.

Chapter 1 presents a nontechnical overview of the main features of the solar neutrino problem, including the theory that is being tested, experiments that show that the measured flux of neutrinos from the Sun is different from the calculated value, a description of some new experiments that are being prepared to solve this problem, and a brief summary of some of the possible explanations of the discrepancy. The last section contains answers to questions that are frequently asked following a general colloquium on solar neutrinos.

The discussion in this chapter, and in most of the book, concentrates on solar neutrinos, because the Sun is the only observable steady source of neutrinos. However, the experimental techniques are the same for studying solar and supernova neutrinos and many of the theoretical considerations are similar. Therefore, Chapter 15 applies the results of previous chapters to the elucidation of the neutrino astronomy of stellar collapse, with particular emphasis on the recent supernova in the Large Magellanic Cloud, SN1987A.

1.1   Where do we stand?

The **solar neutrino problem** can be stated simply. Both the theoretical and the observational results are expressed in terms of the solar neutrino unit, **SNU**, which is the product of a characteristic calculated solar neutrino flux (units: cm^{-2} s^{-1}) times a theoretical cross section for neutrino absorption (unit: cm^2). A SNU has, therefore, the units of events per target atom per second and is chosen for convenience equal to 10^{-36} s^{-1}.

The predicted rate for capturing solar neutrinos in a ^{37}Cl target is

$$\text{Predicted rate} = (7.9 \pm 2.6) \text{ SNU}, \tag{1.1a}$$

where the indicated uncertainty represents the total theoretical range including three standard deviation (3σ) uncertainties for measured input parameters. The rate observed by Davis and his associates in a chlorine radiochemical detector is

$$\text{Observed rate} = (2.1 \pm 0.9) \text{ SNU}, \tag{1.1b}$$

where the error is again a 3σ uncertainty.

The disagreement between the predicted and the observed rates constitutes the solar neutrino problem. There is no generally accepted solution to the problem although a number of interesting possibilities have been proposed.

The discrepancy between calculation and observation has recently been confirmed by an independent technique using the Japanese detector of neutrino–electron scattering, Kamiokande II. The preliminary Kamiokande II result is

$$\frac{\phi_{\text{observed}}}{\phi_{\text{predicted}}} = 0.45 \pm 0.15, \tag{1.2}$$

where the neutrino flux, ϕ, is from the rare ^8B solar neutrinos discussed below (see Table 1.1, Section 1.4) and the quoted error is the 1σ uncertainty.

The predictions used in Eqs. (1.1) and (1.2) are valid for the combined **standard model**, that is, the standard model of electroweak theory (of Glashow, Weinberg, and Salam) and the standard solar model. Throughout this book, the unmodified expression "standard model" will mean the combined standard model.

1.2   Why does anyone care?

The central question for solar neutrino research is easily stated. Is the solar neutrino problem caused by unknown properties of neutrinos or by a lack of understanding of the interior of the Sun? In other words, is this a case of new physics or faulty astrophysics?

Many physicists are interested in solar neutrino experiments because these observations provide a special opportunity to explore the weak interactions with a unique sensitivity, corresponding to differences in the squares of neutrino masses of order $\Delta m^2 \geq 10^{-12}$ eV^2. Experiments that use the Sun as a source of neutrinos are more sensitive to mixing among quantum states than are corresponding laboratory experiments because of the large distance between the source in the Sun and the target on Earth and because of the relatively low-energy of solar neutrinos (~ 1 MeV to 10 MeV). In solar neutrino experiments, the weak interactions have a longer proper time to make their effects felt than in terrestrial experiments. The potential increase in sensitivity over feasible laboratory or atmospheric experiments is about seven orders of magnitude in Δm^2.

In the standard model of electroweak interactions, all neutrino masses are exactly zero; widely differing values for the masses are possible in extensions of the standard model. Estimates have been made of neutrino masses based upon unified models for the strong, weak, and electromagnetic interactions. The published values range from $\Delta m^2 \sim 10^2$ eV^2 to $\sim 10^{-12}$ eV^2. Over much of this range, the only known way to test the theoretical models is by solar neutrino experiments.

From the astronomical point of view, solar neutrino experiments test in a rigorous way the theories of nuclear energy generation in stellar interiors and of stellar evolution. These tests are independent of many of the uncertainties that complicate the comparison of the theory with observations of stellar surfaces. For example, convection and turbulence are generally believed to be important near stellar surfaces but unimportant in the solar interior. Furthermore, the effects of rotation and magnetic fields, which are hard to model accurately, further complicate the analysis of observations of stellar surfaces.

Observations of solar neutrinos constitute critical tests of the theory of stellar evolution. We know more about the Sun than about any other star. We know its mass, its luminosity, its surface chemical composition, and its age much more accurately than we can determine these crucial parameters for any other star. Moreover, the Sun is still in the simplest stage of stellar evolution: According to standard theory, it is a middle-aged **main sequence** star, calmly burning hydrogen without violent or rapid evolution. Thus we expect to be able to calculate what the Sun is doing more accurately than we can predict the behavior of less familiar stars that are evolving rapidly.

Modifications of the theory of stellar evolution have been proposed as possible solutions of the solar neutrino problem. None of these solutions is fully consistent with well-established physics. However, these modifications show what is at issue for astronomy. The published astronomical solutions change conventional ideas about how stars evolve, ideas that astronomers use every day in their research. In interpreting astronomical observations (made by detecting photons), and in constructing astronomical theories, we use the theory of stellar evolution to determine the ages of stars, to interpret their compositions, to infer the evolution of galaxies, and to place limits on the chemical composition of the primordial material of the universe. Each of these basic astronomical industries is called into question by at least one of the proposed solutions of the solar neutrino problem.

In order that the reader may begin to form an opinion on these questions, the next three sections summarize in a qualitative way the aspects of stellar evolution theory that are necessary to predict the expected event rate in a solar neutrino experiment.

1.3   Stellar evolution

In order to understand the solar neutrino problem, one needs to know the main ingredients of the theory of stellar evolution that are used in the construction of solar models. Chapter 2 summarizes stellar evolution theory and Chapter 4 describes the standard solar model. This section presents some of the highlights from Chapters 2 and 4. Nonstandard solar models are discussed in Chapter 5.

The Sun is assumed to be spherical and to have evolved quasistatically (from one approximately equilibrium configuration to another approximately equilibrium configuration) for a period of about 5×10^9 yr. Evolution is manifested by the loss of photons from the surface of the star, which in turn is balanced by the burning of protons into α-particles in the core of the Sun. The overall reaction can be represented symbolically by the relation

$$4\mathrm{p} \to \alpha + 2\mathrm{e}^+ + 2\nu_\mathrm{e} + 25 \text{ MeV}. \tag{1.3}$$

Protons are converted to α-particles, positrons, and neutrinos, with a release of about 25 MeV of thermal energy for every four protons burned. Each conversion of four protons to an α-particle is known as a **termination** of the chain of energy-generating reactions that accomplishes the nuclear fusion. The thermal energy that is supplied by nuclear fusion ultimately emerges from the surface of the Sun as sunlight.

Energy is transported in the deep solar interior mainly by photons, which means that the opacity of matter to radiation is important. The pressure that supports the Sun is provided largely by the thermal motions of the electrons and ions.

Some of the principal approximations used in constructing standard solar models deserve special emphasis because of their fundamental roles in the calculations. These approximations have been investigated carefully for possible sources of departure from the standard scenario.

(1). **Hydrostatic equilibrium.** The Sun is assumed to be in hydrostatic equilibrium, that is, the radiative and particle pressures of the model exactly balance gravity. Observationally, this is known to be an excellent approximation since a gross departure from hydrostatic equilibrium would cause the Sun to collapse (or expand) in a free-fall time of less than an hour.

(2). **Energy transport by photons or by convective motions.** In the deep interior, where neutrinos are produced, the energy transport is primarily by photon diffusion; the calculated radiative opacity is a crucial ingredient in the construction of a model and has been the subject of a number of recent detailed studies.

(3). **Energy generation by nuclear reactions.** The primary energy source for the radiated photons and neutrinos is nuclear fusion, although the small effects of gravitational contraction (or expansion) are also included.

(4). **Abundance changes caused solely by nuclear reactions.** The initial solar interior is presumed to have been chemically homogeneous. In regions of the model that are stable to matter convection, changes in the local abundances of individual isotopes occur only by nuclear reactions.

A **standard solar model** is the end product of a sequence of models. The calculation of a model begins with the description of a main sequence star that has a homogeneous composition. Hydrogen burns in the stellar core, supplying both the radiated luminosity and the thermal pressure that supports the star against the force of gravity. Successive models are calculated by allowing for composition changes caused by nuclear reactions, as well as the mild evolution of other parameters, such as the surface luminosity and the temperature distribution inside the star. The models that describe later times in an evolutionary sequence have inhomogeneous compositions. In the model for the current Sun, the innermost mass fraction of hydrogen is about half the surface (initial) value.

A satisfactory solar model is a solution of the evolutionary equations that satisfies boundary conditions in both space and time. One seeks a model with a fixed mass $M_\odot$, with a total luminosity (in photons) $L_\odot$, and with an outer radius $R_\odot$ at an elapsed time of 4.6×10^9 yr (the present age of the Sun, determined accurately from meteoritic ages). The assumed initial values of chemical composition and entropy are iterated until an accurate description is obtained of the Sun at the present epoch. The solution of the evolution equations determines the initial values for the mass fractions of hydrogen, helium, and heavy elements, the present distribution of physical variables inside the Sun, the spectrum of acoustic oscillation frequencies observed on the surface of the Sun, and the neutrino fluxes.

The physical conditions in the solar interior where the neutrinos are produced are different from the conditions of everyday life but they are not so different as to suggest that the relevant physics will contain important surprises. The central temperature is 15×10^6 K (corresponding to a kinetic energy of a little more than a keV) and the central density is about 150 g cm^{-3}. As far as is known, the physics of the gaseous (largely nondegenerate) solar interior is relatively simple. Calculations of the standard solar model include corrections to the equation of state that arise from electron degeneracy and plasma effects, but these corrections are small and do not significantly affect the predicted neutrino fluxes.

The initial chemical composition is assumed to have been approximately the same everywhere in the Sun; the present-day surface composition is assumed to reflect the initial ratio of heavy elements to hydrogen. The surface of the Sun is too cool for nuclear reactions to have significantly altered the composition. Several authors have discussed a possible solution of the solar neutrino problem in which the initial chemical composition was very nonuniform. However, no one has yet succeeded in constructing a theoretical model, consistent with all the available observations, in which the initial composition was strongly inhomogeneous.

The calculation of a standard solar model begins with an estimate of the initial fraction of the total stellar mass that was in the form of hydrogen. The surface measurements give only the ratios relative to hydrogen of the abundances of elements heavier than helium; the abundance of helium on the solar surface cannot be measured accurately. The model parameters are evolved quasistatically, taking account of the composition changes and the energy released by nuclear reactions. The accuracy with which the interior calculations must be carried out is much higher than for most other applications of stellar evolution theory, because the neutrino measurements relate directly to processes occurring in the solar interior and because the calculated fluxes are sensitive to the physical conditions.

The luminosity boundary condition states that the model luminosity at the present age must equal the observed solar luminosity; this condition has a strong effect on the calculated neutrino fluxes. The reason is that both the luminosity and the neutrino fluxes are produced by nuclear reactions in the solar core.

Evolutionary codes have been developed by a number of different research groups; when adjusted for differences in input parameters, the codes yield individual solar neutrino fluxes that are the same to an accuracy of about 10% or better.

Both for the Sun and for other stars, the theory of stellar evolution is in satisfactory agreement with conventional astronomical observations. Some of the principal results from the calculations with a standard solar model include a model luminosity that increases by about 40% over the five billion years the Sun has been shining, a calculated initial fractional helium abundance by mass of $Y = 0.27$, and a complete spectrum for the p-mode (pressure) oscillations (observed in a variety of ways on the solar surface). The standard solar model gives a satisfactory (although necessarily incomplete) account of what is known about the Sun from photons.

The adopted model for convection of material does not significantly affect the predicted solar neutrino fluxes since the solar interior is calculated to be in radiative equilibrium and the physical conditions in the deep interior of the model are practically independent of the description of the solar surface. Observations made with photons are often sensitive to what is happening in the outer (less well-understood) convective layers of the Sun.

The theory of stellar evolution has a number of successes to its credit. The most basic achievement is the theoretical relation between the mass and the photon luminosity of stars that is in agreement with observation over almost two orders of magnitude in mass (six orders of magnitude in luminosity). In addition, the theory successfully accounts for the distribution of known stars in the luminosity–temperature or luminosity–color plane. Since most of the observed plane (known technically as the **Hertzsprung–Russell diagram** or H–R diagram) is empty, the representation of the positions of the known stars by conventional models is a major triumph. The calculated frequencies of the surface oscillations of the Sun agree with the observed values to better than 1%, which constitutes a great success for the theory despite the fact that disagreements exist at the level of a few tenths of a percent (disagreements that might arise from processes that are unimportant for the solar neutrino problem).

The greatest achievement of the theory of stellar evolution is so overwhelming that it is usually overlooked. Astronomers use the theory routinely in interpreting astronomical observations of the physical and chemical characteristics of individual stars of widely different types in a variety of environments, in both nearby and distant galaxies. Stellar evolution theory has been remarkably successful in providing a framework for discussing these traditional astronomical observations without obvious inconsistencies.

The bottom line of this brief survey of the theory of stellar evolution is that only the ^{37}Cl solar neutrino experiment, and most recently the Kamiokande II experiment, are inconsistent with the standard theory. Even this inconsistency could be just apparent. A popular scenario for explaining the solar neutrino problem (which has many variations) supposes that solar neutrinos are indeed produced in the quantity predicted by the standard solar model, but that they do not reach the Earth in the form in which they are emitted at the Sun. If something happens to the neutrinos in the solar interior or on the way to the Earth, then the standard solar model may be correct and the solar neutrino experiments could be telling us something new about neutrino propagation over large distances.

1.4   Nuclear energy generation and neutrino fluxes

The Sun shines by converting protons into α-particles. About 600 million tons of hydrogen are burned every second to supply the solar luminosity. Nuclear physicists have worked for half a century to determine the details of this transformation.

The main nuclear burning reactions in the Sun are shown in Table 1.1, which represents the energy-generating **pp chain**. This table also indicates the relative frequency with which each reaction occurs in the standard solar model. Chapter 3 describes in more detail both the pp chain and the less important (for the Sun) **CNO cycle**. In the CNO cycle, the fusion of four protons to form an α-particle is achieved through reactions involving carbon, nitrogen, and oxygen.

The fundamental reaction in the solar energy-generating process is the proton–proton (pp) reaction. In the pp reaction, a proton β-decays in the vicinity of another proton forming a bound system, deuterium (^{2}H). This reaction (number 1a in Table 1.1) produces the great majority of solar neutrinos; however, these pp neutrinos have energies below the detection thresholds for the ^{37}Cl and Kamiokande II experiments. Experiments with ^{71}Ga that are in progress in the Soviet Union and in Europe are sensitive primarily to neutrinos from the pp reaction. More rarely, a three-body reaction involving two protons and an electron initiates the reaction chain. While this reaction (number 1b in Table 1.1) occurs with a relative frequency of only one in 250, the resulting neutrino energy is larger by the equivalent of two electron masses, raising it above the threshold in the chlorine experiment. The deuteron produced by either of the initiating reactions is burned quickly by a (p,γ) reaction that forms ^{3}He (reaction 2 in Table 1.1). Reactions 1a and 2 occur in essentially all terminations of the pp chain in the Sun; reaction 1b occurs only rarely, in approximately 0.4% of all pp terminations. The richness and complications of the pp cycle begin at the next stage.

Most of the time, 85% in the standard solar model, the proton–proton chain is terminated by two ^{3}He nuclei fusing to form an

Table 1.1. The pp chain in the Sun. The average number of pp neutrinos produced per termination in the Sun is 1.85. For all other neutrino sources, the average number of neutrinos produced per termination is equal to (the termination percentage/100).

Reaction	Number	Termination[†] (%)	ν energy (MeV)
$p + p \rightarrow {}^2H + e^+ + \nu_e$	1a	100	≤ 0.420
or			
$p + e^- + p \rightarrow {}^2H + \nu_e$	1b (pep)	0.4	1.442
${}^2H + p \rightarrow {}^3He + \gamma$	2	100	
${}^3He + {}^3He \rightarrow \alpha + 2p$	3	85	
or			
${}^3He + {}^4He \rightarrow {}^7Be + \gamma$	4	15	
${}^7Be + e^- \rightarrow {}^7Li + \nu_e$	5	15	(90%) 0.861
			(10%) 0.383
${}^7Li + p \rightarrow 2\alpha$	6	15	
or			
${}^7Be + p \rightarrow {}^8B + \gamma$	7	0.02	
${}^8B \rightarrow {}^8Be^* + e^+ + \nu_e$	8	0.02	< 15
${}^8Be^* \rightarrow 2\alpha$	9	0.02	
or			
${}^3He + p \rightarrow {}^4He + e^+ + \nu_e$	10 (hep)	0.00002	≤ 18.77

[†]The termination percentage is the fraction of terminations of the pp chain, $4p \rightarrow \alpha + 2e^+ + 2\nu_e$, in which each reaction occurs. The results are averaged over the model of the current Sun. Since in essentially all terminations at least one pp neutrino is produced and in a few terminations one pp and one pep neutrino are created, the total of pp and pep terminations exceeds 100%.

α-particle plus two protons (reaction 3 of Table 1.1). No additional neutrinos are formed in this dominant mode.

About 15% of the time, a ${}^3He$ nucleus will capture an already existing α-particle to form ${}^7Be$ plus a gamma ray (reaction 4). It is the neutrinos formed after this process that are primarily detected in the ${}^{37}Cl$ experiment. Nearly always, the ${}^7Be$ nucleus will undergo electron capture, usually absorbing an electron from the continuum of ionized electrons (reaction 5). This branch produces neutrinos with energies of 0.9 MeV (90% of the time), which contribute small (but not negligible) fractions of the predicted standard model capture rate in the ${}^{37}Cl$ and ${}^{71}Ga$ experiments. There is no experiment in progress that isolates the contribution of the ^{7}Be neutrinos, although some suggestions for practical detectors have been made.

Most of the predicted capture rate in the ^{37}Cl experiment comes from the rare termination in which ^{7}Be captures a proton to form radioactive ^{8}B (reaction 7). The ^{8}B decays to unstable ^{8}Be, ultimately producing two α-particles, a positron, and a neutrino. The neutrinos from ^{8}B decay have a maximum energy of less than 15 MeV. Although the reactions involving ^{8}B occur only once in every 5000 terminations of the pp chain, the total calculated event rates for the ^{37}Cl and Kamiokande II experiments are dominated by this rare mode.

For ^{37}Cl, the ^{8}B contribution is most important because many of the neutrinos from this source are sufficiently energetic to excite a superallowed transition between the ground state of ^{37}Cl and the analogue excited state of ^{37}Ar (which closely resembles the ground state of ^{37}Cl). None of the more abundant neutrinos have enough energy to cause this strong analogue transition. A theoretical calculation originally showed that the sensitivity for the detection of ^{8}B neutrinos is increased by about a factor of 20 by transitions to excited states. This result has been confirmed by beautiful nuclear physics experiments on the isotopic quartet of the mass 37 nuclear system.

The last reaction in Table 1.1, number 10, is extremely rare, occurring about twice in every 10^7 terminations of the pp chain. Nevertheless, the neutrinos from this reaction may be detectable in some direct counting electronic experiments (with, e.g., deuterium or ^{40}Ar) because they have the highest energies of any of the sources in Table 1.1.

The neutrinos from reaction 1b, which is initiated by three particles, p+e+p, are known as **pep** neutrinos. The neutrinos from the ^{3}He+p reaction are known as **hep** neutrinos.

The neutrino spectrum predicted by the standard model is shown in Figure 1.1, where contributions from both line and continuum sources are included. For Kamiokande II, only the ^{8}B and hep neutrinos (reaction 10) have enough energy to produce recoil electrons above the dominant backgrounds.

The measured event rate in a solar neutrino experiment is the product of neutrino fluxes times the interaction cross sections. The calculation of neutrino cross sections is discussed in Chapter 8.

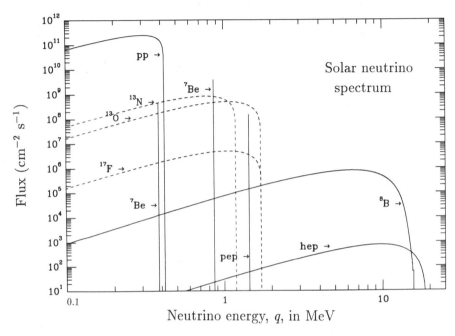

Figure 1.1 Solar neutrino spectrum This figure shows the energy spectrum of neutrinos predicted by the standard solar model. The neutrino fluxes from continuum sources (like pp and ^{8}B) are given in the units of number per cm^2 per second per MeV at one astronomical unit. The line fluxes (pep and ^{7}Be) are given in number per cm^2 per second. The spectra from the pp chain are drawn with solid lines; the CNO spectra are drawn with dotted lines. Chapter 6 discusses the neutrinos that are believed to be produced in the Sun.

1.5 Uncertainties in the predictions

Is there really a solar neutrino problem? The answer is yes if the difference between the predicted and the measured capture rates exceeds the range of the uncertainties. The answer is no if the uncertainties exceed the discrepancy between theory and observation.

Table 1.2.   Calculated solar
neutrino fluxes.

Source	Flux (10^{10} cm^{-2} s^{-1})
pp	6.0 (1 ± 0.02)
pep	0.014 (1 ± 0.05)
hep	8×10^{-7}
^{7}Be	0.47 (1 ± 0.15)
^{8}B	5.8×10^{-4} (1 ± 0.37)
^{13}N	0.06 (1 ± 0.50)
^{15}O	0.05 (1 ± 0.58)
^{17}F	5.2×10^{-4} (1 ± 0.46)

A quantitative statement of the uncertainties is at the heart of our
subject and is discussed separately in Chapter 7.

The solar neutrino fluxes calculated from the standard solar model
are shown in Table 1.2. The uncertainties in the calculated neutrino
fluxes are also shown in Table 1.2.

The flux of the basic pp neutrinos can be calculated to an estimated accuracy of 2% using the standard solar model. Thus the
pp flux, the dominant flux of solar neutrinos, can be thought of as
a reliable source, placed at an astronomical distance, which can be
used for physical experiments on the propagation of neutrinos.

The production rate for the rare neutrinos from ^{8}B β-decay is sensitive to conditions in the solar interior, because of the relatively
high Coulomb barrier for the ^{7}Be(p, γ)^{8}B reaction ($\sim$ 10 MeV compared to a mean thermal energy of 1 keV). The calculated flux of
^{7}Be electron capture neutrinos is intermediate in sensitivity between
the pp and the ^{8}B neutrinos.

The calculated uncertainties are described in this book in terms
of a **total theoretical range**. It would be wonderful to be able to
calculate a *true* "three standard deviation level of confidence," but
this cannot be done because the probability distribution is unknown
for parameters that must be calculated, not measured (e.g., radiative opacity or higher-order corrections to neutrino cross sections).
In practice, the meaning of the total theoretical range is that, if the

true value lies outside this range, someone who has determined an input parameter (experimentally or theoretically) has made a mistake.

For measured quantities (e.g., nuclear reaction rates), my colleagues and I use standard 3σ limits to estimate the uncertainties. For theoretical quantities, some authors calculate the uncertainties together with a best estimate, but more often they must be assessed independently. In practice, we usually take the uncertainties in quantities that are calculated to be equal to the range in values in published state-of-the-art calculations, especially when this range exceeds (as it usually does) the published estimates of uncertainties. (Of course, the theory could be wrong in some fundamental way that would not be reflected in scatter in the values obtained by different treatments.) Quantities for which only one calculation is available require a more delicate judgment. For example, we have chosen to multiply the value of higher-order corrections to neutrino capture cross sections by three and call this the total uncertainty. It is possible that we assign relatively larger errors for experimentally determined parameters (for which the errors are more easily quantifiable) than we do for the calculated parameters such as the opacity. However, the adopted procedure is as objective as any we can think of and has the advantage of simplicity.

In the final analysis, the method for estimating errors is defined by the examples that are described in Chapter 7. In all cases, the procedures and assumptions we use to obtain the final uncertainties are stated explicitly. The reader with a better (or different) way of estimating the errors can easily recalculate the uncertainties in all the predicted event rates using the results and prescriptions described in this book.

In order to predict the event rate in a given detector, we must combine theoretical uncertainties with statistical (and systematic) errors in measured quantities. In doing so, we have adopted one cardinal rule: Errors associated with different sources are combined incoherently, that is, the total uncertainty is the square root of the sum of the squares of the individual uncertainties.

How reliable are the theoretical predictions? Do the error estimates really describe the total theoretical range? There is an informal but objective way of answering these questions that complements the quantitative error estimates described in Chapter 7.

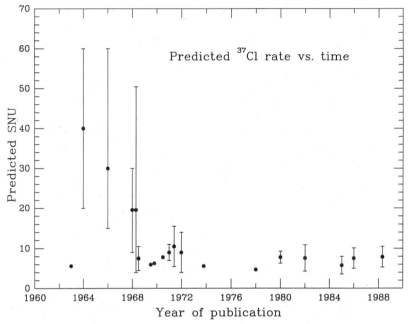

Figure 1.2   Predicted capture rates as a function of time   The published predictions of the author and his collaborators of neutrino capture rates in the ^{37}Cl experiment are shown as a function of the date of publication.

Many theorists use as their private estimate of the uncertainties in experimental parameters the dispersion of measured values reported by respected experimental groups. This estimate is used if the dispersion among the published values exceeds, as it usually does, the quoted experimental errors. (Experimentalists probably do much the same thing with theoretical calculations, using the dispersion among what theorists calculate as a measure of the uncertainty.) One can do something analogous for the solar neutrino problem.

Figure 1.2 shows the predicted capture rates for the ^{37}Cl experiment as a function of the date of publication for each paper in which my collaborators and I have published a calculated value (1963 to 1988). The original error bars are shown for every case in which they were published. All 14 values published since 1968 are consistent with the range given in Eq. (1.1a).

The discussion up to this point has emphasized the theory of neutrino emission from the Sun. Fortunately, this theory can be tested observationally.

Table 1.3. Solar neutrino experiments.

Target	Source(s)	Operating group(s)
^{37}Cl	^{8}B (^{7}Be)	Univ. Penn
ν_e-e	^{8}B (hep)	Kamiokande II
		LVD
		Univ. Sydney
		Canada/USA/UK
		ICARUS
^{98}Mo	^{8}B	LANL
^{2}H	^{8}B (hep)	Canada/USA/UK
^{40}Ar	^{8}B (hep)	ICARUS
^{81}Br	^{8}B, ^{7}Be	Penn/Tennessee
^{71}Ga	pp (^{7}Be)	GALLEX; USSR
^{115}In	pp	European
^{11}B	^{8}B	Bell Lab.
^{4}He	pp	Brown Univ.
(ν_e-e)		

1.6 Experiments are required for progress

Progress in solar neutrino research requires new experiments. There are a plethora of interesting theoretical explanations of the solar neutrino problem that are consistent with the existing data. New measurements are necessary to determine which solution nature has adopted. Observations must be made with different detectors and techniques in order to test theoretical predictions and to eliminate the possibility that systematic uncertainties influence the interpretations. Some of the most promising experiments are discussed in the following sections. For convenience, Table 1.3 lists the major solar neutrino experiments that are being developed or are in progress. The last column lists the operating group(s) primarily responsible for each experiment. All of the experiments are discussed in more detail in Chapters 10 to 14.

1.7 The ^{37}Cl experiment

The beautiful ^{37}Cl experiment of Davis and his collaborators was for two decades the only operating solar neutrino detector. The reaction

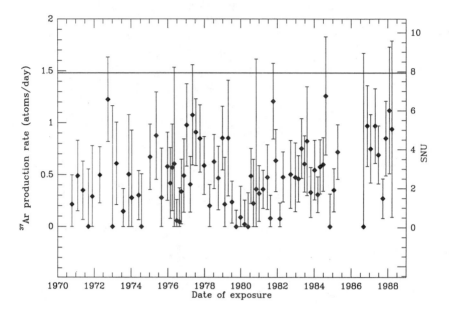

Figure 1.3 ^{37}Cl observations   The observed rate in the ^{37}Cl solar neutrino experiment (from observations by Davis, Cleveland, and Rowley). The line at 7.9 SNU across the top of the figure represents the prediction of the standard model.

that is used for the detection of the neutrinos is:

$$\nu_e + {}^{37}\text{Cl} \Rightarrow e^- + {}^{37}\text{Ar}, \tag{1.4}$$

which has a threshold energy of 0.8 MeV. The target is a tank containing 10^5 gallons of C_2Cl_4 (perchloroethylene, a cleaning fluid), deep in the Homestake Gold Mine in Lead, South Dakota. The underground location is necessary in order to avoid background events from cosmic rays. Every few months, for about 15 years, Davis and his collaborators extracted a small sample of ^{37}Ar, typically of order 15 atoms, out of the total of more than 10^{30} atoms in the tank. The ^{37}Ar produced in the tank is separated chemically from the C_2Cl_4, purified, and counted in low background proportional counters. The typical background counting rate for the counters corresponds to about one radiative decay of an ^{37}Ar nucleus a month! Experiments have been performed to show that ^{37}Ar produced in the tank is extracted with more than 90% efficiency.

Table 1.4. Predicted capture rates
for a ^{37}Cl detector.

Neutrino source	Capture rate (SNU)
pp	0.0
pep	0.2
hep	0.03
^{7}Be	1.1
^{8}B	6.1
^{13}N	0.1
^{15}O	0.3
^{17}F	0.003
Total	7.9 SNU

Figure 1.3 shows all the experimental data that were reported by
Rowley, Cleveland, and Davis at the 1984 Solar Neutrino Conference in Lead, South Dakota, together with their 1σ error bars. The
average rate at which ^{37}Ar is produced is

$$\text{Production rate} = 0.462 \pm 0.04 \text{ atoms day}^{-1}, \qquad (1.5a)$$

of which a small part is background (from cosmic ray events),

$$\text{Background rate} = 0.08 \pm 0.03 \text{ atoms day}^{-1}. \qquad (1.5b)$$

Subtracting the known background rate from the production rate
yields the capture rate

$$\text{Capture rate} = (2.05 \pm 0.3) \text{ SNU}, \qquad (1.5c)$$

which is due to solar neutrinos if all of the significant contributions
to the background have been recognized. The errors quoted in these
three observed rates are all 1σ uncertainties. The nine experiments
made in 1986 to 1988 are two standard deviations [(3.2 ± 0.7) SNU]
higher than the average for 1970–1988.

Table 1.4 shows the contribution to the total capture rate predicted by the combined standard model. Of the total eight SNU,
about 75% (6 SNU) is contributed by the ^{8}B neutrinos. The difference between the average measured value shown in Figure 1.3 and
eight SNU (the dotted line) is "the solar neutrino problem."

We can make two important inferences about stellar evolution provided that neutrinos reach us from the Sun without being altered in transit, that is, assuming the correctness of the standard model of the electroweak theory. First, the CNO cycle is unimportant for the Sun. If the CNO cycle were the dominant source of energy production, then the neutrino fluxes could be calculated to satisfactory accuracy without constructing a detailed solar model. To a good approximation, each 25 MeV of energy would be accompanied by one ^{13}N and one ^{15}O neutrino. The theoretical capture rate implied if the CNO cycle is dominant is 29 SNU. The result that the observed rate is much less than 29 SNU is a not-often-discussed triumph for stellar evolution theory. Bethe originally suggested that the CNO cycle was the primary energy source for the Sun, but detailed calculations of stellar models showed that a star must be slightly more massive than the Sun in order for the CNO reactions to dominate. The second inference is that the standard solar model is wrong in some important respect. Unfortunately, we cannot conclude anything from the ^{37}Cl experiment about the basic proton–proton (or pep) reaction. The pp reaction is below the energy threshold for the ^{37}Cl detector and the pep reaction contributes only 0.25 SNU, which is a small fraction of the observed capture rate.

The ^{37}Cl experiment is discussed in more detail in Chapter 10.

1.8   The Kamiokande II experiment

The Kamiokande II experiment, which is located in the Japanese Alps, detects Cerenkov light emitted by electrons that are scattered in the forward direction by solar neutrinos. The reaction by which the neutrinos are observed is

$$\nu + e \rightarrow \nu' + e', \qquad (1.6)$$

where the primes on the outgoing particle symbols indicate that the momentum and energy of each particle can be changed by the scattering interactions. This is the first of several solar neutrino experiments that are planned or are in progress in which ν–e scattering will be studied; other experiments are discussed in Chapter 13. For the higher-energy neutrinos (> 5 MeV, i.e., ^{8}B and hep neutrinos only) that can be observed by this process using available techniques, the scattering provides additional information not available with a radiochemical detector. Neutrino–electron scattering experiments furnish information about the incident neutrino energy spectrum (from measurements of the recoil energies of the scattered electrons), determine the direction from which the neutrinos arrive, and record the precise time of each event.

The Kamiokande experiment was originally designed as a three kiloton underground water Cerenkov detector in order to study proton decay. In late 1984, improvements were begun to make possible the detection of the relatively low-energy events that are expected to be produced by solar neutrinos. This detector also provided (together with the IMB detector, a water Cerenkov detector located in a salt mine in Ohio) the first observation of neutrinos from a supernova, SN1987A; the supernova neutrinos happen to have energies in about the same range as those expected from solar ^{8}B neutrinos. Fortunately, an upgrade of Kamiokande II to a full-time solar neutrino detector had been completed just a few months before the neutrinos from the supernova explosion, SN1987A (which occurred in a nearby galaxy, the Large Magellanic Cloud), reached the Earth.

The preliminary results from the Kamiokande II detector yield a ^{8}B neutrino flux that is approximately 0.45 of the standard model flux, about 3σ away from zero and from the standard model value. This result applies for recoil electrons with a minimum total energy of 9.3 MeV. A significant forward peaking of the recoil electrons is observed along the direction of the Earth–Sun axis. This result is of great importance since all of the previous observational results on solar neutrinos came from a single ^{37}Cl experiment.

1.9 A geochemical experiment: ^{98}Mo

A geochemical experiment has been developed over the past several years at Los Alamos National Laboratory, in which neutrinos are detected via their absorption by molybdenum atoms shielded from atmospheric phenomena by the covering of a deep mine. The absorption of neutrinos produces an unstable but long-lived isotope of technetium that would not be present in a steady state situation in which there were no high-energy solar neutrinos. The method uses the neutrino capture reaction

$$\nu_e + {}^{98}\text{Mo} \rightarrow {}^{98}\text{Tc}^* + e^-. \tag{1.7}$$

Neutrino absorption populates a variety of excited states of technetium that can only be populated by the ^{8}B (and much less importantly, hep) neutrinos.

The ^{98}Tc has a mean life of 4.2 million years; hence, the present abundance reflects the average production rate during a period of several million years. The lifetime is too long for the isotope to be detected from its natural radioactivity; instead, enough technetium is accumulated so that a sufficient number of atoms ($N \approx 10^7$) are available for the ^{98}Tc to be counted by an ultra sensitive mass-spectrometer. This experiment will test the constancy of the flux of ^{8}B neutrinos over the past several million years.

Standard ideas about the time scale for solar evolution (estimated to be $\sim 10^{10}$ yr) imply that the time-averaged flux of ^{8}B neutrinos measured with the ^{98}Mo experiment will be the same to within 1% (much less than the experimental uncertainties) as the contemporary flux determined from the ^{37}Cl and Kamiokande II detectors.

The experimentalists use the Henderson molybdenum mine in Colorado, in which the ore is recovered at great depth (1500–1800 m). The sources of background have been carefully evaluated and are believed to be sufficiently low to permit the observation of ^{98}Tc production from solar neutrinos. The largest uncertainty in the predicted rate, about a factor of two, comes from the neutrino absorption cross section.

Chapter 12 discusses several radiochemical and geochemical solar neutrino experiments.

1.10   Gallium detectors

Two radiochemical solar neutrino experiments using ^{71}Ga are under way, one by a primarily Western European collaboration (GALLEX) and the second by a group which will work in the Soviet Union (Institute for Nuclear Research, Moscow). The GALLEX collaboration plans to use 30 tons of gallium in an aqueous solution; the detector will be located in the Gran Sasso Laboratory in Italy. The Soviet experiment will use about 60 tons of gallium metal as a detector in a solar neutrino laboratory constructed underneath a high mountain in the Baksan Valley in the Caucasus Mountains of the Soviet Union. The scale of both of these experiments is impressive considering that, at the time the experimental techniques were developed, the total world production of gallium was only 10 tons per year!

Table 1.5. Predicted capture rates for a ^{71}Ga detector.

Neutrino source	Capture rate (SNU)
pp	70.8
pep	3.0
hep	0.06
^{7}Be	34.3
^{8}B	14.0
^{13}N	3.8
^{15}O	6.1
^{17}F	0.06
Total	132^{+20}_{-17} SNU

The gallium experiments can furnish unique and fundamental information about nuclear processes in the solar interior and about neutrino propagation. The neutrino absorption reaction is:

$$\nu_e + {}^{71}\text{Ga} \Rightarrow e^- + {}^{71}\text{Ge}. \tag{1.8}$$

The germanium atoms are removed chemically from the gallium and the radioactive decays of ^{71}Ge (half-life 11.4 days) are measured in small proportional counters. The threshold for absorption of neutrinos by ^{71}Ga is 0.233 MeV, which is well below the maximum energy of the pp neutrinos. No other solar neutrino experiment has a demonstrated capability to detect the low-energy neutrinos from the basic pp reaction (reaction 1a of Table 1.1).

Table 1.5 shows the calculated contribution from individual neutrino sources to the predicted capture rate. Neutrinos from the basic pp reaction are expected, according to the standard model, to produce approximately half of the computed total capture rate. The other main contributors are ^{7}Be neutrinos, about one-quarter of the total rate, and ^{8}B neutrinos, about 10%.

The initial chemical extraction is different in the GALLEX and the Soviet experiments, but the final chemical procedures and the methods of low-level counting will be similar for both collaborations. The comparison of the results from the two experiments can provide a valuable check on any possible systematic errors. The GALLEX collaboration will employ gallium in the chemical form of an aqueous solution of gallium chloride and hydrochloric acid. The radioactive ^{71}Ge is removed from this solution by purging with gas. The GALLEX collaboration will use a single vessel to contain the gallium chloride solution. The extracted ^{71}Ge will be converted to germane (GeH_4) and measured in a miniature gas proportional detector. This group hopes to receive their full amount of gallium by late 1989 and to begin measurements in 1990.

The Soviet experiment, with American collaborators, is sometimes referred to as SAGE (Soviet-American-Gallium-Experiment). The SAGE collaboration will use gallium metal as the target material. The germanium extraction process involves mixing the metal with a dilute hydrochloric acid solution. The ^{71}Ge is removed from the acidic solution by purging with gas collected in water, a procedure similar to the primary ^{71}Ge separation used in the gallium chloride process. The process used by the Soviet group is slower and more cumbersome, but is also capable of obtaining a high recovery yield. The ^{71}Ge radioactivity will be measured in small proportional counters similar to those developed for the ^{37}Cl experiment. The Soviet experimentalists have 60 tons of gallium available and hope to begin taking data sometime in late 1988 or early 1989.

The gallium experiments may indicate which class of solution is correct for the solar neutrino problem, faulty astrophysics or new physics. Most nonstandard solar models predict event rates that are not very different from the standard solar model. The minimum rate that is consistent with the assumption that nuclear fusion currently balances the solar luminosity is about 60% of the standard model value, provided that no physics beyond the standard electroweak model affects the propagation of the neutrinos. Some explanations of the solar neutrino problem that involve particle physics imply that the event rate in the gallium experiments will be much less than the standard model prediction, perhaps no more than 10% of the standard value.

Chapter 11 discusses gallium experiments in more detail.

1.11   Next-generation experiments

Two powerful new detectors are being developed as next-generation experiments. One is a proposed one kiloton heavy water experiment (D_2O), to be placed in an INCO nickel mine near Sudbury,

Ontario (Canada). The other is a three kiloton detector of liquid argon, to be placed in the Gran Sasso underground laboratory in central Italy. The deuterium experiment is a collaboration between Canadian, American, and British scientists; the argon detector is primarily an Italian experiment with (so far) limited American participation. The deuterium detector measures the energy and direction of recoil electrons by observing their Cerenkov light with photomultipliers. The argon detector forms a three-dimensional electronic image of the positions of recoil electrons produced by neutrino absorption or scattering, by drifting the electrons in a homogeneous electric field.

These experiments are sensitive to ^{8}B and hep neutrinos, but the other solar neutrinos will be below the energy thresholds that are set at several MeV in order to avoid numerous lower-energy background events. Hep neutrinos can produce higher-energy recoil electrons than can ^{8}B neutrinos; the highest-energy hep neutrino events can be distinguished experimentally from the more frequent events produced by ^{8}B neutrinos. Fortunately, ^{8}B and hep neutrinos are produced in different regions of the Sun and are therefore affected differently by some of the most often discussed "nonstandard" solar models. The two sources provide complementary information about the Sun. The fact that there are two very different sources increases one's assurance that a solar neutrino signal will be detected. The expected rates vary from several hundred to several thousand events per year in each detector, depending upon the adopted explanation of the solar neutrino problem.

Both the argon and the deuterium experiments utilize multipurpose detectors, which can study solar neutrinos by neutrino capture and by electron–neutrino scattering. Neutrino absorption is sensitive only to solar ν_e whose type (flavor) is unchanged in transit to the Earth. For electron–neutrino scattering, the cross section for ν_μ or ν_τ is about one-seventh the cross section for ν_e. The solar origin of the events can be tested using the angular distribution (with respect to the Earth–Sun direction) of the electrons scattered by the neutrinos.

Absorption reactions make possible the measurement of individual neutrino energies, E_ν, using the relation $E_\nu = E_e + \text{constant}$, where E_e is the energy of the electron that is produced and the constant is equal to the difference of initial and final nuclear masses. This simple relation is valid because nuclei absorb momentum but very little energy in capturing solar neutrinos, since nuclei are much heavier than electrons and neutrinos. The measurement of individual energies of neutrinos will constitute a test of the predicted shape of the energy spectrum.

The deuterium experiment may include a detection mode that is equally sensitive to all three types of neutrinos, ν_e, ν_μ, and ν_τ. In this **neutral-current mode**, deuterium nuclei are disintegrated into their constituent neutrons and protons without changing the charge of the nucleons. Neutral currents make possible the ideal Equal Opportunity Detector. Other methods of detection have zero sensitivity for all but one type of neutrino (as in neutrino absorption) or are much more sensitive to neutrinos of one type than any other (as in neutrino–electron scattering). The measurement of the neutral-current disintegration of deuterium can provide a determination of the total flux of solar neutrinos; radiochemical experiments (with ^{37}Cl, ^{71}Ga, and ^{98}Mo detectors) detect only the component of the flux that is in the form of electron neutrinos. The ratio of event rates for the charged-current (absorption) process and the neutral-current (disintegration) process could be a sensitive indicator of otherwise inaccessible effects of neutrino transformations.

The interaction cross sections applicable to these experiments can be calculated accurately. Therefore, almost all of the total uncertainties in the predicted rates are determined by the uncertainties in the solar production rate of ^{8}B and hep neutrinos and by the unknown weak interaction effects on neutrino propagation.

Most recently, an experiment has been proposed that would utilize ^{11}B as a detector and could, in principle, observe simultaneously the neutral-current, electron-scattering, and capture (charged-current) reactions. Different detector environments are being investigated.

Chapter 14 discusses several direct counting experiments.

1.12 Electronic detection of pp neutrinos

Several detectors are being developed to observe the basic (low-energy) pp neutrinos by electronic methods.

Two of the detectors are being designed to observe neutrino–electron scattering at low temperatures. A crystalline silicon detector could function as a low-temperature (milliKelvin) bolometer measuring the energy deposited by individual neutrinos. A table-top size detector is currently being developed in order to test the design concepts. A superfluid helium detector has been proposed to study the excitations produced in the fluid when neutrinos are scattered by individual electrons.

In addition, an ^{115}In detector is being developed by several European groups. This detector would work as a liquid scintillator, registering neutrinos that are captured by ^{115}In. The planned detector is in principle capable of measuring accurately the incident spectrum of pp neutrinos (see Figure 1.1), although natural radioactive backgrounds remain a serious problem.

All of these experiments are discussed in more detail in Chapter 12.

1.13   Coherent scattering detectors

Several European and American groups are involved in ambitious programs to use the **coherent scattering** of neutrinos by nuclei at low temperatures to develop sensitive detectors of solar neutrinos and dark matter. This process can be represented by the reaction

$$\nu + A \rightarrow \nu + A, \tag{1.9}$$

where the nucleus contains A neutrons and protons. The cross section for coherent scattering is approximately proportional to N^2, the square of the total number of neutrons in the target nucleus. The name "coherent scattering" reflects the fact that all of the neutrons act together. Coherent scattering is equally sensitive to all neutrinos described by the standard electroweak model, providing another method (in addition to the disintegration of deuterium and the excitation of ^{11}B) of achieving an Equal Opportunity Detector.

The rate of coherent scattering per target nucleus is much larger than the rate per target particle for the incoherent processes discussed previously (neutrino absorption, neutrino–electron scattering), because of the occurrence of the factor of N^2 in the expression for the cross section. The large cross section implies that relatively small quantities of material ($\gtrsim$ 0.1 kton) can be used for a solar neutrino experiment; the possibilities of using different medium or heavy nuclei are being explored. The detectors could be calibrated conveniently using antineutrinos from a reactor. The major difficulty in constructing a neutral-current detector is that the observable signal is from nuclear recoil, which has a low energy (typically keV or lower, hence the low temperature of the suggested detectors) and lacks specificity (the deposition of energy by nuclear recoil does not provide a unique signature). In principle, coherent scattering could be used to detect all of the solar neutrino sources. In practice, the ^{8}B neutrinos, which may have higher energies, may be the easiest to detect since both the coherent scattering cross section and the magnitude of the nuclear recoil are proportional to the square of the neutrino energy.

1.14   Some explanations of the solar neutrino problem

There are two classes of "solutions" to the solar neutrino problem, those that modify the astrophysical description of the Sun in order to obtain a nonstandard solar model and those that hypothesize properties of neutrinos in order to obtain a nonstandard theory of electroweak interactions.

Nonstandard solar models are constructed by changing something, either the physics or the input data, from our current best estimate to something that is less plausible. Some nonstandard models invoke enhanced element diffusion, convective instabilities, or a solar interior greatly deficient in heavy elements. Most of the nonstandard models that have been published were invented in order to reduce the calculated flux of ^{8}B neutrinos. Many nonstandard models are described in Chapter 5.

One suggested solution has been made that might solve simultaneously the famous "dark matter" problem in astronomy and the solar neutrino problem. In this suggestion, weakly interacting massive particles, **WIMPs**, are produced in the early universe in an abundance that accounts for the unseen matter in large astronomical systems. The properties of the so-far undiscovered WIMPs could be just such that they redistribute some of the heat in the solar interior, smoothing out the temperature gradient, and reducing the flux of ^{8}B neutrinos to be consistent with the experimental limit.

In the other class of solutions (described in Chapter 9), the flux of neutrinos in the interior of the Sun is predicted correctly by the standard solar model, but these neutrinos are transformed from ν_e to another type of neutrino on their way to the terrestrial detector. For nearly all proposed solutions, this transformation requires a nonzero neutrino mass.

A massive neutrino may have a magnetic moment. If this moment were as large as 10^{-10} to $10^{-11} \mu_B$ (Bohr magnetons), it is possible that the familiar left-handed ν_e could have its spin flipped to a right-handed ν_e in passing through magnetic fields of a few thousand gauss in the outer part of the Sun. Right-handed neutrinos do not have normal weak interactions and would therefore not be detected. This solution was proposed many years ago and has been revived recently by Soviet scientists to account for an apparent anti-correlation of the neutrino capture rate in the ^{37}Cl experiment with the sunspot cycle. The greatest change, a strong decrease in the observed capture rate, occurred near the time of the onset of solar cycle 21 in 1979 and 1980. The large magnetic moment needed for this solution, while barely consistent with existing laboratory limits of the order of $10^{-10}\mu_B$, is much larger than is obtained in calculations with conventional electroweak models. An alternative explanation of the apparent anticorrelation of SNUs with spots is that it is due to a rather improbable coincidence. Observations of the neutrino capture rate with the ^{37}Cl experiment in 1990 and 1991 will help to resolve this question.

No conclusive evidence exists that any of the neutrinos have a nonzero mass. Direct experiments, such as the measurement of the electron energy spectrum from ^{3}H decay, have yielded, with certainty, only upper limits on the masses. The mass of a $\bar{\nu}_e$ is known to be less than 16 eV from observations of the supernova, SN1987A (see Chapter 15). There are also arguments based upon cosmology that suggest that none of the neutrinos has a mass greater than 80 eV.

The standard model of **electroweak interactions** is normally presented in a form in which neutrinos are massless. However, in most extensions of the model, neutrinos have a mass. Of particular interest are **grand unified theories (GUTs)** which unify the different types of interaction (weak, electromagnetic, and strong) into a single interaction at very high energies, or equivalently at a very large mass scale, M_X. Most GUTs require that neutrinos have a mass.

In GUTs, the neutrino masses are inversely proportional to the mass scale M_X. For many models, the neutrino mass, m_ν, is $\sim M_{eW}^2/M_X$, where M_{eW} is a characteristic electroweak mass of order 10^2 GeV. For M_X as large as 10^{15} GeV, as required in many GUTs, all neutrinos have masses well below 1 eV. Furthermore, one expects $m(\nu_e) << m(\nu_\mu) << m(\nu_\tau)$, so that all masses are many orders of magnitude smaller than the range accessible by direct terrestrial experiments.

The physics community was electrified in 1985 when an elegant solution for the solar neutrino problem was proposed that is consistent with expectations from GUT theories of neutrino mass. According to this solution, a ν_e created in the solar interior is almost completely converted into ν_μ or ν_τ as the neutrino passes through the Sun. This conversion reflects the enhancement by the matter in the Sun of the probability that a neutrino of an electron type oscillates into a neutrino of a different type; it is usually referred to as the Mikheyev–Smirnov–Wolfenstein **MSW effect** in honor of its discoverers (see the discussion in Chapter 9).

In order for the MSW effect to occur, the **flavor eigenstates** ν_e, ν_μ, and ν_τ must be different from the **mass eigenstates**. The flavor eigenstates are created in weak decays and have weak interactions with their associated charged leptons (electron, muon, and tau) that can be written in a simple (diagonal) form. The mass eigenstates, which have diagonal mass matrices, are the states in which neutrinos propagate in a vacuum. The mass eigenstates are often denoted by ν_1, ν_2, and ν_3. For a simplified description in terms of two eigenstates, the relation between flavor and mass eigenstates in vacuum is described by a single **mixing angle** θ_V, where $\tan\theta_V$ is the relative amplitude of ν_1 and ν_2 in the ν_e wave function ($\nu_e = \cos\theta_V \nu_1 + \sin\theta_V \nu_2$).

Neutrino oscillations occur because of the variation with time of the relative phases of different components of the neutrino wave function. The ν_e wave function has contributions from two or more mass eigenstates; the relative phases of the mass eigenstates change with time because different mass eigenstates move with different speeds. The changing phases cause the initial ν_e wave function to look sometimes like an eigenstate of a different flavor.

In a medium with a nonzero density of electrons and nucleons, the index of refraction of neutrinos is proportional to the density of scatterers times the forward scattering amplitude. Electron neutrinos have a different interaction with electrons than do neutrinos of any other flavor, an almost obvious but profound fact that is the essence of the MSW effect. The heavier mass eigenstate acquires a tiny amount of extra mass as a result of the special interaction of ν_e with electrons in the center of the Sun. This extra mass is proportional to the weak interaction coupling constant and to the ambient electron density.

Because the electron density (and hence the extra mass) varies, the relative phases of the different mass eigenstates change from what they are in the original ν_e wave function. The ν_e in vacuum most nearly resembles the lower mass eigenstate ν_1, but at the large ambient electron densities that exist in the solar interior the newly created ν_e most nearly resembles a heavier mass eigenstate ν_2. If the electron density changes sufficiently slowly (the adiabatic approximation) in the solar interior, ν_e may remain very similar to the mass eigenstate ν_2, which in vacuum is primarily ν_μ or ν_τ (and is not detected in the ^{37}Cl experiment). Even for very small vacuum mixing angles, the electrons in the Sun can change the phases of a ν_e wave so that it looks almost exactly like a neutrino wave of a different flavor.

For a large range of neutrino parameters, electron neutrinos created in the solar interior will pass through a region where the electron density is such that the energies of two mass eigenstates become nearly equal and the oscillation amplitude is maximized. The separation in energy is of order 10^{-20} GeV at the density at which two mass eigenstates are most nearly equal.

The large suppression of the ^{8}B flux observed in the ^{37}Cl experiment can be explained if the mixing angle, $\theta \gtrsim 0.6°$. The mass of the heavier neutrino can lie between 10^{-4} eV and 10^{-2} eV.

This range of mixing angles and masses appears natural in the theoretical framework of GUTs. Typically, in GUTs the heaviest neutrino, ν_τ, has a mass of the order $(100 \text{ GeV})^2/M_X$, where M_X is a large mass scale. In the simplest versions of some grand unified models, the value of M_X is of the order 10^{15} GeV, the unification scale. Thus $m(\nu_\tau) \sim 10^{-2}$ eV and $\Delta m^2 \sim 10^{-4}$eV^2, so that ν_e transforms into ν_τ, provided the mixing angle θ is not smaller than 10^{-2} and, as expected, $m(\nu_e) << m(\nu_\tau)$. In some recent theories, the value of M_X is given by an intermediate mass scale of order 10^{11} to 10^{12} GeV. In this case the mass of ν_τ is too large to affect the solar neutrino problem, although it could be of great importance in cosmology. However, the mass of ν_μ could be of the order of 10^{-2} eV so that the solar ν_e would transform into ν_μ.

The study of solar neutrinos is the low-energy frontier of high-energy physics. If the MSW solution is correct, then information about the grand unification mass scale, $M \sim 10^{15}$ GeV, is being obtained from an interaction that is driven by a neutrino mass difference of order 10^{-20} GeV. How marvelous and awesome a possibility!

The MSW effect is an attractive solution to the solar neutrino problem that can be tested by carrying out additional experiments on solar neutrinos.

1.15   Discriminating among explanations

Future experiments can discriminate between solutions to the solar neutrino problem based on nonstandard solar models and those
based on new neutrino physics. In nonstandard solar models, the
shape of the neutrino spectrum from individual neutrino sources,
such as the ^{8}B spectrum, will be the same as in Figure 1.1. The
total flux from a given source can be altered by astrophysical considerations but the shape is fixed by nuclear physics. On the other
hand, the transformation of ν_e to ν'_μ or ν_τ due to the MSW effect
would be energy dependent and therefore the spectrum would be
distorted. Nearly all proposed nonstandard solar models predict the
same flux as the standard model for the predominant ν_e from the pp
reaction. In contrast, the MSW effect can, for certain parameters,
decrease significantly the flux of ν_e pp neutrinos.

   In most oscillation scenarios, the missing ν_e arrive at Earth as ν_μ or
ν_τ. Detectors based on neutrino–electron scattering or pure neutral-
current interactions should be able to detect the missing neutrinos.

1.16   The next decade

The first quarter century of solar neutrino astronomy has produced a
well-defined "solar neutrino problem." Experiments to be performed
in the next decade may reveal the solution to this problem and point
us toward either a more complete theory of stellar energy generation
or of neutrino propagation. If we are lucky, solar neutrino experiments might do both.

1.17   Questions and answers

Certain questions are often asked following lectures on solar neutrinos, or at informal gatherings of physicists, chemists, or astronomers.
This section contains a representative sample of the questions that
are most frequently asked (or that are most instructive), together
with my answers. The answers presume in some cases a background
knowledge that is given later in the book or an expertise that is pos-
sessed by the asker of the question. Readers who prefer to acquire
information in a more orderly or formal manner may skip the following section since the discussion in the subsequent chapters does
not depend on these questions and answers. Some colleagues have

suggested that this colloquial section does not belong in a scholarly book. They are correct. This book is intended to stimulate and to inform, rather than to impress with scholarship.

The justification for giving the answers in the present chapter is that the questions have occurred to many people after only a brief introduction to the subject (e.g., in a one hour colloquium), similar to the overview given here. I have used personal pronouns throughout this section to indicate the informal nature of both the questions and the answers.

Question 1. What do you think is the cause of the solar neutrino problem?

> **Answer.** Simplicity. Our models of the solar interior and of neutrino propagation are not strongly constrained by experimental data. My guess is that a decade of new experiments will show that we need more sophisticated theoretical models, astrophysical and physical. I am impressed by the fact that all of the suggested solutions involving nonstandard solar models are in conflict with something we think we understand, whereas some suggestions of new physics, for example the MSW effect, do not suffer from this defect. Of course, the lack of conflict with solutions involving new physics may just reflect the difficulty of testing the proposed nonstandard physical theories except by solar neutrino experiments.
>
> Incidentally, I think the MSW effect is a beautiful idea, whether or not it is correct.

Question 2. Isn't it presumptuous to claim that you can calculate the equation of state of matter in the solar interior to sufficient accuracy to predict the neutrino fluxes?

> **Answer.** No. The equation of state in the interior is relatively simple. At the temperatures (keV) and densities (10^2 g cm^{-3}) of interest, the matter is fully ionized and is reasonably close to a perfect gas. Corrections for electron degeneracy and for plasma screening affect the calculated fluxes by a few percent or less.

Question 3. Are there any nuclear physics experiments that should be done in support of the proposed solar neutrino experiments?

> **Answer.** Yes. Calibration experiments for (p,n) reactions. These reactions are used in some cases to estimate neutrino cross sec-

tions. For example, the cross sections for neutrino absorption to excited states in the ^{71}Ga experiments are estimated by studying the reaction ^{71}Ga(p,n)^{71}Ge, where the ^{71}Ge is left in different excited states. Measurements should be made for many cases in which the matrix elements are known accurately from β-decay experiments to calibrate the ratio of weak interaction matrix elements to (p,n) cross sections.

The ^{3}He(^{3}He,2p)^{4}He, the ^{3}He(^{4}He,γ)^{7}Be, and the ^{7}Be(p,γ)^{8}B cross sections should be remeasured if anyone has a good idea how the experiments could be done differently or better. The ^{7}Be+p reaction is especially important; a precision measurement with a 1σ accuracy of better than 5% would be a breakthrough. The thermal neutron capture cross section on ^{3}He should be measured since the calculated standard model flux of hep neutrinos is approximately proportional to this quantity.

Question 4. Which of the nonstandard solar models do you think has the best chance of being correct?

Answer. I don't have a favorite. But, I am impressed with the discussion by Press of the nonlinear focusing of internal gravity waves in the solar core. There is probably not enough energy in the relevant convective frequencies for this proposal to be correct, but the theoretical analysis indicates the complexity and sophistication that may be required if we are to change significantly the standard solar models.

Question 5. Don't measurements of p-mode solar oscillations make solar neutrino experiments unnecessary? Aren't oscillation measurements providing the same information at much less cost?

Answer. No. The oscillation measurements and the neutrino experiments are complementary. The p-mode oscillations are most sensitive to the outer regions of the Sun; the solar neutrinos to the innermost regions. Consider a hypothetical example. Suppose that for some nuclear physics reason the ^{7}Be(p,γ)^{8}B reaction does not occur at all, that it has a cross section of zero. This would not have a measurable effect on any solar oscillation measurements, but solar neutrino experiments would be greatly affected. Even setting the ^{3}He+^{4}He reaction rate to zero, which affects the structure of the calculated model star, does not affect significantly the calculated p-mode oscillation frequencies. Both oscillation and neutrino experiments are necessary to understand the Sun.

Question 6. Why was there only one experiment in 25 years?

    Answer. I don't know. Both Ray Davis and I tried for many years to help get other experiments started. But each solar neutrino experiment is difficult and expensive; each one ultimately takes at least a decade of work. Few people are willing to put in that amount of time and work on one experiment, an experiment that is neither pure physics, nor pure astronomy, nor pure anything. Just hard work. In addition, it has been difficult for funding agencies in the United States to figure out how to handle solar neutrino experiments; they don't fit into any of the standard categories of particle physics, nuclear physics, chemistry, or astronomy. They are a mixture, and that is hard to fund out of budgets that are designed for specific disciplines. The bottom line is: Everyone is in favor of interdisciplinary projects but no one wants the money to come out of his or her discipline's budget.

Question 7. Why should the ^{37}Cl solar neutrino experiment be continued? Hasn't it already told us about all it is going to?

    Answer. The ^{37}Cl experiment should operate during the era of the Kamiokande II, ^{71}Ga, and the other next-generation experiments in order to provide cross-checks on unusual events or runs. Certainly, the ^{37}Cl measurements must be continued in order to settle observationally the controversial question of whether or not there is a dependence of counting rate on phase in the sunspot cycle. Also, the background in the ^{37}Cl experiment can and should be measured accurately in order to refine the inferences about solar neutrino fluxes.

Question 8. I have just talked with Professor X who reports that he has recalculated the standard solar model and has found a value for the neutrino capture rate that is very different from what you have given. Doesn't this mean you have underestimated the errors in the solar models?

    Answer. No. It's easy to get the wrong answer if you apply a stellar evolution program written for another purpose to the solar neutrino problem. The accuracy required in physical description and in input parameters is much higher for solar neutrino work than for most other stellar evolution research. I am sure that his answer will agree with the standard one after we have a chance to compare computer codes in detail. There have already been many such cross-checks made.

Question 9. The value you have obtained for the initial helium abundance seems too low to me. It is not significantly greater than what I think is the best value for the amount of helium produced in the Big Bang. Shouldn't you change the model in some way to give a better value?

> **Answer.** That's not the way we play the game. If the result from stellar evolution is in conflict with a cosmological inference, then either stellar evolution or cosmology is wrong. If you try to force stellar evolution to give the cosmologically preferred solution, then you are elevating cosmology to astronomical theology, which (in my opinion) is not healthy for either cosmology or stellar evolution.

Question 10. Isn't the standard solar model in trouble because of the large predicted increase of luminosity with solar age? Wouldn't this luminosity evolution have seriously affected the Earth and its inhabitants?

> **Answer.** I don't think so. Global ecology is not understood well enough to use the history of the Earth as a sensitive test of solar evolution. My impression is that we are not able to predict with certainty in which direction a luminosity increase would influence life on Earth. The physics of the atmosphere is too complicated, with too many unknowns. One must also know the albedo of the Earth as a function of stellar age, which involves additional significant uncertainties. It would be interesting to turn the question around: Can one find at least one plausible geophysical model that is consistent with the evolution predicted by the standard model and with everything else we know about the history of the Earth?

Question 11. Do you really think you can calculate the response of the ^{37}Cl detector to an accuracy of 10%?

> **Answer.** Yes. The original calculations were based in part on a theoretical nuclear model. The same model predicted that ^{37}Ca would decay by positron emission and could be used to determine empirically the nuclear matrix elements. The appropriate nuclear experiments were done. In agreement with the predictions of the model, ^{37}Ca was found to decay by positron emission, and the measurements were used to refine the neutrino cross sections.

Question 12. Which of the nuclear reactions determines the observed solar luminosity or the rate of nuclear energy generation? Is it the slow initial pp reaction?

    Answer. No. The solar luminosity is determined by atomic and gravitational physics, not by nuclear physics.

    The rates of the nuclear reactions, which are sensitive to temperature and density, adjust themselves so that the net flow of energy equals what is required by macroscopic physics. You can read the opposite (and wrong) assertion in many review papers. Don't be misled. A lot was understood about stellar structure before the nuclear source of energy production was identified.

Question 13. What caused the big decrease in the uncertainty in the predicted capture rates from 1964 to 1968?

    Answer. Improvements in the measured low-energy nuclear reaction rates, especially the ^{3}He–^{3}He cross section.

Question 14. I have read that the pp neutrino flux is independent of solar models. Is that because the flux is related closely to the observed luminosity?

    Answer. The claim has been made often in the literature that the pp flux is independent of the solar model. This claim is wrong. If the central temperature in the standard solar model were slightly higher, then most of the predicted neutrinos would be from the CNO cycle, not the pp chain. Even assuming the pp chain is dominant, the number of pp neutrinos is uncertain. Depending on which termination of the pp chain is dominant, reaction 3 (the ^{3}He+^{3}He reaction) or reaction 4 (the ^{3}He+^{4}He reaction) of Table 1.1, the calculated flux can vary by a factor of two. One pp neutrino and one ^{7}Be (or ^{8}B) neutrino are produced for each 25 MeV of fusion energy if the ^{3}He+^{4}He reaction is much more frequent than the ^{3}He+^{3}He reaction, a situation that would imply a pp neutrino flux equal to about half the standard value. If reaction 3 occurs, as it does for most of the terminations in the standard solar model, two pp neutrinos are required to complete the fusion chain of Eq. (1.3) and obtain 25 MeV fusion energy. The relative frequency of reactions 3 and 4 can be determined theoretically only with the aid of an accurate solar model. Fortunately, the input parameters for the solar model permit a relatively precise calculation of the pp flux. One can make a model-independent estimate of the pp

neutrino flux that is accurate to a factor of two by assuming
that each 25 MeV of photon energy that reaches the Earth from
the Sun is accompanied by either one or two pp neutrinos.

Question 15. How can you be certain that hydrodynamic phenomena, perhaps analogous in complexity to the red spot on Jupiter, are
not important in the solar interior?

    Answer. I can't. Hydrodynamical or turbulent phenomena are
    not well understood in astronomical contexts. However, many
    suggestions made over the past two decades have been investi-
    gated. None of the proposed mechanisms for driving major hy-
    drodynamical phenomena seem very plausible with our current
    understanding. The standard solar model is stable to spherical
    and nonspherical thermal perturbations. The energy required
    to mix the Sun enough to affect appreciably the neutrino fluxes
    is of order a few percent of the gravitational binding energy of
    the Sun, thousands of times larger than the present solar rota-
    tional energy. No energy source has been identified that could
    drive hydrodynamic phenomena in the solar interior. Measure-
    ments of the acoustic oscillation frequencies indicate that, to a
    depth of order 0.3 of the solar radius, there are no large depar-
    tures from a standard, relatively quiescent Sun.

Question 16. You have made the errors so large that it looks like
nothing can be learned. Why do you quote 3σ errors? Isn't there
anything you can do to reduce those errors. I have always quoted
1σ errors for my solar neutrino experiment.

    Answer. I wish the errors were smaller! But, I give my best
    judgment as to what the uncertainties are so that people can
    decide how to think about the difference between your obser-
    vations and my calculations, about whether the difference is
    big enough to justify building new experiments or exploring
    new theories. The most important single measurement that
    one could perform to reduce the uncertainties would be a preci-
    sion measurement, accurate to 3% (1σ), of the low-energy cross
    section factor for the $^7\mathrm{Be}(p,\gamma)^8\mathrm{B}$ reaction, but even this would
    not reduce greatly the total theoretical uncertainty. The total
    uncertainty is contributed by many different quantities, no one
    of which is dominant. You can of course divide the errors I
    quote by three if you prefer dealing with effective 1σ errors.

Question 17. How do you rank, in order of likelihood, the various
particle physics solutions of the solar neutrino problem?

Answer. The MSW solution is the most probable in my view because it requires the minimum extension of the successful standard electroweak theory. All the other explanations require something that is *a priori* surprising: vacuum oscillations (fine-tuning, large mixing angle), spin flip (large magnetic moment), decay (large coupling constant), and WIMPs (new particle, new interactions).

Question 18. The measured capture rates in the ^{37}Cl experiment show a lot of fluctuations. Is there evidence for variability?

Answer. In the data obtained prior to 1985, there was no variability I believed was significant. The data obtained after the experiment was resuscitated with new pumps in 1987 and early 1988 are higher than the previous average by two standard deviations. I do not know what to make of this. Measurements that will be made with the ^{37}Cl and Kamiokande II detectors over the next several years are required in order to settle the question of time dependence. It is obviously of great importance.

Question 19. I think you are being unfair to the WIMP hypothesis. This is the only known explanation that can solve both the solar neutrino problem and the dark matter problem. What's more, it accounts naturally for the tiny but persistent discrepancy between calculation and observation of the frequency separation between some nearly degenerate acoustic wave oscillations, the p-modes. Why don't you agree that this is a beautiful idea, just as beautiful as the MSW effect?

Answer. You are right; WIMPs are beautiful. But, I think they are unlikely to exist with the properties necessary to solve the solar neutrino problem. The scattering cross section off of protons must be of order 10^3 times as large as the cross section for ordinary Z^0 exchange. The annihilation cross section must be strongly suppressed relative to the scattering cross section or there must be a cosmological asymmetry between WIMPs and anti-WIMPs. The mass of the WIMPs must be in a narrow window between 2 GeV and 10 GeV. There is not a shred of laboratory evidence that WIMPs exist with these remarkable properties. Also, the small discrepancy between calculation and observation of the solar acoustic modes may have a purely astronomical explanation, for example, it could be due to poorly understood phenomena associated with the outer regions of the

Sun or to enhanced chemical inhomogeneities due to element
diffusion.

Question 20. Isn't the solar neutrino problem overblown? As a
practicing astronomer, I know of many examples in which calculations do not agree with observations to better than a factor of two.
If you take your lowest allowed value, about 5.3 SNU, and Davis's
largest allowed value, about 3.0 SNU, the difference is even less than
a factor of two. What's the big deal?

> **Answer.** You are right that the difference between theory and ob-
> servation is not a large factor, but it is the ratio of the difference
> to the uncertainties that is significant. The standard calcula-
> tions and the observations do not overlap even if you push the
> theory and the observations simultaneously to their extreme
> limits set by known uncertainties, as you have done in compar-
> ing 5.3 SNU with 3.0 SNU. The solar neutrino problem differs
> from many puzzles in astronomy in that we know more about
> the Sun than about any other star and we believe that there is a
> fundamental theory, stellar evolution, that permits an accurate
> calculation. Solar evolution is the simplest and best understood
> case of stellar evolution. If we can't calculate accurately how
> the Sun shines, then we may have to be cautious in claiming
> we understand more complicated astronomical systems.

Question 21. What do you think is the most important quantity
to measure in order to solve the solar neutrino problem?

> **Answer.** The immediate key to the solar neutrino problem is
> the neutrino energy spectrum. The standard solar model and
> all non-standard solar models (without radically new particle
> physics) predict the same energy spectra from individual neu-
> trino sources, like the ^{8}B neutrinos or the pp neutrinos. Er-
> rors in the solar model can change the number of neutrinos
> from a particular source (like ^{8}B neutrinos) but cannot affect,
> within measurable accuracy, the shape of the spectrum of ener-
> gies emitted by a single source. The spectrum of energies from a
> specific neutrino source is determined by nuclear physics and is
> the same no matter what is going on in the Sun, provided noth-
> ing happens to the neutrinos after they are created in the solar
> interior. Explanations that involve particle physics beyond the
> standard electroweak model generally predict observed energy
> spectra at Earth that are different from what is expected on
> the basis of the standard model. In the nonstandard particle

physics explanations, what happens to the neutrinos depends upon their energies.

Experiments that are in progress or that are being developed can perform some of the required measurements. The ^{8}B neutrinos are the experimentally most accessible source to study for determining the shape of the energy spectrum and this will be done in neutrino-electron scattering experiments and in future experiments using neutrino absorption. A measurement of the shape of the spectrum of the fundamental low-energy pp neutrinos is of crucial importance (since some of the particle-physics effects depend strong on energy) but is much more difficult. A measurement of the shape of the ^{7}Be neutrino line could determine directly the average solar interior temperature, from the width of the line, and is the ultimate challenge to experimentalists.

Techniques are being developed to measure neutral current interactions, which determine the total flux of neutrinos independent of their type (flavor). These measurements will provide a sharp discrimination between the large class of solar neutrino explanations in which particle physics causes neutrinos to change flavors but conserves the total number of neutrinos and the opposite explanations in which the Sun produces a smaller number of neutrinos than is predicted by the currently standard solar model.

2

Stellar structure and evolution

Summary

Stellar evolution is the prototype of a good astronomical theory. Astronomers use the theory of how stars evolve to interpret observations in many different fields, for example, calculating the ages of stars, making quantitative inferences regarding chemical composition, unraveling the evolution of galaxies, and deciphering the universal Big Bang (using the inferred primordial helium abundance and the ages of the oldest stars as major clues). Each of these applications, employed routinely by astronomers in their everyday work, is called into question by at least one of the suggested explanations of the solar neutrino problem (see Chapter 5). A lot is at stake for astronomers in solar neutrino tests of the theory of stellar evolution.

The Sun is in the simplest state of stellar evolution, the long-lived quasistatic **main sequence** (or hydrogen burning) phase. Most astronomers study more exotic astronomical systems (in which things "happen" more rapidly) and nearly all astronomers concentrate on photons (not neutrinos) that are emitted from the outermost regions of stars where complicated nonlinear effects such as turbulence and convection are important. One might expect that these more dynamic astronomical situations would also be more difficult to understand. After all, we know much more about the Sun than we do about any other star. Moreover, the Sun's interior is believed to be in a quiescent state and therefore the relevant physics is simple.

However, after decades of work, we are still not certain whether the physics of solar energy generation is as simple as expected. We have to await more diagnostic solar neutrino experiments for a final decision.

The study of the physics of the interior of the Sun requires a different approach than the statistical investigations that are appropriate for most research in stellar evolution. Astronomers usually predict or measure the relative numbers of stars with different masses, colors, ages, or chemical compositions. In studying the Sun, we focus instead on the detailed physical processes of an individual object.

This chapter outlines the general principles of the theory of stellar evolution and summarizes the overall comparison with observations. The first section, 2.1, describes the life history of a star in terms of the separate time scales for the gravitational, electromagnetic, and strong forces. The following section, 2.2, presents the basic equations of stellar evolution and uses these equations to obtain order of magnitude estimates for the interior temperature of the Sun and for the central pressure. Two of the principal triumphs of stellar evolution theory are the prediction of the mass–luminosity relation and the explanation of the Hertzsprung–Russell diagram, the locus of observed stars in the plane of luminosity versus color. Section 2.3 describes the basis for these theoretical results and compares the predictions with observations of galactic stars. The final section, 2.4, summarizes the status of stellar evolution theory. Standard calculations give a satisfactory account of many observations made with photons and are used routinely by astronomers to interpret their results. Only solar neutrino experiments are apparently in direct conflict with the standard theory.

In order to understand what is being tested by solar neutrino experiments, one needs to know at least as much about stellar evolution as is presented in this brief chapter. To do research in stellar evolution, one must know much more. Annotated references to some of the classical textbooks and review articles in stellar evolution are given at the end of this chapter [see works by Chandrasekhar (1939), Salpeter (1957), Schwarzschild (1958), Sears (1964), Iben (1967, 1974), Cox and Giuli (1968), Tayler (1970), Clayton (1983), and VandenBerg (1985)].

2.1   Life history of a star

The life history of a star is an interplay between the gravitational, the electromagnetic (or atomic), and the strong nuclear forces. These three forces determine the characteristic time scales in a star's life. The characteristic time scales are derived below for stars like the Sun using order of magnitude estimates. Here $M_\odot$, $R_\odot$, and $L_\odot$ are, respectively, the solar mass, radius, and luminosity (numerical values given in Table 4.1).

First, the **gravitational time scale** is the ratio of the gravitational energy to the total luminosity.

$$t_{\text{gravity}} \approx GM_\odot^2/R_\odot L_\odot \approx 10^7 \,\text{yr}. \qquad (2.1)$$

The gravitational time scale is also approximately equal to the thermal time scale (or **Kelvin–Helmholtz time scale**), the rate at which the stored thermal energy is radiated: $E_{\text{thermal}}/L_\odot$. The equivalence of these two estimates for the gravitational time scale is guaranteed by the virial theorem, which relates kinetic and potential energies.

The **atomic time scale** is the time for a photon to random-walk out of the solar interior. The total number of steps is the square of the distance traveled, $R_\odot$, to the typical photon mean path, λ. The time to go one step is of order (λ/c). Thus:

$$t_{\text{atomic}} \sim (R_\odot/\lambda)^2 \frac{\lambda}{c} \approx 10^4 \,\text{yr}. \qquad (2.2)$$

A typical mean free path is $\sim 1/\kappa \langle \rho \rangle$, ~ 1 cm, where κ is the product of the Thomson cross section and Avogadro's number and $\langle \rho \rangle \sim 1$ gm cm^{-3}.

Some authors have mistakenly assumed that $t_{\text{atomic}} \approx t_{\text{gravity}}$. The two characteristic times would be equal if most of the thermal energy were in the form of photons. However, in the solar interior most of the thermal energy is in the form of random motions of the electrons and ions, while only $\sim$ a few tenths of a percent resides in photons. The **nuclear time scale** is:

$$t_{\text{nuclear}} \approx \epsilon \times 0.1 \times M_\odot c^2/L_\odot \sim 10^{10} \,\text{yr}. \qquad (2.3)$$

Here 0.1 represents the approximate fraction of a stellar mass that is exhausted before a star leaves the main sequence and becomes a giant star that evolves more rapidly. The quantity ϵ represents the fraction (0.7%) of the rest mass that is converted to thermal energy in the burning of protons into α-particles. The nuclear time scale is the characteristic epoch over which a star evolves in the main sequence.

The Sun is the only star close enough for us to use the fourth known force, the weak force, to look inside a stellar core and to explore, with the aid of neutrinos, a stellar interior that is steadily burning nuclear fuel. The photons that we see are last scattered (or are produced) in the outer 0.05% of a solar radius, whereas neutrinos come to us directly from the solar core. In fact, essentially all of the observations of stellar structure and evolution with which astronomers are familiar rely on light emitted from the outer layers of a star.

2.2   Basic equations

The basic equations of stellar evolution and structure involve simple physics. They are given explicitly here so that we can see what is most important for solar interior models. Spherical symmetry and quasistatic equilibrium without material convection are assumed in order to obtain equations that are relatively transparent. Both of these assumptions are quantitatively accurate for constructing models of the solar interior. Rotation, mass loss, and magnetic fields are not considered here; the possible effects of these phenomena are discussed in Chapter 5 in connection with nonstandard models.

The first equation is the condition for **hydrostatic equilibrium**. Let $P(r)$ and $\rho(r)$ be the pressure and density at a distance r from the center of the star and let $M(r)$ be the enclosed mass. The balance between the gravitational and pressure forces implies that

$$\frac{\mathrm{d}P(r)}{\mathrm{d}r} = -\frac{GM(r)\rho(r)}{r^2}. \qquad (2.4)$$

Equation (2.4) is satisfied to high accuracy since the Sun would collapse in less than an hour if the equality were significantly violated. The great simplicity that results from the assumption of quasistatic equilibrium is justified, in the case of the Sun, by evidence that life has existed on the Earth for more than a billion years and hence that the solar luminosity has not changed drastically during that period.

The temperature gradient between the interior and the surface gives rise to a net flux of energy from the central regions of the Sun. The equation governing **energy transport** is:

$$L_r = -4\pi r^2 (ac/3) \frac{1}{\kappa\rho} \frac{\mathrm{d}T^4}{\mathrm{d}r}. \qquad (2.5)$$

Here L_r is the energy per unit time that passes through a sphere of radius r and T is the temperature. The total **opacity** κ is the combination of a radiative and a conductive opacity: $\kappa^{-1} \equiv \kappa_{\mathrm{rad}}^{-1} + \kappa_{\mathrm{cond}}^{-1}.$

For solar interior conditions, the radiative opacity dominates the total opacity. The pressure is the sum of the radiative and the (dominant) thermal pressure:

$$P(r) = \frac{a}{3}T^4 + \frac{1}{\mu}\frac{k\rho T}{m_{\mathrm{H}}}\left(1 + D\right).$$ (2.6)

Here k is Boltzmann's constant and D represents easily calculated corrections for the degeneracy of electrons and for Debye–Hückel modifications to the equation of state. Let X, Y, and Z be the fractions by mass of the material that is in the form, respectively, of hydrogen, helium, and heavier elements. The **mean molecular weight** μ is, in terms of these parameters: $\mu^{-1} = 2X + (3/4)Y + (1/2)Z$.

The rate at which the luminosity is produced in spherical shells is the sum of nuclear and mechanical energy generation:

$$\frac{dL_r}{dr} = \rho(4\pi r^2)\left(\epsilon_{\mathrm{nuclear}} - T\frac{dS}{dt}\right),$$ (2.7)

where S is the stellar entropy. For main sequence stars like the Sun, nuclear energy is generated at a much greater rate than mechanical energy. The observed luminosity is the integral over the star of the local energy generation:

$$L_\odot = \int_0^{R_\odot} \left(\frac{dL_r}{dr}\right) dr,$$ (2.8)

assuming that the star is in an approximate steady state.

In order to complete the description of the equations of stellar structure, we must specify the boundary conditions at the outer edge of the star. The boundary conditions have both a good and a bad aspect. The bad aspect is that the outer layers of stars like the Sun are strongly influenced by convection and by turbulence, physical processes that are not well understood in the stellar context. The good aspect is that the outer boundary conditions do not matter much for processes that occur in the deep interior. In practice, for stars like the Sun, it is sufficient to assume that the outer regions are in convective equilibrium and to apply well-known approximate relations in this region to determine the dependence of pressure on temperature. Detailed calculations, changing the assumptions and the parameters used, show that the conventional atmospheric approximations have no significant effect on the calculated solar neutrino fluxes. However, the uncertain physics in stellar atmospheres often does complicate the comparison between theory and observations made with photons emitted from the surface layers. For example, interpretations of solar p-mode oscillations involve detailed models of the outer layers of the Sun.

An order of magnitude estimate for the central pressure and temperature can be made in the same spirit as the estimates of the characteristic time scales in Eqs. (2.1) to (2.3). The central pressure can be estimated with the aid of Eq. (2.4). Assume that the Sun can be represented by three regions: a central core, a midway surface, and an outer surface (where the pressure is zero). Apply Eq. (2.4) at the midway surface. Then the left-hand side can be set equal to the central pressure and the density can be taken to be equal to the mean solar density (1 g cm^{-3}). With these crude approximations, the central pressure is:

$$P_c \sim 2\rho G M_\odot / R_\odot \sim 6 \times 10^{15} \text{ dynes cm}^{-2}. \tag{2.9}$$

Using the equation of state of a perfect gas, the central temperature of the Sun is about:

$$T_c \sim \frac{m_H P_c}{2k\rho_c} \sim 10^7 \text{ K}. \tag{2.10}$$

These estimates for the central pressure and temperature are in satisfactory agreement with the more accurate values obtained by detailed calculations (see Chapter 4).

2.3   Principal results

The basic equations lead to two powerful results: (1) a relation between stellar mass and luminosity; and (2) a description of what kinds of stars can exist. We begin with the mass–luminosity relation. Dimensional analysis is the simplest way to derive the shape of the **mass–luminosity relation**. Assuming, as the crudest approximation, that the opacity is independent of temperature and density (e.g., the opacity is caused only by electron scattering), one can use the equation for energy transport from the previous section to obtain the proportionalities:

$$L_r \propto \frac{R^2 T^4}{\langle \kappa \rangle \rho R} \propto \frac{R^4 T^4}{\langle \kappa \rangle M}. \tag{2.11}$$

In addition, the equation of state indicates that

$$T \propto \frac{P}{\rho} \propto \frac{M\rho}{R\rho} \propto \frac{M}{R}. \tag{2.12}$$

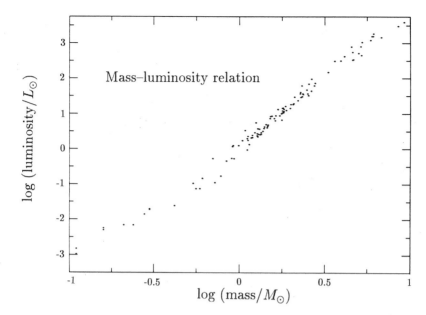

Figure 2.1   Mass–luminosity relation   This figure shows the observed relation between mass and luminosity for binary stars whose masses and luminosities are accurately measured. The data are taken from Popper (1980), Tables 2, 4, 7, and 8.

Combining Eqs. (2.11) and (2.12), one obtains the simplest form of a mass–luminosity relation:

$$L \propto M^3. \tag{2.13}$$

This relation represents the general trend of how the observed luminosities of stars depend on their masses, although the exact value of the exponent obtained with detailed stellar models varies systematically with mass, in large part because of the dependence of opacity on temperature and density.

Figure 2.1 shows the observed mass–luminosity relation for stars in the disk of our Galaxy. It is difficult to measure stellar masses accurately; this can only be accomplished in special systems (such as well-separated binary stars). Total luminosities are also difficult to measure since one must determine the distance to the star, as well as add to the visual radiation all the energy that is emitted in other wavelength regions. The Sun is the only star close enough that we can measure its total luminosity or mass to an accuracy of a percent (or better). The data in Figure 2.1 are therefore necessarily somewhat uncertain. However, a tight relation between luminosity and mass is apparent.

Theoretical calculations reproduce the main features of the observed mass–luminosity relation, exhibiting different power-law dependences in different mass ranges. This agreement between calculation and observation constitutes a major success of the theory of stellar structure and evolution.

Most other triumphs of the theory of stellar evolution are connected with the famous **H–R diagrams** (**Hertzsprung–Russell** diagrams), an example of which is shown in Figure 2.2. An H–R diagram shows the logarithm of the stellar luminosity as a function of some property of the stellar surface (such as color, type of spectrum, or effective temperature, defined below).

Astronomers often display their results in an H–R diagram that shows bolometric magnitude, M_{bol}, versus $B–V$ (blue–visual) color. The **bolometric magnitude**, M_{bol}, is defined in terms of the total luminosity, L, by:

$$M_{bol} = 4.63 - 2.5 \, \text{Log}_{10}(L/L_\odot). \tag{2.14}$$

The $B–V$ color is 2.5 times the logarithm of the ratio of the received fluxes in two broad wavelength bands, visual (V) and blue (B). The value of the solar luminosity, $L_\odot$, is 3.86×10^{33} ergs^{-1}. The **effective temperature** is sometimes used in H–R diagrams instead of the color or spectral type. By definition, luminosity $= 4\pi R^2 \sigma T_{eff}^4$. The color $B–V$ is related to the effective surface temperature of a star by accurate calibrations.

The most striking feature of an H–R diagram is its emptiness: Stars only inhabit a small fraction of the two-dimensional parameter space (cf. Figure 2.2).

Figure 2.2 shows the H–R diagram for stars in the solar neighborhood. Like all observational H–R diagrams, Figure 2.2 is expressed in terms of the luminosity within a given wavelength band (here the violet, V, band). The visual absolute magnitude is defined by an equation analogous to Eq. (2.14) except that all quantities refer to luminosities emitted in the V-band and the constant 4.63 is replaced by 4.79.

For the nearby stars shown in Figure 2.2, which are a mixture of ages and compositions, the so-called "main sequence" is the band

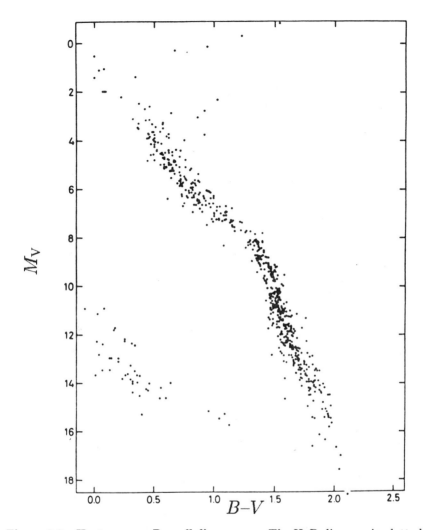

Figure 2.2   Hertzsprung–Russell diagram   The H–R diagram is plotted for 535 stars in the solar neighborhood whose distances and photon fluxes in two colors (B and V) are accurately known. The vertical axis shows the absolute visual magnitude, M_V, which is proportional to the logarithm of the luminosity emitted in the visual wavelength band. The horizontal axis, B–V, is proportional to the logarithm of the ratio of received fluxes in two broad wavelength bands, blue (B) and visual (V). The Sun is plotted at $M_V = 4.8$ and B–$V = 0.6$. This figure is taken from Gliese (1983).

stretching from the upper left-hand corner of the H–R diagram to the lower right-hand corner. The stars with the highest surface temperatures have the smallest B–V colors and are on the left-hand side of the diagram; the most luminous stars are at the top. The stars on the right-hand side of the main sequence strip have lifetimes on the main sequence that are longer than the current age of the Galaxy. The lower right-hand corner contains the least massive known stars, with masses $\lesssim 0.1 M_{\odot}$, which are very faint red objects known as M dwarfs and brown dwarfs. The upper left-hand part of the diagram contains the rare short-lived blue stars that are much more massive than the Sun. To the right and above the main sequence are the evolved red giants, and in the lower left-hand corner are the faint white dwarfs, dying stars cooling through the last stages of stellar evolution. The paucity of observed giant stars is the result of the short amount of time spent in the giant phase.

The Sun is a typical main sequence star with $M_V = 4.8$ and B–$V = 0.6$.

To a first approximation, the theory of stellar evolution is an explanation of the H–R diagram. According to current understanding, an embryonic solar-mass star evolves almost vertically down in this diagram until it reaches the main sequence. Then the star sits quietly on the main sequence burning hydrogen until the proton fuel of its core is exhausted. When the core hydrogen is used up, the star begins to evolve more rapidly towards the giant branch, where its subsequent history is the result of a succession of episodes involving nuclear energy generation and gravitational contraction. After one or more phases of significant mass loss, the final stages of evolution of the remnant core can lead to the formation of a **white dwarf**, a dense star that is supported by electron degeneracy pressure and which radiates thermal energy accumulated by contraction. For stars that are more massive than some value which is not known well (but which may be of order $8 M_{\odot}$), remnant cores cannot be supported by electron degeneracy pressure, but instead collapse to a much denser star, a neutron star, in which neutron degeneracy and nuclear forces provide the support. The formation of a neutron star can give rise to a spectacular explosion, a supernova (discussed in Chapter 15). For still more massive stars, black holes may be formed.

2.4 Present status

Computer codes that calculate stellar evolution are described in a number of excellent review articles. Many well-tested codes are in existence around the world, although most of the programs are not suitable for calculating precise solar neutrino fluxes. Codes written for other purposes usually do not evaluate with the accuracy and detail required of solar-neutrino calculations some of the physics of the nuclear reactions, of the opacity, of the radiative transfer, or of the equation of state. But the principles are the same.

Many of the existing codes can reproduce well the statistical features of the mass–luminosity relation and of the H–R diagram described in the preceding section. Stellar evolution codes also show why the upper left-hand corner of the diagram is empty for older stars that populate globular clusters (these massive stars evolve rapidly and have all left the main sequence in globular clusters). The theory is also able to account approximately for the difference in the main sequence luminosities between the relatively young disk stars that are the most common representatives in the solar neighborhood and the older globular cluster stars. The older stars have fewer metals than the young disk stars, which formed after the interstellar medium was polluted by heavy elements created in the explosions of first generation stars. This relative lack of metals changes the radiative opacity, which is reflected in the observed and calculated H–R diagrams of globular clusters and in their calculated (but not yet observed) mass–luminosity laws.

The theory of stellar evolution gives a satisfactory account of many observations and is routinely used by astronomers to interpret their results. Some problems remain. The ages of stars in the oldest globular clusters appear to be dangerously close to exceeding the estimated age of the universe. Fewer faint white dwarf stars are observed than are predicted by simple calculations. There is not yet a comprehensive theory of star formation. Some effects, such as magnetic fields, rotation, or turbulence, are difficult to model realistically and accurately.

Nevertheless, only the ^{37}Cl and Kamiokande II experiments are in direct conflict with the standard theory. Even for the neutrino experiments, there is a conflict with stellar evolution theory only if the standard model of electroweak interactions is correct.

Bibliographical Notes

Chandrasekhar, S. (1939) *An Introduction to Stellar Structure* (Chicago: University of Chicago Press). Authoritative account of the mathematical theory of stellar structure prior to the introduction of modern numerical methods.

Clayton, D.D. (1983) *Principles of Stellar Evolution and Nucleosynthesis* (Chicago: University of Chicago Press). The standard textbook from which students (and professors) have learned the fundamentals of the subject for two decades (first edition 1968). This edition contains a new preface which lists many of the modern references.

Cox, J.P. and R.T. Giuli (1968) *Principles of Stellar Structure* (New York: Gordon & Breach) Vols. I and II. Comprehensive summary of the theory.

Eddington, A.S. (1926) *The Internal Constitution of the Stars* (Cambridge: Cambridge University Press). The first systematic treatment of radiative equilibrium inside stars.

IAU Symposium 105 (1984) *Observational Tests of the Stellar Evolution Theory*, edited by A. Maeder and A. Renzini (Dordrecht: Reidel). A recent summary of stellar evolution theory with only a passing reference to the solar neutrino problem.

Iben, I. Jr. (1967) *Ann. Rev. Astron. Astrophys.*, **5**, 571; (1974) *Ann. Rev. Astron. Astrophys.*, **12**, 215. Comprehensive discussion of the evolution of stars based upon numerical models.

Salpeter, E.E. (1957) *Rev. Mod. Phys.*, **29**, 244. A concise, clear explanation of the physical principles of stellar energy sources and stellar structure by a master of both subjects.

Schwarzschild, M. (1958) *Structure and Evolution of the Stars* (Princeton: Princeton University Press). Classical description of the theory of stellar evolution with emphasis on physical understanding. The clearest book ever written on the subject.

Sears, R.L. (1964) *Ap. J.*, **140**, 477. One of the earliest detailed models of the Sun.

Tayler, R.J. (1970) *The Stars: Their Structure and Evolution* (Wykeham Science Series, No. 10). A lucid elementary introduction to the physical principles.

VandenBerg, D.A. and R.A. Bell (1985) *Ap. J. Supp.*, **58**, 561; VandenBerg, D.A. (1985) *Ap. J. Supp.*, **58**, 711. State-of-the-art numerical models for stars with masses comparable to the Sun.

3

Nuclear fusion reactions

Summary

The Sun shines by burning hydrogen fuel. As the fuel is consumed, the star evolves. Stellar evolution is accompanied by nuclear fusion, which incidentally produces neutrinos that can be used to test the theory of how stars evolve.

The Sun and other main sequence stars evolve slowly by adjusting their temperatures so that the average thermal energy of a nucleus is small compared to the Coulomb repulsion an ion feels from potential fusion partners. The large Coulomb repulsion slows the nuclear reaction rates to astronomically long time scales. The determination of nuclear reaction rates in stars is a difficult task because the relevant energies are so low that the laboratory event rates are small. Fortunately, after decades of intense laboratory and theoretical research, the fusion cross sections that are most important for the solar neutrino problem are known to satisfactory accuracy.

Main sequence stars fuse protons to gain energy via the proton–proton chain or the CNO cycle, two distinct sets of nuclear reactions. Observations of neutrinos of different energies can provide direct evidence of the number and type of fusion reactions that occur in the solar interior.

This chapter summarizes the role of nuclear reactions in main sequence stars. The first section, 3.1, shows why nuclear fusion is the only known energy source that can supply the luminosity of a star for a stellar lifetime. The following section, 3.2, provides a summary of the principal relations that are used in calculating the rates of nuclear reactions in main sequence stars. The next section, 3.3, reviews the reactions in the pp chain, giving accurate values for the cross section parameters and their uncertainties. Section 3.4 presents the relations between neutrino fluxes, which are measurable, and the frequencies of reactions in the pp chain, which must be calculated with a stellar model. These relations can be used to determine the relative frequency of some nuclear reactions which do not produce any neutrinos as well as those which do emit neutrinos. For the CNO cycle, Section 3.5 gives accurate values for the cross section parameters. Section 3.6 shows how to calculate the rate at which thermal energy is produced in a star using the reaction rates given in the previous sections. Finally, Section 3.7 describes a difficult challenge for nuclear experimentalists, the reduction of the uncertainties in the reaction rates to a level at which the estimated errors in the rates do not influence the interpretation of the solar neutrino experiments.

Nuclear reactions in stars have been discussed extensively in the literature. The reader is referred to the references cited in the main text of the chapter for details that are omitted here, see especially Bethe (1939), Burbidge *et al.* (1957), Cameron (1957, 1958), Salpeter (1957), Parker, Bahcall, and Fowler (1964), Fowler, Caughlan, and Zimmerman (1975), Barnes (1981), Clayton (1983), Fowler (1984), Parker (1986), and Rolfs and Rodney (1988), as well as the more specific references cited in the text of Tables 3.2 and 3.3.

3.1   The essentials

The Sun has been shining for several billion years. Evidence to support this proposition comes from different fields of science. For example, fossils of primitive organisms have been found on Earth that are more than a billion years old. In addition, the oldest known rocks (in Southwest Greenland) are of order 3.8 billion years old and meteoritic ages are typically found to be about 4.5 billion years, with **very** small dispersion. All of the available evidence suggests that the solar system was formed, and the Sun began to shine, several billion years ago.

The amount of energy that is required to maintain sunshine for such a long time is orders of magnitude larger than the available chemical or gravitational energy. We saw in Chapter 2 [cf. Eq. (2.1)] that gravitational energy could support the solar luminosity for only about ten million years, hundreds of times too short to encompass the history of the solar system. The available chemical energy is even less, insufficient to maintain the Sun's luminosity for the approximately ten thousand years of recorded history.

The fusion of hydrogen nuclei to form helium is the only known process that is sufficient to provide the required amount of energy for the long times required by the biological, geological, and astronomical evidence. The arguments by which physicists and astronomers reached this conclusion were developed in the early 1900s by a number of prominent scientists. The story of the struggle to develop a quantitative theory of nuclear energy fusion is told in brilliant and fascinating works by Russell (1919), Eddington (1920, 1926), Gamow (1938), Chandrasekhar (1939), Bethe (1939), and Fowler (1984) that are listed at the end of this chapter under Bibliographical Notes.

Looking back over the past century, the development of the theory of nuclear energy generation contains brilliant insights and false starts, both supported initially by about equal amounts of observational information. As a result of the many factual clarifications that have been wrested from difficult laboratory experiments and from increasingly more detailed calculations, it is now possible to describe the status of the theory in a concise and coherent form.

The luminosity of the Sun is determined by the atomic opacity, the chemical composition, and the balance of gravitational and pressure forces, not by nuclear physics. We derived in Chapter 2 the approximate luminosity of a star in terms of these quantities without referring to the source of stellar energy [see Eq. (2.11)]. Atomic and gravitational physics set the scale of stellar luminosities, which corresponds to a slow rate of nuclear fusion.

What is the characteristic luminosity, L_{ch}, for the Sun that is obtained from an order of magnitude estimate based solely on the nuclear energy release per reaction? Ignoring the implications of atomic and gravitational physics, the characteristic luminosity may be written

$$L_{ch} \sim \frac{\epsilon N \Delta E}{\tau}, \tag{3.1}$$

where ϵ is the fraction of the total number of solar nuclei, N ($\sim 10^{57}$), that participate in fusion and $\Delta E \sim 25$ MeV. The characteristic lifetime, τ_{ch}, is

$$\tau_{ch} \sim \frac{1}{n\sigma v}, \tag{3.2}$$

where the central number density is $n \sim 10^{26}$ cm^{-3} and the ion velocities are $v \sim 10^9$ cm s^{-1}. A typical nuclear cross section for energies above the Coulomb barrier is $\sigma \sim 1$ millibarn $\sim 10^{-27}$ cm^{-3}. These numbers yield a short characteristic lifetime, $\tau_{ch} \sim 10^{-8}$ s, much less than the astronomical time scale of 10^{10} yr. Assuming a plausible value of $\epsilon \sim 10^{-2}$, one finds

$$L_{ch} \sim 10^{20} L_{\odot}. \tag{3.3}$$

This order of magnitude calculation yields a luminosity that is 20 orders of magnitude too large! To achieve the appropriate luminosity, a star must carry out nuclear fusion at energies (i.e., temperatures) that are well below the Coulomb barrier so that the effective fusion cross sections are reduced many orders of magnitude below the values that pertain at MeV kinetic energies.

In order to maintain the slow rate of energy generation corresponding to the macroscopic physics, a star adjusts its temperature so that the average thermal energy of a nucleus is small compared to the maximum repulsive potential between two nuclear charges of the same sign, the **Coulomb barrier**. When the ratio between the Coulomb barrier and the thermal energy is large (it is typically greater than a thousand for the Sun, see below), the reaction rates are slowed to the required macroscopic time scales. A direct consequence of this slowing of the reaction rates is that most of the energy production of a star comes from a narrow effective energy range. The effective energy is large compared to the thermal energy but small compared to the barrier height. Therefore, the Maxwell–Boltzmann factor and the Coulomb barrier penetration factor are both small in the effective energy range but their product is a maximum.

Since the kinetic energy that is most effective in producing stellar nuclear fusion is low, the cross sections usually cannot be measured at the relevant energies in the laboratory. The fact that stars insist on hoarding their energy and expending it at rates which are unnaturally low by laboratory physics standards makes experiments in nuclear astrophysics difficult.

The mean lifetime is about 10 billion years (the age of the universe) for a proton to be fused together with another proton to form deuterium in the solar center. This fact should be kept in mind when gauging the difficulty of simulating solar nuclear reactions on the time scale of a single graduate student thesis.

The overall transformation that occurs as the result of the reactions discussed below can be represented symbolically by the relation

$$4p \to \alpha + 2e^+ + 2\nu_e, \tag{3.4}$$

that is, four protons are fused to form an α-particle, two positrons, and two neutrinos of the electron "flavor."[†] The positrons annihilate with free electrons providing the star with $2m_e c^2$ ($= 1.02$ MeV) of energy in addition to the nuclear and kinetic energies of the fusing particles. The total energy released in Eq. (3.4) is $\simeq 26.731$ MeV, of which, on the average, only a small amount (0.6 MeV for the current standard solar model) is carried away by neutrinos. It is customary to refer to the process represented by Eq. (3.4) as **hydrogen burning** and to speak of each completed conversion as a **termination** of the proton–proton chain or the CNO cycle.

3.2 Reaction rates

The **reaction rate** (number of reactions per unit volume per unit of time) between two nuclear species is described by the formula

$$R_{12} = \frac{n(1)n(2)}{(1 + \delta_{12})} \langle \sigma v \rangle_{12}, \tag{3.5}$$

where $n(1), n(2)$ are the number densities of particles of type 1 and 2, σ is their interaction cross section, v the magnitude of their relative velocity, and the Kronecker delta prevents double counting of identical particles. A simpler notation is convenient for performing calculations. One can write

$$R_{12} \equiv \frac{\langle 1, 2 \rangle \, n(1)n(2)}{(1 + \delta_{12})}, \tag{3.6}$$

where the angular bracket represents, by comparison with the previous equation, the average product of cross section times velocity.

The rates for nuclear fusion in the solar interior are dominated by Coulomb barriers. Equation (2.10) shows that the typical thermal energy of particles in the solar interior is only a few keV, which is much smaller than the typical height of Coulomb barriers – a few MeV– among light elements. Nuclear fusion is possible in main

---

[†]In some cases, the precise transformation is $4p + e^- \to \alpha + e^+ + 2\nu_e$, which is almost equivalent to Eq. (3.4).

sequence stars because of the quantum mechanical effect of tunneling through a potential barrier first discussed by Gamow in connection with α-decay [see Gamow (1938)].

The energy dependence of the fusion cross sections can be represented most simply by a formula in which the geometrical factor, the De Broglie wavelength squared, and the barrier penetration factor have both been removed, leaving a function $S(E)$ that varies smoothly in the absence of resonances. The conventional definition is:

$$\sigma(E) \equiv \frac{S(E)}{E} \, \exp(-2\pi\eta), \qquad (3.7)$$

where

$$\eta = Z_1 Z_2 (e^2/\hbar v). \qquad (3.8)$$

The quantity $\exp(-2\pi\eta)$ is known appropriately as the **Gamow penetration factor**. The value of $S(E)$ at zero energy is know as the **cross section factor**, S_0.

To a large extent, the possibility of performing laboratory experiments that are relevant to nuclear fusion in main sequence (and giant) stars depends upon the simplicities that are expressed by the definitions given in Eqs. (3.7) and (3.8). The Gamow penetration factor is responsible for the decrease of nuclear fusion cross sections by many orders of magnitude between energies that are practical for laboratory experiments and the much lower solar energies. Typically, the laboratory rates are conveniently large at energies of order several MeV but rapidly become too small to measure at energies of order of 100 keV or below. Fortunately, the complicated aspects of nuclear physics are embodied in the experimentally accessible function, $S(E)$, which for light nuclei is not expected (except in the presence of rare low energy resonances that are treated separately) to change much on scales that are significantly less than an MeV. The basic strategy adopted by all workers [following the pioneering investigation of Bethe (1939)] is to measure the cross section for energies that are accessible in the laboratory and to use Eq. (3.7) (sometimes guided by theoretical arguments about the shape of the extrapolation to lower energies) to parametrize the cross sections at thermal energies.

At the temperatures and densities relevant for the solar interior, the interacting particles reach a Maxwellian distribution in a time that is infinitesimal compared to the mean lifetime for a nuclear reaction. Thus the average over velocities indicated in Eq. (3.5) can be carried out using a Maxwell–Boltzmann distribution. The average cross section times velocity can then be written as a convenient function of energy in the following form:

$$\langle \sigma v \rangle = \left(\frac{8}{\pi \mu (kT)^3} \right)^{1/2} f_0 \int_0^\infty dE \, S(E) \exp(-2\pi\eta - E/kT), \quad (3.9)$$

where μ is the reduced mass of the interacting particles. The quantity f_0 is a factor, calculated accurately by Salpeter (1954), which enhances the reaction rate and is caused by the screening of bare nuclear charges by electrons from the solar plasma. In the weak screening approximation that is appropriate for the Sun, one can write $f_0 = \exp(0.188 Z_1 Z_2 \zeta \rho^{1/2} T_6^{-3/2})$, where T_6 is the temperature in units of 10^6 K, $\zeta = [\sum_i (X_i Z_i^2 / A_i + X_i Z_i / A_i)]^{1/2}$, and X_i is the mass fraction of nuclei of type i. A small correction to Salpeter's expression has been derived by Carraro, Schäfer, and Koonin (1988) [for screening effects in low-energy laboratory experiments see Assenbaum, Langanke, and Rolfs (1987) and Engstler et al. (1988)].

The integration over kinetic energies can be accomplished by performing a power series expansion in terms of the inverse of a large quantity, τ, where:

$$\tau = 3E_0/kT = 42.487 \left(Z_1^2 Z_2^2 A T_6^{-1} \right)^{1/3}. \quad (3.10)$$

The expansion is carried out in the vicinity of a most probable energy of interaction, E_0,

$$\begin{aligned} E_0 &= \left[(\pi \alpha Z_1 Z_2 kT)^2 (mAc^2/2) \right]^{1/3} \\ &= 1.2204 (Z_1^2 Z_2^2 A T_6^2)^{1/3} \text{ keV}. \end{aligned} \quad (3.11)$$

Here m is the atomic mass unit and $A = A_1 A_2 / (A_1 + A_2)$ is the reduced mass in atomic mass units. The power series expansion in τ^{-1} converges rapidly since τ is large (typically 15 to 40) for all non-resonant reactions of interest in the pp chain or the CNO cycle. For the most important reactions in the Sun, E_0 is between 6 keV and 10 keV.

The average product of cross section times velocity can be written in the compact form

$$\langle \sigma v \rangle = 1.3005 \times 10^{-15} \left[\frac{Z_1 Z_2}{A T_6^2} \right]^{1/3} f_0 S_{\text{eff}} \exp(-\tau) \text{ cm}^3 \text{ s}^{-1}, \quad (3.12)$$

when S_{eff} is expressed in keV–barns. To first order in τ^{-1} [Bahcall (1966a)],

$$S_{\text{eff}} = S(E_0) \left\{ 1 + \tau^{-1} \left[\frac{5}{12} + \frac{5 S' E_0}{2S} + \frac{S'' E_0^2}{S} \right]_{E=E_0} \right\}. \qquad (3.13)$$

Here $S' = \mathrm{d}S/\mathrm{d}E$. In most analyses in the literature, the values of S and associated derivatives are quoted at zero energy, not at E_0. In order to relate Eq. (3.13) to the usual formulae, one must express the relevant quantities in terms of their values at $E = 0$. The appropriate connection is

$$S_{\text{eff}}(E_0) \simeq$$

$$S(0) \left[1 + \frac{5}{12\tau} + \frac{S'\left(E_0 + \frac{35}{36}kT\right)}{S} + \frac{S'' E_0}{S}\left(\frac{E_0}{2} + \frac{89}{72}kT\right) \right]_{E=0} \qquad (3.14)$$

If the last term in Eq. (3.14), proportional to S'', is set equal to zero, then the above expression reduces to the usually quoted formula.

The temperature dependence of a nonresonant thermonuclear reaction is determined by the above equations. It is sometimes instructive to define an approximate power-law dependence on temperature by the equation

$$\text{Rate} \propto T^n. \qquad (3.15a)$$

In terms of the characteristic ratio, τ, defined above, the exponent n is

$$n = (\tau - 2)/3. \qquad (3.15b)$$

The "effective" exponent by which a nuclear reaction rate or neutrino flux depends upon temperature will differ from the value given in Eq. (3.15b), in which τ is assumed to be calculated at a specific point in the star (e.g., the center). The effective exponent results from an average over the temperature–density profile of the star (see Section 6.2).

The lifetime for a nucleus of type 1 in the presence of nuclei of type 2 is

$$T_1 = \frac{1}{n_2 \langle \sigma v \rangle_{12}}. \qquad (3.16)$$

Equation (3.16) has been used to calculate the lifetimes under typical solar conditions that are quoted in the following section.

3.3 The pp chain

A *The principal reactions*

Table 3.1 summarizes the principal nuclear reactions that are important in main sequence stars like the Sun, the **proton–proton chain** (or **pp chain**). Alternative nuclear reactions are unimportant under solar conditions [see, e.g., Parker, Bahcall, and Fowler (1964), Bahcall and Wolf (1964), Parker (1972), and Hardie *et al.* (1984)]. The table also shows, for the standard solar model, in what percentage of the terminations of the pp chain each reaction occurs and what are the energies of the neutrinos, if any, that are emitted. The percentage of terminations are computed using the equilibrium equations from Section 3.3C and the neutrino fluxes from Table 6.5.

The first reaction shown is the basis for the whole pp chain (hence the name). Qualitatively, the **pp reaction** (number 1a) resembles the β-decay of the neutron; in this case, a proton decays in the vicinity of another proton to form a bound deuteron. The rate for this primary reaction is too slow to be measured in the laboratory at relevant energies since the transformation proceeds via the weak interaction. However, the rate can be calculated accurately using the theory of low-energy weak interactions and the measured properties of proton–proton scattering and the deuteron. In more than 99% of the terminations of the proton–proton chain in the standard solar model an electron neutrino is emitted with a maximum energy of 0.420 MeV.

Detecting the neutrinos from the pp reaction is one of the primary goals of solar neutrino astronomy. The radiochemical ^{71}Ga experiments which are discussed in Chapter 11 are sensitive to pp neutrinos, together with other neutrino sources that also contribute to the overall detector response (see Table 11.1). Some electronic experiments have been proposed that could isolate the pp neutrinos for detailed study (see the discussion of low-temperature detectors in Section 13.8 and of ^{115}In in Section 14.3).

The next reaction in Table 3.1 is known as **pep**. The pep reaction is important in principle since the ratio of pep to pp neutrinos is practically independent of solar models. Thus pep neutrinos contain essentially the same information about the rate of the basic fusion reactions as do the lower-energy pp neutrinos. Moreover, the 1.4 MeV (line) neutrinos from pep are above the threshold energy for absorption in the ^{37}Cl experiment (0.814 MeV threshold) although

Table 3.1. The pp chain in the Sun. The average number of pp neutrinos produced per termination in the Sun is 1.85. For all other neutrino sources, the average number of neutrinos produced per termination is equal to (the termination percentage/100).

Reaction	Number	Termination[†] (%)	ν energy (MeV)
$p + p \rightarrow {}^2H + e^+ + \nu_e$	$1a$	100	≤ 0.420
or			
$p + e^- + p \rightarrow {}^2H + \nu_e$	$1b$ (pep)	0.4	1.442
${}^2H + p \rightarrow {}^3He + \gamma$	2	100	
${}^3He + {}^3He \rightarrow \alpha + 2p$	3	85	
or			
${}^3He + {}^4He \rightarrow {}^7Be + \gamma$	4	15	
${}^7Be + e^- \rightarrow {}^7Li + \nu_e$	5	15	(90%) 0.861 (10%) 0.383
${}^7Li + p \rightarrow 2\alpha$	6	15	
or			
${}^7Be + p \rightarrow {}^8B + \gamma$	7	0.02	
${}^8B \rightarrow {}^8Be^* + e^+ + \nu_e$	8	0.02	< 15
${}^8Be^* \rightarrow 2\,\alpha$	9	0.02	
or			
${}^3He + p \rightarrow {}^4He + e^+ + \nu_e$	10 (hep)	0.00002	≤ 18.77

[†]The termination percentage is the fraction of terminations of the pp chain, $4p \rightarrow \alpha + 2e^+ + 2\nu_e$, in which each reaction occurs. The results are averaged over the model of the current Sun. Since in essentially all terminations at least one pp neutrino is produced and in a few terminations one pp and one pep neutrino are created, the total of pp and pep terminations exceeds 100%.

those from the basic pp reaction are not. The rate for the pep reaction can be calculated accurately using weak interaction theory in terms of the rate of the pp reaction. Unfortunately, the reaction is sufficiently rare ($\sim 0.4\%$ of the terminations in the standard solar model) that the expected signal is comparable with the background for the ${}^{37}Cl$ experiment.

The sum of the pp and pep contributions in column 3 of Table 3.1 exceeds 100% of the terminations. In essentially all of the terminations, at least one pp neutrino is produced. Most often, two pp neutrinos are created. In about 0.4% of the terminations, one pep neutrino and one pp neutrino are produced. About twenty terminations in a million produce two pep neutrinos and no pp neutrinos.

The ^{2}H+p reaction (reaction 2 in Table 3.1) is so fast that its rate is unimportant. It always occurs, but with no observable signature. Historically, the rate of the ^{3}He+^{3}He reaction (reaction 3 of Table 3.1) constituted the principal uncertainty in the initial prediction of the solar neutrino fluxes (see the discussion of Figure 1.2 in Chapter 1). Fortunately, the reaction has by now been well studied and its rate at thermal energies is rather accurately known. This reaction completes the pp chain in about 85% of the terminations in the standard solar model, with no additional neutrino emission.

The ^{3}He+^{4}He reaction (reaction 4 of Table 3.1) leads to the two important neutrino-producing reactions involving ^{7}Be. The cross section for the ^{3}He+^{4}He reaction was first measured by Holmgren and Johnston (1958, 1959) and found to be much larger than was expected. Fowler (1958) immediately pointed out the importance of the Holmgren–Johnston result for solar neutrino astronomy. The reaction has been studied extensively in the past few years and the rate is now well determined (see Table 3.2).

In the Sun, ^{7}Be is almost always destroyed by electron capture (reaction 5 of Table 3.1), usually from free electrons in the solar plasma. The rate for this process can be calculated accurately using weak interaction theory.

The ^{7}Be+p reaction (reaction 7 of Table 3.1) occurs only rarely in the standard solar model, in about 1 out of every 5000 terminations of the pp chain. (The ^{7}Be electron capture reaction is about a thousand times more probable.) Nevertheless, the ^{7}Be+p reaction is of crucial importance since it leads to ^{8}B neutrinos with energies as high as 14 MeV that are more easily detected than the more abundant lower-energy (pp, pep, and ^{7}Be) neutrinos. Despite their rarity, neutrinos from ^{8}B decay dominate the predicted capture rate in the ^{37}Cl experiment and in most of the other proposed solar neutrino experiments (see Table 1.3 and Chapters 10 to 14). Unfortunately, there are still significant uncertainties in the low-energy cross section for this reaction [see Parker (1986), Bahcall and Ulrich (1988), and the references in Section 7.2A].

The final reaction in Table 3.1 is the hep reaction, which produces the highest-energy solar neutrinos. The neutrinos from this reaction are extremely rare, but they may be measurable in some of the electronic experiments that are discussed in Chapter 14 (e.g., with the D_2O and liquid argon detectors).

B   The reaction parameters

Table 3.2 presents the reaction parameters for the most important reactions in the pp chain. For each reaction, the table lists the total thermal energy release – **Q-value** – including the $2m_e c^2$ following positron annihilation, exclusive of neutrino energy losses. The average neutrino energy loss, $\langle q \rangle$, is also shown; this quantity was computed by averaging over the spectrum of neutrinos that are emitted. To measure one of the low-energy cross section factors, S_0, usually requires years of work by expert nuclear experimentalists. The final column in Table 3.2 is the calculated lifetime in the solar interior of the target (first-listed) nucleus.

The caption to the table contains references to some of the principal work on each reaction. More complete references to this research can be found in Fowler, Caughlan, and Zimmerman (1967, 1975), Kavanagh (1972), Rolfs and Trautvetter (1978), Barnes (1981), Bahcall *et al.* (1982), Fowler (1984), Parker (1986), Bahcall and Ulrich (1988), and Rolfs and Rodney (1988). The reader should consult especially the discussion in Bahcall *et al.* (1982) and Parker (1986) for a critical analysis of the rates and their uncertainties.

Two reactions that are important in the pp chain do not involve nuclear fusion, but instead electron capture. These electron capture reactions are (numbers 2 and 6) the pep and ^{7}Be–e$^-$ reactions. The rates for these reactions can be calculated accurately using weak interaction theory and the local physical conditions of the solar plasma. The results are:[†]

$$R_{\mathrm{p+e+p}} \cong 1.102 \times 10^{-4} (\rho/\mu_e) T_6^{-1/2} (1 + 0.02T_6) R_{\mathrm{pp}}, \qquad (3.17)$$

and

$$R_{(^7\mathrm{Be+e^-})} = 5.54 \times 10^{-9} (\rho/\mu_e) T_6^{-1/2} [1 + 0.004(T_6 - 16)]\,\mathrm{s}^{-1}. \qquad (3.18)$$

Here $\mu_e \simeq [2/(1 + X)]$ is the mean molecular weight per electron.

[†]The evaluation of the ^{7}Be electron capture rate was the first calculation the author did in solar neutrino theory, see Bahcall (1962b). R. Davis, Jr., at the suggestion of W.A. Fowler, wrote Bahcall about the possibility of calculating the rate.

Table 3.2. Reaction parameters for the proton–proton chain. The uncertainties for the cross section factors are indicated in parentheses; they correspond to 3σ errors and have been calculated using the data given in the original experimental papers [see Parker (1986) and Bahcall and Ulrich (1988) for discussions]. Some of the principal references are given below for each numbered reaction.

Reaction	No.	Q (MeV)	$<q_{\nu_e}>$ (MeV)	S_0 (keV barns)	dS/dE (barns)	t (yr)
^{1}H(p,e$^+\nu_e$)^{2}H	1	1.442	0.265	4.07(1±0.051)E−22	4.52E−24	10^{10}
^{1}H(p,e$^-$,ν_e)H	2	1.442	1.442	[see Eq. (3.17)]		10^{12}
^{2}H(p,γ)^{3}He	3	5.494		2.5E−04	7.9E−06	10^{-8}
^{3}He(^{3}He,2p)^{4}He	4	12.860		5.15(1±0.17)E+03	−0.9	10^5
^{3}He(^{4}He,γ)^{7}Be	5	1.586		0.54(1±0.06)	−3.1E−04	10^6
^{7}Be(e$^-$,ν_e)^{7}Li	6	0.862	0.862	[see Eq. (3.18)]		10^{-1}
		0.384	0.384			
^{7}Li(p,α)^{4}He	7	17.347		52(1 ± 0.5)	0	10^{-5}
^{7}Be(p,γ)^{8}B	8	0.137		0.0243(1 ± 0.22)	−3E−05	10^2
^{8}B(e$^+\nu_e$)^{8}4Be*	9	17.980	6.710			10^{-8}
^{8}Be*(α)^{4}He						
^{3}He(p,e$^+\nu_e$)^{4}He	10	19.795	9.625	8E−20		10^{12}

References: (1) Bahcall and May (1969) and Bahcall *et al.* (1982); (2) Bahcall and May (1969); (3) Griffiths, Lal, and Scarfe (1963); (4) Bahcall and Ulrich (1988) and Krauss, Becker, Trautvetter, and Rolfs (1987); (5) Alexander, Ball, Lennard, Geissel, and Mak (1984), Parker (1986), and Hilgemeier *et al.* (1988); (6) Bahcall and Moeller (1969); (7) Rolfs and Kavanagh (1986); Fowler (1987); (8) Filippone, Elwyn, Davids, and Koetke (1983) and Parker (1986); (9) Wilkinson and Alburger (1971) and Warburton (1986); (10) Werntz and Brennan (1973) and Tegner and Bargholtz (1983).

Note that the rate of the pep reaction is proportional to the rate of the pp reaction. Tests with a number of solar models have shown that the proportionality factor is practically independent of the details of the stellar model; the variation in the proportionality constant from one stellar model to another is $\lesssim 10\%$.

3.4 Neutrino fluxes and terminations of the pp chain

The basic process in the pp chain is the burning of four protons to form an α-particle, two positrons, and two neutrinos. The isotopes of ^{2}H, ^{3}He, ^{7}Be, ^{7}Li, and ^{8}B are collaborating spectators whose local number densities do not change when the chain is in equilibrium. Thus one can derive convenient and informative relations giving the relative abundances of the participating ions by considering the steady state equations balancing creation and destruction of species whose total number density is constant.

For example, by considering the reactions involving ^{2}H one immediately writes down

$$n(^2\text{H}) = \frac{\langle ^1\text{H}, ^1\text{H} \rangle \, n(^1\text{H})}{2 \, \langle ^1\text{H}, ^2\text{H} \rangle}. \qquad (3.19)$$

Similarly, the balancing of reactions involving ^{3}He implies that

$$\langle ^3\text{He}, ^3\text{He} \rangle \, n(^3\text{He})^2 + \langle ^3\text{He}, ^4\text{He} \rangle \, n(^3\text{He})n(^4\text{He})$$
$$= \frac{\langle ^1\text{H}, ^1\text{H} \rangle \, n(^1\text{H})^2}{2}, \qquad (3.20)$$

where the small contribution of the pep reaction is omitted from the right-hand side of Eq. (3.20). The equilibrium relations for ^{7}Be require that

$$n(^7\text{Be}) = \frac{\langle ^3\text{He}, ^4\text{He} \rangle}{\langle e^-, ^7\text{Be} \rangle \, n(e) + \langle ^1\text{H}, ^7\text{Be} \rangle \, n(^1\text{H})} n(^3\text{He})n(^4\text{He}). \qquad (3.21)$$

These relations can be used, together with the reaction rates discussed above, to calculate the relative abundances of each of the ions of interest in the pp chain. However, a word of caution is in order. There is a region in the Sun where the temperature is high enough to produce ^{3}He (via reactions 1 and 2 of Table 3.1) but where the temperature is too low to burn ^{3}He (via reactions 4 or 5 of Table 3.1). In this nonequilibrium region, Eq. (3.20) does not apply. Fortunately, this temperature domain is not important for our further considerations.[†]

The equilibrium relations given in Eq. (3.19), (3.20), and (3.21) can be used to compute, in terms of the emitted neutrino fluxes, ϕ, the fractions, F, of terminations of the pp chain that occur via the ^{3}He+^{3}He reaction and the ^{3}He+^{4}He reaction. Each conversion of

---

[†]In the region outside $0.2R_\odot$, the ^{3}He that is produced is not burned because the temperature is too low. This results in a nonequilibrium abundance for ^{3}He that increases as the Sun ages, see discussion of Figures 7 and 8 in Bahcall and Ulrich (1988).

four protons to an α-particle is known as a **termination** of the energy-generating chain. In deriving the expressions, one must remember that there is a factor of $1/2$ that occurs in the expression for the rate at which terminations are accomplished by the ^{3}He+^{3}He reaction, a factor which is absent in the expression for the rate of destruction of ^{3}He nuclei. The $1/2$ is present in the termination rate because the interacting particles are identical. Since two particles are destroyed for each termination by the ^{3}He+^{3}He reaction, the factor of two cancels out of the destruction rate.

The fraction of terminations, F, can be written as

$$F(i,j) \equiv \frac{\langle i,j \rangle\, n(i)n(j)}{(1+\delta_{12})[R_{33}+R_{34}]}, \qquad (3.22)$$

where the term in square brackets in the denominator of Eq. (3.22) represents the total rate of the terminations from the ^{3}He+^{3}He reaction (R_{33}) and the ^{3}He+^{4}He reaction (R_{34}). The total rate is proportional to one-half of the sum, Sum, of the fluxes

$$\text{Sum} = [\phi(\text{pp}) + \phi(\text{pep}) + \phi(^7\text{Be}) + \phi(^8\text{B})]. \qquad (3.23)$$

For simplicity, the tiny contribution of $\phi(\text{hep})$ to the sum has been omitted.

For all neutrino sources except pp and hep, the termination fraction is equal to the number of neutrinos produced per termination. Equation (3.22) must be generalized somewhat for pp and pep neutrinos since more than one pp or pep neutrino can be produced if a termination occurs via the ^{3}He+^{3}He reaction.

The fraction of terminations that occur via the ^{3}He+^{3}He reaction is

$$F(^3\text{He},^3\text{He}) = \frac{[\phi(\text{pp}) + \phi(\text{pep}) - \phi(^7\text{Be}) - \phi(^8\text{B})]}{[\phi(\text{pp}) + \phi(\text{pep}) + \phi(^7\text{Be}) + \phi(^8\text{B})]}, \qquad (3.24)$$

and the fraction that occurs via the ^{3}He+^{4}He reaction is

$$F(^3\text{He},^4\text{He}) = \frac{2\,[\phi(^7\text{Be}) + \phi(^8\text{B})]}{[\phi(\text{pp}) + \phi(\text{pep}) + \phi(^7\text{Be}) + \phi(^8\text{B})]}. \qquad (3.25)$$

The ratio, averaged over the Sun, of the number of pep reactions to the number of pp reactions is

$$N(\text{pep})/N(\text{pp}) = \phi(\text{pep})/\phi(\text{pp}), \qquad (3.26)$$

and the average ratio of ^{7}Be+p reactions to ^{7}Be electron capture is

$$N(^8\text{B})/N(^7\text{Be}) = \phi(^8\text{B})/\phi(^7\text{Be}). \qquad (3.27)$$

The average number of pp neutrinos produced per termination is

$$N(\text{pp}) = \frac{2\phi(\text{pp})}{[\phi(\text{pp}) + \phi(\text{pep}) + \phi(^7\text{Be}) + \phi(^8\text{B})]}, \tag{3.28}$$

which is about 1.85 for the standard solar model. The average number of pep neutrinos per terminations is much smaller (0.004),

$$N(\text{pep}) = \frac{2\phi(\text{pep})}{[\phi(\text{pp}) + \phi(\text{pep}) + \phi(^7\text{Be}) + \phi(^8\text{B})]}. \tag{3.29}$$

The fraction of terminations that involve ^{7}Be neutrinos is

$$F(^7\text{Be}) = \frac{2\phi(^7\text{Be})}{[\phi(\text{pp}) + \phi(\text{pep}) + \phi(^7\text{Be}) + \phi(^8\text{B})]}, \tag{3.30}$$

while the average number of ^{8}B neutrinos is tiny,

$$F(^8\text{B}) = \frac{2\phi(^8\text{B})}{[\phi(\text{pp}) + \phi(\text{pep}) + \phi(^7\text{Be}) + \phi(^8\text{B})]}. \tag{3.31}$$

The even smaller number of hep neutrinos per termination is

$$F(\text{hep}) = \frac{2\phi(\text{hep})}{[\phi(\text{pp}) + \phi(\text{pep}) + \phi(^7\text{Be}) + \phi(^8\text{B})]}. \tag{3.32}$$

The fraction of terminations in which pp neutrinos participate can be calculated from

$$F(\text{pp}) = (1 - \epsilon^2)R_{33} + (1 - \epsilon)R_{34}, \tag{3.33a}$$

where

$$\epsilon = \frac{\phi(\text{pep})}{\phi(\text{pp})}. \tag{3.33b}$$

The corresponding fraction for pep neutrinos is

$$F(\text{pep}) = (2\epsilon - \epsilon^2)R_{33} + \epsilon R_{34}. \tag{3.34}$$

Since in some terminations both pp and pep neutrinos participate, the sum of the fractions $F(\text{pp})$ and $F(\text{pep})$ exceeds unity,

$$F(\text{pp}) + F(\text{pep}) = 1 + 2\epsilon(1 - \epsilon)R_{33}. \tag{3.35}$$

The above relations for the fractional terminations and average numbers of neutrinos produced per termination show how measurements of different solar neutrino fluxes can reveal details about the operation of nuclear fusion reactions in the Sun. The termination fractions that are given in Tables 1.1 and 3.1 were derived using the above equations.

Table 3.3. The CNO cycle. The main reactions are shown for the carbon–nitrogen–oxygen cycle in the Sun.

Reaction	ν energy (MeV)
$^{12}C + p \rightarrow \,^{13}N + \gamma$	
$\quad ^{13}N \rightarrow \,^{13}C + e^+ + \nu_e$	$\lesssim 1.199$
$^{12}C + p \rightarrow \,^{14}N + \gamma$	
$^{14}N + p \rightarrow \,^{15}O + \gamma$	
$\quad ^{15}O \rightarrow \,^{15}N + e^+ + \nu_e$	$\lesssim 1.732$
$^{15}N + p \rightarrow \,^{12}C + \alpha$	

3.5 The CNO cycle

Table 3.3 shows the main reactions in the **carbon–nitrogen–oxygen cycle (CNO cycle)**. In this set of reactions, originally discussed by Bethe (1939), the overall conversion of four protons to form an α-particle, two positrons, and two neutrinos [Eq. (3.4)] is achieved with the aid of ^{12}C, the most abundant heavy isotope in normal stellar conditions. The total energy release – before deducting neutrino losses (see below) – is the same as for the pp chain (26.7 MeV).

At higher temperatures than are relevant for the solar interior, the CNO cycle involves infrequently traversed side chains [see, e.g., Caughlan and Fowler (1962), Parker (1986), and Rolfs and Rodney (1988)]. These side chains do not significantly affect the energy production via the CNO cycle, which itself only constitutes a small contribution to the total solar luminosity (of order 1.5% in the standard solar models). The additional high-temperature reactions are therefore not shown in Table 3.3.

Table 3.4 summarizes the main reaction parameters for the CNO cycle, including the S factors, the Q-values, and the mean neutrino losses ($\langle q \rangle$-values, computed by averaging over the neutrino spectra). For convenience, Table 3.4 also gives the cross section for the $^{16}O(p,\gamma)^{17}F$ reaction since the neutrinos produced by ^{17}F β-decay provide in principle a way for measuring the initial oxygen abundance in the solar interior [see Bahcall et $al.$ (1982)], a quantity of great interest for nuclear astrophysics and cosmology. At present

Table 3.4.   Reaction parameters for the CNO cycle. The uncertainties for the cross section factors are indicated in parentheses; they correspond to 1σ errors and have been calculated using the data given in the original experimental papers [see Bahcall *et al.* (1982) and Parker (1986)].

Reaction	Q (MeV)	$\langle q \rangle$ (MeV)	$S(0)$ (MeV barns)	$S'(0)$ (barns)	$S''(0)$ ($\frac{\text{barns}}{\text{MeV}}$)	Refs.
$^{12}C(p,\gamma)^{13}N$	1.943		1.45(1±0.15)E−03	2.45E−03	6.80E−02	1
$^{13}N(e^+,\nu_e)^{13}C$	2.221	0.7067				
$^{13}C(p,\gamma)^{14}N$	7.551		5.50(1±0.15)E−03	1.34E−02	9.87E−02	2
$^{14}N(p,\gamma)^{15}O$	7.297		3.32(1±0.12)E−03	−5.91E−03	9.06E−03	3
$^{15}O(e^+\nu_e)^{15}N$	2.754	0.9965				
$^{15}N(p,\gamma)^{16}O$	12.128		6.4(1±0.09)E−02	3E−02	4.0	4
$^{15}N(p,\alpha)^{12}C$	4.966		78.0(1±0.17)	351	1.11E+04	5
$^{16}O(p,\gamma)^{17}F$	0.600		9.4(1±0.16)E−03	−2.3E−02	6.0E−02	6
$^{17}F(e^+,\nu_e)^{17}O$	2.762	0.9994				

References: (1) Rolfs and Azuma (1974); (2) Fowler, Caughlan, and Zimmerman (1967); (3) Fowler Caughlan, and Zimmerman (1975) and Schröder *et al.* (1987); (4) Rolfs and Rodney (1974); (5) Zyskind and Parker (1979); (6) Rolfs (1973).

the measurement of the small ^{17}F neutrino flux does not appear to be feasible.

The $^{14}N(p,\gamma)^{15}O$ reaction is the slowest process in the chain. Below about 10^7 K, only the first three reactions in Table 3.4 occur sufficiently rapidly to take place in a solar lifetime. For regions of the Sun below this temperature, the ^{14}N is largely not burned and some ^{12}C is converted to ^{14}N. This incompleteness in the CNO cycle results in a moderate difference between the calculated ^{13}N and ^{15}O solar neutrino fluxes (see Table 6.5).

3.6   Energy generation

The rate at which nuclear reactions contribute energy to a star can be computed from the relation

$$\epsilon = \sum_r (Q - \langle q \rangle)_r R_r, \qquad (3.36)$$

where the values of Q and $\langle q \rangle$ are given in Table 3.2 for each reaction in the pp chain and in Table 3.4 for reactions in the CNO cycle. The reaction rates, R_r, are described in Section 3.2 with the appropriate cross section factors being listed in Tables 3.2 and 3.3. Bahcall and Ulrich (1988) (see their Appendix A and Table XXI) describe a convenient numerical scheme for treating the reactions in the pp chain and the CNO cycle, including the temperature dependence of the neutrino losses.

The average thermal energy released in the CNO cycle is 25.027 MeV. The average amount of energy released to the star by the ^{3}He+^{3}He termination is $\simeq 26.196$ MeV. (This result includes a small correction, calculated with the aid of the standard solar model of Chapter 4, for the neutrino energy loss by pep.) The average thermal energy release via the reactions that involve ^{7}Be electron capture (reactions 5 to 7 of Table 3.1) is 25.649 MeV and is only 19.754 MeV for the termination that proceeds through ^{8}B (reactions 8 and 9).

For the pp chain, the energy production rate can be written conveniently in the following form:

$$\epsilon(\mathrm{pp}) = 26.731\chi\langle^1\mathrm{H},{}^1\mathrm{H}\rangle\left[n(^1\mathrm{H})^2/2\right] \text{ MeV cm}^{-3}\,\text{s}^{-1}. \quad (3.37)$$

The quantity χ depends upon the frequency with which the chain is completed by the different terminations discussed above. For relatively low temperatures ($T \sim 10^6$ K), all of the terminations proceed via the ^{3}He+^{3}He reaction and $\chi \simeq 0.49$ (when neutrino energy losses are included). For temperatures more characteristic of the solar interior, $\chi \lesssim 0.6$.

The ratio of pp energy generation to CNO energy generation can be written in terms of the observable neutrino fluxes as follows:

$$\frac{L(\mathrm{CNO})}{L(\mathrm{pp})} =$$

$$\frac{[11.01\phi(^{13}\mathrm{N}) + 14.02\phi(^{15}\mathrm{O})]F(3,4)}{6.671\phi(\mathrm{pp})F(3,4) + [12.86F(3,3) + 18.98F(3,4)]\phi(^7\mathrm{Be})}. \quad (3.38)$$

Here the fractional terminations, $F(3,3)$ and $F(4,4)$, are defined, respectively, in terms of the neutrino fluxes in Section 3.4. For simplicity, the small ratios of $\phi(^8\mathrm{B})/\phi(^7\mathrm{Be})$ and $\phi(\mathrm{pep})/\phi(\mathrm{pp})$ are neglected in writing the above equation.

3.7 What remains to be done?

Over the past two decades, heroic experiments and laborious calculations have been performed to determine the rates of individual reactions in the pp chain and the CNO cycle. This work has determined the nuclear reaction rates to an accuracy that shows that the solar neutrino problem cannot be accounted for by recognized uncertainties in the nuclear fusion parameters. The results of this collective achievement, the work of hundreds of scientists in institutions distributed around the world, are summarized in Tables 3.2 and 3.3.

The challenge for the next two decades is to reduce the uncertainties in each of the reaction rates to a level where the errors in determining the nuclear fusion parameters do not influence the interpretation of solar neutrino experiments, which roughly translates into the goal of achieving a 3σ uncertainty in the predicted neutrino fluxes of less than 10%. Refinements in our knowledge of low-energy cross sections that will permit this accuracy in the predictions will be difficult and expensive in terms of time, thought, and accelerator facilities, but are required in order to keep the uncertainties resulting from nuclear fusion parameters from limiting the usefulness of solar neutrino experiments.

The most important reaction for solar neutrino astronomy, the ^{7}Be(p,γ)^{8}B reaction, is, with the existing data, the most susceptible to systematic uncertainties. Parker (1986) has analyzed six independent experiments and finds a 1σ uncertainty of 7.4%. The accumulated data for this reaction represents more than two decades of imaginative and difficult work. This crucial reaction is more difficult to study, compared to other important pp-chain reactions, because the target, ^{7}Be, is radioactive. Barker and Spear (1986) have drawn attention to systematic uncertainties in the normalization of the rate via the ^{7}Li(d,p)^{8}Li reaction, the stopping cross section of protons in lithium, the relative weighting of different experiments, and the extrapolation of the cross section to stellar energies. They suggest that the appropriate cross section factor could be 25% lower than derived by Parker (1986), who does not agree with some of the Barker–Spear comments. More experimental work is required because the predictions for many solar neutrino experiments (e.g., those using detectors of ^{37}Cl, deuterium, liquid argon, boron, molybdenum, and neutrino–electron scattering) are proportional to, or approximately proportional to, the value of the low-energy cross section factor for this reaction. The ultimate goal is to reduce the 1σ uncertainty to about 3% – a Herculean task – so that it not limit the interpretation of the solar neutrino experiments.

A precise measurement of the rate of the ^{7}Be(p, γ)^{8}B reaction represents the most important challenge for experimental nuclear astrophysics. The parameters for the important reactions ^{3}He(^{3}He,2p)^{4}He and ^{3}He(α, γ)^{7}Be are relatively well determined [Bahcall and Ulrich (1988) and Parker (1986)] after two decades of experimental struggles. The recognized uncertainties (see Table 3.2) of these reaction parameters imply 3σ uncertainties in the interpretation of experiments that are sensitive to ^{8}B neutrinos that are about 18% and 13%, respectively (see the discussion of power-law relations in Section 7.3). Additional improvements in the accuracy of the measurements for these reactions will be difficult but important.

The 3σ uncertainty in the pp rate also causes an appreciable uncertainty, about 14%, in the predicted ^{8}B neutrino flux. The main contributor is the meson exchange correction which has a total uncertainty of 4% [based upon the spread among the results of different authors, see Bahcall et al. (1982) for references] although some authors have estimated uncertainties that are a factor of two smaller. The nuclear matrix element is rather well determined [effective 3σ uncertainty of only 2.5%, Bahcall and May (1969)]. The recent determinations of the ratio of axial vector to vector weak coupling constants [e.g., Bopp et al. (1986)] correspond to a total 3σ uncertainty of only 2%.

Bibliographical Notes

Bethe, H.A. (1939) Phys. Rev., **55**, 434. The authoritative and comprehensive discussion of nuclear interactions among light elements that underlies all subsequent investigations of nuclear fusion in stars [see also Weizsacker (1937, 1938)].

Burbidge, E.M., G.R. Burbidge, W.A. Fowler, and F. Hoyle (1957) Rev. Mod. Phys., **29**, 547. Epochal paper that established the research agenda in this field for three decades. [Bahcall (1962a) pointed out that many weak interaction rates tabulated in the Appendix were inapplicable because β-decay rates are different in stars.]

Cameron, A.G.W. (1957) Chalk River Report CRL-41 [see also Cameron, A.G.W. (1958) Ann. Rev. Nucl. Sci., **8**, 299]. Contains some of the fundamental work on the physics of nuclear astrophysics. Original and far reaching.

Chandrasekhar, S. (1939) *An Introduction to Stellar Structure* (Chicago: University of Chicago Press). Chapter 11 is a fascinating description of the history and understanding of stellar energy generation immediately before Bethe's epochal papers.

Clayton, D.D. (1983) *Principles of Stellar Evolution and Nucleosynthesis* (Chicago: University of Chicago Press). The standard textbook from which students (and professors) have learned the fundamentals of the subject for almost two decades. This edition contains a new preface which lists many of the modern references.

Eddington, A.S. (1920) *Observatory*, **43**, 341. A review of the theory of internal constitution of the stars. Breathtakingly beautiful and insightful. No one can be a serious student of the subject without reading this paper. The paper also contains an astonishingly prophetic nonscientific statement: *If, indeed, the sub-atomic energy in the stars is being freely used to maintain their great furnaces, it seems to bring a little nearer to fulfillment our dream of controlling this latent power for the well-being of the human race – or for its suicide.*

Eddington, A.S. (1926) *The Internal Constitution of the Stars* (Cambridge: Cambridge University Press). A beautifully written summary of the early theory of stellar evolution. Chapter 8 contains a fascinating account of the first gropings toward understanding the source of stellar energy generation.

Fowler, W.A. (1984) *Rev. Mod. Phys.*, **56**, 149. A physical and fun account of nuclear astrophysics, with special emphasis upon the crucial role of laboratory measurements. By the man who made it happen.

Gamow, G. (1938) *Phys. Rev.*, **53**, 595. A pioneering investigation of nuclear energy sources and stellar evolution. The historical importance of Gamow's contributions is sometimes not appropriately recognized.

Rolfs, C. and W. Rodney (1988) *Cauldrons in the Cosmos* (Chicago: University of Chicago Press). A valuable review of nuclear astrophysics with special emphasis on laboratory experiments that determine the cross sections.

Russell, H.N. (1919) *Publ. Astronomical Soc. Pacific,* **31**, 129. Set the astronomical constraints for the source of stellar energy.

Salpeter, E.E. (1957) *Rev. Mod. Phys.*, **29**, 244. A concise, clear explanation of the physical principles of stellar energy sources and stellar structure by a master of both subjects.

Schatzman, E. (1951) *C. R.*, **232**, 1740. Suggested the termination of the pp chain by the ^{3}He$+^3$He reaction. An important contribution that is sometimes forgotten.

4

The standard solar model

Summary

The Sun is an astronomical laboratory. Because of its proximity to
the Earth, we are able to obtain information about the Sun that is
not accessible for other stars. We can measure with precision solar
parameters that are known to only one or two significant figures for
any other star. For example, we can determine precise values for the
solar mass, radius, geometric shape, photon spectrum, total luminosity, surface chemical composition, and age. In addition, astronomers
have measured accurate frequencies for thousands of acoustic oscillation modes that are observed at the solar surface. These frequencies
contain information about the solar interior. We are beginning to
measure the spectrum of neutrinos produced by nuclear reactions
in the solar interior. The geological records, as well as the plan-
ets, comets, and meteorites, all provide information about the past
history of the Sun. Taken together, this treasure of experimental
information provides a unique opportunity to test theories of stellar
structure and evolution.

This chapter describes the main input parameters used in standard
solar models, the logic of the calculations, and some of the important results. The most significant input parameters are the chemical
abundances, the radiative opacity, and the equation of state, as well
as the nuclear reaction rates which are discussed in Chapter 3. The
calculational procedure is straightforward, although the most highly
developed computer codes are complicated because numerical precision is important.

Details of the standard solar model are useful in different applications and are presented in the form of tables and figures. **Astroseismology**, the determination of characteristics of stars from their observed surface oscillation frequencies, is the experimental frontier of stellar evolution. It is hoped that precision observations over the next several years will reveal oscillation frequencies in many stars, establishing astroseismology as a standard and key technique in stellar diagnostics. **Helioseismology**, the study of solar seismology, already provides unique observational constraints on solar properties and is reviewed in this chapter.

The presentation given here is based upon publications written over the course of more than two decades of collaboration with close associates, especially Roger Ulrich [see in particular Bahcall, Bahcall, and Ulrich (1969), Bahcall and Ulrich (1971), Bahcall and Sears (1972), Bahcall, Huebner, Lubow, Parker, and Ulrich (1982), Ulrich (1986), and especially Bahcall and Ulrich (1988), as well as the references cited in the caption to Figure 10.1].

Different stellar evolution codes give results for solar neutrino fluxes that are in agreement (to $\sim 10\%$) with the results obtained using the standard models in the Bahcall–Ulrich series, when account is taken of the greater detail that is included in the Bahcall–Ulrich codes and of differences in input parameters. Some comparisons of the results obtained with different codes are described by Abraham and Iben (1971), Bahcall and Sears (1972), Chitre, Ezer, and Stothers (1973), Wheeler and Cameron (1975), Filippone and Schramm (1982), Cassè, Cahen, and Doom (1986), Cahen, Doom, and Cassé (1986), and others. The most recent and detailed comparison, by Sienkiewicz, Bahcall, and Paczyński (1989), uses an independent computer code and gives results that agree to better than 3% for all the pp neutrino fluxes (including the sensitive ^{8}B neutrino flux) with the Bahcall and Ulrich (1988) values.

The first section, 4.1, describes the main input parameters that are used in calculating a standard solar model, while Section 4.2 outlines the calculational method. Tables 4.4 and 4.5 of Section 4.3 present a detailed numerical representation of the standard solar model. Section 4.3 also gives some global characteristics of the standard model. The last section, 4.4, discusses helioseismology, emphasizing the complementarity of p-mode oscillation studies and solar neutrino experiments.

Table 4.1. Some important solar quantities. The measured parameters are: photon luminosity, mass, radius, oblateness, and age. All other quantities are calculated with the aid of the standard solar model.

Parameter	Value
Photon luminosity ($L_\odot$)	3.86×10^{33} erg s^{-1}
Neutrino luminosity	$0.023 L_\odot$
Mass ($M_\odot$)	1.99×10^{33} g
Radius ($R_\odot$)	6.96×10^{10} cm
Oblateness	$\leq 2 \times 10^{-5}$
$[(R_{\text{equatorial}}/R_{\text{polar}}) - 1)]$	
Effective (surface) temperature	5.78×10^3 K
Moment of inertia	7.00×10^{53} g cm^2
Age	$\approx 4.55 \times 10^9$ yr
Initial helium abundance by mass	0.27
Initial heavy element abundance by mass	0.020
Depth of convective zone	$0.26 R_\odot (0.015 M_\odot)$
Central density	148 g cm^{-3}
Central temperature	15.6×10^6 K
Central hydrogen abundance by mass	0.34
Neutrino flux from pp reaction	6.0×10^{10} cm^{-2} s^{-1}
Neutrino flux from ^{8}B decay	6×10^6 cm^{-2} s^{-1}
Fraction of energy from pp chain	0.984
Fraction of energy from CNO cycle	0.016

Table 4.1 lists some of the main physical characteristics of the Sun. Of special importance for the solar neutrino problem are the accurately determined luminosity and mass, the initial heavy element to hydrogen ratio, and an upper limit on the intrinsic solar oblateness [Dicke, Kuhn, and Libbrecht (1985)].

4.1   The input parameters

The major input parameters or functions that are used in a standard solar model are: nuclear parameters (discussed in Chapter 3), solar luminosity, solar age, equation of state, elemental abundances, and radiative opacity. The adopted luminosity and age are listed in Table 4.1; the data for these two quantities have not changed significantly since the extensive discussion of Bahcall *et al.* (1982).

A   *Chemical abundances*

The chemical abundances of the elements affect the computed radiative opacity and hence the temperature–density profile of the solar interior. Many astronomers working in wavelength regions ranging from the ultraviolet to the infrared have contributed to the difficult observational and theoretical tasks of determining accurate elemental abundances for the solar surface, a strenuous effort of the astronomical community that has been in progress for more than 50 years. This joint effort of many different researchers has been ably summarized in two reviews (completed approximately at the same time) by Grevesse (1984) and Aller (1986). Bahcall and Ulrich (1988) adopted the abundances listed by Grevesse (1984) as the best estimates that were used in determining the radiative opacity for the standard solar model that is discussed in this chapter.

The present composition of the solar surface is presumed, in standard solar models, to reflect the initial abundances of all of the elements that are at least as heavy as carbon. The fractional abundance by mass of elements heavier than helium is called the **heavy element abundance** and is traditionally denoted by Z. The corresponding abundances by mass of hydrogen and helium are denoted by X and Y.

The initial ratio by mass of elements heavier than helium relative to hydrogen, Z/X, is one of the crucial input parameters in the determination of a solar model. The fractional abundances of each of the elements are also important in determining the stellar opacity, which is closely linked to the predicted neutrino fluxes.

Table 4.2 lists the individual fractional abundances of the heavy elements that are recommended by Grevesse (1984) and Aller (1986). The two studies are in excellent agreement. The best estimate value of Z/X has increased considerably from the value of 0.0228 of the

Table 4.2. Fractional abundances of heavy
elements.

Element	Number fraction [Grevesse (1984)]	Number fraction [Aller (1986)]
C	0.29661	0.27983
N	0.05918	0.05846
O	0.49226	0.49761
Ne	0.06056	0.06869
Na	0.00129	0.00125
Mg	0.02302	0.02552
Al	0.00179	0.00198
Si	0.02149	0.02672
P	0.00017	0.00018
S	0.00982	0.01040
Cl	0.00019	0.00019
Ar	0.00230	0.00227
Ca	0.00139	0.00134
Ti	0.00006	0.00007
Cr	0.00028	0.00035
Mn	0.00017	0.00016
Fe	0.02833	0.02382
Ni	0.00108	0.00114
Total	1.000	1.000

Ross–Aller (1976) mixture that was used by Bahcall *et al.* (1982), in an earlier study of the standard solar model. The Grevesse (1984) value is $(Z/X)_{\mathrm{Grevesse}} = 0.02765$ and for the Aller (1986) mixture $(Z/X)_{\mathrm{Aller}} = 0.02739$. The difference between the Ross–Aller value of Z/X and the current value of Grevesse (1984) and Aller (1986) is about 19%.

In order to employ these surface abundances in stellar interior calculations, two important but quantitatively plausible assumptions are made. First, the Sun is assumed to be chemically homogeneous when it arrives on the main sequence. Pre-main sequence models of solar type stars are convectively mixed [see Hayashi (1961, 1966)]. Second, the composition of the present solar surface is assumed to reflect the initial abundances of all elements at least as heavy as carbon. Nuclear burning, for material presently confined to the outer parts of the Sun, is negligible because the temperatures within the present convective zone are relatively low.

B   The radiative opacity

The transport of energy in the central regions of the Sun is primarily through photon radiation, although electron conduction contributes somewhat in the innermost regions and convection dominates near the surface. The calculated radiative opacity depends upon the chemical composition and upon the modeling of complex atomic processes. The calculations require, for the solar interior, the use of large computer codes in order to include all of the known statistical mechanics and atomic physics [see Huebner (1986)]. The primary source for accurate astrophysical opacities has been, for many years, the Los Alamos National Laboratory codes, presumably developed for related thermonuclear applications.

Table 4.3 gives a numerical representation of the Los Alamos opacities for the Grevesse mixture; this tabulation can be useful to different workers in stellar evolution in testing their codes. The table covers the parameter range that is relevant for the calculation of standard solar models. The units of density are g cm^{-3} and the units of temperature are 10^6 K.

The opacities were calculated with the programs of the Los Alamos Opacity Library and were made available to Bahcall and Ulrich (1988) (see their Table III and the discussions in Section II.C and Appendix A) through the generous cooperation of W. Huebner. A detailed description of the physics that underlies these calculations is given in Huebner (1986).

Because the opacity determines in large part the temperature profile (see Section 2.2), the adopted opacity constitutes an important source of uncertainty for solar neutrino calculations [see especially the discussions in Section 7.2C, as well as in Bahcall *et al.* (1982), Huebner (1986), and Bahcall and Ulrich (1988)]. Bahcall *et al.* (1982) compared two different opacity calculations in the region of interest for solar interior calculations and concluded that the typical uncertainty is less than 10% (the maximum difference between the independent Los Alamos and Livermore Laboratory codes in the region of interest, see Table 7.1).

One can understand qualitatively why the opacity, despite the complexity of the most precise calculations, does not constitute a

Table 4.3. Rosseland mean opacities (cm^2 gm^{-1}). The first column give the temperature, T_6, in units of 10^6 K. The remaining columns give the opacity at different values of ρ/T_6^3, where ρ is the mass density in cgs units.

ρ/T_6^3 T_6	2.818E-2	3.981E-2	5.623E-2	2.818E-2	3.981E-2	5.623E-2
1.000	4.659E+1	5.802E+1	7.116E+1	4.407E+1	5.557E+1	6.892E+1
1.218	4.122E+1	5.168E+1	6.421E+1	3.823E+1	4.850E+1	6.075E+1
1.483	3.931E+1	4.931E+1	6.083E+1	3.592E+1	4.528E+1	5.681E+1
1.807	3.262E+1	4.050E+1	5.009E+1	2.989E+1	3.694E+1	4.555E+1
2.200	2.728E+1	3.286E+1	3.889E+1	2.464E+1	3.004E+1	3.588E+1
2.680	1.921E+1	2.214E+1	2.496E+1	1.760E+1	2.048E+1	2.335E+1
3.264	1.267E+1	1.433E+1	1.606E+1	1.160E+1	1.326E+1	1.494E+1
3.975	8.107E+0	9.145E+0	1.036E+1	7.404E+0	8.393E+0	9.451E+0
4.841	5.236E+0	5.940E+0	6.731E+0	4.703E+0	5.358E+0	6.115E+0
5.896	3.527E+0	4.017E+0	4.597E+0	3.137E+0	3.584E+0	4.118E+0
7.181	2.528E+0	2.895E+0	3.341E+0	2.205E+0	2.553E+0	2.957E+0
8.746	1.941E+0	2.240E+0	2.604E+0	1.680E+0	1.952E+0	2.279E+0
10.652	1.620E+0	1.875E+0	2.192E+0	1.397E+0	1.624E+0	1.910E+0
12.973	1.434E+0	1.666E+0	1.938E+0	1.237E+0	1.434E+0	1.672E+0
15.800	1.306E+0	1.484E+0	1.719E+0	1.122E+0	1.285E+0	1.474E+0
19.243	1.130E+0	1.285E+0	1.429E+0	9.754E−1	1.086E+0	1.242E+0
23.436	9.526E−1	1.053E+0	1.176E+0	8.077E−1	8.932E−1	1.000E+0
	$X=0.7300, Z=0.0195$			$X=0.3500, Z=0.0195$		

major source of uncertainty for solar neutrino calculations. At the center of the Sun, the opacity, in all the mixtures discussed, is dominated by inverse bremsstrahlung. In fact, more than half of the radiative opacity in the central regions ($T > 10^7$ K) is produced by photon scattering on free electrons and by inverse bremsstrahlung in the presence of completely ionized hydrogen and helium, processes that can be calculated with the aid of elementary quantum mechanics and statistical mechanics to an accuracy of better than 10%. Numerical experiments show that a hypothetical fractional change in the radiative opacity, for temperatures above 6×10^6 K, typically gives rise to a comparable fractional change in the computed neutrino fluxes [see Bahcall, Bahcall, and Ulrich (1969)].

Table 4.3 (con't).

ρ/T_6^3 T_6	2.818E-2	3.981E-2	5.623E-2	2.818E-2	3.981E-2	5.623E-2
1.000	4.886E+1	6.085E+1	7.463E+1	4.609E+1	5.813E+1	7.208E+1
1.218	4.326E+1	5.424E+1	6.738E+1	4.002E+1	5.078E+1	6.359E+1
1.483	4.136E+1	5.188E+1	6.399E+1	3.772E+1	4.752E+1	5.963E+1
1.807	3.434E+1	4.267E+1	5.280E+1	3.140E+1	3.883E+1	4.791E+1
2.200	2.879E+1	3.470E+1	4.109E+1	2.595E+1	3.167E+1	3.786E+1
2.680	2.030E+1	2.342E+1	2.641E+1	1.859E+1	2.164E+1	2.468E+1
3.264	1.340E+1	1.515E+1	1.698E+1	1.226E+1	1.402E+1	1.579E+1
3.975	8.553E+0	9.644E+0	1.092E+1	7.814E+0	8.855E+0	9.966E+0
4.841	5.502E+0	6.238E+0	7.062E+0	4.947E+0	5.633E+0	6.422E+0
5.896	3.687E+0	4.197E+0	4.798E+0	3.284E+0	3.749E+0	4.303E+0
7.181	2.627E+0	3.008E+0	3.468E+0	2.295E+0	2.655E+0	3.074E+0
8.746	2.007E+0	2.315E+0	2.689E+0	1.739E+0	2.020E+0	2.356E+0
10.652	1.668E+0	1.930E+0	2.256E+0	1.440E+0	1.674E+0	1.968E+0
12.973	1.473E+0	1.712E+0	1.991E+0	1.272E+0	1.475E+0	1.720E+0
15.800	1.340E+0	1.523E+0	1.762E+0	1.153E+0	1.321E+0	1.514E+0
19.243	1.156E+0	1.313E+0	1.459E+0	1.000E+0	1.114E+0	1.270E+0
23.436	9.707E−1	1.072E+0	1.196E+0	8.246E−1	9.111E−1	1.019E+0
		$X=0.7300,\ Z=0.0208$			$X=0.3500,\ Z=0.0208$	

C   The equation of state

The equation of state, the relation between pressure and density, must include accurately the effects of radiation pressure and electron degeneracy [see, e.g., Rakavy and Shaviv (1967) or Schwarzschild (1958)], and screening interactions [according to the Debye–Hückel theory, see footnote 15 of Bahcall and Shaviv (1968)]. All of these effects can be included without unusual complications in a stellar interior code; the remaining recognized uncertainties do not significantly affect the calculated solar structure or the neutrino fluxes [see Bahcall *et al.* (1982) and Ulrich (1982)]. However, numerical experiments show that the computed neutrino fluxes are sensitive to hypothetical localized changes in the equation of state when the perturbations are introduced near 8×10^6 K [Bahcall, Bahcall, and Ulrich (1969)].

4.2   General method

A   The ingredients

The **standard solar model** is calculated using the best physics and
input parameters that are available at the time the model is constructed. Thus the set of numbers that correspond to the standard
solar model vary with time, hopefully (nearly) always getting closer
to the "true" standard model. In the quarter of a century that standard solar models have been used to compute neutrino fluxes, there
have been many hundreds of improvements in the input parameters
and in the description of the physics. A few seemingly esoteric up-
grades of the codes made noticeable differences in the predictions
of neutrino fluxes, but a number of the most difficult and careful
investigations of new physics or input parameters resulted in little
change in the calculated fluxes.

Some of the principal approximations used in constructing standard models deserve special attention since they have been investigated particularly thoroughly or often for possible sources of departure from the standard scenario.

(1) **Hydrostatic equilibrium.** The Sun is assumed to be
in hydrostatic equilibrium; the radiative and particle pressures of the model exactly balance gravity. Observationally,
this is known to be an excellent approximation since a gross
departure from hydrostatic equilibrium would cause the Sun
to collapse in a free-fall time, which is less than an hour.
Pulsation, rotation, and pressure due to magnetic fields are
all estimated to be unimportant for purposes of calculating
solar neutrino fluxes.[†]

(2) **Energy transport by photons or convective motions.** In the deep interior, which is most important for the
solar neutrino problem, the energy transport is primarily
by photon diffusion and is described in terms of the Rosseland mean opacity. For regions that are unstable against
convective motions, the temperature gradient is taken to

[†]There are detailed discussions of the validity of these approximations, as well as
estimates of the small expected effects of departures from the standard assumptions, in the references listed in the caption to Figure 10.1 and in the discussion
of nonstandard models in Chapter 5.

be the adiabatic gradient except near the surface (important for the helioseismological calculations) where mixing length theory is used. Additional transport due to acoustic or gravity waves is negligible in the standard solar model. The possibility of energy transport by weakly interacting massive particles (WIMPs) is discussed in Chapter 5.

(3) **Energy generation by nuclear reactions.** The primary energy source for the radiated photons and neutrinos is nuclear fusion, although small effects of contraction or expansion are included in the standard solar model. The standard codes include departures from nuclear equilibrium that are caused by the fusion processes themselves, for example, in the abundance of ^{3}He (see Chapter 3). Chapter 5 discusses some examples of energy generation by processes other than hydrogen fusion.

(4) **Abundance changes caused solely by nuclear reactions.** The primordial solar interior is chemically homogeneous in the standard model. Changes in the local abundances of individual isotopes occur only by nuclear reactions in those regions of the model that are convectively stable. Thermal and gravitational diffusion are not included at present, because they are estimated to be small over the lifetime of the Sun [see Cox, Guzik, and Kidman (1989)]. The numerical formalism required to include conveniently diffusive effects has been developed and will be included in future standard solar models.

B   Calculational procedure

A standard solar model is the end product of a sequence of models. One begins with a main sequence star that has a homogeneous composition. Hydrogen burns in the deep interior of the model, supplying both the radiated luminosity and the local heat (thermal pressure) which supports the star against gravitational contraction. Successive models are calculated by allowing for composition changes caused by nuclear reactions, as well as the mild evolution of other parameters; the integration of the nuclear abundance equations involves some numerical complications that can be handled best by specialized techniques. The nuclear interaction rates are interpolated between the previous and new models and multiplied by a time step (usually of order 5×10^8 or 10^9 yr., in order to determine the new chemical composition as a function of mass fraction included. The model at the advanced time is computed using the new composition. The models in an evolutionary sequence have inhomogeneous compositions; in the model for the present epoch, the innermost mass fraction of hydrogen is about one-half the surface (initial) value.

The stellar evolution models are constructed by integrating from the center outward and from the surface inward, requiring that the two solutions match at a convenient point that is typically at about $0.2M_\odot$. Only a relatively crude treatment of the solar atmosphere is required for computing accurate values for solar interior parameters. Even a 10% change in the outer radius of the model does not significantly affect the calculated neutrino fluxes [see Sears (1964) or Bahcall and Shaviv (1968), Eq. (3)]. The difference between the most careful and the crudest treatment of the solar convection zone corresponds to at most a 2% change in the calculated solar neutrino fluxes [see Bahcall and Ulrich (1988), Section X.D]. Bahcall and Sears (1972) describe and compare many of the earlier standard solar models and the methods by which they were derived.[†]

How does one proceed in practice? Guess an initial set of parameters; march the models along in time using difference equations to represent the equations of stellar evolution; calculate the predicted characteristics of the present Sun; and then iterate the results until good numerical agreement is obtained between the model and the observed Sun. In the models that Bahcall and Ulrich construct, one begins by guessing initial values of X, the original homogeneous hydrogen abundance, and S, an entropy-like variable.[‡] Typically, an evolutionary sequence requires of order five to seven solar models of progressively greater ages to match the luminosity and radius to the desired one part in 10^5.

The initial helium abundance of the model is determined in the process of iteration. The other two composition parameters are fixed

---

[†]Details regarding the codes that Ulrich and Bahcall use are given in the appendices of Bahcall *et al.* (1982) and Bahcall and Ulrich (1988).

[‡]Appendix A of Bahcall *et al.* (1982) defines S and discusses the initial steps in the construction of the model. S determines the adiabat of the convection zone. In earlier treatments of the problem, one adjusted the constant $K = P/T^{2.5}$, which gives the relation between pressure and temperature in the convective envelope [see Sears (1964) or Bahcall and Shaviv (1968)].

by the surface (initial) ratio of Z/X that is taken from observations and by the fact that the sum of all the mass fractions is equal to unity, that is, $X + Y + Z = 1.0$.

A satisfactory solar model is a solution of the evolutionary equations that satisfies boundary conditions in both space and time. One seeks a model with a fixed mass $M_{\odot}$ and with a total luminosity (in photons) equal to $L_{\odot}$ and an outer radius $R_{\odot}$ at an elapsed time of 4.6×10^9 yr, the present age of the Sun. The initially assumed values of X (the hydrogen mass fraction) and S (the entropy-like variable) are iterated until an accurate description is obtained of the Sun at the present epoch. Empirical relations have been derived which can be used as guides in this iterative process [see Eq. (15) of Bahcall et al. (1982)]. The solution of the evolution equations determines the values for the mass fractions of hydrogen, helium, and heavy elements, the present complete run of physical variables inside the Sun, the spectrum of acoustic oscillation frequencies observed on the surface of the Sun, and the neutrino fluxes.

The luminosity boundary condition has an especially strong effect on the calculated neutrino fluxes. The reason is that both the luminosity and the neutrino fluxes are produced by nuclear reactions in the deep solar interior. If there were no competition between different branches of the pp chain (see Table 1.1), then holding the luminosity constant would imply that the computed fluxes would not change as one varied parameters or physical assumptions. The reason that neutrino fluxes are a good diagnostic of the solar interior is because of the sensitivity of the branching ratios of the competing reactions of the pp chain. The strong coupling between the luminosity and the neutrino fluxes has resulted in some groups deriving incorrect results because they have not iterated the models sufficiently.

What are nonstandard solar models? "Nonstandard" solar models (which are discussed in Chapter 5) are, by definition, constructed by changing something, physics or input data, from the current best-guess to something that is less plausible. Most of the nonstandard solar models that have been published were invented in order to "solve" the solar neutrino problem. [Systematic investigations and reviews of many of the nonstandard models have been given by Bahcall, Bahcall, and Ulrich (1969), Rood (1978), Boyd et al. (1985), Roxburgh (1985a,b), Schatzman (1985), and Newman (1986).] If the physics or input parameters that are used in the nonstandard models were believed to be correct, then the "nonstandard" features would have been incorporated into the standard model.

4.3   Some characteristics of the standard model

There are a number of characteristics of the standard model that are of general interest. For example, the fraction of the photon luminosity that originates in the pp chain is 0.984; the corresponding fraction for the CNO cycle is 0.016. The net expansion at the present epoch corresponds to a luminosity fraction of -0.0003. The convection zone terminates at 1.92×10^6 K, corresponding to a radius of about $0.74R_{\odot}$ and a density of 0.12 g cm^{-3}; the convection zone comprises the outer 1.5% of the solar mass.[†] One-half of the photon luminosity (or the flux of pp neutrinos) is produced within the inner $0.09M_{\odot}$ ($R \leq 0.11R_{\odot}$); 95% of the photon luminosity is produced within the inner $0.36M_{\odot}$ ($R \leq 0.21R_{\odot}$). The neutrino luminosity is 2.3% of the photon luminosity, which corresponds to an average of 0.572 MeV lost in neutrinos per termination of the pp chain. The pp chain is terminated 85.5% of the time by the ^{3}He–^{3}He reaction (number 3 of Table 1.1) and 14.5% of the time by the ^{3}He–^{4}He reaction (number 4 of Table 1.1).

Tables 4.4 and 4.5 provide a detailed numerical description of the solar interior of the standard model.

Table 4.4 lists the principal physical and chemical characteristics of the standard solar model. The first six columns present the physical variables that define the model: the mass included in the current and all inner zones (in units of $M_{\odot}$), the radius (in units of $R_{\odot}$), the temperature (in K), the density (in g cm^{-3}), the pressure (in ergs cm^{-3}), the luminosity integrated up to and including the current zone (in units of $L_{\odot}$), and S_{crit} [which is related to the ratio of adiabatic and radiative gradients, see Bahcall and Ulrich (1988)]. The last seven columns give the most important isotopic abundances, the fractions by mass of ^{1}H, ^{3}He, ^{4}He, ^{7}Be, ^{12}C, ^{14}N, and ^{16}O. The initial heavy element abundance Z is 0.01961.

[†]The precise parameters for the convective zone are unimportant for the solar neutrino problem although they are important for the calculation of the p-mode oscillation frequencies.

Table 4.4   Characteristics of the standard solar model

M/M☉	R/R☉	T	ρ	P	L/L☉	S_crit	X(^{1}H)	X(^{3}He)	X(^{4}He)	X(^{7}Be)	X(^{12}C)	X(^{14}N)	X(^{16}O)
0.0	0.0	1.56E+07	1.48E+02	2.29E+17	0.0	-0.127	0.34111	7.74E-06	0.63867	1.65E-11	2.61E-05	6.34E-03	8.48E-03
0.00001	0.0039	1.56E+07	1.48E+02	2.29E+17	0.000	-0.124	0.34103	7.73E-06	0.63875	1.65E-11	2.61E-05	6.34E-03	8.48E-03
0.00005	0.0083	1.56E+07	1.47E+02	2.28E+17	0.000	-0.127	0.34317	7.88E-06	0.63661	1.64E-11	2.60E-05	6.33E-03	8.50E-03
0.00017	0.0120	1.56E+07	1.46E+02	2.27E+17	0.001	-0.130	0.34546	8.04E-06	0.63432	1.62E-11	2.59E-05	6.31E-03	8.52E-03
0.00040	0.0158	1.56E+07	1.45E+02	2.26E+17	0.003	-0.132	0.34885	8.29E-06	0.63092	1.59E-11	2.56E-05	6.29E-03	8.54E-03
0.00078	0.0197	1.55E+07	1.44E+02	2.24E+17	0.007	-0.134	0.35328	8.63E-06	0.62649	1.56E-11	2.54E-05	6.26E-03	8.58E-03
0.00135	0.0237	1.55E+07	1.42E+02	2.21E+17	0.012	-0.137	0.35868	9.06E-06	0.62108	1.52E-11	2.50E-05	6.22E-03	8.62E-03
0.00214	0.0277	1.54E+07	1.40E+02	2.18E+17	0.018	-0.141	0.36499	9.58E-06	0.61476	1.47E-11	2.47E-05	6.18E-03	8.66E-03
0.00320	0.0317	1.53E+07	1.37E+02	2.15E+17	0.027	-0.144	0.37217	1.02E-05	0.60758	1.41E-11	2.43E-05	6.14E-03	8.71E-03
0.00456	0.0358	1.52E+07	1.35E+02	2.12E+17	0.038	-0.147	0.38016	1.09E-05	0.59958	1.34E-11	2.38E-05	6.10E-03	8.76E-03
0.00625	0.0400	1.51E+07	1.32E+02	2.08E+17	0.051	-0.150	0.38890	1.18E-05	0.59084	1.27E-11	2.34E-05	6.06E-03	8.81E-03
0.00832	0.0442	1.50E+07	1.29E+02	2.03E+17	0.067	-0.155	0.39833	1.28E-05	0.58140	1.19E-11	2.29E-05	6.02E-03	8.86E-03
0.01080	0.0484	1.49E+07	1.26E+02	1.99E+17	0.085	-0.160	0.40839	1.39E-05	0.57133	1.11E-11	2.24E-05	5.98E-03	8.90E-03
0.01373	0.0528	1.48E+07	1.23E+02	1.94E+17	0.106	-0.163	0.41903	1.52E-05	0.56069	1.02E-11	2.19E-05	5.94E-03	8.94E-03
0.01715	0.0572	1.46E+07	1.19E+02	1.89E+17	0.130	-0.167	0.43017	1.67E-05	0.54954	9.33E-12	2.13E-05	5.91E-03	8.98E-03
0.02109	0.0616	1.45E+07	1.16E+02	1.83E+17	0.157	-0.171	0.44176	1.85E-05	0.53794	8.44E-12	2.08E-05	5.88E-03	9.01E-03
0.02560	0.0662	1.43E+07	1.12E+02	1.78E+17	0.186	-0.176	0.45428	2.05E-05	0.52542	7.55E-12	2.03E-05	5.86E-03	9.04E-03
0.03071	0.0708	1.42E+07	1.08E+02	1.72E+17	0.217	-0.179	0.46672	2.28E-05	0.51297	6.69E-12	1.98E-05	5.84E-03	9.06E-03
0.03645	0.0756	1.40E+07	1.05E+02	1.66E+17	0.251	-0.183	0.47942	2.54E-05	0.50026	5.86E-12	1.92E-05	5.82E-03	9.08E-03
0.04287	0.0804	1.38E+07	1.01E+02	1.60E+17	0.287	-0.188	0.49233	2.85E-05	0.48735	5.08E-12	1.87E-05	5.81E-03	9.10E-03
0.05000	0.0853	1.37E+07	9.70E+01	1.53E+17	0.325	-0.192	0.50536	3.20E-05	0.47431	4.35E-12	1.82E-05	5.80E-03	9.11E-03
0.05769	0.0902	1.35E+07	9.33E+01	1.47E+17	0.363	-0.196	0.51817	3.60E-05	0.46150	3.70E-12	1.76E-05	5.79E-03	9.12E-03
0.06538	0.0948	1.33E+07	8.99E+01	1.41E+17	0.400	-0.200	0.52988	4.02E-05	0.44978	3.16E-12	1.71E-05	5.78E-03	9.13E-03
0.07308	0.0992	1.31E+07	8.68E+01	1.36E+17	0.434	-0.205	0.54066	4.46E-05	0.43900	2.71E-12	1.67E-05	5.78E-03	9.14E-03
0.08077	0.1033	1.30E+07	8.40E+01	1.31E+17	0.466	-0.208	0.55064	4.93E-05	0.42902	2.33E-12	1.62E-05	5.78E-03	9.14E-03

0.08846	0.1073	1.28E+07	8.13E+01	1.26E+17	0.497	-0.213	0.55990	5.42E-05	0.41975	2.01E-12	1.58E-05	5.78E-03	9.14E-03
0.09615	0.1111	1.27E+07	7.88E+01	1.22E+17	0.525	-0.216	0.56853	5.95E-05	0.41111	1.74E-12	1.54E-05	5.77E-03	9.14E-03
0.10385	0.1147	1.25E+07	7.64E+01	1.18E+17	0.553	-0.219	0.57659	6.51E-05	0.40304	1.50E-12	1.50E-05	5.77E-03	9.15E-03
0.11154	0.1182	1.24E+07	7.42E+01	1.14E+17	0.579	-0.224	0.58414	7.10E-05	0.39549	1.30E-12	1.47E-05	5.77E-03	9.15E-03
0.11923	0.1217	1.22E+07	7.20E+01	1.10E+17	0.603	-0.228	0.59121	7.73E-05	0.38841	1.13E-12	1.40E-05	5.77E-03	9.15E-03
0.12692	0.1250	1.21E+07	7.00E+01	1.06E+17	0.626	-0.233	0.59785	8.40E-05	0.38177	9.86E-13	1.43E-05	5.77E-03	9.15E-03
0.13462	0.1283	1.20E+07	6.81E+01	1.03E+17	0.648	-0.237	0.60409	9.11E-05	0.37552	8.59E-13	1.39E-05	5.77E-03	9.15E-03
0.14231	0.1315	1.18E+07	6.63E+01	9.92E+16	0.668	-0.241	0.60996	9.86E-05	0.36964	7.50E-13	1.45E-05	5.77E-03	9.15E-03
0.15000	0.1346	1.17E+07	6.45E+01	9.60E+16	0.688	-0.246	0.61549	1.07E-04	0.36410	6.56E-13	1.68E-05	5.77E-03	9.15E-03
0.15600	0.1370	1.16E+07	6.32E+01	9.36E+16	0.702	-0.250	0.61958	1.13E-04	0.36001	5.91E-13	2.14E-05	5.76E-03	9.15E-03
0.16200	0.1393	1.15E+07	6.20E+01	9.12E+16	0.716	-0.254	0.62349	1.20E-04	0.35610	5.33E-13	3.07E-05	5.75E-03	9.15E-03
0.16800	0.1417	1.14E+07	6.07E+01	8.89E+16	0.729	-0.259	0.62722	1.27E-04	0.35236	4.81E-13	4.76E-05	5.73E-03	9.15E-03
0.17400	0.1440	1.13E+07	5.95E+01	8.67E+16	0.742	-0.264	0.63079	1.35E-04	0.34879	4.34E-13	7.59E-05	5.70E-03	9.15E-03
0.18000	0.1462	1.13E+07	5.84E+01	8.46E+16	0.754	-0.268	0.63420	1.43E-04	0.34537	3.92E-13	1.20E-04	5.65E-03	9.15E-03
0.18600	0.1485	1.12E+07	5.72E+01	8.25E+16	0.766	-0.275	0.63747	1.51E-04	0.34211	3.54E-13	1.84E-04	5.57E-03	9.15E-03
0.19200	0.1507	1.11E+07	5.61E+01	8.05E+16	0.777	-0.281	0.64059	1.60E-04	0.33899	3.20E-13	2.72E-04	5.47E-03	9.15E-03
0.19800	0.1529	1.10E+07	5.50E+01	7.85E+16	0.788	-0.287	0.64359	1.69E-04	0.33600	2.90E-13	3.87E-04	5.34E-03	9.15E-03
0.20400	0.1551	1.09E+07	5.40E+01	7.66E+16	0.798	-0.295	0.64646	1.79E-04	0.33315	2.63E-13	5.28E-04	5.17E-03	9.15E-03
0.21000	0.1572	1.08E+07	5.30E+01	7.47E+16	0.807	-0.304	0.64922	1.89E-04	0.33041	2.38E-13	6.94E-04	4.98E-03	9.15E-03
0.21600	0.1594	1.08E+07	5.20E+01	7.29E+16	0.817	-0.312	0.65185	2.00E-04	0.32779	2.16E-13	8.82E-04	4.76E-03	9.15E-03
0.22200	0.1615	1.07E+07	5.10E+01	7.11E+16	0.826	-0.321	0.65438	2.11E-04	0.32529	1.96E-13	1.09E-03	4.52E-03	9.15E-03
0.22800	0.1636	1.06E+07	5.00E+01	6.94E+16	0.834	-0.329	0.65681	2.22E-04	0.32289	1.78E-13	1.30E-03	4.26E-03	9.15E-03
0.23400	0.1657	1.05E+07	4.91E+01	6.77E+16	0.842	-0.337	0.65913	2.34E-04	0.32058	1.61E-13	1.53E-03	4.00E-03	9.15E-03
0.24000	0.1678	1.04E+07	4.82E+01	6.61E+16	0.850	-0.346	0.66136	2.47E-04	0.31838	1.47E-13	1.75E-03	3.74E-03	9.15E-03
0.24600	0.1699	1.04E+07	4.73E+01	6.45E+16	0.857	-0.352	0.66349	2.60E-04	0.31627	1.33E-13	1.98E-03	3.48E-03	9.15E-03
0.25200	0.1719	1.03E+07	4.64E+01	6.29E+16	0.865	-0.364	0.66550	2.73E-04	0.31429	1.22E-13	2.18E-03	3.24E-03	9.15E-03
0.25800	0.1740	1.02E+07	4.55E+01	6.14E+16	0.872	-0.373	0.66746	2.87E-04	0.31235	1.11E-13	2.39E-03	3.00E-03	9.15E-03
0.26400	0.1760	1.02E+07	4.47E+01	5.99E+16	0.878	-0.381	0.66934	3.03E-04	0.31049	1.01E-13	2.58E-03	2.78E-03	9.15E-03
0.27000	0.1781	1.01E+07	4.38E+01	5.84E+16	0.885	-0.390	0.67113	3.18E-04	0.30871	9.22E-14	2.75E-03	2.58E-03	9.15E-03
0.27600	0.1801	1.00E+07	4.30E+01	5.70E+16	0.891	-0.399	0.67285	3.35E-04	0.30700	8.40E-14	2.91E-03	2.39E-03	9.15E-03

Table 4.4 (*continued*).

$M_r/M_\odot$	$R/R_\odot$	T	ρ	P	$L_r/L_\odot$	S_{crit}	$X(^1H)$	$X(^3He)$	$X(^4He)$	$X(^7Be)$	$X(^{12}C)$	$X(^{14}N)$	$X(^{16}O)$
0.28200	0.1821	9.95E+06	4.22E+01	5.56E+16	0.896	-0.406	0.67450	3.52E-04	0.30536	7.65E-14	3.06E-03	2.21E-03	9.15E-03
0.28800	0.1841	9.88E+06	4.14E+01	5.42E+16	0.902	-0.414	0.67607	3.71E-04	0.30379	6.97E-14	3.19E-03	2.06E-03	9.15E-03
0.29400	0.1861	9.81E+06	4.06E+01	5.29E+16	0.907	-0.423	0.67758	3.90E-04	0.30228	6.36E-14	3.31E-03	1.92E-03	9.15E-03
0.30000	0.1881	9.74E+06	3.99E+01	5.16E+16	0.912	-0.430	0.67902	4.10E-04	0.30084	5.80E-14	3.42E-03	1.80E-03	9.15E-03
0.31000	0.1914	9.63E+06	3.86E+01	4.95E+16	0.919	-0.446	0.68129	4.45E-04	0.29856	4.98E-14	3.57E-03	1.62E-03	9.15E-03
0.32000	0.1948	9.52E+06	3.74E+01	4.75E+16	0.926	-0.459	0.68339	4.84E-04	0.29644	4.27E-14	3.69E-03	1.48E-03	9.15E-03
0.33000	0.1981	9.41E+06	3.63E+01	4.55E+16	0.933	-0.473	0.68535	5.26E-04	0.29445	3.67E-14	3.79E-03	1.37E-03	9.15E-03
0.34000	0.2014	9.31E+06	3.51E+01	4.36E+16	0.939	-0.488	0.68716	5.71E-04	0.29261	3.15E-14	3.86E-03	1.28E-03	9.15E-03
0.35000	0.2047	9.20E+06	3.40E+01	4.18E+16	0.945	-0.502	0.68885	6.20E-04	0.29088	2.71E-14	3.92E-03	1.21E-03	9.15E-03
0.36000	0.2080	9.10E+06	3.29E+01	4.01E+16	0.950	-0.517	0.69042	6.74E-04	0.28927	2.33E-14	3.97E-03	1.16E-03	9.15E-03
0.37000	0.2113	9.00E+06	3.18E+01	3.84E+16	0.954	-0.533	0.69187	7.32E-04	0.28776	2.00E-14	4.01E-03	1.11E-03	9.15E-03
0.38000	0.2146	8.90E+06	3.08E+01	3.67E+16	0.959	-0.548	0.69322	7.95E-04	0.28636	1.72E-14	4.04E-03	1.08E-03	9.15E-03
0.39000	0.2179	8.80E+06	2.98E+01	3.52E+16	0.963	-0.564	0.69447	8.64E-04	0.28504	1.48E-14	4.06E-03	1.05E-03	9.15E-03
0.40000	0.2212	8.70E+06	2.88E+01	3.36E+16	0.966	-0.580	0.69563	9.38E-04	0.28381	1.27E-14	4.08E-03	1.03E-03	9.15E-03
0.41000	0.2246	8.60E+06	2.78E+01	3.22E+16	0.970	-0.589	0.69670	1.02E-03	0.28266	1.09E-14	4.09E-03	1.02E-03	9.15E-03
0.42000	0.2279	8.51E+06	2.69E+01	3.08E+16	0.973	-0.600	0.69770	1.11E-03	0.28158	9.38E-15	4.10E-03	1.00E-03	9.15E-03
0.43000	0.2313	8.41E+06	2.60E+01	2.94E+16	0.976	-0.610	0.69862	1.21E-03	0.28056	8.06E-15	4.11E-03	9.93E-04	9.15E-03
0.44000	0.2347	8.32E+06	2.51E+01	2.81E+16	0.978	-0.622	0.69946	1.32E-03	0.27960	6.91E-15	4.12E-03	9.86E-04	9.15E-03
0.45000	0.2381	8.22E+06	2.42E+01	2.68E+16	0.981	-0.633	0.70024	1.44E-03	0.27871	5.93E-15	4.12E-03	9.80E-04	9.15E-03
0.46000	0.2415	8.13E+06	2.34E+01	2.56E+16	0.983	-0.644	0.70096	1.57E-03	0.27786	5.08E-15	4.12E-03	9.76E-04	9.15E-03
0.47000	0.2450	8.04E+06	2.25E+01	2.44E+16	0.985	-0.658	0.70161	1.72E-03	0.27705	4.35E-15	4.13E-03	9.73E-04	9.15E-03
0.48000	0.2485	7.95E+06	2.17E+01	2.32E+16	0.987	-0.667	0.70221	1.88E-03	0.27629	3.72E-15	4.13E-03	9.70E-04	9.15E-03
0.49000	0.2520	7.86E+06	2.09E+01	2.21E+16	0.989	-0.682	0.70275	2.07E-03	0.27557	3.18E-15	4.13E-03	9.68E-04	9.15E-03
0.50000	0.2555	7.76E+06	2.01E+01	2.10E+16	0.990	-0.694	0.70324	2.26E-03	0.27488	2.71E-15	4.13E-03	9.67E-04	9.15E-03
0.51000	0.2591	7.67E+06	1.94E+01	2.00E+16	0.992	-0.707	0.70368	2.47E-03	0.27423	2.29E-15	4.13E-03	9.66E-04	9.15E-03
0.52000	0.2628	7.58E+06	1.86E+01	1.90E+16	0.993	-0.721	0.70409	2.69E-03	0.27361	1.93E-15	4.13E-03	9.65E-04	9.15E-03

0.53000	0.2664	7.49E+06	1.79E+01	1.81E+16	0.994	-0.736	0.70446	2.90E-03	0.27303	1.60E-15	4.13E-03	9.64E-04	9.15E-03
0.54000	0.2702	7.41E+06	1.72E+01	1.71E+16	0.995	-0.751	0.70480	3.08E-03	0.27251	1.31E-15	4.13E-03	9.64E-04	9.15E-03
0.55000	0.2739	7.32E+06	1.65E+01	1.62E+16	0.996	-0.766	0.70512	3.22E-03	0.27204	1.05E-15	4.13E-03	9.64E-04	9.15E-03
0.58500	0.2876	7.01E+06	1.42E+01	1.34E+16	0.998	-0.815	0.70621	3.18E-03	0.27100	4.01E-16	4.13E-03	9.63E-04	9.15E-03
0.62000	0.3020	6.70E+06	1.20E+01	1.09E+16	0.999	-0.854	0.70723	2.49E-03	0.27066	1.15E-16	4.14E-03	9.63E-04	9.15E-03
0.65500	0.3176	6.39E+06	1.01E+01	8.69E+15	1.000	-0.896	0.70806	1.74E-03	0.27059	2.74E-17	4.14E-03	9.63E-04	9.15E-03
0.69000	0.3344	6.08E+06	8.34E+00	6.82E+15	1.000	-0.946	0.70866	1.15E-03	0.27058	5.69E-18	4.14E-03	9.63E-04	9.15E-03
0.72500	0.3529	5.76E+06	6.75E+00	5.22E+15	1.000	-0.996	0.70907	7.40E-04	0.27058	1.04E-18	4.14E-03	9.63E-04	9.15E-03
0.76000	0.3737	5.44E+06	5.32E+00	3.88E+15	1.000	-1.029	0.70934	4.67E-04	0.27058	1.61E-19	4.14E-03	9.63E-04	9.15E-03
0.79500	0.3975	5.09E+06	4.06E+00	2.77E+15	1.001	-1.065	0.70952	2.95E-04	0.27058	2.06E-20	4.14E-03	9.63E-04	9.15E-03
0.83000	0.4255	4.73E+06	2.96E+00	1.88E+15	1.001	-1.101	0.70962	1.93E-04	0.27058	2.11E-21	4.14E-03	9.63E-04	9.15E-03
0.86500	0.4597	4.33E+06	2.03E+00	1.18E+15	1.001	-1.111	0.70967	1.38E-04	0.27057	1.58E-22	4.14E-03	9.63E-04	9.15E-03
0.90000	0.5038	3.88E+06	1.27E+00	6.59E+14	1.001	-1.093	0.70970	1.12E-04	0.27058	6.87E-24	4.14E-03	9.63E-04	9.15E-03
0.92373	0.5431	3.53E+06	8.42E-01	3.98E+14	1.001	-1.048	0.70970	1.04E-04	0.27058	4.48E-25	4.14E-03	9.63E-04	9.15E-03
0.94183	0.5818	3.20E+06	5.72E-01	2.45E+14	1.001	-0.974	0.70970	1.02E-04	0.27058	2.90E-26	4.14E-03	9.63E-04	9.15E-03
0.95563	0.6195	2.91E+06	3.96E-01	1.55E+14	1.000	-0.874	0.70970	1.01E-04	0.27058	1.76E-27	4.14E-03	9.63E-04	9.15E-03
0.96616	0.6559	2.64E+06	2.81E-01	9.94E+13	1.000	-0.700	0.70970	1.00E-04	0.27058	9.11E-29	4.14E-03	9.63E-04	9.15E-03
0.97419	0.6906	2.38E+06	2.05E-01	6.53E+13	1.000	-0.499	0.70970	1.00E-04	0.27058	3.55E-30	4.14E-03	9.63E-04	9.15E-03
0.98032	0.7230	2.11E+06	1.54E-01	4.37E+13	1.000	-0.167	0.70970	1.00E-04	0.27059	2.23E-30	4.14E-03	9.63E-04	9.15E-03
0.98499	0.7523	1.82E+06	1.22E-01	2.98E+13	1.000	0.285	0.70970	1.00E-04	0.27059	2.23E-30	4.14E-03	9.63E-04	9.15E-03
0.98855	0.7783	1.57E+06	9.81E-02	2.06E+13	1.000	0.556	0.70970	1.00E-04	0.27059	2.23E-30	4.14E-03	9.63E-04	9.15E-03
0.99127	0.8015	1.36E+06	7.92E-02	1.44E+13	1.000	0.708	0.70970	1.00E-04	0.27059	2.23E-30	4.14E-03	9.63E-04	9.15E-03
0.99334	0.8221	1.19E+06	6.42E-02	1.01E+13	1.000	0.801	0.70970	1.00E-04	0.27059	2.23E-30	4.14E-03	9.63E-04	9.15E-03
0.99492	0.8406	1.03E+06	5.22E-02	7.19E+12	1.000	0.867	0.70970	1.00E-04	0.27059	2.23E-30	4.14E-03	9.63E-04	9.15E-03
0.99612	0.8573	9.04E+05	4.26E-02	5.12E+12	1.000	0.910	0.70970	1.00E-04	0.27059	2.23E-30	4.14E-03	9.63E-04	9.15E-03
0.99704	0.8722	7.91E+05	3.48E-02	3.65E+12	1.000	0.937	0.70970	1.00E-04	0.27059	2.23E-30	4.14E-03	9.63E-04	9.15E-03
0.99775	0.8858	6.92E+05	2.84E-02	2.60E+12	1.000	0.956	0.70970	1.00E-04	0.27059	2.23E-30	4.14E-03	9.63E-04	9.15E-03
0.99828	0.8981	6.04E+05	2.31E-02	1.84E+12	1.000	0.969	0.70970	1.00E-04	0.27059	2.23E-30	4.14E-03	9.63E-04	9.15E-03
0.99869	0.9093	5.25E+05	1.87E-02	1.29E+12	1.000	0.978	0.70970	1.00E-04	0.27059	2.23E-30	4.14E-03	9.63E-04	9.15E-03
0.99900	0.9197	4.54E+05	1.50E-02	8.95E+11	1.000	0.984	0.70970	1.00E-04	0.27059	2.23E-30	4.14E-03	9.63E-04	9.15E-03
1.00000	1.0000	5.77E+03	0.0	0.0	1.000	-1.000	0.70970	1.00E-04	0.27059	2.23E-30	4.14E-03	9.63E-04	9.15E-03

Table 4.5   Neutrino and energy production in the standard model.

R	T6	ln(ρ_c)	d(mass)	d(energy)	dφ(pp)	dφ(^{8}B)	dφ(^{13}N)	dφ(^{15}O)	dφ(^{17}F)	dφ(^{7}Be)	dφ(pep)	dφ(hep)
0.001953	15.6443	1.9939	0.000006	0.000055	0.000038	0.000497	0.000386	0.000450	0.000462	0.000196	0.000059	0.000015
0.006088	15.6305	1.9931	0.000044	0.000384	0.000265	0.003419	0.002651	0.003095	0.003177	0.001364	0.000411	0.000107
0.010122	15.6073	1.9916	0.000119	0.001037	0.000721	0.009025	0.006981	0.008150	0.008381	0.003655	0.001114	0.000294
0.013896	15.5773	1.9897	0.000231	0.002005	0.001407	0.016947	0.013073	0.015262	0.015731	0.006999	0.002167	0.000581
0.017775	15.5360	1.9870	0.000381	0.003273	0.002327	0.026547	0.020413	0.023830	0.024634	0.011270	0.003564	0.000974
0.021711	15.4836	1.9837	0.000569	0.004822	0.003483	0.037078	0.028423	0.033181	0.034405	0.016305	0.005298	0.001484
0.025693	15.4206	1.9796	0.000794	0.006628	0.004876	0.047718	0.036493	0.042602	0.044307	0.021905	0.007356	0.002122
0.029718	15.3474	1.9749	0.001056	0.008664	0.006505	0.057648	0.044046	0.051419	0.053611	0.027844	0.009720	0.002898
0.033788	15.2640	1.9695	0.001356	0.010899	0.008364	0.066110	0.050574	0.059040	0.061660	0.033868	0.012362	0.003825
0.037907	15.1708	1.9635	0.001694	0.013299	0.010445	0.072499	0.055686	0.065008	0.067924	0.039720	0.015249	0.004912
0.042079	15.0679	1.9567	0.002069	0.015829	0.012731	0.076400	0.059124	0.069022	0.072039	0.045144	0.018337	0.006168
0.046308	14.9557	1.9494	0.002481	0.018449	0.015201	0.077623	0.060771	0.070945	0.073828	0.049901	0.021571	0.007600
0.050601	14.8344	1.9414	0.002931	0.021120	0.017826	0.076219	0.060647	0.070802	0.073306	0.053788	0.024891	0.009208
0.054961	14.7040	1.9327	0.003419	0.023800	0.020570	0.072448	0.058888	0.068748	0.070658	0.056647	0.028227	0.010990
0.059393	14.5650	1.9235	0.003944	0.026447	0.023390	0.066743	0.055721	0.065053	0.066204	0.058372	0.031504	0.012940
0.063905	14.4173	1.9135	0.004506	0.029041	0.026263	0.059667	0.051456	0.060075	0.060390	0.058920	0.034673	0.015060
0.068501	14.2614	1.9029	0.005106	0.031528	0.029121	0.051769	0.046398	0.054171	0.053651	0.058302	0.037636	0.017324
0.073187	14.0973	1.8917	0.005744	0.033838	0.031877	0.043597	0.040865	0.047712	0.046434	0.056603	0.040281	0.019689
0.077968	13.9252	1.8798	0.006419	0.035949	0.034491	0.035656	0.035170	0.041065	0.039174	0.053944	0.042559	0.022137
0.082850	13.7454	1.8673	0.007131	0.037819	0.036899	0.028317	0.029583	0.034544	0.032220	0.050486	0.044404	0.024634
0.087781	13.5601	1.8543	0.007692	0.038493	0.038128	0.021407	0.023817	0.027814	0.025313	0.045402	0.044701	0.026491
0.092547	13.3782	1.8413	0.007692	0.036286	0.036410	0.014958	0.017765	0.020748	0.018410	0.037788	0.041595	0.026296
0.097020	13.2053	1.8288	0.007692	0.034240	0.034727	0.010548	0.013367	0.015615	0.013513	0.031599	0.038698	0.026014
0.101255	13.0402	1.8168	0.007692	0.032332	0.033088	0.007497	0.010132	0.011838	0.009994	0.026531	0.035997	0.025663
0.105291	12.8817	1.8050	0.007692	0.030546	0.031499	0.005364	0.007727	0.009030	0.007440	0.022354	0.033479	0.025258

0.109159	12.7290	1.7936	0.007692	0.028869	0.029964	0.003860	0.005923	0.006924	0.005569	0.018892	0.031132	0.024812
0.112881	12.5815	1.7825	0.007692	0.027289	0.028483	0.002791	0.004559	0.005332	0.004188	0.016010	0.028945	0.024332
0.116478	12.4386	1.7715	0.007692	0.025799	0.027058	0.002028	0.003523	0.004122	0.003162	0.013600	0.026907	0.023825
0.119964	12.2999	1.7608	0.007692	0.024392	0.025690	0.001479	0.002731	0.003197	0.002396	0.011579	0.025008	0.023299
0.123353	12.1651	1.7502	0.007692	0.023062	0.024378	0.001082	0.002127	0.002487	0.001822	0.009877	0.023238	0.022758
0.126654	12.0337	1.7398	0.007692	0.021803	0.023121	0.000794	0.001673	0.001939	0.001389	0.008441	0.021589	0.022206
0.129878	11.9056	1.7295	0.007692	0.020612	0.021918	0.000585	0.001355	0.001516	0.001061	0.007226	0.020053	0.021647
0.133032	11.7806	1.7193	0.007692	0.019483	0.020769	0.000432	0.001185	0.001187	0.000813	0.006195	0.018622	0.021083
0.135789	11.6715	1.7103	0.006000	0.014453	0.015435	0.000257	0.000928	0.000745	0.000501	0.004217	0.013595	0.016052
0.138164	11.5778	1.7025	0.006000	0.013830	0.014792	0.000204	0.001061	0.000616	0.000408	0.003747	0.012828	0.015708
0.140507	11.4856	1.6947	0.006000	0.013235	0.014171	0.000161	0.001341	0.000510	0.000332	0.003332	0.012102	0.015365
0.142820	11.3949	1.6869	0.006000	0.012665	0.013574	0.000128	0.001789	0.000421	0.000271	0.002965	0.011416	0.015024
0.145104	11.3055	1.6792	0.006000	0.012120	0.012999	0.000102	0.002404	0.000347	0.000221	0.002640	0.010767	0.014684
0.147362	11.2176	1.6716	0.006000	0.011598	0.012445	0.000081	0.003168	0.000285	0.000181	0.002352	0.010153	0.014347
0.149596	11.1310	1.6639	0.006000	0.011099	0.011913	0.000064	0.004035	0.000233	0.000148	0.002097	0.009572	0.014013
0.151808	11.0457	1.6563	0.006000	0.010622	0.011401	0.000051	0.004942	0.000190	0.000121	0.001870	0.009024	0.013682
0.153998	10.9618	1.6486	0.006000	0.010164	0.010910	0.000041	0.005820	0.000154	0.000099	0.001669	0.008505	0.013355
0.156168	10.8791	1.6410	0.006000	0.009725	0.010438	0.000033	0.006602	0.000124	0.000081	0.001491	0.008015	0.013032
0.158320	10.7977	1.6334	0.006000	0.009304	0.009985	0.000026	0.007234	0.000099	0.000067	0.001333	0.007552	0.012713
0.160455	10.7175	1.6258	0.006000	0.008900	0.009550	0.000021	0.007683	0.000079	0.000055	0.001192	0.007115	0.012398
0.162574	10.6386	1.6182	0.006000	0.008511	0.009133	0.000017	0.007934	0.000062	0.000045	0.001066	0.006702	0.012089
0.164679	10.5607	1.6105	0.006000	0.008137	0.008733	0.000013	0.007992	0.000049	0.000037	0.000955	0.006312	0.011784
0.166769	10.4840	1.6029	0.006000	0.007777	0.008349	0.000011	0.007875	0.000038	0.000030	0.000855	0.005943	0.011484
0.168847	10.4082	1.5952	0.006000	0.007432	0.007980	0.000009	0.007611	0.000030	0.000025	0.000766	0.005595	0.011188
0.170916	10.3365	1.5876	0.006000	0.007109	0.007636	0.000007	0.007263	0.000024	0.000021	0.000689	0.005272	0.010903
0.172973	10.2658	1.5799	0.006000	0.006798	0.007306	0.000006	0.006834	0.000018	0.000017	0.000620	0.004967	0.010623
0.175019	10.1931	1.5722	0.006000	0.006492	0.006981	0.000005	0.006323	0.000014	0.000014	0.000556	0.004674	0.010343
0.177055	10.1212	1.5645	0.006000	0.006198	0.006669	0.000004	0.005786	0.000011	0.000012	0.000499	0.004397	0.010068
0.179084	10.0501	1.5567	0.006000	0.005916	0.006370	0.000003	0.005243	0.000009	0.000010	0.000448	0.004135	0.009797
0.181104	9.9799	1.5490	0.006000	0.005646	0.006083	0.000002	0.004712	0.000007	0.000008	0.000403	0.003888	0.009532

Table 4.5   *(continued)*.

R	T6	ln (ρ_c)	d(mass)	d(energy)	dϕ(pp)	dϕ(^{8}B)	dϕ(^{13}N)	dϕ(^{15}O)	dϕ(^{17}F)	dϕ(^{7}Be)	dϕ(pep)	dϕ(hep)
0.183118	9.9105	1.5412	0.006000	0.005387	0.005807	0.000002	0.004203	0.000005	0.000007	0.000362	0.003654	0.009273
0.185125	9.8418	1.5334	0.006000	0.005140	0.005544	0.000001	0.003727	0.000004	0.000005	0.000325	0.003434	0.009018
0.187127	9.7739	1.5256	0.006000	0.004903	0.005291	0.000001	0.003286	0.000003	0.000005	0.000292	0.003227	0.008768
0.189788	9.6845	1.5151	0.010000	0.007671	0.008284	0.000002	0.004596	0.000004	0.000006	0.000422	0.004947	0.014072
0.193106	9.5743	1.5019	0.010000	0.007085	0.007657	0.000001	0.003658	0.000003	0.000004	0.000353	0.004453	0.013416
0.196415	9.4658	1.4886	0.010000	0.006541	0.007073	0.000001	0.002886	0.000002	0.000003	0.000296	0.004005	0.012783
0.199718	9.3590	1.4752	0.010000	0.006036	0.006529	0.000001	0.002261	0.000001	0.000002	0.000247	0.003599	0.012173
0.203018	9.2537	1.4617	0.010000	0.005568	0.006024	0.000000	0.001761	0.000001	0.000002	0.000207	0.003231	0.011585
0.206317	9.1500	1.4480	0.010000	0.005133	0.005554	0.000000	0.001366	0.000001	0.000001	0.000173	0.002898	0.011019
0.209618	9.0477	1.4342	0.010000	0.004730	0.005117	0.000000	0.001055	0.000000	0.000001	0.000145	0.002597	0.010476
0.212923	8.9469	1.4203	0.010000	0.004356	0.004711	0.000000	0.000812	0.000000	0.000001	0.000121	0.002324	0.009954
0.216234	8.8473	1.4062	0.010000	0.004010	0.004335	0.000000	0.000624	0.000000	0.000000	0.000102	0.002078	0.009453
0.219555	8.7491	1.3920	0.010000	0.003690	0.003986	0.000000	0.000478	0.000000	0.000000	0.000085	0.001856	0.008973
0.222887	8.6520	1.3776	0.010000	0.003393	0.003661	0.000000	0.000365	0.000000	0.000000	0.000071	0.001656	0.008513
0.226232	8.5557	1.3630	0.010000	0.003118	0.003360	0.000000	0.000278	0.000000	0.000000	0.000059	0.001476	0.008073
0.229593	8.4602	1.3483	0.010000	0.002864	0.003081	0.000000	0.000211	0.000000	0.000000	0.000049	0.001313	0.007653
0.232972	8.3653	1.3333	0.010000	0.002629	0.002822	0.000000	0.000160	0.000000	0.000000	0.000041	0.001167	0.007251
0.236372	8.2712	1.3182	0.010000	0.002412	0.002582	0.000000	0.000121	0.000000	0.000000	0.000034	0.001035	0.006869
0.239794	8.1777	1.3029	0.010000	0.002211	0.002359	0.000000	0.000091	0.000000	0.000000	0.000029	0.000917	0.006504
0.243241	8.0848	1.2874	0.010000	0.002026	0.002154	0.000000	0.000068	0.000000	0.000000	0.000024	0.000811	0.006157
0.246716	7.9925	1.2716	0.010000	0.001856	0.001964	0.000000	0.000051	0.000000	0.000000	0.000020	0.000716	0.005827
0.250222	7.9008	1.2556	0.010000	0.001698	0.001788	0.000000	0.000038	0.000000	0.000000	0.000016	0.000631	0.005510
0.253760	7.8096	1.2394	0.010000	0.001551	0.001626	0.000000	0.000028	0.000000	0.000000	0.000013	0.000555	0.005203
0.257334	7.7189	1.2229	0.010000	0.001411	0.001477	0.000000	0.000021	0.000000	0.000000	0.000011	0.000487	0.004898
0.260946	7.6287	1.2061	0.010000	0.001277	0.001340	0.000000	0.000016	0.000000	0.000000	0.000009	0.000427	0.004585

0.264600	7.5389	1.1890	0.010000	0.001144	0.001213	0.000000	0.000000	0.000011	0.000000	0.000000	0.000007	0.000374	0.004253
0.268298	7.4496	1.1716	0.010000	0.001011	0.001097	0.000000	0.000000	0.000008	0.000000	0.000000	0.000006	0.000326	0.003895
0.272044	7.3607	1.1539	0.010000	0.000879	0.000991	0.000000	0.000000	0.000006	0.000000	0.000000	0.000005	0.000284	0.003504
0.280751	7.1626	1.1135	0.035000	0.002080	0.002746	0.000000	0.000000	0.000010	0.000000	0.000000	0.000008	0.000724	0.008788
0.294808	6.8550	1.0458	0.035000	0.001073	0.001869	0.000000	0.000000	0.000003	0.000000	0.000000	0.000002	0.000429	0.004266
0.309802	6.5466	0.9729	0.035000	0.000598	0.001237	0.000000	0.000000	0.000001	0.000000	0.000000	0.000001	0.000244	0.001665
0.325982	6.2360	0.8935	0.035000	0.000357	0.000791	0.000000	0.000000	0.000000	0.000000	0.000000	0.000000	0.000133	0.000569
0.343676	5.9215	0.8062	0.035000	0.000210	0.000486	0.000000	0.000000	0.000000	0.000000	0.000000	0.000000	0.000068	0.000176
0.363339	5.5994	0.7091	0.035000	0.000115	0.000283	0.000000	0.000000	0.000000	0.000000	0.000000	0.000000	0.000032	0.000049
0.385619	5.2643	0.5998	0.035000	0.000054	0.000154	0.000000	0.000000	0.000000	0.000000	0.000000	0.000000	0.000014	0.000012
0.411497	4.9113	0.4741	0.035000	0.000018	0.000076	0.000000	0.000000	0.000000	0.000000	0.000000	0.000000	0.000005	0.000003
0.442571	4.5314	0.3262	0.035000	-0.000002	0.000033	0.000000	0.000000	0.000000	0.000000	0.000000	0.000000	0.000002	0.000001
0.481707	4.1077	0.1459	0.035000	-0.000012	0.000012	0.000000	0.000000	0.000000	0.000000	0.000000	0.000000	0.000000	0.000000
0.523422	3.7037	-0.0483	0.023730	-0.000010	0.000003	0.000000	0.000000	0.000000	0.000000	0.000000	0.000000	0.000000	0.000000
0.562430	3.3650	-0.2219	0.018099	-0.000009	0.000001	0.000000	0.000000	0.000000	0.000000	0.000000	0.000000	0.000000	0.000000
0.600638	3.0595	-0.3867	0.013804	-0.000007	0.000000	0.000000	0.000000	0.000000	0.000000	0.000000	0.000000	0.000000	0.000000
0.637722	2.7785	-0.5420	0.010528	-0.000006	0.000000	0.000000	0.000000	0.000000	0.000000	0.000000	0.000000	0.000000	0.000000
0.673285	2.5126	-0.6863	0.008030	-0.000005	0.000000	0.000000	0.000000	0.000000	0.000000	0.000000	0.000000	0.000000	0.000000
0.706835	2.2482	-0.8173	0.006124	-0.000004	0.000000	0.000000	0.000000	0.000000	0.000000	0.000000	0.000000	0.000000	0.000000
0.737676	1.9661	-0.9300	0.004671	-0.000001	0.000000	0.000000	0.000000	0.000000	0.000000	0.000000	0.000000	0.000000	0.000000
0.765309	1.6944	-1.0287	0.003563	0.0	0.0	0.0	0.0	0.0	0.0	0.0	0.0	0.0	0.0
0.789883	1.4664	-1.1237	0.002717	0.0	0.0	0.0	0.0	0.0	0.0	0.0	0.0	0.0	0.0
0.811795	1.2740	-1.2160	0.002072	0.0	0.0	0.0	0.0	0.0	0.0	0.0	0.0	0.0	0.0
0.831385	1.1102	-1.3064	0.001581	0.0	0.0	0.0	0.0	0.0	0.0	0.0	0.0	0.0	0.0
0.848946	0.9695	-1.3955	0.001206	0.0	0.0	0.0	0.0	0.0	0.0	0.0	0.0	0.0	0.0
0.864737	0.8476	-1.4838	0.000919	0.0	0.0	0.0	0.0	0.0	0.0	0.0	0.0	0.0	0.0
0.878987	0.7412	-1.5721	0.000701	0.0	0.0	0.0	0.0	0.0	0.0	0.0	0.0	0.0	0.0
0.891906	0.6476	-1.6610	0.000535	0.0	0.0	0.0	0.0	0.0	0.0	0.0	0.0	0.0	0.0
0.903687	0.5644	-1.7515	0.000408	0.0	0.0	0.0	0.0	0.0	0.0	0.0	0.0	0.0	0.0
0.914518	0.4898	-1.8451	0.000311	0.0	0.0	0.0	0.0	0.0	0.0	0.0	0.0	0.0	0.0

For some purposes, it is useful to have approximate relations between the physical variables. The relation between the density ρ (in g cm^{-3}) and temperature T_6 (in units of 10^6 K) is close to a polytrope of index 3 in the inner 65% by mass of the Sun,

$$\rho = 0.041(1 \pm 0.07)T_6^3, \tag{4.1}$$

for $T_6 > 6.5$.

Table 4.5 gives the local production rates in the Sun for nuclear energy and for individual neutrino fluxes. The first four columns list the radius, the temperature (in units of 10^6 K), the logarithm (to the base 10) of the electron density, ρ_e (in units of cm$^{-3}/N_A$, where N_A is Avogadro's number), and the mass of the zone (in units of $M_\odot$). The fifth column gives the fraction of the total photon luminosity that is generated in the zone. The last eight columns give the fraction of each neutrino flux that is produced in the zone. Tables 4.4 and 4.5 differ in that the quantities in Table 4.4 refer to the end points of each zone whereas the temperature, radius, and electron density in Table 4.5 refer to the zone center. The remaining quantities in Table 4.5 are all integrals over the individual zones. The quantities in Table 4.5 can be summed to yield the total fluxes from the neutrino-producing reactions. The constraint on the total luminosity for the converged solar model has been enforced through the trapezoidal rule and the reported total neutrino fluxes have also been calculated with this algorithm. Consequently, Table 4.5 is the appropriate source to obtain local neutrino fluxes for studying the MSW effect (which is discussed in Chapter 9).

Figure 4.1 illustrates some of the most interesting physical characteristics of the standard solar model. Figure 4.1a shows the fraction of the energy generation that is produced at different positions in the Sun. The energy generation peaks at a radius of $0.09R_\odot$, which corresponds to about $0.06M_\odot$; the half peak values of the energy generation extend from $0.04R_\odot$ to $0.16R_\odot$, that is, from the inner $0.007M_\odot$ to $0.23M_\odot$. Figures 4.1b and 4.1c illustrate the distributions of temperature and density; the central values are, respectively, 15.6×10^6 K and 148 g cm^{-3}. The density decreases much more rapidly than the temperature; the half peak value for the density occurs at $0.12R_\odot$, whereas the half peak value for the temperature occurs at $0.25R_\odot$. The peak of the energy generation occurs at a temperature of about 14×10^6 K and a density of about 95 g cm^{-3}.

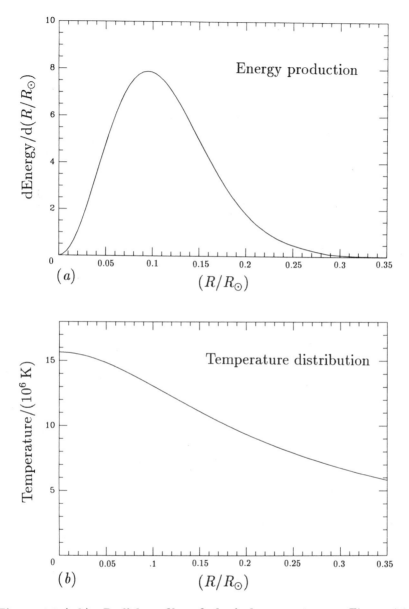

Figure 4.1 (a,b)   Radial profiles of physical parameters   Figure 4.1*a* shows the fraction of the energy generation that is produced at each position in the standard solar model. Figure 4.1*b* illustrates the temperature distribution in the standard solar model.

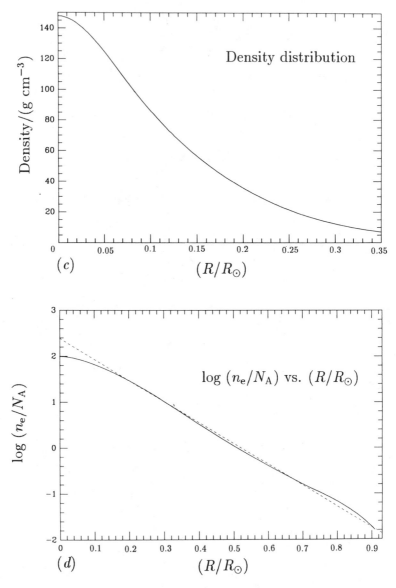

Figure 4.1 (c,d)   Radial profiles of physical parameters (con't)   Figure 4.1c illustrates the density distribution in the standard solar model. Figure 4.1d shows as a solid line the logarithm of the electron number density, N_e, divided by Avogadro's number, N_A, as a function of solar radius. The dotted line is an exponential fit to the density distribution, the parameters of which are given in the text [from Bahcall and Ulrich (1988)].

Figure 4.1d shows the dependence of the electron number density upon solar radius. The way that the electron density depends upon radius is crucial for the MSW effect. The dotted line in Figure 4.1d shows the fit of a linear function to the logarithm of the electron density of the standard solar model. The equation for the straight line in Figure 4.1d is $\log (n_e/N_A) = 2.39 - 4.58x$ or

$$n_e/N_A = 245 \exp(-10.54x) \text{ cm}^{-3}, \tag{4.2}$$

where as before $x = R/R_\odot$. Note that the linear fit is not exact and the parameters depend upon where the fitting is done. In particular, the formula given in Eq. (4.2) gives a value of n_e that is too large by about a factor of 2.5 at the solar center.

Figure 4.2 shows the dependence of the mass fractions upon position in the Sun. Figure 4.2a shows how helium is increased in abundance with respect to hydrogen by nuclear burning in the solar interior. The most remarkable distribution is shown in Figure 4.2b, which illustrates the strongly peaked abundance (on a linear scale) of ^{3}He. In the innermost region, the ^{3}He abundance is small because ^{3}He is burned rapidly by reactions 3 and 4 of Table 1.2. In the outermost region, no ^{3}He is produced by proton burning. There is a sharp peak in the ^{3}He abundance near $0.28R_\odot$. In this region, ^{3}He is produced by the hydrogen burning reactions (numbers 1 and 2 of Table 1.2), but is mostly not burned because reactions 3 and 4 of Table 1.2 require higher temperatures.

Table 4.6 presents some properties of the model as a function of time. The model of the present Sun has a luminosity that has increased by 41% from the nominal zero-age model (when the model Sun first reached quasistatic equilibrium on the main sequence) and the effective temperature has increased by 3%. The flux of ^{8}B neutrinos has increased dramatically; the contemporary flux is a factor of 41 times larger than the zero-age value.

How would phenomena on Earth be affected by the luminosity and temperature changes indicated in Table 4.6? The answer to this question is not known and has not been investigated with the aid of modern computer codes that have been developed for studying contemporary atmospheric and geophysical phenomena. New uncertainties, including the albedo of the Earth as a function of time, are introduced into the already complicated geophysical problems if one tries to calculate phenomena on the time scale of billions of years. But, perhaps, the following more limited question could be

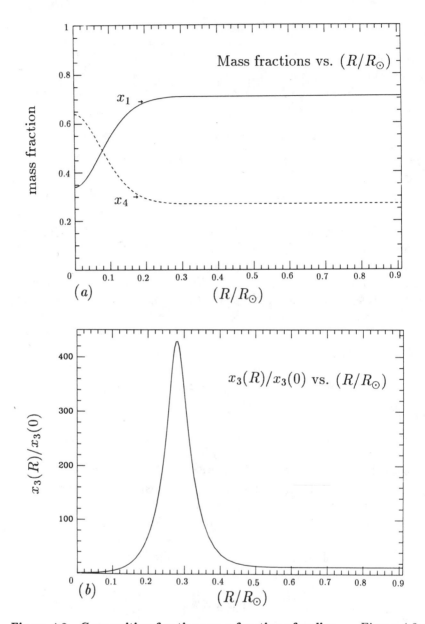

Figure 4.2   Composition fractions as a function of radius   Figure 4.2a shows the logarithm of the hydrogen, X_1, and helium, X_4, mass fractions as a function of position in the standard solar model. Figure 4.2b illustrates the dependence of the ^{3}He abundance upon position [from Bahcall and Ulrich (1988)].

Table 4.6.   Properties of the solar model as a function of time.

Time (10^9 yr)	Luminosity $(L_\odot)$	T_e (K)	Radius $(R_\odot)$	T_c (10^7 K)	$\phi(^8\text{B})$ $(10^6 \text{ cm}^{-2} \text{ s}^{-1})$
0.00	0.7095	5625	0.89	1.34	0.14
0.06	0.7198	5642	0.89	1.34	0.14
0.28	0.7339	5655	0.89	1.35	0.17
0.50	0.7457	5660	0.90	1.36	0.21
1.50	0.7955	5688	0.92	1.40	0.45
2.50	0.8517	5718	0.94	1.44	0.98
3.50	0.9163	5745	0.97	1.49	2.24
4.60	1.0000	5772	1.00	1.56	5.75

answered: Is it possible to find at least one plausible model of the terrestrial atmosphere that is consistent with the evolution indicated in Table 4.6 and with everything we know about the Earth? If it is possible, how specialized is the model?

The helium mass fraction is highest in the interior as the result of hydrogen burning, while the heavy element abundance is constant everywhere, by assumption. In all of the modern calculations, the core of the Sun is convectively stable, although not by much.

The sensitivity of the inferred initial helium abundance to changes in input parameters was evaluated by Bahcall et al. (1982) with the aid of a series of standard solar models. The logarithmic derivatives of Y with respect to the assumed solar age, luminosity, and the most important nuclear parameters are listed in their Table 10. The largest recognized contribution to the uncertainty in the inferred helium abundance is caused by the uncertainty in the initial value of Z/X and is of order a few percent. Standard solar models yield a well-defined value for the initial helium abundance:

$$Y = 0.27 \pm 0.01. \tag{4.3}$$

This initial solar value of helium represents an upper limit to the primordial helium abundance at the beginning of the Big Bang (if there was one). Three determinations of the helium abundance are in satisfactory agreement: the initial solar helium abundance, the present-day abundance of helium in the Galaxy's interstellar medium, and the preferred abundance based on cosmological considerations. All three quantities are equal to within the errors of their determinations, which are at least a few percent.

4.4 Helioseismology

Helioseismology, like terrestrial seismology, provides information about the interior of the body under study by using observations of slight motions on the surface. The technique is analogous to striking a bell and using the frequencies of the emitted sound to make inferences about the bell's constitution. Leighton, Noyes, and Simon (1962) first discovered solar oscillations by studying the velocity shifts in absorption lines formed in the solar surface. To their surprise, they found – instead of the anticipated purely chaotic motions – that the surface of the Sun is filled with patches that oscillate intermittently with periods of the order of 5 minutes and velocity amplitudes of order 0.5 $km s^{-1}$. The oscillatory motion was subsequently detected in measurements of the solar intensity. The oscillations typically persist for several periods with a spatial coherence of order a few percent of the solar diameter.

The correct explanation for the observed solar motions was a long time in coming. We now know [Ulrich (1970) and Leibacher and Stein (1971)] that the Sun acts as a resonant cavity. Sound waves known as **p-modes** (or pressure modes) are largely trapped between the solar surface and the lower boundary of the convection zone. The waves bounce back and forth between spherical-shell resonant cavities bounded on the outside by the reflections due to the density gradient near the solar surface and on the inside by refractions due to the increasing sound speed. The observed motions result from the superposition of several million resonant modes with different periods and horizontal wavelengths. Individual modes may have velocity amplitudes of as much as 20 cm s^{-1} and vertical excursions as large as a meter or two.

In order for a mode to resonate in the solar acoustic cavity, a half-integral number of waves must fit along the path leading from the solar surface to the base of the cavity. The depth of the cavity is fixed by the condition that horizontal wavenumber equals the total wavenumber (i.e., the vertical wavenumber becomes equal to zero), at which point the wave is refracted back towards the surface. The vertical wavenumber decreases with increasing depth in the Sun because the temperature rises in the inner regions. For many of the waves that have been most intensively studied, the base of the resonant cavity is close to the base of the convective zone.

For a given horizontal wavelength only certain periods will correspond to a resonance in the solar cavity. It was therefore predicted [Ulrich (1970)] and subsequently observed [Deubner (1975) and Rhodes, Ulrich, and Simon (1977)] that the strongest solar oscillations fall in a series of narrow bands when the results are displayed in a two-dimensional power spectrum that shows amplitude as a function of both period and horizontal wavelength. Just as in a musical instrument, the largest amplitudes correspond to standing waves that constructively interfere at the boundaries of the cavity. The simplest example is provided by oscillations in an organ pipe. The information about the conditions in the cavity (in the Sun or in the organ pipe) is contained in the observed frequency spectrum since the rate of propagation depends upon the physical conditions of the medium in which the wave travels. However, unlike most musical instruments with which we are familiar, the Sun oscillates in three dimensions, giving rise to an especially rich spectrum of tones or frequencies. These frequencies provide information about the temperature and density distribution within the Sun and its chemical composition. Solar rotation breaks the symmetry between otherwise degenerate modes and enables observers and theorists working together to make important inferences about the rate at which interior regions of the Sun are rotating.

Observations by Claverie et al. (1979) and by Grec, Fossat, and Pomerantz (1980), which utilized the integrated light from the entire solar disk, showed that the oscillations are globally coherent. The modes observed by these techniques provide the most important information currently available for the study of the deep solar interior since they penetrate most deeply toward the solar center (see discussion below).

What causes the excitation and the damping of solar oscillations? This question is difficult to answer both theoretically and observationally. Fortunately, the growth and decay of the modes are not particularly important for the purposes of this book. We need only make the plausible assumption that frequencies are not significantly influenced by the damping or excitation processes. The stability of the oscillations sets a limit on the precision of the helioseismological constraints that we can impose on the solar models. Present observations indicate that the frequencies of the oscillations can be

measured to an accuracy of about two parts in ten thousand, which
provides strong constraints. Indeed, some of the earliest detailed
observational results led to the conclusion that the depth of the solar convection zone was somewhat greater than previously believed
[Gough (1977) and Rhodes, Ulrich, and Simon (1977)].

Several of the nonstandard solar models that are discussed in
Chapter 5 can be tested with the aid of observations of the p-modes.

This section focuses on the relation between the oscillation frequencies and the neutrino fluxes. More detailed reviews of the
method have been given by Deubner and Gough (1984), Leibacher
et al. (1985), Christensen-Dalsgaard, Gough, and Toomre (1985),
Toomre (1986), and Bahcall and Ulrich (1988).

Consider a hypothetical drastic change in the nuclear energy generation: Set the cross section for the ^{3}He $+$ ^{4}He reaction (reaction
4 of Table 1.2) equal to zero. This results in a model with no ^{8}B
and ^{7}Be neutrino fluxes. The predicted solar neutrino capture rates
are reduced by an order of magnitude for most of the experiments
discussed in this book (although not for the ^{71}Ga experiment). The
structure of the Sun is also affected since approximately 15% of the
energy generated in the standard solar model involves this reaction.

The p-mode frequencies are insensitive to even this drastic change
in nuclear energy generation. The characteristic change in p-mode
frequencies caused by switching off the ^{3}He $+$ ^{4}He reaction (and all
the higher-energy neutrino fluxes) is less than 0.01% [see Bahcall
and Ulrich (1988)].

The frequencies of the **g-mode** [or gravity waves], which penetrate
deeply into the stellar interior, exhibit a small sensitivity ($\sim 0.2\%$)
to the hypothetical change in nuclear energy generation. There have
been several reports suggesting that g-modes may have been detected
in the Sun, but these claims are controversial.

The calculations using the standard solar model represent well
the quantitative features of the solar p-mode frequency spectrum.
However, there are small ($\sim$ a few tenths of a percent) discrepancies
between observations and calculations of typical p-mode splittings,
which are of order 10^2 μHz. The most significant discrepancy is
the difference between the calculated and observed value of δ_{02}, the
small (~ 10 μHz) frequency separation (for radial nodes n differing
by 1) between the modes with spherical harmonic degrees $l = 0$ and
$l = 2$. There are theoretical reasons for believing that this frequency
separation is less susceptible to uncertain surface phenomena than

are the much larger frequency splittings. Bahcall and Ulrich (1988) estimate the discrepancy to be at approximately the 3σ level of significance. The observed value for δ_{02} is between 8.9 μHz and 9.9 μHz [Pallé et al. (1987)], depending upon the pairs of radial nodes chosen. The value calculated with the standard solar model is 10.6 μHz.

A small gradient in the initial helium abundance can modify the calculated oscillation frequencies significantly and in the correct sense to improve the agreement with observations [see rows 14 through 17 of Table XX of Bahcall and Ulrich (1988)]. The calculated ^{8}B neutrino flux is increased by about 15% by the specified ad hoc assumption regarding the composition gradient. Thus changes in the interior structure of the Sun sufficient to account for the observed p-mode oscillation frequencies can change the predicted neutrino fluxes by amounts that are smaller than the currently estimated uncertainties in calculating the neutrino fluxes. Hence the explanation of some of the most significant discrepancies between observation and calculation of p-mode oscillation frequencies may require a composition gradient that is slightly different from what is assumed in the standard solar model (in which the composition gradient is due solely to in situ nuclear burning).

Solar models that involve WIMPs (see Section 5.13) can also improve agreement with p-mode oscillation measurements [see Faulkner, Gough, and Vahia (1986) and Däppen, Gilliland, and Christensen-Dalsgaard (1986)]. In this case, however, the predicted ^{8}B neutrino fluxes are reduced by a large factor with respect to the flux predicted by the standard solar model.

The low Z models, which have often been considered as a possible solution to the solar neutrino problem, result in a larger discrepancy between the observed and calculated values of δ_{02} than for the standard solar model. The discrepancy is of order 2 μHz for the low Z models and of order 1 μHz for the standard model, out of a total observed value of about 9 μHz.

The histogram of the fractional contributions to the observed p-mode splitting is shown in Figure 4.3 for mass fractions from $0.05M_\odot$ to $1.0M_\odot$, corresponding to radial intervals from $0.08R_\odot$ to $1.0R_\odot$. The histograms for the production of neutrinos from ^{8}B decay and the generation of the solar luminosity (which is nearly the same as the histogram for the production of neutrinos from the pp reaction) are also displayed in Figure 4.3.

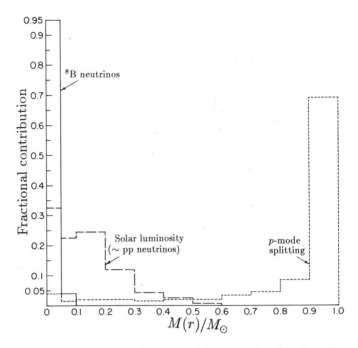

Figure 4.3   Histogram of fractional contributions to p-mode splitting, the flux of neutrinos from ^{8}B decay, and the flux of neutrinos from the pp reaction   Here $M(r)/M$ is the fraction of the solar mass interior to the point r. In order to resolve the ^{8}B neutrino emission, the width of the inner two histogram points is $0.05M(r)/M$, not $0.1M(r)/M$.

Figure 4.3 shows that the three observational quantities, p-mode oscillations, the solar luminosity (or pp neutrinos), and ^{8}B neutrinos, are primarily determined in different regions. Nearly all of the neutrinos from ^{8}B decay originate in the inner 5% of the solar mass. Almost 70% of the p-mode splitting comes from the outer 10% of the solar mass. The important regions for the generation of the solar luminosity, and the flux of neutrinos from the pp reaction, are intermediate in distribution between those for the p-mode splitting and those for the flux of ^{8}B neutrinos. About 33% of the solar luminosity is produced in the inner $0.05M_\odot$ from which the neutrinos from ^{8}B decay also originate but a majority of the nuclear energy is produced in the intermediate region between $0.05M_\odot$ and $0.4M_\odot$. The fractional decomposition shown in Figure 4.3 is essentially independent of any plausible changes in nuclear cross sections.

The comparison between calculated oscillation frequencies and observed p-modes has led to several conclusions that are relevant for nonstandard solar models (see Chapter 5). The depth of the convection zone has been a focus of investigation because p-mode frequencies corresponding to high spherical harmonics l are sensitive to the ratio of mixing length to pressure scale height. Models constructed with a low initial heavy element abundance, which have been suggested as a possible solution of the solar neutrino problem, yield a convection zone that is too shallow to agree with the observations of high-degree 5 minute oscillations [Christensen-Dalsgaard, Gough, and Morgan (1979) and Ulrich and Rhodes (1977)]. Rotation of the Sun breaks the degeneracy of the normal modes with respect to the azimuthal quantum number m; the apparent patterns of oscillation are modified by the different rates at which modes with separate values of m are carried around the solar surface. Observations so far have revealed that the angular velocity does not change rapidly with depth in the solar interior to radii of order $0.3R_\odot$ [see Duvall and Harvey (1984) and Christensen-Dalsgaard et $al.$ (1985)]. The velocity of sound can also be inferred from measurements of the oscillation frequencies and yields values in satisfactory agreement with model calculations for solar radii between 0.4 and $0.9R_\odot$ [Christensen-Dalsgaard et $al.$ (1985)].

The studies of p-mode oscillations and of solar neutrinos are largely complementary. Both techniques are required in order to understand the solar interior and both kinds of studies have influenced work in the complementary field.

The existing solar oscillation frequencies are among the most precisely known of all astrophysical quantities, having a precision equal to 2×10^{-5} in many cases. However, the interpretation of the results obtained from any given site is complicated by the periodicities introduced by the observing cycle, which introduce harmonics of the frequency $(1 \text{ day})^{-1}$. Scientists from more than 60 institutions (many different countries) are establishing a network of at least six observing sites around the world, which will provide coverage for more than 90% of the sidereal day. The network is know as the **GONG project** and is managed by the U.S. National Solar Observatory/National Optical Astronomical Observatory. The network will record 256×256 pixel images of the Sun in velocity, intensity, and line strength (using the Ni I line at 6768 Å) about once a minute. The ultimate accuracy is expected to reach 1 ms^{-1} on the solar surface. The network will produce of order 10^9 bytes of precision data each day.

The GONG project will provide continuous astronomical data of unprecedented quality and precision, making possible analyses of the Sun that rival in precision and in detail the most advanced laboratory studies of terrestrial materials. The data from the GONG project may lead to revolutionary advances in the understanding of the Sun and of stellar evolution.

Bibliographical Notes

Bahcall, J.N., W.F. Huebner, S.H. Lubow, P.D. Parker, and R.K. Ulrich (1982) *Rev. Mod. Phys.*, **54**, 767. A detailed discussion of the principles and practices of constructing a standard solar model. Includes stellar modeling, nuclear physics, and stellar opacity, as well as evaluations of the uncertainties in input parameters.

Bahcall, J.N. and R.K. Ulrich (1988) *Rev. Mod. Phys.*, **60**, 297. State-of-the-art calculations of solar models, neutrino fluxes, and helioseismological frequencies. The numerical results in Chapter 4 are based on this paper.

Cowling, T.G. (1941) *Mon. Not. R. Astron. Soc.*, **101**, 367; Pekeris, C.L. (1939) *Ap. J.*, **88**, 189. Two of the theoretical papers that founded the subject of helioseismology. Clear and exciting to read.

Deubner, F.-L. (1975) *Solar Phys.*, **44**, 371. Demonstrated that the lowest ridges resolved in the average k, ω diagram agreed well with Ulrich's predictions [and later calculations by Ando and Osaki (1975)], establishing the basic correctness of the interpretation of the 5 minute oscillations as standing acoustic waves.

Ezer, D. and A.G.W. Cameron (1965) *Can. J. Phys.*, **43**, 1497. An early and comprehensive discussion of solar evolution by one of the pioneering collaborations in solar neutrino studies.

Grevesse, N. (1984) *Physica Scripta*, **T8**, 49. A critical and comprehensive review of elemental abundances that are determined from accurate atomic data, careful modeling, and thorough analyses of the photospheric observations.

Leibacher, J.W., R.W. Noyes, J. Toomre, and R.K. Ulrich (1985) *Sci. Am.*, **253**, 48. Clear, semipopular account of helioseismology.

Leighton, R.B., R.W. Noyes, and G.W. Simon (1962) *Ap. J.*, **135**, 474. The discovery paper describing observations of the 5 minute oscillations in vertical Doppler velocities and in the brightness fluctuations of spec-troheliograms. Inaugurated the field of solar seismology. The first descrip-tion of the technique and results were given in a fascinating conference talk by Leighton that is published in *Nuovo Cimento Suppl.*, **XXII**, 321 (1961).

Sears, R.L. (1964) *Ap. J.*, **140**, 477. This was one of the first modern treatments of a standard solar model. The paper begins with a wonderful apology: "Theoretical models of the internal structure of the Sun are no longer at the frontier of the theory of stellar structure and evolution."

Ulrich, R.K. (1970) *Ap. J.*, **162**, 993. Explained the 5 minute oscillations as acoustic waves trapped below the photosphere and predicted that stand-ing waves would only be observed along discrete lines in the diagnostic diagram of horizontal wavelength versus frequency. An exciting paper to read because of the many important observations that are described by this epochal insight.

5

Nonstandard solar models

Summary

Faced with a persistent discrepancy between theory and observation, astrophysicists have explored speculative solutions to the solar neutrino problem, **nonstandard solar models**. Nearly every solution alters the standard solar model in a way that is designed to reduce the calculated flux of ^{8}B neutrinos. All the proposals change some quantity or physical relation from something that is generally accepted to something else that is generally regarded as unlikely or impossible.

Nonstandard solar models are stimulating and important. The different models make predictions that can be tested by future solar neutrino experiments or by optical measurements of the acoustic oscillations (p-modes) observed on the surface of the Sun. The models clarify what aspects of stellar evolution are tested by solar neutrino experiments, emphasizing changes in model characteristics that affect the calculated neutrino fluxes. Some nonstandard models that have been proposed imply revolutionary consequences for stellar physics or cosmology, and even in a few cases, for particle physics. Finally, and perhaps most importantly, nonstandard models highlight assumptions or aspects of stellar evolution theory that can profitably be subjected to a more intense scrutiny.

Nearly all of the nonstandard solar models in which the astrophysics is changed were proposed in the decade beginning in the late 1960s, shortly after the recognition of the discrepancy between the predictions of the standard solar model and the observations of the solar neutrino capture rate in the ^{37}Cl experiment. In this chapter, 17 nonstandard models are discussed; 13 of the 17 were proposed more than a decade ago. Of the remaining four, two involve new particle physics (WIMPs and Q-nuclei) and one and one-half (massive mass loss and half of the mechanisms discussed under hydrodynamical phenomena) increase, rather than decrease, the ^{8}B neutrino flux. The action has shifted over the past decade from astrophysical nonstandard solar models (discussed in this chapter) to new particle physics (discussed in Chapter 9).

The great majority of the published nonstandard solar models yield a smaller ^{8}B neutrino flux than the flux that is predicted by the standard solar model. Of the models discussed in this chapter, almost 90% predict a decreased flux. Presumably, the same fertile minds that invented models that decrease the ^{8}B neutrino flux could have invented with equal ease models that increase the predicted neutrino flux; the corresponding ideas were not published because of the lack of experimental justification. This bias is unfortunate since we need to know what might be wrong with the standard solar model, independent of whether the correction leads to an increased or a decreased neutrino flux. The difference between what the Sun produces in neutrinos and what is observed in experiments on Earth is ascribed, in a variety of theories that are discussed in Chapter 9, to neutrino interactions in the Sun and on the way to the Earth. If the solar production of neutrinos is incorrectly estimated in either direction, then incorrect neutrino parameters will be inferred.

This chapter describes a representative selection of some of the most stimulating nonstandard solar models. Much further work remains to be done by stellar evolution theorists. Precise solar models should be calculated for each of the suggested nonstandard models. The predictions of the models should then be tested against the available and future observations, including the frequencies of the solar acoustic oscillations. The implications of the suggested changes in solar evolution for the evolution of other stars should be investigated quantitatively and compared with observations. A significant fraction of the models discussed in this chapter may be in conflict with existing data, but in many cases the time-consuming calculations that are required for an accurate comparison of predictions with observations have not yet been done.

The reader may wish to consult other discussions for further details and references as well as different viewpoints. The reviews by Fowler

(1972), Kuchowicz (1976), Rood (1978), Bahcall and Davis (1982), Roxburgh (1985a,b), Schatzman (1985), Newman (1986), and Press (1986) are of special interest. The nonstandard models that are discussed in this chapter are based upon ideas that include: a low central heavy element abundance (Section 5.1), iron precipitation out of the gaseous phase (Section 5.2), a burnt-out central core (Section 5.3), rapid rotation in the solar interior (Section 5.4), a strong magnetic field (Section 5.5), a mixed-up Sun (Section 5.6), turbulent diffusion (Section 5.7), thermal instabilities (Section 5.8), hydrodynamic phenomena (Section 5.9), nucleo-thermal destabilization of g-modes (Section 5.10), a high mass loss rate (Section 5.11), a central black hole (Section 5.12), WIMPs (Section 5.13), Q-nuclei (Section 5.14), non-Maxwellian velocity distribution (Section 5.15), no formation of ^{7}Be (Section 5.16), and a pure CNO cycle (Section 5.17).

5.1   The low Z model

The temperature gradient in the solar interior is proportional to the radiative opacity [see Eq. (3.5)] and therefore one can lower the central temperature by decreasing the opacity. Two methods of achieving this lower opacity are by supposing that the solar interior has a lower abundance of heavy elements, the low Z model that is described in this section, or by assuming that the iron is precipitated out of the gas phase, a possibility that is discussed in the following section (Section 5.2).

About half of the opacity of the solar interior is contributed by heavy elements if the interior abundance ratio of heavy elements to hydrogen, Z/X, is the same as on the solar surface [see last column of Table V of Bahcall et al. (1982)]. The opacity in the solar interior can be reduced by assumption if one supposes [Bahcall and Ulrich (1971)] that the interior abundance of heavy elements is sufficiently small so as not to contribute significantly to the radiative opacity. If the heavy element abundance in the interior is reduced by a factor of 10 or more compared to the surface abundance, then essentially all of the opacity is caused by electron scattering and inverse bremsstrahlung on hydrogen and helium nuclei. By hypothesizing that the heavy elements are under-represented in the solar interior, the computed neutrino fluxes can be made consistent with the rate observed in the ^{37}Cl and the Kamiokande II experiments.

For example, Bahcall and Ulrich (1988) find values for the ^{37}Cl experiment of between 1.4 SNU and 1.6 SNU for low Z models with a composition discontinuity at different fractions (0.56 to 0.68) of the solar radius. The value of the ^{8}B flux found for their assumed low value of the heavy element abundance, $Z/X \approx 2.7 \times 10^{-3}$, is

$$\phi(^8\text{B}) = 0.7 \times 10^6 \text{ cm}^{-2} \text{ s}^{-1}. \tag{5.1}$$

The second row of Table 6.6 shows the neutrino fluxes calculated by Bahcall and Ulrich (1988) for a low Z model with a predicted ^{37}Cl capture rate of 1.6 SNU. The ^{8}B flux is reduced by about a factor of seven with respect to the standard solar model value, while the ^{7}Be flux is decreased by about a factor of 2.5. The CNO fluxes are decreased dramatically, by more than an order of magnitude, while the pp and pep fluxes are practically unchanged and the hep flux is increased by about 35%.

Sienkiewicz, Paczyński, and Ratcliff (1988) have calculated the effect of assuming a low abundance of heavy elements in a core whose boundary was varied from a small fraction of the solar radius to the full Sun. They found that agreement with the ^{37}Cl experiment could be achieved with a low Z core confined to a radius as small as $0.3R_\odot$.

Gough (1981) and Bahcall and Ulrich (1988) have calculated p-mode oscillation frequencies using low Z models and have concluded that the nonstandard models exacerbate the disagreement between the observed and calculated frequencies. This result makes low Z models seem less attractive, but cannot be used to reject them absolutely. The discrepancies might be explained by some as yet unrecognized process in the convective zone that – when included – results in agreement between theory and observation when a low Z abundance is assumed.

5.2 Iron precipitation

Ironlike elements contribute a significant fraction, $\sim 25\%$, of the opacity in the solar interior. Several authors have suggested that the plasma corrections to the Gibbs free energy of a H, He, and Fe mixture result in a phase transition in which nearly all the Fe precipitates out of the H and He gas [see Pollock and Alder (1978), Ruff and Liszi (1985), and Ruff, Liszi, and Gombos (1985)]. Some refined calculations of the miscibility of iron ions in the solar plasma have not confirmed this suggestion [see especially the review by Ichimaru, Iyetomi, and Tanaka (1987) and references quoted therein]. Assuming that all the iron does precipitate out of the gaseous mixture, Dearborn, Marx, and Ruff (1987) calculated the solar neutrino fluxes. These authors assumed that Fe does not contribute to the radiative opacity. The stellar evolution code that was used by Dearborn et al. (1987) and the physics that it contains were not described in detail. However, the authors reported a 40% reduction in the calculated capture rate for the ^{37}Cl experiment, down from 6.5 SNU for their standard model to 4.0 SNU for an iron-free radiative opacity. If further investigations support the suggestion of iron precipitation, it would be important to construct a detailed solar model with the best available physics to investigate the effect of removing ironlike elements from the opacity.

5.3 A burnt-out core

Prentice (1973, 1976) hypothesized a different chemical inhomogeneity, pointing out that a reduced ^{8}B neutrino flux could be achieved if the Sun possesses a small burnt-out core in which the hydrogen has been exhausted. This model was based upon a number of assumptions.

Prentice supposed that resistive segregation of grain material during the early stages of star formation was followed by a turbulent phase of planetary formation that led to an initial chemically inhomogeneous Sun consisting of a metalrich core, with a mass of a few percent of a solar mass. Strong convection was hypothesized to make the core sufficiently large that it affected the solar neutrino fluxes. The conversion from a hot core burning hydrogen via the CNO cycle to a burnt-out core must have occurred in the recent solar past in order that the observable characteristic of the star appear today to be in agreement with those of a "standard" main sequence model.

No modern detailed models have been constructed based upon this suggestion, but approximate analytic models that are calibrated against standard numerical models exhibit the suggested effects. The CNO fluxes were not computed in these calculations; the suggested [Prentice (1976)] enhancement of the CNO elements in the proposed isothermal core might lead to a significant counting rate from these neutrinos [see Eqs. (6.6) to (6.8)].

5.4   Rapid rotation

Many authors have considered the possibility that rapid rotation in the solar core would reduce the thermal pressure required to support the star against gravity and therefore decrease the emitted solar neutrino flux [see, e.g., Ulrich (1969), Bartenwerfer (1973), Demarque, Mengel, and Sweigart (1973), Rood and Ulrich (1974), and Roxburgh (1974)]. In order to estimate the amount of rotation that is required, consider the equation of hydrostatic equilibrium when an extra term is added that represents the additional support provided by the centrifugal force. By performing a spherical average of the radial component of the centrifugal force, one finds

$$\frac{\mathrm{d}P}{\mathrm{d}r} = -\frac{G\rho(r)M(r)}{r^2} + \frac{2\rho(r)\Omega^2(r)r}{3}, \qquad (5.2)$$

where $\Omega(r)$ is the rotational frequency at a point r in the Sun. In order to decrease significantly the pressure gradient and therefore the neutrino fluxes, the magnitude of the centrifugal support must be at least a percent or so of the gravitational attraction.

Let ϵ be the ratio of centrifugal repulsion to gravitational attraction,

$$\epsilon \equiv \frac{2\Omega^2(r)r^3}{3GM(r)}, \qquad (5.3)$$

or

$$\epsilon \cong 1.6 \times 10^{-8} \left(\frac{27\ \mathrm{d}}{\tau}\right)^2 \left(\frac{r}{0.05R_\odot}\right)^3 \left(\frac{0.01M_\odot}{M(r)}\right). \qquad (5.4)$$

Here τ is the average rotational period in the region in which the ^{8}B neutrino flux is produced, [$\sim 0.05R_\odot$ or $\sim 0.01M_\odot$]. The surface period is 27 days.

If $\epsilon \gtrsim 0.01$, then the rotation in the deep interior must be more than a thousand times faster than on the surface (i.e., the interior period must be of order an hour or less). In order that the calculated oblateness of the Sun resulting from the rotation not exceed the observed value, the outer regions must rotate progressively less rapidly as the Sun ages [see Demarque *et al.* (1973)], although even an hypothesized progressive spin down does not remove all of the difficulties [Rood and Ulrich (1974)].

No detailed models have been constructed that include rapid rotation and take account of the more stringent oblateness limit imposed by recent solar measurements [see Table 4.1 or Dicke, Kuhn, and Libbrecht (1985)].

The analysis of solar acoustic (p-mode) oscillations indicates [see Duvall *et al.* (1984) and Brown and Morrow (1987)] that the angular velocity Ω is remarkably uniform with depth, being within 30% of its surface value for radii as small as $0.3R_\odot$. There is no hint of an increase by the required factor of order of a thousand. Unfortunately, the existing inferences from the available data probably cannot be relied upon interior to about $0.3R_\odot$ [see, e.g., Gough (1985)], which is outside the region in which the ^{8}B neutrinos are produced (see Figure 6.1). It is possible, however, that a more specific analysis of the existing data could provide a useful limit on ϵ in the solar core. Future observations of the rotational splittings of the low-degree acoustic modes will almost certainly strongly constrain ϵ in the relevant region.

5.5 A strong magnetic field

How would a strong magnetic field affect the predicted fluxes? As a first approximation, several groups have investigated the effect of including magnetic pressure in the equation of hydrostatic equilibrium (similar to what was done above for the centrifugal force). One has in this case [Abraham and Iben (1971), Bahcall and Ulrich (1971), Bartenwerfer (1973), and Chitre, Ezer, and Strothers (1973)],

$$\frac{dP_{\text{gas}}}{dr} = -\frac{G\rho(r)M(r)}{r^2} - \frac{1}{8\pi}\frac{d}{dr}\left[B^2(\mathbf{r})\right]. \qquad (5.5)$$

Thus the effect of including a magnetic field is to increase the pressure gradient if the magnetic field decreases toward the surface. The fractional increase, β, in the pressure is

$$\beta \equiv \frac{B^2}{8\pi P_{\text{gas}}} \cong 0.2 \left(\frac{B(r)}{10^9 \text{ gauss}}\right)^2 \left(\frac{P_{\text{gas}}(0)}{P_{\text{gas}}(r)}\right). \qquad (5.6)$$

An interior field of order 10^9 gauss is required in order to change significantly the calculated neutrino fluxes.

The equations describing the normal modes for a magnetic field inside the Sun are given by Cowling (1945) and were solved numerically, together with the equations of stellar evolution, by Bahcall and Ulrich (1971). The lifetime of the first Cowling mode was found to be about 2.5×10^{10} yr, suggesting that a primordial magnetic field could survive in this mode. The influence of a possible primordial field on the mixing of regions with different elemental abundances is an interesting problem for future research.

The calculated effect of including a magnetic field with a central value of 10^9 gauss was to increase by a factor of two the predicted event rate for the ^{37}Cl experiment [Bahcall and Ulrich (1971)]. Since the effect had the opposite sign from what was desired, no further investigation of the Cowling modes seems to have been made. Also, Cowling (1965) has summarized earlier arguments that the interior field must be less than about 10^8 gauss in order to be stable.

Several authors have considered instead strong magnetic fields that are confined to the core of the Sun [see, e.g., Bartenwerfer (1973) and Chitre, Ezer, and Strothers (1973)]. The decay lifetime for such fields would be expected to be much less than the age of the Sun [$\sim 10^8$ yr $(R_{\text{core}}/0.1R_\odot)$] and therefore some unspecified mechanism must be involved that continuously regenerates the fields.

There is a decisive argument which shows that magnetic fields confined to the solar interior cannot significantly influence the calculated solar interior fluxes. A magnetic flux tube is essentially a bubble – although a complicated one – in which the interior density is less than the ambient density, causing the field to float to the surface. Parker (1974) presented an elegant discussion in which he showed that magnetic buoyancy would bring a strong interior field to the solar surface in less than 10^8 yr.

Brown and Morrow (1987) used observations of the quadratic splittings of p-mode oscillations to set an upper limit of about 10^7 gauss on a toroidal magnetic field that is confined to the solar interior.

5.6   A mixed-up Sun

The central temperature of the Sun increases as a function of time on the main sequence in order to compensate for the gradual depletion of the hydrogen abundance and the increasing luminosity of the star (see Chapter 4, especially Table 4.6). The calculated flux of ^{8}B neutrinos therefore increases monotonically with solar age. Ezer and Cameron (1968) suggested that the flux of ^{8}B neutrinos could be suppressed in the stellar models if the central hydrogen abundance could be maintained near its initial value. They proposed that extensive mixing of different regions of the Sun has been operative throughout solar history. This suggestion has spawned an industry

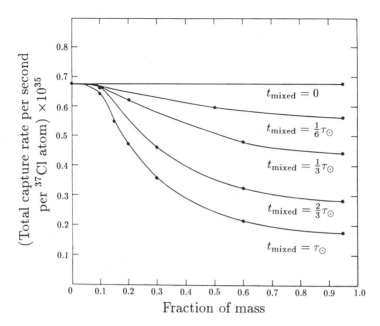

Figure 5.1   Dependence of ^{37}Cl capture rate upon mixed fraction and duration   The total capture rate is shown as a function of the fraction of the solar mass that is mixed and the duration of mixing. Here $\tau_\odot$ is the solar lifetime.

devoted to investigating the effects of composition mixing on solar and stellar evolution, and on the calculated neutrino fluxes [see, e.g., Bahcall, Bahcall, and Ulrich (1968), Shaviv and Salpeter (1968) and Schatzman (1969), and for recent reviews Roxburgh (1985a,b) and Spruit (1987)].

Figure 5.1, which is reproduced from Bahcall, Bahcall, and Ulrich (1968), shows that in order to obtain agreement with the ^{37}Cl experiment the mixing must be extensive: lasting for practically the entire prior history of the Sun and extending over at least 60% of the solar mass. In the numerical experiments that are illustrated in Figure 5.1 the entire region that was mixed was assumed to be continuously homogenized, a maximum-mixing scenario.

No known mechanism has been found that could supply the energy necessary to keep the Sun thoroughly mixed. In fact, three arguments have been proposed that make the "mixed-up" solution appear unlikely. The first argument [described in Bahcall, Bahcall, and Ulrich (1968)] is based upon the mean molecular weight gradient that is built up as a function of time due to nuclear burning and the second argument [Shaviv and Salpeter (1968)] is based upon the increased lifetime that would result for stars if homogenization were a common phenomenon. The third argument is based upon energetics [see Roxburgh (1985a,b) or Spruit (1987)].

The mean molecular weight μ is defined by the relation

$$n = \frac{\rho}{\mu m_{\mathrm{H}}}, \tag{5.7}$$

where n is the number density, m_{H} is the mass of a hydrogen atom, and ρ is the mass density. The molecular weight increases as a function of time in the solar interior as hydrogen is gradually converted to helium. Since the hydrogen conversion is most efficient in the high-temperature core, the mean molecular weight in the standard solar model is highest in the solar interior and decreases gradually with increasing radius. For the standard model described in Chapter 4, the molecular weight at the present solar age is 0.85, 0.79, 0.70, and 0.61 at $R/R_{\odot} = 0.0, 0.05, 0.2$, and 1.0, respectively. In order to drive mixing against the slow buildup of mean molecular weight (ignoring thermal stratification), an effective pressure gradient of order

$$\frac{\partial P(\mu)}{\partial r} = -\left(\frac{\partial \mu}{\mu \partial r}\right) P \simeq -\left(\frac{\Delta \mu}{\mu}\right) \frac{GM(r)\rho(r)}{r^2} \tag{5.8}$$

must be maintained. The rotational velocity of the Sun has usually been invoked as the primary source of such a pressure gradient. In order for rotation to overcome the mean molecular weight gradient, the ratio of the gradient of the effective pressure from the mean molecular weight to the gradient of the centrifugal pressure must be less than or of order unity. The ratio S can be described in terms of macroscopic properties by the relation [Bahcall *et al.* (1969)]

$$S = \left|\frac{\partial \mu}{\partial r}\right| \bigg/ \frac{\partial P(\Omega)}{\partial r} \simeq \frac{\Delta \mu}{\mu} \frac{GM(r)}{\Omega^2(r) r^3}. \tag{5.9}$$

For the standard solar model, the rate at which the mean molecular weight increases with time is given approximately by

$$\frac{\Delta \mu}{\mu \Delta t} \approx (6 \pm 2) \times 10^{-2} \text{ per } 10^9 \text{ yr}, \tag{5.10}$$

where the largest increase applies at the center of the Sun and the smallest is for $R = 0.1 R_{\odot}$. Combining the previous two equations, one finds a time, $t_{\rm grad}$, over which a molecular weight gradient sufficient to prevent a rotational instability builds up,

$$t_{\rm grad} \approx 10^4 \text{ yr} \left(\frac{27 \text{ d}}{P}\right)^2.$$

(5.11)

Here P is the rotational period in the solar interior. The time allowed for a rotational instability to develop, Eq. (5.11), is three orders of magnitude shorter than the shortest characteristic time in the solar interior, the Kelvin–Helmholtz time scale (see Section 2.1). This argument suggests the rotational instabilities will be quelled by the mean molecular weight gradient before the instability has a chance to develop a large amplitude.

Measurements of the rotational splittings observed in the frequencies of the acoustic p-modes appear to rule out [Duvall et al. (1984)] rotational periods short enough to drive the mixing over an appreciable fraction of the solar age. Taken together with the results summarized in Figure 5.1, the estimate given above for $t_{\rm mixed}$ implies that the mean molecular weight gradient inhibits mixing to an extent that prevents mixing from greatly affecting the calculated neutrino fluxes.

The mixing hypothesis might exacerbate the well-known problem that the inferred ages of the oldest globular clusters exceed the age of the Universe unless the cosmological parameters are chosen carefully [see Shaviv and Salpeter (1968)]. Recent estimates for the ages of globular clusters usually fall in the range 15–20×10^9 yr, which requires a Hubble constant in the range of 50 km s^{-1} Mpc^{-1} [cf., e.g., Sandage (1982), Paczyński (1984), and VandenBerg (1983)]. An unmixed star evolves from the main sequence into a red giant when about 0.1 of its mass has been converted into ^{4}He. For a mixed star, the mass fraction burned and the main sequence lifetime are likely to increase by a factor of order (mixed fraction)/0.1. The mixed fraction cannot be large without exacerbating the age problem, provided the same mixing process occurs in globular cluster stars as in the Sun. Detailed stellar models must be constructed to investigate this conjecture since under some circumstances the burning rate might be increased sufficiently to more than compensate for the large mass of hydrogen that is available to be burned [see Joseph (1984)].

A large amount of energy is required to mix the interior against the stable gradient of helium [see Spruit (1987) and Roxburgh (1985a,b)]. To decrease the composition gradient that is built up in the standard model by a factor of order unity requires an amount of energy that is of order a few percent of the present gravitational binding energy of the Sun (since helium is a significant fraction of the total solar mass and most of the helium is, in the standard model, in the stellar core). The required energy is about five orders of magnitude more than the total present rotational energy if the angular velocity is constant with radius.

The combined force of these three arguments suggests that it is unlikely that mixing can affect the interior of the Sun sufficiently to change significantly the neutrino fluxes.

5.7   Turbulent diffusion

The effects of a specific model of mixing have been considered by Schatzman (1969, 1985) and his collaborators [e.g., Schatzman and Maeder (1981)]. In this model, the diffusion coefficient, D, is assumed to be related to the microscopic viscosity, ν, by the relation

$$D = \text{Re}^* \nu. \tag{5.12}$$

Here Re^* is a dimensionless constant which must be of the order of 10^2 in order to reduce the computed flux of ^8B neutrinos to a value that is consistent with observation. There is no convincing microscopic justification for assuming such a large value of Re^*, but Schatzman and his colleagues have argued that the assumption of a large effective diffusion coefficient allows one to understand better the abundances of ^3He, Li, and ^{13}C that are observed in the atmospheres of different stars [see Schatzman (1985) for a review of this work and for the relevant references; some reservations about the validity of these arguments have been expressed by Michaud (1985)].

The spectrum of the low-degree acoustic (p-mode) oscillations computed assuming $\text{Re}^* \sim 10^2$ seems to be in significant disagreement with observations of accurately measured splittings of low-degree modes [see Ulrich and Rhodes (1983) and Cox, Kidman, and Newman (1985)]. Also, an investigation of meridional circulations suggests that for the values of Re^* that solve the neutrino problem, the solar interior should be rotating several times faster than the solar surface [see Tassoul and Tassoul (1984a,b) and Michaud (1985)], a result that is in apparent conflict with inferences about solar rotation from recent measurements of p-mode splittings [see Duvall *et al.* (1984)].

The relation between models with turbulent viscosity and the observed spectrum of p-mode oscillations should be clarified over the next few years, when accurately computed solar models are compared in detail with existing and future observations of p-mode splittings.

5.8   Thermal instabilities

What could be wrong with the quasistatic description of stellar evolution presented in Chapter 2? One obvious possibility is that the solar interior is unstable and that conventional calculations such as those described in Chapters 2 and 4 give misleading results because they neglect instabilities that drive macroscopic processes. In fact, even for spherical models of giant stars there is a well-known instability in which there is a thermal runaway in a thin nondegenerate shell that is burning a highly temperature-sensitive nuclear energy source [Schwarzschild and Härm (1965)].

Could there be a thermal instability in the Sun that modifies the physical description of the standard solar model? Numerical calculations show that the standard solar model is stable against both spherical thermal instabilities [Schwarzschild and Härm (1973)] and nonspherical thermal instabilities [Rosenbluth and Bahcall (1973)].

Schwarzschild and Härm (1973) performed a linear perturbation analysis of the spherical equations of stellar evolution that are summarized in Section 2.2. No unstable modes were found despite an extensive search of all the interesting time scales. The standard solar model does not have eigenvalues corresponding to unstable modes with a real part corresponding to an e-folding time of 10 years or longer and an imaginary part (corresponding to an oscillatory instability, if not zero) of at least 10 years. The roots of the lowest modes are complex conjugates with characteristic (damped) frequencies that correspond, as expected, to periods of order 10^7 years [cf., Eq. (2.1)]. The energy generation rate was assumed in this analysis to be represented by a single power of the temperature, an obvious oversimplification (cf., Chapter 3), but the temperature exponent was varied over a wide range including values chosen to maximize the opportunity for an instability.

Rosenbluth and Bahcall (1973) considered the more complicated case of nonspherical thermal instabilities in the presence of chemically inhomogeneous layers. The treatment included general forms for the opacity and the chemical composition (as well as their gradients). Their analysis also took account of the interdependence, and different time scales, which exist in the pp chain. The main result was a necessary and sufficient condition for stability to nonspherical perturbations, a condition that was shown to be satisfied for the standard model of the Sun.

These results are not surprising, although there was no rigorous proof prior to the explicit calculations that major thermal instabilities do not occur in the main sequence Sun. Intuitively, one expects that a nondegenerate star will be stable to thermal instabilities in simple configurations since a small temperature increase will produce a large energy deposition that will cause the perturbed element to expand and therefore to cool. A similar effect operates in the opposite direction for a temperature decrease. Despite the intuitive expectation of stability to thermal perturbations, it is useful, because of the solar neutrino discrepancy, to have explicit calculations that demonstrate the anticipated solar resiliency.

5.9 Hydrodynamic phenomena

Are there hydrodynamic phenomena that alter significantly the conventional quasistatic picture of main sequence stellar evolution? Not enough is known to answer this question with certainty, but pioneering investigations by Press (1981) and Press and Rybicki (1981) have shown that the struggle to find an answer involves complicated and challenging physics questions.

The particular process that was investigated in some detail by Press (1981) is the hydrodynamic coupling by internal waves (g-modes of high radial wavenumbers) between turbulent fluid motions at the base of the convective zone and shear fluid motions closer to the stellar center. The induced fluid shear motions tend to become nonlinear in amplitude near the center of the star, possibly causing hydrodynamic and coupled radiative-hydrodynamic effects. These phenomena have not been investigated thoroughly in the stellar context, although internal gravity waves occur in the deep ocean and have been extensively studied in the oceanographic environment [see, e.g., Garrett and Munk (1979)]. Press suggested that internal waves can lead to turbulence that causes species mixing, possibly reducing the predicted ^8B neutrino flux.

The study of internal gravity waves in stars requires mathematical and physical treatments that are more sophisticated than is usual in the theory of stellar evolution. Press and Rybicki (1981) manage to show, despite the inherent complexities of the problem, that hydraulically enhanced diffusion of entropy can in appropriate circumstances increase the effective radiative opacity of a medium, such as the solar interior, that is stably stratified. This effect, if it occurs, would increase the flux of ^{8}B neutrinos that is predicted by the standard solar model.

Do the internal waves generated in the convective zone penetrate to the solar interior? Not very much, if the standard model of the present-day Sun is a reliable guide. Most of the energy that is created in g-modes at the base of the convective zone will be in the form of waves that are highly damped as they propagate toward the solar interior [Press (1981)]. However, the "safety factor" is not large. Press pointed out that only 2% of the mechanical flux available at the base of the convective zone need be introduced into waves with frequencies of order 10 times the convective overturn frequency in order to drive nonlinear internal waves all the way into the solar core. Spruit (1987) argues that Press's estimates are optimistic, but no one has carried out a self-consistent calculation of a solar model in which internal waves are allowed to play a significant role in the evolution of the star.

The first investigations of hydrodynamical phenomena in the solar interior have raised a number of fundamental questions in the theory of stellar evolution. Further study is required in order to decide with confidence whether or not the associated phenomena affect to a significant extent the calculated neutrino fluxes, although existing estimates suggest that the influence will be small. In any event, these investigations [Press (1981) and Press and Rybicki (1981)] constitute an exemplary demonstration of what may be required in terms of new ideas and technical developments if the solution of the solar neutrino problem lies in the realm of nonstandard solar models.

5.10 ^{3}He instability

Dilke and Gough (1972) suggested, in an influential paper that stimulated much subsequent work, that certain low-order g-modes may be unstable at particular epochs of the Sun's main sequence evolution. They proposed that the solar interior contains growing amplitude oscillations driven by the gradient of ^{3}He that is built up by the burning of hydrogen and by the temperature sensitivity of the nuclear reactions. Subsequent investigations by a number of authors [see especially Christensen-Dalsgaard, Dilke, and Gough (1974) and Roxburgh (1985a,b)] have confirmed that this possibility exists, although the suggested instability of the modes depends upon details of uncertain interactions with turbulence in the convective zone.

The original Dilke–Gough proposal was that the instability would trigger a finite amplitude, but episodic, destabilization of g-modes in the central regions. The amplitude of the unstable mode might grow large enough to cause a significant mixing of the material in the core, reducing the expected ^{8}B neutrino flux (cf. Section 5.6) and providing a simultaneous explanation of the terrestrial ice ages. Other authors [see the reviews by Roxburgh (1985a,b)] have suggested that the instability could lead to steady diffusive mixing or that the Sun settles down to finite amplitude g-mode oscillations. This latter possibility has been advocated by Roxburgh (1985a,b).

In a pioneering study, Rood (1978) showed that a transient phase of mixing, such as was originally conjectured by Dilke and Gough (1982) and by Fowler (1972), produces a complex result that initially increases the ^{8}B neutrino flux as fresh ^{3}He is introduced into the solar core. A low contemporary neutrino flux can be achieved in a stellar model if the mixing is assumed to occur on the correct time scale and over a sufficiently large fraction of the star. However, such a star would be undergoing massive structural changes today; Rood estimated that the luminosity would be changing currently by a factor of two on a time scale of a few million years. Is recent mixing of a major fraction of the mass of the Sun consistent with the observed acoustic mode observations? The required combination of stellar modeling and analysis of p-mode oscillations that is necessary to answer this question has not yet been performed.

If an instability has important consequences for the Sun, then its implications for other stars should be investigated. Rood (1978) has suggested that any episodic mechanism for extensive mixing will spread out or make clumpy the main sequences in galactic and globular clusters, lengthen the estimated age of globular clusters, and possibly destroy the agreement between calculation and observation of the observed gap near turnoff in the H–R diagram of old stellar clusters [see also Shaviv and Salpeter (1971)]. Rood's suggestions have not been tested by constructing detailed stellar models.

Press (1986) has stressed the importance of damping, pointing out that a tiny amount of dissipation at the level of $\gtrsim 10^{-12}$ of fractional energy loss per cycle would tilt the balance from growth to net damping. One example of a damping mechanism is nonlinear wave–wave (or mode–mode) interactions, which Press suggests would limit any instability of the Dilke–Gough sort to a tiny mode amplitude that could not drive significant mixing of the solar interior fluid. Another source of possible damping could be the coupling of the evanescent extension of the wave with the overlying convective zone and the stellar atmosphere. Press concludes that unmodeled sources of loss at the $\gtrsim 10^{-12}$ level could easily be present.

The basic reason for Press's cautionary remarks is that sources of mode growth are few, requiring as they do a source of free energy, whereas sources of mode damping are likely to be found whenever the idealizations of the instability model are examined in detail.

Further work on the Dilke–Gough mechanism should be carried out in the context of detailed solar models and of the recent observations of p-modes that extend into the solar core. It will be difficult, however, to prove that an unaccounted for dissipation mechanism does not exist at the level of 10^{-12}.

5.11   Massive mass loss

Stars are observed to lose mass in different stages of their lives. However, the effect of mass loss on the internal structure of the Sun is not included in the standard solar model. Could this be a significant omission? It seems unlikely, but there are at least two suggestions in the literature that propose huge mass loss scenarios that would affect the computed neutrino fluxes.

The mass loss rate today from the Sun is about

$$\dot{M}_{\odot}(\text{today}) \sim 10^{-14} M_{\odot} \ \text{yr}^{-1}. \tag{5.13}$$

There has not yet been a study of how neutrino fluxes depend upon the mass loss rate or integrated mass loss during the solar lifetime. One can make a preliminary guess of the effect using the mass–luminosity relation (see Chapter 2) and the calculated sensitivities of the neutrino fluxes to the assumed solar luminosity (Table 7.2). It is plausible that a total mass loss of order a percent or so of $M_{\odot}$ will be found necessary in order to affect significantly the calculated neutrino fluxes. Therefore, the current mass loss rate is orders of magnitude below the estimated minimum mass loss rate for affecting neutrino emission.

Willson, Bowen, and Struck-Marcell (1987) have suggested that in the early life of the Sun the mass loss rate was five to six orders of magnitude larger than the current mass loss rate. They propose that the Sun may have begun life as a star of mass $2M_\odot$ and shed the outer half of its mass during solar childhood. Willson *et al.* (1987) motivate this suggestion by describing astronomical observations that can be interpreted with the aid of a massive mass loss rate.

The energy required to drive a large mass loss is much greater than is available from stellar rotation, but Willson *et al.* (1987) point out that the necessary energy is only of order a few percent of the total solar luminosity. They justify assuming that a significant fraction of the stellar luminosity is converted to mechanical energy by reference to observations of young F and G stars that have observed X-ray luminosities in the range of 10^{-2} to 10^{-3} of the total stellar luminosity. No specific theoretical mechanism has been suggested for converting, with the required efficiency, the thermal energy released in the stellar core to mechanical energy of mass loss.

Mass loss increases the computed ^{8}B neutrino flux over what is found with the standard solar model. The bigger stellar mass requires a larger luminosity than in the standard solar model, resulting in a higher rate of conversion of hydrogen to helium. Thus the model with a large mass loss rate has a relatively lower hydrogen content today, implying that a higher central temperature will be necessary to produce the presently observed solar luminosity. This higher central temperature results in a larger computed ^{8}B neutrino flux. Guzik, Willson, and Brunish (1987) calculated detailed solar models with massive mass loss, using nuclear reaction rates and opacities obtained from the 1970's literature and an early version of Iben's stellar evolution program; they found ^{8}B neutrino fluxes that are up to a factor of two larger than in the standard solar model.

This nonstandard solar model is difficult to reconcile with the observed abundances of Li, Be, and B on the surface of the Sun. If the presently observed outer region of the Sun were previously buried under a much larger overburden of material, the higher temperature at this early stage would have destroyed by nuclear burning all of the light elements. For example, lithium is burned at temperatures in excess of 3×10^6 K. Willson *et al.* (1987) acknowledge this difficulty and suggest that spallation or some other mechanism may account for the observed surface abundances of the light elements. A large amount of energy is required to produce the observed light elements by spallation and no source of this energy is known.

5.12 A central black hole

Hawking (1971) suggested that a small accreting black hole in the center of the Sun could solve the solar neutrino problem. His idea was that a significant fraction of the solar luminosity could be supplied by gravitational energy produced by accretion onto a black hole, removing at least partially the necessity for nuclear fusion with the related neutrino emission. Numerical solar models based upon Hawking's proposal have been computed by Strothers and Ezer (1973) and by Clayton, Newman, and Talbot (1975a).[†] The treatment by Clayton and his colleagues is the most detailed and is the basis for the discussion given below.[‡]

Hawking's original proposal and the subsequent numerical implementations do not discuss the details of some challenging questions. For example, how does one explain the observed mass–luminosity relation (Figure 2.1) that is predicted by conventional stellar evolution (without central black holes that supply additional energy)? Why are the H–R diagrams (see Figure 2.2 and Section 2.3) for the distribution of stars in color and temperature well accounted for by standard solar models? Why are there not anomalous stars (harboring black holes) in the H–R diagrams? Is it unusual for a star, like the Sun, that appears normal, to contain a black hole? How many seed black holes are there in the interstellar medium and how does a star form around a black hole?

In the black hole solar models, the accretion luminosity is

$$L_{\text{accretion}} = f \dot{M}_{\text{star}} c^2, \qquad (5.14)$$

where f is an efficiency factor for converting mass accretion from the gas star (mass M_{star}) to observed (visual) luminosity. No detailed calculation has been made of f but Clayton et al. (1975a) speculate

---

[†]For a more dramatic rendering of this idea, see Clayton (1986).

[‡]The implications of this model for helioseismology have not been worked out; it seems unlikely that such a drastic change from the standard model would be in agreement with the observed p-mode oscillation frequencies.

that the value of f may be large, perhaps even as big as 0.1. The black hole gains mass by accretion according to the relation

$$\frac{\mathrm{d}m_{\mathrm{BH}}}{\mathrm{d}t} = (1-f)\dot{M}_{\mathrm{star}}. \tag{5.15}$$

The accretion is supposed to supply energy at the maximum luminosity possible, the so-called Eddington luminosity in which the radiation pressure on the accreting material of the emitted light just balances the gravitational attraction of the black hole. The accretion luminosity is therefore

$$L_{\mathrm{accretion}} = \frac{4\pi G m_{\mathrm{BH}} c}{\kappa}, \tag{5.16}$$

where κ is the radiative opacity. In order for the accretion luminosity to supply today an appreciable fraction of the solar luminosity, the mass of the central black hole must be of order of $10^{-5} M_{\odot}$,

$$m_{\mathrm{BH}} \, (\mathrm{today}) = \left(\frac{\kappa}{0.34}\right) \left(\frac{L_{\mathrm{accretion}}}{L_{\odot}}\right) \times 10^{-5} M_{\odot}. \tag{5.17}$$

The illustrative value of the radiative opacity used in the above equation, $0.3 \ \mathrm{cm^2 \, gm^{-1}}$, corresponds to the scattering of low-energy photons in a pure ionized hydrogen gas [cf. Clayton et $al.$ (1975a)].

The mass of the black hole increases exponentially with time according to the relation

$$m_{\mathrm{BH}}(t) = m_0 \exp\left(+t/\tau_0\right), \tag{5.18}$$

in which the e-folding time is determined by Eqs. (5.7) to (5.9) to be

$$\tau_0 = \frac{f\kappa c}{(1-f)4\pi G} \simeq 4\left(\frac{f}{0.1}\right)\left(\frac{\kappa}{0.34}\right) \times 10^7 \ \mathrm{yr}. \tag{5.19}$$

The remaining mass of the gaseous star at a time Δt in the future is

$$M_{\mathrm{star}}(t) = M_{\odot} \left[1 - \left(\frac{m_{\mathrm{BH}}}{M_{\odot}}\right) \exp\left(\frac{\Delta t}{\tau_0}\right)\right]. \tag{5.20}$$

The gaseous star will disappear into the black hole in a time Δt that is small compared to the age of the Sun,

$$\frac{\Delta t}{t_{\mathrm{age}}} = \left(\frac{\tau_0}{t_{\mathrm{age}}}\right) \ln\left(\frac{M_{\odot}}{m_{\mathrm{BH}}}\right) \cong 0.1 \left(\frac{f}{0.1}\right). \tag{5.21}$$

The black hole model makes the present epoch special. We happen to live in the period just before the Sun disappears entirely.

5.13   WIMPs

Weakly interacting massive particles, **WIMPs,** have been proposed as a simultaneous solution to the missing mass problem and to the solar neutrino problem [see, e.g., Press and Spergel (1985), Faulkner and Gilliland (1985), and Spergel and Press (1985)]. If exotic weakly interacting particles make up the galactic missing mass and are accreted in sufficient numbers by the Sun, then they can modify the energy transport in the solar core just enough to reduce the calculated event rates to the values observed in the ^{37}Cl and Kamiokande II experiments, provided that the scattering cross section and the mass for the WIMPs are chosen to lie in an appropriate range (see below). There are difficulties in making this scenario consistent with any known particle. The most stringent requirements are [Krauss *et al.* (1986)] that the scattering cross section must be large compared to typical weak interaction cross sections and that the annihilation cross section must be small compared to the scattering cross section: $\sigma_{\text{annihilation}} \lesssim 10^{-4}\sigma_{\text{scattering}}$. However, several authors have discussed new particles that could satisfy the required conditions [see, e.g., Gelmini, Hall, and Lin (1987), Raby and West (1987), and Griest and Seckel (1987)].

WIMPs can carry energy from the central (highest temperature) regions of the Sun to slightly further out, lowering the temperature gradient and the ^{8}B neutrino flux [see especially Spergel and Press (1985), Faulkner and Gilliland (1985), Gilliland, Faulkner, Press, and Spergel (1986), and Griest and Seckel (1987)]. Because they provide a new channel for transporting energy, WIMPs have a similar effect on the neutrino fluxes as lowering the radiative opacity.

What are the ideal properties of a particle that can perform this task? The particles should be confined to the innermost region of the Sun in which the ^{8}B neutrinos are produced in order that their effects not be discernible in conventional astronomical measurements. For a particle of mass m_{W} that is in thermal equilibrium with the plasma in the center of the Sun, the radius out to which it can reach is approximately

$$r \approx 0.13 R_{\odot} \left(m_{\text{proton}}/m_{\text{W}}\right)^{1/2}. \qquad (5.22)$$

Thus in order to confine the WIMPs to the region in which the ^{8}B neutrinos are produced, their mass must be larger than about 2 GeV. A higher limit ($\gtrsim$ 4 GeV) can be obtained by requiring that WIMPs not evaporate from the Sun [see Gould (1987a) and references therein], although it is possible to circumvent this limit if the scattering cross section is sufficiently high (see discussion below). In order that the WIMPs reach out to the region of ^8B neutrino production, $m_W \lesssim 10$ GeV.

An optimum scattering cross section for transporting energy would allow the WIMPs to interact about 40 times per orbit [see Nauenberg (1987)], which gives a cross section per baryon of

$$\sigma_{\text{optimum}} \approx 5 \times 10^{-36} \, \text{cm}^2. \tag{5.23}$$

This cross section corresponds to a mean path that is in order of magnitude equal to the radius of the Sun. Thus, in addition to being the *optimal* cross section for heat transport, this value is also the *minimum* cross section for saturating the capture probability [see Spergel and Press (1985)]. That is, for cross sections larger than σ_{optimum} most WIMPs which intersect the Sun are captured, while below σ_{optimum} the capture probability declines linearly with cross section. With this optimal cross section, a tiny number, N_W, of WIMPs can transport a significant fraction of the solar luminosity. The fraction of the solar abundance of protons,

$$\frac{N_W}{N_{\text{protons}}} \approx 10^{-11}, \tag{5.24}$$

is sufficient to reduce significantly the ^8B neutrino flux if the scattering cross section has the optimal value.

The optimum cross section that is given in Eq. (5.23) is the weighted average of the scattering cross sections for hydrogen and helium. If the scattering cross section by helium is negligible for WIMPs, then the cross section for scattering by hydrogen should be $\sim 15 \times 10^{-36}$ cm^2. A cross section this large would cause the optical depth of WIMPs in the Sun to be so large as to prevent evaporation of WIMPS down to about 3 GeV.

For scattering cross sections that are much larger or much smaller than the optimal value, fractional abundances much larger than 10^{-11} are required. The parameters can be adjusted so as to solve simultaneously the solar neutrino and missing mass problems [see Press and Spergel (1985) and the work by Faulkner and Gilliland that is summarized in Steigman *et al.* (1978)]. The WIMP hypothesis may improve the agreement between observation and calculation of the p-mode oscillation frequencies [see Faulkner, Gough, and Vahia

(1986) and Däppen, Gilliland, and Christensen-Dalsgaard (1986)] although the significance of the improvement is unclear [see Bahcall and Ulrich (1988), Section X].

A major subroutine must be added to conventional stellar evolution codes in order to include nonlocal energy transport by WIMPs. An approximate description of the energy transport has been given by Spergel and Press (1985), Faulkner and Gilliland (1985), and Gilliland *et al.* (1986) [see also Nauenberg (1986)]. The problem is complicated by the many orders of magnitude over which the WIMP parameters can vary. Moreover, the WIMPs are not in local thermal equilibrium with their surroundings.

There is not yet available a definitive calculation of the predicted neutrino fluxes if WIMPs are the correct solution of the solar neutrino problem. In particular, there is no published firm lower bound above which the ^{8}B flux must lie.

Gilliland *et al.* (1986) have estimated the expected neutrino fluxes for an illustrative case, using nuclear reaction rates, an equation of state, and opacities that were determined in the 1970s. Without performing a detailed *ab initio* calculation, one cannot rigorously correct the calculated fluxes of Gilliland *et al.* (1986) to take account of the improvements since the 1970s in reaction rates, opacities, and equation of state.

However, Bahcall and Ulrich (1988) have estimated the neutrino fluxes to be expected on the WIMP hypothesis by assuming that for each neutrino source the ratio of the flux on the WIMP hypothesis to the best estimate for the flux on the basis of the standard model (described in Chapter 4) is the same as the ratio of WIMP to standard fluxes calculated by Gilliland *et al.* (1986). Explicitly, they assume: $\phi_W(i)/\phi_{Stnd.}(i) = \phi_{W,G.}(i)/\phi_{Stnd.,G.}$, where the subscript "G." stands for Gilliland *et al.* (1986).

Table 6.6 lists, in the last row, the neutrino fluxes determined by scaling. The important flux of ^{8}B neutrinos should be given reasonably accurately by this procedure since it depends most strongly upon the cross section for the rare (p, γ) reaction on ^{7}Be, which does not affect the structure of the Sun. The fluxes of pp and pep neutrinos are also probably well determined since they differ by only 2% and 7%, respectively, in the WIMP and standard models. The flux of ^{7}Be neutrinos is not accurately determined; it is about 20% higher in the standard model than in the WIMP model. The CNO fluxes cannot be reliably evaluated by the scaling procedure; their neutrino fluxes are about a factor of two larger for the standard model than the WIMP model.

The main result is that the flux of ^{8}B neutrinos is reduced for the illustrative case considered by Gilliland *et al.* (1986) by about a factor of four relative to the standard model. The calculated capture rate is decreased by about the same factor of four for electron–neutrino scattering and for ^{2}H, ^{40}Ar, ^{81}Br, and ^{98}Mo detectors. The reduction is about a factor of two for a ^{7}Li detector and almost a factor of three for ^{37}Cl. For ^{71}Ga the reduction is only about 25% and is even smaller for an ^{115}In detector.

Two results are likely to be unchanged by future improvements in the solar model or by the choice of WIMP parameters. First, the pp fluxes will be similar in the standard and WIMP models; they differ by only a few percent according to the existing calculations. Second, the ratio of ^{7}Be to ^{8}B neutrino fluxes will probably remain the same to within a factor of two for fixed input nuclear physics parameters.

The WIMP hypothesis has a great advantage over nearly all the other nonstandard models discussed in this chapter: The WIMP explanation can be verified or disproved by feasible laboratory experiments. If WIMPs have the properties required to explain the solar neutrino problem, then they can be detected in dark matter searches using established technology [see, e.g., Drukier, Freese, and Spergel (1986) and Sadoulet *et al.* (1988)]. In addition, the WIMP hypothesis predicts a smaller value than the standard solar model for P_0, the normalized asymptotic period difference for the g-mode pulsation frequencies [cf. Faulkner, Gough, and Vahia (1986)]. Most other nonstandard solar models predict a larger value for P_0 than the standard solar model.

5.14   Q-nuclei

Does modern particle physics permit other schemes for deriving energy from nuclear fusion than the traditional pp chain and CNO cycle that are discussed in Chapter 3? Boyd and his collaborators [see Boyd *et al.* (1983), Sur and Boyd (1985), Boyd *et al.* (1985), and Takahashi and Boyd (1988)] have proposed that a tiny fraction ($\sim 10^{-15}$) of the nuclei in the center of the Sun contain embedded particles called Q-particles that have properties which enable them to solve the solar neutrino problem. The Q-nuclei provide an additional source of energy.

Free quarks are the prototype for Q-particles, although other possibilities have been suggested [see, e.g., Takahashi and Boyd (1988)]. The simplest proposal [Boyd et al. (1985)] involves a helium nucleus plus an embedded Q-particle catalyzing the burning of four protons to form an α-particle, similar to the way ^{12}C catalyzes proton fusion in the CNO cycle (see Section 3.4). Just as in the CNO cycle, the catalysis is driven by (p,γ) reactions punctuated by two β-decays. In order to make this scheme work, Q-nuclei must be somewhat more tightly bound than normal nuclei so that proton capture by a pregnant helium nucleus produces a stable mass 5 nucleus (also with an embedded Q-particle). An additional constraint on the physics of Q-nuclei is that ^{8}B β-decay be only a rare mode for the proposed Q-cycle [Sur and Boyd (1985)], that is, that a pregnant ^{7}Be nucleus positron-decay before it can capture a proton to form a ^{8}B nucleus plus an embedded quark. The effect of Q-nuclei on stellar structure produces some novel effects [see Joseph (1984)].

The Q-nucleus hypothesis is almost unique among nonstandard models in increasing the predicted event rate for a ^{71}Ga detector (see Chapter 11). The calculated capture rate [Sur and Boyd (1985)] is between 1.5 and 3 times as large as the standard rate (132 SNU), depending upon the unknown end-point energies for the decay of quark nuclei.

Nonstandard solar models involving Q-nuclei have not been investigated by other groups, presumably because one has to postulate new particles [see, however, de Rújula, Giles, and Jaffe (1978) or Cahn and Glashow (1981)] and to adjust their physical interactions so as to satisfy the constraints mentioned above.

5.15   Non-Maxwellian velocity distribution

The rate at which neutrinos are produced in the solar interior depends sensitively upon the number of relatively high-energy ions that are present and can penetrate the Coulomb barrier (see, e.g., the discussion in Section 3.1). Clayton and his colleagues [see Clayton (1974), Clayton et al. (1975b), and also Kocharov (1972)] have speculated that the observed deficit in ^{8}B neutrinos may reflect a departure from the assumed Maxwellian velocity distribution in the solar plasma. Questions regarding the validity under solar conditions of the Maxwellian distribution have been raised informally from time to time in seminars and in scientific discussions of the solar neutrino problem. The inevitability of the Maxwellian distribution can be appreciated by estimating the orders of magnitude of the physical quantities. For simplicity, the results are given for the case of a pure hydrogen plasma, although the reader can immediately generalize the formulae to the appropriate chemical mixture.

The Coulomb cross section for large-angle scattering of ions of relative energy E is

$$\sigma\,(\geq \pi/2) \;=\; \frac{1}{8}\,\frac{e^4}{E^2}\,. \tag{5.25}$$

Each such large-angle scattering changes significantly the individual ion energies.

The time, τ_{Coulomb}, between significant energy exchanges is therefore

$$\tau_{\text{Coulomb}} \;=\; \frac{m_{\text{H}}}{\rho\sigma v} \;=\; 10^{-12}\;\text{s}\;\left(\frac{E}{20\;\text{keV}}\right)^{3/2}\left(\frac{150\;\text{g cm}^{-3}}{\rho}\right), \tag{5.26}$$

where ρ is the ion density and v the relative velocity corresponding to E. The characteristic time for the nuclear reactions of interest ranges from 10^2 years to 10^{12} years (see the last column of Table 4.1). Therefore the ratio, R, of the time between strong Coulomb collisions that tend to restore the Maxwellian distribution and the available time between fusions is enormous:

$$
\begin{aligned}
R \;&=\; \frac{\tau_{\text{nuclear}}}{\tau_{\text{Coulomb}}} \\[4pt]
&=\; 3 \times 10^{20}\left(\frac{\tau_{\text{nuclear}}}{10^2\;\text{yr}}\right)\left(\frac{20\;\text{keV}}{E}\right)^{3/2}\left(\frac{\rho}{150\;\text{g cm}^{-3}}\right).
\end{aligned} \tag{5.27}
$$

Maxwell (1890) showed that the low-order moments of the distribution function relax to their equilibrium values in a few mean collision times, given by the expression above for τ_{Coulomb} with $E \sim 1\,\text{keV}$. Therefore collisions are adequate by more than 22 orders of magnitude to guarantee that the lowest moments are Maxwellian. But, what about the tail of the Maxwellian distribution?

The time to fill the Maxwellian tail out to an energy E has been shown, by explicit calculations in which the Boltzmann equation was solved with different approximations and energy-dependent cross sections, to be approximately equal to the self-collision time for particles of energy E [see MacDonald, Rosenbluth, and Chuck (1957) and Krook and Wu (1976)]. A Maxwellian distribution accurately describes the collision rate if $R(E)$ is much greater than unity. Thus even for the tail of the energy distribution that is most important for the neutrino-producing reactions, corresponding to energies of order 20 keV (see Section 3.2), the criterion for the validity of a Maxwellian law is satisfied by 20 orders of magnitude.

5.16   No ^{8}B produced or $S_{34} = 0$

The results that are discussed in this and the following section are based upon assumptions that are in conflict with laboratory measurements of nuclear reaction rates. The models in these sections are not intended to be solutions of the solar neutrino problem or serious competitors to the standard solar model, but are included in order to illustrate the interrelationship between different aspects of theory and observation.

Suppose that the cross section for the rare reaction that produces ^{8}B (number 7 of Table 3.1) were identically zero. What would be the consequences for solar neutrino experiments and for stellar evolution?

The consequences for solar neutrino studies would be disastrous. The proposed experiments using detectors of ^{2}H, ^{11}B, ^{40}Ar, ^{98}Mo, or neutrino–electron scattering at energies above 5 MeV would yield only upper limits or much reduced event rates due to the extremely rare hep neutrinos. Only the ^{71}Ga experiment (see Chapter 11) and the next-generation experiments designed to detect pp neutrinos (see Sections 13.5D and 14.3) would have signals comparable to those expected on the basis of the standard model. Nevertheless, the predicted capture rate for the ^{37}Cl experiment would be in agreement with observation,

$$\sum_i \phi_i \sigma_i = 1.8 \text{ SNU}. \tag{5.28}$$

The assumption that the rare ^{8}B reaction is entirely suppressed would have no significant consequences for stellar evolution. No measurement that can be performed with photons could reveal this drastic – for solar neutrinos – departure from the standard model.

The reaction ^{3}He(^{4}He,γ)^{7}Be (reaction 4 of Table 3.1) is of much greater significance for stellar evolution. About 15% of the terminations of the pp chain proceed through this reaction, which is proportional to the cross section factor S_{34}. The interior structure of the solar model is affected significantly by setting S_{34} equal to zero. In particular, the central hydrogen abundance is increased by 10% and the central ^{3}He abundance by 54%, relative to the standard solar model. The central pressure is increased by 3%.

The consequences for solar neutrino experiments would be even more disastrous than only suppressing the ^{8}B neutrinos [see Bahcall and Ulrich (1988)]. Setting $S_{34} = 0$ would also eliminate the important line neutrinos resulting from electron capture on ^{7}Be (see Table 3.1). The event rates in the ^{2}H, ^{11}B, ^{40}Ar, ^{98}Mo, and neutrino–electron scattering experiments would all be greatly reduced or made undetectable. In addition, the predicted event rate in the ^{37}Cl experiment would be reduced to

$$\sum_i \phi_i \sigma_i = 0.6 \text{ SNU}. \tag{5.29}$$

The predicted capture rate for the ^{71}Ga experiment would be reduced by 35%, from the standard model value of 132 SNU to 86 SNU.

The p-mode oscillation frequencies are insensitive to even this drastic change in nuclear energy generation. Let $\Delta\nu$ be the characteristic difference in frequencies caused by switching off the ^{3}He–^{4}He reaction. Bahcall and Ulrich (1988) have shown that

$$\frac{\Delta\nu}{\nu_{\text{standard}}} \lesssim 10^{-4}. \tag{5.30}$$

The fractional changes in the central hydrogen abundance and central pressure are more than two orders of magnitude larger than the fractional change in p-mode oscillation frequencies.

The g-mode frequencies exhibit a small sensitivity ($\sim 0.2\%$) to the hypothetical change in nuclear energy generation.

This numerical experiment demonstrates the complementarity of neutrino flux measurements and helioseismology.

5.17   CNO reaction dominant

According to the standard theory of stellar evolution, stars only slightly heavier than the Sun burn hydrogen via the CNO cycle. What would one observe if this prediction of the standard theory were incorrect and the dominant source of hydrogen fusion in the Sun were the CNO cycle [as originally suggested by Bethe (1939)]?

The solar neutrino spectrum can be computed simply if one assumes that the main branch of the CNO cycle supplies the entire solar luminosity. In this case, one finds [Bahcall (1969)] that

$$\phi(^{13}N) = \phi(^{15}O) = 3.4 \times 10^{10} \text{ cm}^{-2} \text{s}^{-1}. \tag{5.31}$$

Hence the calculated ^{37}Cl capture rate is

$$\sum_i \phi_i \sigma_i = 28 \text{ SNU}, \tag{5.32}$$

and the CNO model prediction for the ^{71}Ga experiment is

$$\sum_i \phi_i \sigma_i = 610 \text{ SNU}. \tag{5.33}$$

The ^{7}Li and ^{115}In detectors would also show large event rates. On the other hand, the CNO model predicts a negligible event rate for most of the other experiments discussed in this book, including those with detectors based upon ^{2}H, ^{11}B, ^{40}Ar, ^{98}Mo, and neutrino–electron scattering.

Bibliographical Notes

Dilke, F.W.W. and D.O. Gough (1972) *Nature*, **240**, 262. A stimulating proposal suggesting a nuclear–thermal instability that induces the geological ice ages and temporarily depresses the ^{8}B solar neutrino flux.

Ezer, D. and A.G.W. Cameron (1968) *Astrophys. Lett.*, **1**, 177. The first nonstandard solar model. Founded an industry based on the influence of mixing on the solar neutrino fluxes.

Fowler, W.A. (1972) *Nature*, **238**, 24. An influential paper that proposed two "desperate" solutions to the solar neutrino problem.

Hawking, S. (1971) *Mon. Not. R. Astron. Soc.*, **152**, 75. Original proposal that the universe contains many small black holes. Suggested that a black hole formed at the center of the Sun could explain the solar neutrino problem.

Littleton, J.E., H.M. Van Horn, and H.L. Helfer (1972) *Ap. J.*, **173**, 677. Showed that longitudinal plasma waves carried only a negligible fraction of the total energy transport.

Parker, E. (1974) *Astron. Space Sci.*, **31**, 261. Elegant discussion of magnetic buoyancy.

Press, W.H. (1981) *Ap. J.*, **245**, 286. A pioneering investigation of the effects of internal waves in stellar interiors.

Rood, R.T. (1978) in *Proceedings of Informal Conference on Status and Future of Solar Neutrino Research*, edited by G. Friedlander, (Brookhaven National Labortory) Report No. 50879, Vol. 1, p. 175. A balanced and insightful review of nonstandard solar models during the peak of their popularity.

Roxburgh, I.W. (1985) *Solar Phys.*, **100**, 21. A systematic review of many of the most-discussed nonstandard models, with extensive references.

Schatzman, E. (1969) *Ap. J. Lett.*, **3**, L139. Proposed that turbulent diffusion explains the solar neutrino problem.

Spergel, D.N. and W.H. Press (1985) *Ap. J.*, **296**, 679. A superb physical discussion of how WIMP energy transport can occur and affect neutrino emission. See also the prescient work by Faulkner and Gilliland that is summarized briefly in Steigmann *et al.* (1978) [*A. J.* **83**, 1050], a report of an idea before its time had come.

Spruit, H.C. (1987) in *Internal Solar Angular Velocity*, edited by B.R. Durney and S. Sofia (Dordecht: Reidel) p. 185. A clear discussion of angular momentum transport in the radiative interior of the Sun.

Ulrich, R.K. (1969) *Ap. J.*, **158**, 427. Opened the Pandora's box of rapid rotation.

6

The neutrinos

Summary

What nuclear reactions produce solar neutrinos? How many neutrinos are expected? What are the shapes of the calculated neutrino energy spectra? What information do neutrinos carry about the interior regions of the Sun?

There are eight principal nuclear reactions or decays that produce solar neutrinos; six of the eight nuclear sources create neutrinos with continuous energy spectra and the other two (pep and ^{7}Be) produce neutrino lines. The fundamental pp neutrinos are more numerous by more than an order of magnitude than any other source, but they are low energy and therefore difficult to detect. The ^{8}B neutrinos are energetic, but rare. Most of the experiments discussed in this book are primarily sensitive to ^{8}B neutrinos.

The energy spectrum of neutrinos from a particular nuclear reaction or β-decay is determined by nuclear physics. The shape of the energy spectrum is therefore independent of astronomical uncertainties. Unless new weak interaction physics is occurring, the normalized energy spectra from individual sources must have the shapes described in this chapter. The total number can be affected by astronomy, but not the relative number as a function of energy. All nonstandard and standard solar models give the same shape for the energy spectra, at production, of individual neutrino sources.

Experiments that measure the energy spectra of solar neutrinos can distinguish between faulty astrophysics or new physics as the origin of the solar neutrino problem.

Different sources of neutrinos are created in somewhat different parts of the Sun. The ^{8}B neutrinos, for example, are primarily produced in the core of the Sun while the pp and hep neutrinos originate much further out. Each source of neutrinos is most sensitive to specific input parameters and to the physical conditions in the region of that source's production.

The first section, 6.1, discusses the principal neutrino-producing reactions in the pp chain and in the CNO cycle. The neutrino fluxes carry information about the solar interior that is described in Section 6.2. The discussion in this section is based upon three figures, which show where the different neutrino fluxes originate in the standard solar models and which illustrate the temperature dependence of the pp, pep, ^{7}Be, ^{8}B fluxes for 1000 "standard models" with different input parameters. Section 6.3 contains figures and tables of the predicted neutrino spectra. The following section, 6.4, gives numerical values of the solar neutrino fluxes from the standard solar model, as well as from representative nonstandard solar models. Table 6.5 presents the standard model neutrino fluxes and their total uncertainties. The last section, 6.5, estimates neutrino fluxes from galactic stars other than the Sun.

6.1 The neutrino-producing reactions

The most important neutrino-producing reactions in the proton–proton chain are:

$$\text{p} + \text{p} \longrightarrow {}^2\text{H} + \text{e}^+ + \nu_\text{e} \qquad (q \leq 0.420 \text{ MeV}), \qquad (6.1)$$

$$\text{p} + \text{e}^- + \text{p} \longrightarrow {}^2\text{H} + \nu_\text{e} \qquad (q = 1.442 \text{ MeV}), \qquad (6.2)$$

$$^3\text{He} + \text{p} \longrightarrow {}^4\text{He} + \nu_\text{e} \qquad (q \leq 18.773 \text{ MeV}), \qquad (6.3)$$

$$^7\text{Be} + \text{e}^- \longrightarrow {}^7\text{Li} + \nu_\text{e}$$
$$(q = 0.862 \text{ MeV}, 89.7\%; 0.384 \text{ MeV}, 10.3\%), \qquad (6.4)$$

$$^8\text{B} \longrightarrow {}^7\text{Be}^* + \text{e}^+ + \nu_\text{e} \qquad (q \lesssim 15 \text{ MeV}). \qquad (6.5)$$

Here q is the neutrino energy.

The **energy spectrum** is the relative number of neutrinos of different energies from a given reaction or reactions. The pp, hep, and ^{8}B reactions [Eqs. (6.1), (6.3), and 6.5)] produce neutrinos with all energies from zero up to a maximum energy that is shown in each case in parentheses; the neutrinos from each of these sources has a different **continuous spectrum** that is discussed in Section 6.3. The pep and ^{7}Be neutrinos have well-defined energies and are known as **neutrino lines**.

The neutrinos from the pp reaction, Eq. (6.1) are in some sense the "most fundamental," since this reaction initiates the chain [Eq. (3.1)] that supplies nearly all (about 99%) of the solar luminosity. Moreover, the pp neutrinos are overwhelmingly the most abundant. Unfortunately, the pp neutrinos have low energies and are therefore difficult to detect. Neutrino cross sections increase rapidly with energy and the backgrounds from other sources decrease rapidly with energy.

The easiest neutrinos to detect are the relatively high-energy neutrinos from the hep (^{3}He+p) and ^{8}B reactions, Eqs. (6.3) and (6.5), respectively. Unfortunately, neutrinos from both of these sources are scarce, the hep neutrinos being much fewer in number than the rare ^{8}B neutrinos. Therefore, most of the experiments discussed in this book are based upon the detection of ^{8}B neutrinos.

Two of the neutrino sources listed above produce neutrino lines: the pep and ^{7}Be sources [Eqs. (6.2) and (6.3)]. The line shape of these sources contains in principle detailed information about the thermal structure of the solar interior [see Eqs. (6.14) and (6.13) below]. Unfortunately, none of the experiments in progress can distinguish these relatively low-energy neutrino lines.

The most important reactions that are believed to produce solar neutrinos within the CNO cycle are:

$$^{13}N \longrightarrow {}^{13}C + e^+ + \nu_e \qquad (q \leq 1.199 \text{ MeV}), \qquad (6.6)$$
$$^{15}O \longrightarrow {}^{15}N + e^+ + \nu_e \qquad (q \leq 1.732 \text{ MeV}), \qquad (6.7)$$

and

$$^{17}F \longrightarrow {}^{17}O + e^+ + \nu_e \qquad (q \leq 1.740 \text{ MeV}). \qquad (6.8)$$

The ^{13}N and ^{15}O reactions [Eqs. (6.6) and (6.7)] provide information about the CNO cycle (see Table 3.2). The ^{17}F neutrinos [Eq. (6.8)] are potentially of great interest since their flux is a measure of the initial oxygen abundance in the solar interior. However, all three of the CNO neutrino sources are expected to be difficult to detect because neither the calculated fluxes nor the characteristic energies are large.

Two laboratory sources of neutrinos are of special interest since they can be used to calibrate some of the solar neutrino detectors. The first, ^{51}Cr, will be used to calibrate the ^{71}Ga solar neutrino experiment and could be used to calibrate the ^{115}In experiment [see Raghavan (1978b)]. This isotope decays by capturing an electron and emitting a neutrino line:

$$e^- + {}^{51}Cr \longrightarrow {}^{51}V + \nu_e$$
$$(q = 0.746 \text{ MeV}, 90.1\%; 0.426 \text{ MeV}, 9.9\%). \quad (6.9)$$

Fortunately, ^{51}Cr can be obtained in sufficient quantity, megacuries, to make practical a collaboration of the gallium experiment. The neutrino energy for chromium is, however, below threshold for the ^{37}Cl experiment. Therefore, for the chlorine detector, one would have to use ^{65}Zn [Alvarez (1973)], an isotope that is much more difficult to handle experimentally. The decay modes of ^{65}Zn are:

$$e^- + {}^{65}Zn \longrightarrow {}^{65}Cu + \nu_e$$
$$(q = 0.227 \text{ MeV}, 50.75\%; 1.343 \text{ MeV}, 47.8\%), \quad (6.10a)$$
$${}^{65}Zn \longrightarrow {}^{65}Cu + e^+ + \nu_e \quad (q \leq 0.330 \text{ MeV}, 1.45\%). \quad (6.10b)$$

The neutrino line from electron capture on ^{37}Ar (the inverse of the detection reaction in the ^{37}Cl experiment) has some advantages as a calibrator of transitions whose cross sections can not be calculated accurately. These advantages include the similarity in energies of the solar ^{7}Be line and the laboratory ^{37}Ar neutrino line and the absence of associated gamma radiation [see Haxton (1988b)]. The decay process is

$$e^- + {}^{37}Ar \longrightarrow {}^{37}Cl + \nu_e \quad (q = 0.814 \text{ MeV}). \quad (6.11)$$

The maximum neutrino energy, q_{max}, that is given above for the continuum sources is the difference between the nuclear masses minus the mass of the positron that is produced. The values of q_{max} can be calculated from the tabulated mass differences, Δ, of neutral atoms using the relation

$$q_{max} = \Delta - 2m_e + |\delta_{atomic}|. \quad (6.12)$$

Equation (6.12) takes account of the fact that the neutrino sources in the Sun are ionized while the neutral atoms whose masses are tabulated have electrons with binding energy δ_{atomic}.

6.2   Information content

What information about the solar interior do the individual neutrino fluxes carry?

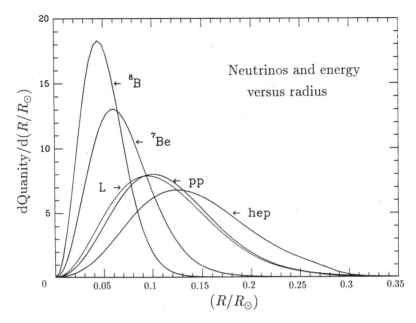

Figure 6.1 Neutrino and energy production as a function of radius
The fraction of neutrinos that originate in each fraction of the solar radius is $[\mathrm{dFlux}/\mathrm{d}(R/R_\odot)]\mathrm{d}(R/R_\odot)]$. The figure illustrates the production fraction for ^{8}B, ^{7}Be, pp, and hep neutrinos for the standard solar model that is described in Chapter 4. The fraction of the solar electromagnetic luminosity that is produced at each radius is denoted by L.

The predicted neutrino fluxes (or neutrino luminosities) are derived by integrating the production rate, R, over the volume (mass) of the Sun. The individual production rates are sensitive functions of the local variables: temperature, density, and chemical composition.

$$L_{\nu_i} \equiv \int \mathrm{d}M\, R_i\left[\rho(M), T(M), \{X_i\}\right]. \qquad (6.13)$$

Figure 6.1 shows where in the Sun the pp, ^{8}B, ^{7}Be, and hep neutrinos originate. For comparison, the figure also shows the fraction of the solar photon luminosity, L, that is generated in each radius interval and which leaks out from the surface. The information that is imprinted on the neutrinos of each nuclear source depends upon the local physical conditions, which are discussed in Chapter 4.

The region in which the pp flux is produced is very similar to the region in which most of the photon luminosity is produced; the peak occurs in both distributions at essentially the same position, $0.09R_\odot$ and $T = 14 \times 10^6$ K. The half peak positions for the pp production are shifted outward with respect to the energy generation by only about $0.01R_\odot$.

Because of its strong temperature dependence, the ^{8}B production is peaked at much smaller radii, $0.05R_\odot$ and $T = 15 \times 10^6$ K, and is generated in a narrower region, $0.02R_\odot$ to $0.07R_\odot$. The region of ^{7}Be production is intermediate between the ^{8}B and pp regions, peaking at $0.06R_\odot$ and spreading (half peak range) from $0.03R_\odot$ to $0.10R_\odot$.

The hep distribution is the most extended, ranging (half peak points) from $0.06R_\odot$ to $0.21R_\odot$ with a peak at $0.13R_\odot$ and $T = 12 \times 10^6$ K. The hep neutrinos are produced at relatively large radii in the solar core because the ^{3}He abundance increases as one goes outward from the center, peaking at $0.27R_\odot$. At the lower temperatures on the outer edge of the outer solar core, the pp chain produces ^{3}He (by the proton burning reactions 1 and 2 of Table 3.1), but the rate for burning ^{3}He (by reactions 3, 4, and 10 of Table 3.1) is slower than the production rate. The burning is slow because of the relatively high Coulomb barrier for the reactions that destroy ^{3}He. The difference between the relatively fast ^{3}He production rate and the relatively slow destruction rate creates a large nonequilibrium abundance of ^{3}He outside the highest temperature solar core.

The solar neutrino experiments that measure individual recoil electrons of higher energy, like the natural water electron scattering detectors and the ^{2}H and ^{40}Ar experiments (see Chapter 13), sample neutrinos from two different parts of the solar interior. These detectors are sensitive to both the ^{8}B neutrinos, produced at relatively small radii and high temperatures, and the hep neutrinos, which are generated at larger radii and lower temperatures. Fortunately, it is possible to distinguish on the basis of the energy of the recoil electrons between events that originate from hep neutrinos and those that are caused by ^{8}B neutrinos, allowing one experiment to probe regions of the solar interior that are both inside and outside the main energy-producing region.

The contained mass, $M(\leq r) = \int_0^r dM(r')$, determines to a large extent the local temperature and density (see Chapter 2). Therefore, it is instructive to examine what mass fractions, $M(\leq r)/M_\odot$,

Table 6.1. Neutrino fluxes as a function of mass fraction. The table shows the mass fraction, $M(\leq r)/M_\odot$, within which the specified percentage of each neutrino flux is produced.

Source	25%	50%	75%
pp	0.058	0.135	0.216
pep	0.036	0.104	0.180
hep	0.104	0.186	0.276
^{7}Be	0.011	0.026	0.065
^{8}B	0.005	0.011	0.021
^{13}N	0.006	0.021	0.150
^{15}O	0.005	0.014	0.031
^{17}F	0.005	0.011	0.026

contain, respectively, 25%, 50%, and 75% of the production of each neutrino source.

Table 6.1 shows the mass fractions (25%, 50%, 75%) for each neutrino source. We see from Table 6.1 that the main production of ^{8}B neutrinos is confined to the inner 2% of the solar mass, while the pp and hep neutrinos are produced mostly out to 22% and 28% of $M_\odot$, respectively.

The local rate of production of ^{8}B neutrinos is a sensitive function of temperature; Flux $\propto T^\alpha$, where $\alpha \simeq 24$ at the center of the Sun. The total flux of neutrinos from the standard solar model is an integral over the entire temperature and density profile and has therefore a different temperature dependence than the local (central) flux. Very approximately, the ^{8}B neutrino flux is proportional to $(S_{pp} \times S_{34}^2 \times S_{17}^2/S_{33} \times S_{e,7}^2)^{1/2}$. The S-factors in the above expression correspond, in order of appearance, to reactions 1, 6, 9, 5, and 7 of Table 3.1. Since the temperature dependence is strong, one may write the total rate of ^{8}B neutrino production in the following suggestive form:

$$\phi(^8\text{B}) \approx \text{const.} \times T^\beta. \qquad (6.14a)$$

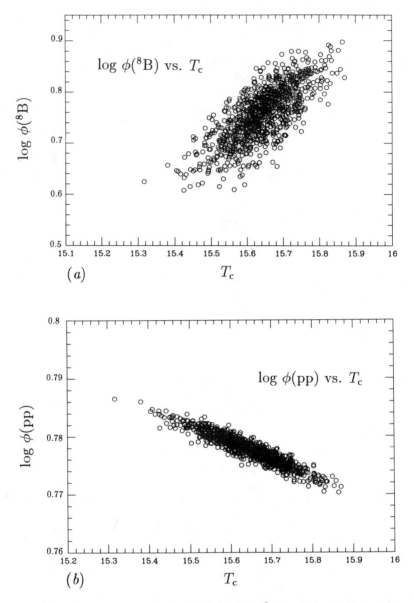

Figure 6.2 Temperature dependence of the ^{8}B and pp neutrino fluxes
In Figure 6.2a, the logarithm of the calculated ^{8}B neutrino flux is shown
as a function of central temperature for the 1000 solar models that were
calculated by Bahcall and Ulrich (1988). For this figure, the temperature is
given in units of 10^6 K and the flux in units of 10^6 cm^{-2} s^{-1}. In Figure 6.2b,
the logarithm of the pp neutrino flux is displayed as a function of central
temperature for the same 1000 solar models. For this figure, the flux is
expressed in units of 10^{10} cm^{-2} s^{-1}; the unit of temperature is the same
as for Figure 6.2a

Figure 6.2a shows the relation between ^{8}B neutrino flux and central temperature for the 1000 solar models that were calculated by Bahcall and Ulrich (1988). With $\beta \sim 18$, Eq. (6.14a) describes approximately the relation between temperature and ^{8}B neutrino flux for the models that were calculated. The precise value of β depends upon how one weighs the contributions corresponding to different models. If one assumes that nothing happens to the neutrinos from the time they are produced until they are observed on Earth and that the nuclear cross sections are reasonably well known, then a measurement of the ^{8}B neutrino flux will determine the central temperature of the Sun to an accuracy of about a percent or better.

The pp neutrino flux is much less sensitive to temperature. Figure 6.2b shows the relation between the pp neutrino flux and the central temperature for the same 1000 solar models. The band of points in Figure 6.2b is described approximately by the equation

$$\phi(\text{pp}) \approx \text{const.} \times T^{-1.2} \ . \qquad (6.14b)$$

The pp neutrino flux decreases with increasing temperature, contrary to naive expectation. The reason for this nonintuitive behavior is that the total number of low-energy neutrinos, from the pp reaction and from ^{7}Be electron capture, is essentially fixed by the condition that the model luminosity at the current epoch equal the observed solar luminosity. As the temperature increases, a greater fraction of the terminations of the pp chain occurs via the ^{3}He–^{4}He reaction (which produces one pp neutrino and one ^{7}Be neutrino) and a smaller fraction occurs via the ^{3}He–^{3}He reaction (which requires two pp neutrinos).

The temperature dependences of the hep and ^{7}Be neutrino fluxes are shown in Figure 6.3. There is much scatter in the relation between the logarithm of the hep flux and the temperature, implying that a simple power law is not a good approximation. For a crude representation, one can write

$$\phi(\text{hep}) \approx \text{const.} \times T^{-\gamma}, \qquad (6.14c)$$

where $\gamma \sim 3$ to 6, depending upon how different points are weighted. The decrease of the flux with temperature reflects, in this case, the fact that the equilibrium abundance of ^{3}He decreases sharply with increasing temperature.

Figure 6.3b shows that the ^{7}Be flux increases with temperature as

$$\phi(^7\text{Be}) \approx \text{const.} \times T^8. \qquad (6.14d)$$

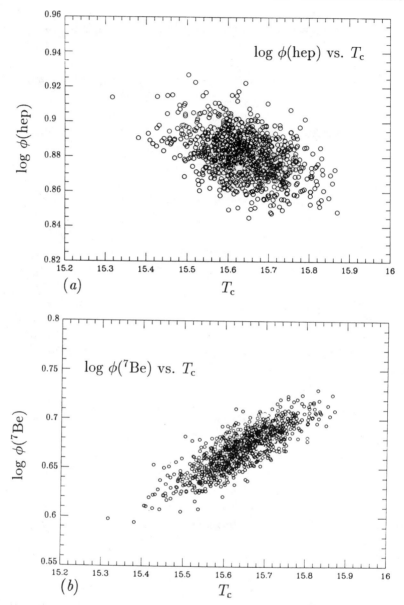

Figure 6.3   Temperature dependence of the hep and ^{7}Be neutrino fluxes
In Figure 6.3a, the logarithm of the calculated hep neutrino flux is shown as
a function of central temperature for the 1000 solar models that were calculated by Bahcall and Ulrich (1988). In Figure 6.3b, the logarithm of the
^{7}Be neutrino flux is displayed as a function of central temperature for the
same 1000 solar models. The unit of temperature is 10^6 K. For Figure 6.3a,
the fluxes are expressed in units of 10^3 cm^{-2} s^{-1} and for Figure 6.3b, in
units of 10^9 cm^{-2} s^{-1}.

The temperature dependence of the ^{7}Be is rather strong, but not as extreme as for the ^{8}B flux.

Each of the models used in constructing Figures 6.2 and 6.3 is the solution of the same complicated set of coupled differential equations with boundary conditions in space and in time that describe the physical evolution of the Sun, together with many input parameters. The iterated solar model that satisfies the boundary conditions and the initial physical assumptions yields a complete temperature and density profile of the solar interior. The central temperature is *not* an adjustable parameter; it is a byproduct. The central temperature was used in plotting Figure 6.2 only for simplicity. The figure could just as well show the ^{8}B neutrino flux as a function of the average temperature in the inner 10% of the solar mass or the temperature at 0.05 $R_\odot$, where the rate of the ^{8}B neutrino flux peaks (see Figure 6.1).

6.3 Neutrino energy spectra

A measurement of the shape of the energy spectrum of neutrinos that reaches us from the Sun will provide a decisive test of whether the solar neutrino problem is caused by our lack of understanding of the solar interior or by new physics. Whatever modification is made in the solar model, the shape of the spectrum of electron neutrinos from each source will remain the same unless nonstandard electroweak effects occur.

This section presents detailed information about the expected neutrino spectra.

Figure 6.4 shows the neutrino spectrum that is predicted for the standard solar model that is discussed in Chapter 4. This spectrum includes all of the important sources discussed in the previous subsection. The neutrino fluxes from continuum sources are given in the units of number per cm^2 per second per MeV at one astronomical unit and the line fluxes are given in number per cm^2 per second.

Each energy spectrum from a specified solar neutrino source has a characteristic shape that is independent of the conditions in the solar interior [up to terms of at least of order (kT/q_{max}); see Appendix A of Bahcall and Ulrich (1988)]. The pp spectrum rises slowly and peaks at about 0.31 MeV (which is $\approx$ 3/4 of the maximum neutrino energy), after which the flux is strongly cut off. The ^{8}B spectrum, on the other hand, is more nearly symmetric, with a peak at 6.4

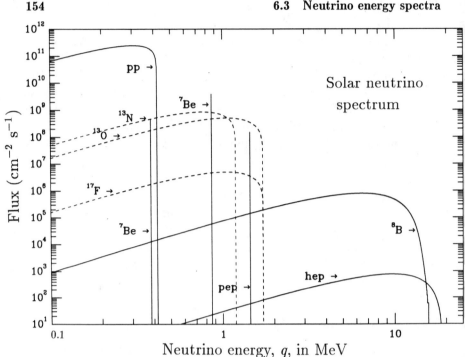

Figure 6.4   Solar neutrino spectrum   This figure shows the energy spectrum of neutrinos that is predicted by the standard solar model discussed in Section 4.3. The neutrino fluxes from continuum sources are given in the units of number per cm^2 per second per MeV at one astronomical unit and the line fluxes are given in number per cm^2 per second. The spectra from the pp chain are drawn with solid lines; the CNO spectra are drawn with dotted lines.

MeV and a somewhat extended tail. The hep spectrum is rather symmetric and peaks at 9.6 MeV. The ^{13}N spectrum, like the pp spectrum, rises slowly and then cuts off sharply; the ^{13}N peak is at 0.76 MeV.

Tables 6.2 to 6.4 present the calculated energy spectra for the pp, ^{8}B, and hep neutrino sources. These tables should be useful in making detailed calculations of the implications of the MSW effect (see Section 9.2). The energy spectrum for the ^{8}B neutrinos is taken from Bahcall and Holstein (1986) and the other spectra are adapted from Bahcall and Ulrich (1988).

For neutrino lines, pep or ^{7}Be, the shape of the neutrino spectrum is produced by the thermal motion of the electrons. The probability

Table 6.2. The pp neutrino spectrum. The normalized pp neutrino energy spectrum, $P(q)$, is given in intervals of 10 keV. The neutrino energy, q, is expressed in MeV and $P(q)$ is normalized per MeV.

q	$P(q)$	q	$P(q)$	q	$P(q)$	q	$P(q)$
0.01	0.0139	0.12	1.45	0.23	3.49	0.34	3.93
0.02	0.0542	0.13	1.64	0.24	3.63	0.35	3.80
0.03	0.119	0.14	1.84	0.25	3.75	0.36	3.63
0.04	0.205	0.15	2.04	0.26	3.86	0.37	3.41
0.05	0.312	0.16	2.24	0.27	3.95	0.38	3.13
0.06	0.436	0.17	2.44	0.28	4.02	0.39	2.76
0.07	0.576	0.18	2.63	0.29	4.07	0.40	2.28
0.08	0.731	0.19	2.82	0.30	4.10	0.41	1.60
0.09	0.897	0.20	3.00	0.31	4.10	0.42	0.00
0.10	1.07	0.21	3.17	0.32	4.08		
0.11	1.26	0.22	3.34	0.33	4.02		

that a neutrino with energy, q, is produced in a region that has a temperature T is:

$$P(q) = \text{const.} \times T^{-1/2} \exp\left[-\left(\frac{m_i c^2}{2kT}\right)\left(\frac{q - q_0}{q_0}\right)^2\right], \quad (6.15)$$

where m_i is the mass of the ion that participates in the neutrino-producing reaction. The shape of the line that emerges from the Sun must be calculated by averaging Eq. (6.15) over the mass distribution of the Sun. Thus the observed shape has the form

$$\langle P(q) \rangle = M_\odot^{-1} \int dM(T) \,[\text{production rate}\,(T)]\, P(q, T). \quad (6.16)$$

The total width of the lines is expected to be ~ 1 keV.

Figures 6.5 and 6.6 show the individual energy spectra for the important pp and ^{8}B neutrino sources, respectively. Figure 6.7 shows the spectra for the hep and ^{13}N sources. The calculation of the ^{8}B spectrum requires special treatment because of the broadness of the ^{7}Be state that is predominantly populated in the β-decay.

Table 6.3.   The ^{8}B neutrino spectrum. The neutrino energy q is in MeV and $P(q)$ is the probability that a neutrino with energy q is emitted between $q \pm 0.5$ MeV.

q	$P(q)$	q	$P(q)$	q	$P(q)$	q	$P(q)$	q	P(q)
0.1	0.00022	3.1	0.07625	6.1	0.13211	9.1	0.09832	12.1	0.02526
0.2	0.00079	3.2	0.07929	6.2	0.13239	9.2	0.09603	12.2	0.02328
0.3	0.00152	3.3	0.08227	6.3	0.13256	9.3	0.09368	12.3	0.02137
0.4	0.00257	3.4	0.08521	6.4	0.13262	9.4	0.09130	12.4	0.01952
0.5	0.00386	3.5	0.08808	6.5	0.13258	9.5	0.08888	12.5	0.01776
0.6	0.00537	3.6	0.09089	6.6	0.13242	9.6	0.08643	12.6	0.01607
0.7	0.00709	3.7	0.09364	6.7	0.13216	9.7	0.08394	12.7	0.01446
0.8	0.00899	3.8	0.09631	6.8	0.13180	9.8	0.08143	12.8	0.01294
0.9	0.01105	3.9	0.09891	6.9	0.13133	9.9	0.07890	12.9	0.01150
1.0	0.01328	4.0	0.10144	7.0	0.13075	10.0	0.07634	13.0	0.01015
1.1	0.01565	4.1	0.10388	7.1	0.13007	10.1	0.07377	13.1	0.00889
1.2	0.01815	4.2	0.10623	7.2	0.12930	10.2	0.07119	13.2	0.00771
1.3	0.02077	4.3	0.10850	7.3	0.12842	10.3	0.06860	13.3	0.00664
1.4	0.02349	4.4	0.11068	7.4	0.12745	10.4	0.06601	13.4	0.00565
1.5	0.02630	4.5	0.11277	7.5	0.12638	10.5	0.06342	13.5	0.00476
1.6	0.02920	4.6	0.11476	7.6	0.12522	10.6	0.06083	13.6	0.00396
1.7	0.03217	4.7	0.11665	7.7	0.12397	10.7	0.05826	13.7	0.00325
1.8	0.03520	4.8	0.11845	7.8	0.12263	10.8	0.05569	13.8	0.00263
1.9	0.03828	4.9	0.12014	7.9	0.12120	10.9	0.05314	13.9	0.00210
2.0	0.04140	5.0	0.12173	8.0	0.11969	11.0	0.05061	14.0	0.00166
2.1	0.04455	5.1	0.12322	8.1	0.11810	11.1	0.04811	14.1	0.00129
2.2	0.04773	5.2	0.12460	8.2	0.11643	11.2	0.04563	14.2	0.00099
2.3	0.05093	5.3	0.12587	8.3	0.11468	11.3	0.04319	14.3	0.00075
2.4	0.05414	5.4	0.12703	8.4	0.11286	11.4	0.04078	14.4	0.00056
2.5	0.05735	5.5	0.12809	8.5	0.11097	11.5	0.03841	14.5	0.00042
2.6	0.06055	5.6	0.12904	8.6	0.10902	11.6	0.03609	14.6	0.00030
2.7	0.06374	5.7	0.12987	8.7	0.10699	11.7	0.03381	14.7	0.00022
2.8	0.06691	5.8	0.13060	8.8	0.10491	11.8	0.03159	14.8	0.00016
2.9	0.07006	5.9	0.13121	8.9	0.10277	11.9	0.02942	14.9	0.00011
3.0	0.07317	6.0	0.13171	9.0	0.10057	12.0	0.02731	15.0	0.00008

6.4   Solar neutrino fluxes

A   Standard model

Table 6.5 gives the best estimates for each of the neutrino fluxes, together with the total theoretical uncertainties. The fluxes were com-

Table 6.4. The hep neutrino spectrum. The normalized neutrino energy spectrum, $P(q)$, is given in intervals of 0.447 MeV. The neutrino energy, q, is expressed in MeV and $P(q)$ is normalized per MeV.

q	$P(q)$	q	$P(q)$	q	$P(q)$	q	$P(q)$
0.447	8.00E−04	5.364	6.29E−02	10.3	9.66E−02	14.8	5.02E−02
0.894	3.05E−03	5.811	6.92E−02	10.7	9.50E−02	15.2	4.32E−02
1.341	6.53E−03	6.258	7.50E−02	11.2	9.25E−02	15.6	3.62E−02
1.788	1.10E−02	6.705	8.03E−02	11.6	8.93E−02	16.1	2.94E−02
2.235	1.64E−02	7.152	8.49E−02	12.1	8.54E−02	16.5	2.29E−02
2.682	2.24E−02	7.599	8.89E−02	12.5	8.08E−02	17.0	1.68E−02
3.129	2.88E−02	8.046	9.22E−02	13.0	7.56E−02	17.4	1.13E−02
3.576	3.56E−02	8.493	9.47E−02	13.4	6.98E−02	17.9	6.62E−03
4.023	4.25E−02	8.940	9.64E−02	13.9	6.36E−02	18.3	2.92E−03
4.470	4.95E−02	9.387	9.73E−02	14.3	5.70E−02	18.8	0.00E+00
4.917	5.63E−02	9.83	9.74E−02				

Table 6.5. Calculated solar neutrino fluxes.

Source	Flux $(10^{10}$ cm^{-2} s$^{-1})$
pp	6.0 (1 ± 0.02)
pep	0.014 (1 ± 0.05)
hep	8×10^{-7}
^{7}Be	0.47 (1 ± 0.15)
^{8}B	5.8×10^{-4} (1 ± 0.37)
^{13}N	0.06 (1 ± 0.50)
^{15}O	0.05 (1 ± 0.58)
^{17}F	5.2×10^{-4} (1 ± 0.46)

puted for the standard solar model that was described in Chapter 4. The total theoretical range for each flux was evaluated using the parameter uncertainties discussed in Chapter 7. The second column of Table 6.5 gives the fluxes and the estimated total uncertainties.

The flux of neutrinos from the proton–proton reaction is known more accurately than for any other sources; the estimated uncertainty in the pp flux is only 2%.

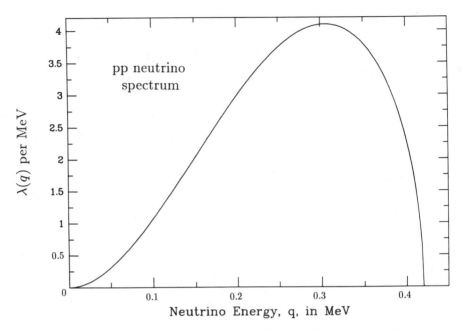

Figure 6.5   The pp neutrino spectrum   This figure shows the energy spectrum, $\lambda(q)$, of pp neutrinos. The integral of $\lambda(q)\mathrm{d}q$ is normalized to unity for $\mathrm{d}q$ measured in MeV.

Many authors have misinterpreted this small uncertainty in the standard model flux and have claimed that the flux of pp neutrino reaction is fixed by the observed solar luminosity, independent of solar models[†]. In fact, the computed flux would be about one-half as large as given in Table 6.5 if ^{3}He were burned primarily by interactions with an α-particle rather than (as in the standard solar model) by interactions with other ^{3}He nuclei. Two pp reactions are required to terminate the chain via reaction 5 of Table 3.1 (^{3}He+^{3}He, which predominates in the standard solar model) whereas only one pp reaction is necessary if reaction 6 (^{3}He+^{4}He) dominates. Fortunately, the cross section factors for reactions 5 and 6 are both relatively well known; they indicate that the ^{3}He+^{3}He reaction oc-

---

[†]There is a model-independent *lower limit* to the pp neutrino flux [see Bahcall, Cleveland, Davis, and Rowley (1985) or Section 11.3].

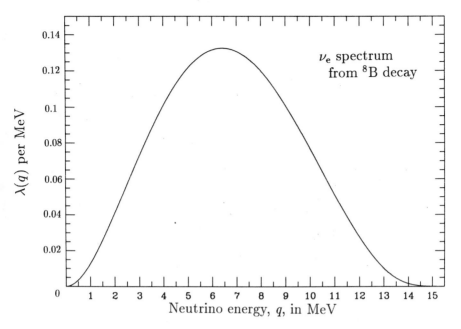

Figure 6.6 The ^{8}B neutrino spectrum   This figure shows the energy spectrum, $\lambda(q)$, of ^{8}B neutrinos and is taken from Bahcall and Holstein (1986). The integral of $\lambda(q)\,dq$ is normalized to unity for dq measured in MeV.

curs about six times as often as the ^{3}He$+^4$He reaction under average solar conditions.

The uncertainty in the flux of ^{7}Be neutrinos is also moderately small, of order 15%.

The uncertainty in the crucial ^{8}B neutrinos is large, 37%. The largest single uncertainties in the calculation of this flux derive from the measurement of the low-energy nuclear cross section factor for the production of ^{8}B and from the influence of the assumed heavy element abundance on the opacity of the solar interior.

The uncertainties are of order 50% for the CNO neutrinos. The largest contributors to their uncertainties are the heavy element abundance and the rate at which ^{14}N is burned.

At present one cannot make a useful estimate of the total uncertainty for the hep neutrino flux because the thermal neutron capture cross section is not well known. All of the other uncertainties influ-

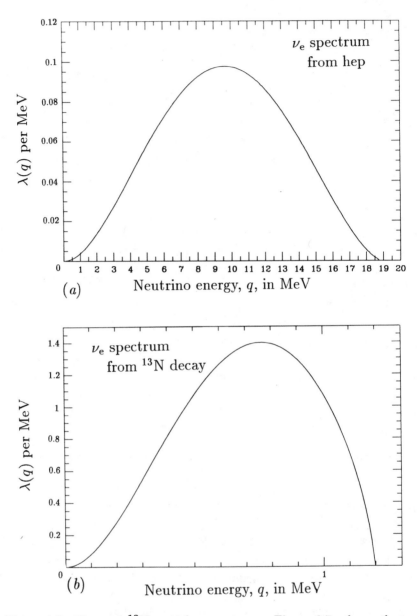

Figure 6.7 Hep and ^{13}N neutrino spectra   Figure 6.7a shows the energy spectrum, $\lambda(q)$, of the hep neutrinos and Figure 6.7b shows the energy spectrum of ^{13}N neutrinos. The integral of $\lambda(q)\,dq$ is normalized to unity for dq measured in MeV.

encing the calculated flux from this source are small [see Bahcall and Ulrich (1988)].

B Nonstandard models

One of the best ways to determine what is at stake for the theory of stellar evolution in solar neutrino experiments is to calculate nonstandard solar models, using different assumptions regarding the underlying astrophysical context or to vary important physical quantities or functions, such as the equation of state, the opacity, or nuclear reaction rates (see Chapter 5 for a review of nonstandard solar models). Systematic investigations and reviews of many of the nonstandard models have been given by Bahcall, Bahcall, and Ulrich (1969), by Rood (1978), by Roxburgh (1985a,b), by Schatzman (1985), and by Newman (1986). Unfortunately, relatively few of the nonstandard models have been calculated with a precision that is equal to that used for standard models; hence it is often difficult to determine exactly what they predict for the different solar neutrino experiments.

Table 6.6 shows the results for a number of representative nonstandard models that Bahcall and Ulrich (1988) calculated with the same precision with which they constructed standard solar models.

The first row of Table 6.6 refers to the standard solar model of Chapter 4. The last row refers to the model with the maximum pp neutrino flux that is consistent with the observed solar luminosity; this model was used by Bahcall, Cleveland, Davis, and Rowley (1985) to calculate the minimum event rate for a ^{71}Ga experiment that is consistent with general ideas of stellar evolution.[†]

The following subsection discusses first models with *ad hoc* inhomogeneities in their chemical compositions (i.e., inhomogeneities that are not due to nuclear fusion) and the subsequent subsection describes two models with nonstandard nuclear physics.

Ad hoc inhomogeneous models. The first class of models has specially concocted inhomogeneities in their initial abundance distribu-

---

[†]The value for the pep flux given here differs slightly from the previously calculated value because the ratio of pep to pp neutrino fluxes obtained from the standard model has decreased by a small amount.

tion. Suppose that segregation of the elements took place while the Sun formed out of the solar nebula or as a result of diffusion during the Sun's early history. The variation of chemical composition from planet to planet shows that at least some elemental segregation was possible early in the Sun's life. There are specific physical processes, involving, for example, magnetic fields or the properties of grains, that are sufficiently poorly understood that we feel justified in at least calculating the effects of possible inhomogeneities.

Two qualitatively different types of variable composition models were considered in detail. In the first set of models, the heavy element abundance, Z, which was assumed to be the same at all points and at all times in the standard model, is assumed to have a smaller value in the interior than in the outer regions. Models of this class are usually referred to in the literature as "low Z" models. In the second set of models, the initial helium abundance, Y, is assumed to be nonuniform, larger in the interior than in the outer regions. The first set of models (low Z) was developed as a possible solution of the solar neutrino problem [Iben (1969) and Bahcall and Ulrich (1971)] and the second set of models (variable initial Y) was considered by Ulrich et al. (1983) as a possible solution of a problem in helioseismology.

Models with variable Z are nonstandard because their opacities are different from those in the standard model, while models with variable initial Y alter directly the mean molecular weight and sound speed. Although both types of variation could be present simultaneously, they have been considered separately in order to isolate their principal effects. Also, in order to limit the number of free parameters, Bahcall and Ulrich (1988) assumed that the abundance changes occur as step functions, which eliminates the necessity of specifying a mass increment over which the variation occurs. For the particular model cited here, the heavy element abundance was reduced by a factor of 8.5 with respect to the surface value for all points interior to $0.56R_\odot$.

The first row of Table 6.6 gives for comparison the fluxes from the standard solar model.

The second row of Table 6.6 shows the neutrino fluxes for a low Z model that agrees reasonably well with the results of the ^{37}Cl experiment (1.6 SNU). The reduction of heavy elements alters the model structure primarily through the change in the heavy element opacity. Hence, the maximum effect on the central temperature and

Table 6.6. **Neutrino fluxes for nonstandard and standard solar models.** The unit of flux is $cm^{-2} s^{-1}$ at the Earth's surface. The first row refers to the "standard" solar model described in Chapter 4. The next five rows refer to nonstandard models with either a low initial heavy element abundance (row 2), a high initial helium abundance in the solar interior (row 3), $S_{17} = 0$ (row 4), $S_{34} = 0$ (row 5), or a cosmologically significant mass density of WIMPs (row 6). The last row gives the fluxes for a model in which the pp neutrino flux has the maximum value consistent with the observed solar luminosity. The last column shows the total capture rate predicted for the ^{37}Cl experiment.

Model	Y	Z	pp (E10)	pep (E8)	hep (E3)	^7Be (E9)	^8B (E6)	^{13}N (E8)	^{15}O (E8)	^{17}F (E6)	^{37}Cl (SNU)
Standard	0.271	0.0196	6.00	1.40	7.58	4.69	5.76	6.09	5.22	5.16	7.9
Low Z	0.127	0.0023	6.38	1.66	10.42	1.83	0.80	0.22	0.12	0.10	1.6
High Y	0.266	0.0196	6.04	1.40	7.87	5.15	6.45	6.56	5.70	5.66	8.8
No ^8B	0.271	0.0196	5.98	1.39	7.90	4.88	0.0	6.08	5.21	5.15	1.8
$S_{34} = 0$	0.250	0.0168	6.49	1.62	8.70	0.0	0.0	3.98	3.14	2.99	0.6
WIMPs	. . .	. . .	6.15	1.5	. . .	3.3	1.4	2.2	1.9	1.9	2.7
Max pp	. . .	. . .	6.53	1.50	0.0	0.0	0.0	0.0	0.0	0.0	0.0

neutrino flux is achieved by placing the composition jump as near to the solar surface as possible so that the opacity is reduced throughout the radiative regions.

The third row of Table 6.6 gives the neutrino fluxes and composition parameters for a model that has evolved with an inhomogeneous initial helium and hydrogen composition. This particular model had an initial helium abundance that was enhanced by 0.025 (and a hydrogen abundance that was decreased by the corresponding amount) for the inner 10% of the mass of the Sun. The motivation for this particular model is that it gives better agreement with observation of the calculated p-mode oscillation frequencies than does the standard model, although the computed capture rate for the ^{37}Cl experiment, 8.8 SNU, is in worse agreement with observation, assuming nothing happens to the neutrinos on their way to Earth from the solar interior.

Ad hoc altered nuclear physics. If the rare reaction that produces ^8B neutrinos, number 8 in Table 3.1, does not occur for some unknown reason, the predicted capture rate for the ^{37}Cl experiment would be

1.8 SNU, in agreement with existing observations. The fluxes for this model are listed in the fourth row of Table 6.6 since the assumption that ^{8}B is not produced is in some ways the "simplest" solution of the solar neutrino problem, although it is contradicted by a series of independent and careful laboratory experiments on the low-energy cross section for the production of ^{8}B.

The model labeled "$S_{34} = 0.0$" was constructed in order to answer the following question: What happens to the calculated p-mode oscillation frequencies if one artificially changes the overall energy generation significantly? This question was answered by constructing a solar model in which the rate of the important ^{3}He $+^4$Hereaction (number 5 of Table 3.1) is set equal to zero ($S_{34} = 0$). For the standard solar model, about 15% of the terminations of the pp chain proceed through this reaction. Therefore, the calculated solar structure is significantly altered from what is obtained with the standard parameters, unlike the previously considered "no ^{8}B model." The model calculated with $S_{34} = 0$ has a significantly different temperature and density profile for this nonstandard model. The fifth row of Table 6.6 gives the calculated neutrino fluxes for the $S_{34} = 0$ model. For most of the solar neutrino detectors discussed in this book, the predicted event rate is reduced by *more than an order of magnitude* compared to what is expected on the basis of the standard solar model. Solar neutrino astronomy would be very much more difficult were this nonphysical hypothesis correct. However, the p-mode frequencies calculated for this model are not significantly different from those evaluated for the standard solar model.

WIMPs. Weakly interacting massive particles (WIMPs) have been proposed as a simultaneous solution to the missing mass problem and to the solar neutrino problem [see, e.g., Press and Spergel (1985), Faulkner and Gilliland (1985), and Spergel and Press (1985)]. If exotic weakly interacting particles make up the galactic missing mass, then they can modify the energy transport in the solar core in just such a way as to reduce the calculated capture rate to the value observed for ^{37}Cl, provided that the scattering cross section and the mass for the WIMPs are chosen to lie in an appropriate range [see Faulkner and Gilliland (1985), Spergel and Press (1985), and Gould (1987a)]. There are difficulties in making this scenario consistent with any already discovered particle, but various authors have discussed new particles that might satisfy the required conditions [see, e.g., Gelmini, Hall, and Lin (1987), Raby and West (1987), and Greist and Seckel (1987)].

There is not yet available a definitive calculation of the predicted neutrino fluxes if WIMPs are the correct solution of the solar neutrino problem. However, Gilliland, Faulkner, Press, and Spergel (1986) have estimated the expected neutrino fluxes using parameters and physical variables that are not the best available. The neutrino fluxes given in the last row of Table 6.6 were determined [Bahcall and Ulrich (1988)] by an approximate scaling of the Gilliland *et al.* (1986) results to the parameter and physics of the standard solar model. The important flux of ^{8}B neutrinos should be given reasonably accurately by the scaling argument since it depends most strongly upon the cross section for the rare (p, γ) reaction on ^{7}Be, which does not affect the structure of the Sun. The fluxes of pp and pep neutrinos are also probably well determined since they differ by only 2% and 7%, respectively, in the WIMP and standard models. The flux of ^{7}Be neutrinos is probably not very well determined; it is about 20% higher in the standard model than in the WIMP model. The CNO fluxes cannot be reliably evaluated by the scaling procedure; their neutrino fluxes are about a factor of two larger for the standard model than the WIMP model. The WIMP hypothesis can be tuned so that it reduces the ^{8}B neutrino flux by a factor of four or by almost any desired reduction factor that may be indicated by the observations. The basic results that are likely to be relatively invariant to future improvements in the solar model and in the choice of WIMP parameters are the relative lack of change in the pp flux from the standard to the WIMP model (only 2.5%) and, to within a factor of order two, the ratio of ^{7}Be to ^{8}B flux (about 2.4×10^3 with the assumed input nuclear physics parameters).

6.5   Fluxes from other stars

The integrated neutrino flux, ϕ_ν, from all of the stars burning hydrogen in the Galaxy except the Sun can be written conveniently in the following form

$$\Phi_\nu \;=\; \left(\frac{\mathrm{d}M}{\mathrm{d}N_\nu}\right)^{-1} \int \frac{\mathrm{d}V(R)\rho(R)}{4\pi|\vec{R} - \vec{R}_\odot|^2}, \tag{6.17}$$

where $\mathrm{d}V(R)$ is the volume element, and $\rho(R)$ is the mass density of stars, at a distance R from the galactic center. The quantity $\mathrm{d}M$ is the stellar mass required to produce dN_ν neutrinos per unit of time. The derivative (dN_ν/dM) can be expressed as

$$\frac{dN_i}{dM} \equiv \frac{\epsilon_i N_{i\odot}}{M_\odot} = \frac{\epsilon_i \left(4\pi\phi_{i\odot}\right) (1 \text{ AU})^2}{M_\odot}, \tag{6.18}$$

where the index i denotes a particular neutrino source (e.g., pp or ^{8}B) and $\phi_{i\odot}$ is the flux of neutrinos (in cm^{-2} s^{-1}) from the Sun [1 AU (one **astronomical unit**) is the average distance of the Sun from the Earth, 1.5×10^{13} cm, 8 light minutes]. The quantities ϵ_i represent the ratios of the rate of emission of neutrinos of type i averaged over all the galactic stars compared to the solar rate.

Combining the above two equations, one obtains

$$\Phi_i = \left[\epsilon_i \frac{(1 \text{ AU})^2}{\langle R \rangle^2} \left(\frac{M_{\text{visible}}}{M_\odot}\right)\right] \phi_{i\odot}. \tag{6.19}$$

For the Bahcall and Soneira (1980) galaxy model, the total mass of visible stars in the Galaxy is $M_{\text{visible}} = 2 \times 10^{10} M_\odot$ and the density-weighted average distance, $\langle R \rangle^{-2}$, is 10 kpc^{-2}. Therefore one can write

$$\Phi_i \cong 5 \times 10^{-9} \epsilon_i \phi_{i\odot}. \tag{6.20}$$

A proper calculation of the ϵ_i would entail computing accurate neutrino fluxes from a representative sample of stars of different masses and then weighing the results according to the percentages of galactic stars with specified masses. These calculations have not been done, but one might guess that $\epsilon(\text{pp}) \lesssim 1$, $\epsilon(^7\text{Be}) \lesssim 10^{-1}$, $\epsilon(^8\text{B}) \lesssim 10^{-1.5}$, and $\epsilon(\text{CNO}) \lesssim 10^{-1}$. A significant fraction of all stars will emit pp neutrinos in an amount proportional to their mass while, for example, the emission of ^{8}B neutrinos will occur only in a relatively narrow range of stellar masses (at lower temperatures, the Coulomb barrier will quench ^{7}Be neutrino emission and at higher temperatures the CNO cycle dominates main sequence hydrogen burning).

For any reasonable values of ϵ_i, the neutrino flux, Φ_i, from stars other than the Sun is many orders of magnitude less than the solar flux, $\phi_{i\odot}$, just as it is for photon fluxes.[†]

[†]Neutrinos and antineutrinos produced by radioactive decays in the interior of the Earth have typical fluxes estimated from the Earth's heat balance of order

Bibliographical Notes

Bahcall, J.N., W.A. Fowler, I. Iben, Jr., and R.L. Sears (1963) *Ap. J.*, **137**, 344. First detailed solar model constructed to calculate solar neutrino fluxes. The predicted number of neutrino captures by ^{37}Cl was disappointingly small because the transition to the analogue state of ^{37}Ar was not considered.

Fowler, W.A. (1958) *Ap. J.*, **127**, 551. Pointed out that the Holmgren–Johnston result could lead to a measurable ^{8}B neutrino flux if the solar interior temperature was sufficiently high.

Holmgren, H.D. and R. Johnston (1958) *Bull. Am. Phys. Soc. II*, **3**, 26; (1959) *Phys. Rev.*, **113**, 1556. Showed experimentally that the cross section for the ^{3}He$(\alpha, \gamma)^7$Be reaction was about 10^3 times larger than had been previously estimated.

10^6 cm^{-2} s^{-1} to 10^7 cm^{-2} s^{-1} and energies that range from ~ 0.3 MeV to somewhat more than 3 MeV [see Ezer (1966), Marx and Lux (1970), and Krauss, Glashow, and Schramm (1984)]. These fluxes are too small to observe with current or planned detectors. The electron capture line at 1.51 MeV from ^{40}K decay is of special interest for future detectors since the well-defined electron recoil energy might be observable in a massive experiment designed to study the neutrino line from ^{7}Be electron capture in the Sun.

7

Theoretical uncertainties

Summary

Is there a "solar neutrino problem?" Yes, provided the errors in the predictions and the observations are less than the discrepancy between theory and experiment. Otherwise, no; the excitement is about a false alarm. In order to decide if there is a problem or not, to determine if something new has been discovered, we must establish the "room for maneuver" in the theoretical calculations. A quantitative evaluation of the errors is at the heart of our subject.

This chapter discusses the uncertainties in predicting event rates in solar neutrino experiments. The total theoretical range for a given experiment is computed using 3σ errors for all of the measured input parameters. Uncertainties for theoretical quantities are more difficult to determine; the theoretical uncertainties are sometimes estimated by comparing the results obtained by different authors.

The method used in calculating errors is best defined by examples. Individual sections in this chapter describe the estimated uncertainties in nuclear reaction cross sections, the heavy element abundance on the surface of the Sun, the radiative opacity, and the equation of state, as well as other quantities. For some important experiments, the neutrino interaction cross sections constitute the dominant uncertainty. A crucial question for the interpretation of these experiments is: How accurately do (p,n) reactions measure neutrino absorption matrix elements? Further work is required to answer this question satisfactorily.

How do individual neutrino fluxes depend on the input parameters? In most cases, the dependences can be represented to the required accuracy by power laws. Monte Carlo simulations show that power-law approximations are satisfactory even near the tail of the distribution of errors.

Section 7.1 describes the logic that is used in determining the errors in the input parameters, while Sections 7.2 and 7.3 present some illustrative examples. In many of the papers on this subject, the total uncertainty in the theoretical predictions for a given experiment has been determined using numerically calculated partial derivatives of individual neutrino fluxes with respect to the input parameters, together with the estimated uncertainties in each parameter. This procedure assumes that the uncertainties from individual parameters add incoherently. Section 7.4 outlines how the partial derivatives of the fluxes are calculated and presents some of the most important dependences in the form of power-law relations.

The calculation of solar neutrino fluxes is a nonlinear problem, involving coupled partial differential equations that must be solved subject to boundary conditions in space and time (the final model must have the observed solar luminosity and radius at the present solar age). Could an extrapolation to large changes of the results obtained with the partial derivatives seriously misrepresent the effects of nonlinear couplings? Could the assumption that individual uncertainties from different parameters combine incoherently significantly underestimate the overall uncertainty? Probably not. Section 7.5 describes a Monte Carlo evaluation of the theoretical uncertainties that is based upon the results of 1000 accurately computed standard solar models, the input parameters for which were drawn from representative statistical distributions. The Monte Carlo simulations are in good agreement with the results obtained by using the partial derivatives and by assuming that the uncertainties from different parameters add incoherently.

Section 7.6 and Table 7.5 summarize the principal uncertainties for solar neutrino experiments with the following detectors: ^{2}H, ^{7}Li, ^{37}Cl, ^{40}Ar, ^{71}Ga, ^{81}Br, ^{98}Mo, and ^{115}In, as well as ν_e-e^- scattering.

The presentation in this chapter is based upon the analysis of Bahcall et al. (1982) and Bahcall and Ulrich (1988).

7.1 Logic

This book presents the calculated uncertainties in terms of a **total theoretical range**. In an earlier attempt to systematize the discussion of all the errors [Bahcall *et al.* (1982)], this same range was referred to as an "effective 3σ level of confidence." This nomenclature has been modified [see Bahcall and Ulrich (1988)], because 3σ suggests a greater statistical rigor than is possible for the calculation of solar neutrino fluxes. It would be very useful to have a *true* "3σ level of confidence," but this is not possible to determine since the probability distribution is unknown for parameters that must be evaluated by theoretical calculations (e.g., radiative opacity or forbidden corrections to neutrino cross sections). In practice, my colleagues and I have tried to make the "total theoretical range" sufficiently broad so that if the true value of some parameter lies outside the estimated range, someone has made a mistake in determining the parameter.

A massive collective effort is required to determine the best parameters and their uncertainties that are to be used in constructing a standard solar model. No one person, or small group of persons, can be expert in all the fields that are involved in the theoretical calculations. The construction of an accurate solar model, and the estimation of the errors, depends upon the work of a great many scientists in different disciplines.

For measured quantities (e.g., nuclear reaction rates or solar luminosity), standard 3σ limits are used.

For theoretical quantities, the original authors have only occasionally evaluated the full uncertainties. More often, my colleagues and I had to make our assessment of the possible errors. In practice, we have often taken uncertainties in quantities that are calculated theoretically, not measured, to be equal to the range in values in published state-of-the-art calculations. We may assign by this procedure relatively larger errors for experimentally determined parameters (for which the errors are more easily quantifiable) than for the calculated parameters, such as the opacity. However, the adopted procedure is as objective as any we can think of and has the advantage of simplicity. In cases where only one calculation is available, we have had to make a more delicate judgment. For example, we have chosen to multiply the value of forbidden corrections to neutrino capture cross sections by three and call this the total uncertainty.

We have sometimes been forced to use a "rule of thumb" for estimating the uncertainties which is based upon a common practice among theoretical physicists. Many theoreticians determine the uncertainties in important experiments by ignoring the formal errors quoted by their colleagues. In the privacy of their offices, these theorists estimate the errors by determining the dispersion in values from experimental groups with good reputations. If only one group has measured a number for a particular quantity, then these skeptics make an estimate of the minimum likely error by comparing the values published as a function of time by the same experimental group. (The discussion of Figure 10.1, which displays the time dependence of the published values of the predicted capture rates for the ^{37}Cl experiment, is an example of the use of this procedure outside the usual range of applications.)

The method for estimating errors is defined by examples of what is done (see examples in Sections 7.2 and 7.3). In all cases, the procedures and assumptions that are used to obtain the uncertainties are stated explicitly. If the reader has a better (or different) way of estimating the errors, then he can recalculate easily the uncertainties in all the predicted event rates using the results and prescriptions described in this chapter and in Bahcall *et al.* (1982) and Bahcall and Ulrich (1988).

Most of the predicted rates and uncertainties that appear in the literature are calculated assuming that the standard solar model is correct and that nothing happens to the neutrinos on the way to the Earth from the Sun. However, there is convincing experimental evidence (from the ^{37}Cl and Kamiokande II experiments, see Chapter 1 or Chapters 10 and 13) that this combined conservative hypothesis is incorrect.

How can we calculate the rates and uncertainties if the incident neutrino energy spectrum is different from the standard spectrum? The procedure is relatively simple if the shape of the energy spectrum for each of the individual neutrino sources is unchanged and only the total number of neutrinos of a specified type is altered. This condition is satisfied for all of the so-called nonstandard solar models (see Chapter 5), including the WIMP hypothesis, since for these models the relative numbers of neutrinos of different kinds are not the same as in the standard model, but the normalized energy spectra of the individual sources are unchanged. In this case, we can use the cross sections (that are given in Chapter 8) together with the appropriate neutrino fluxes from Chapter 6 and the uncertainties specified in Section 7.2. The situation is more complicated if the shapes of the individual energy spectra are altered by physical processes such as the MSW effect. In this case, one must use the individual cross sections as a function of energy (that are given in Chapter 8) to calculate effective absorption cross sections for the specified spectrum shape. The uncertainties can be inferred in most cases from the information given in Section IV.B of Bahcall and Ulrich (1988) (see also Chapter 8), since the fractional uncertainties in the cross section for a given source are, to a satisfactory approximation in most instances, independent of the precise shape of the spectrum.

7.2 Examples

A Nuclear reaction cross sections

It is simple to determine the 3σ error bars for the nuclear reaction cross sections that are measured experimentally since in all the important cases the original data are published or accessible and one can perform as many statistical tests as one likes. Recent reviews of the nuclear reaction data are contained in Parker (1986) and Bahcall and Ulrich (1988). The only question of principle is: How does one treat the occasional cases where one or two apparently very good experiments lie well outside the estimated errors? In practice, we have usually calculated predicted solar neutrino event rates based upon both of the competing experimental values, with the error given by each group. For laboratory experiments, the major controversies about input parameters have all been resolved by new measurements.

The nuclear reaction parameters for the pp and CNO chain are given, together with their estimated uncertainties (1σ limits), in Tables 3.1 and 3.2 of Chapter 3.

The most important reactions to measure accurately are the $^3\text{He} + ^3\text{He}$, $^3\text{He} + ^4\text{He}$, $^7\text{Be} + p$, and $^3\text{He} + p$ and reactions (numbers 3, 4, 7, and 10 of Table 3.1). Of all of the solar nuclear fusion reactions, the least well known and the most crucial for solar neutrinos is the $^7\text{Be} + p$ reaction, number 7, which produces the ^8B neutrinos. The low-energy cross section factor for this reaction is uncertain by 22% (3σ), which translates into an equal uncertainty in the predicted flux of ^8B neutrinos [see Eq. (7.7)]. It will be hard to improve significantly this limit because the statistical and systematic difficulties are great in measuring the small interaction cross section for a target of radioactive ^{7}Be [see, e.g., Parker (1986), Filippone *et al.* (1983), and Kajino, Toki, and Austin (1987)].

The rates of two of the most important reactions must be calculated, not measured. They are the pp and $e^- + {^7}$Be reactions (numbers 1a and 5 of Table 3.1). We have evaluated the uncertainties in those reactions by considering different theoretical calculations [see Section II.A of Bahcall *et al.* (1982) and Table 3.1].

B   Surface heavy element abundance

The present composition of the solar surface is presumed, in standard solar models, to reflect the initial abundances of all of the elements that are at least as heavy as carbon. The effects of diffusion and nuclear reactions on the surface abundance of heavy elements and of hydrogen are expected on the basis of plausible estimates to be negligible.

The initial ratio of heavy elements to hydrogen, Z/X, is one of the crucial input parameters in the determination of a solar model (see Chapter 2). The fractional abundances of the elements are important in determining the stellar opacity, which is closely linked to the predicted neutrino fluxes. Bahcall *et al.* (1982) used the elemental abundances recommended by Ross and Aller (1976) as the standard composition for which the radiative opacities were calculated. Since that work was completed, there has been much observational and theoretical research designed to improve our knowledge of the individual elemental abundances on the solar surface. This collective effort of many different researchers has been ably summarized in two reviews (completed approximately at the same time) by Grevesse (1984) and Aller (1986). In computing the standard solar model discussed in Chapter 4, Bahcall and Ulrich (1988) adopted the surface abundances listed by Grevesse (1984).

The two recent abundance studies are in excellent agreement. The best estimate value of Z/X has increased considerably from the value of 0.0228 of the Ross–Aller (1976) mixture. The Grevesse (1984) value is $(Z/X)_{\text{Grevesse}} = 0.02765$ and for the Aller (1986) mixture $(Z/X)_{\text{Aller}} = 0.02739$. The difference between the Ross–Aller (1976) value of Z/X and the current value of Grevesse (1984) and Aller (1986) is about 19%, almost twice the uncertainty that was estimated in Bahcall and Ulrich (1988). In accordance with the "rule-

Table 7.1. Neutrino fluxes
calculated with Los Alamos and
Livermore opacity codes. The
results are given for the standard
model parameters of Bahcall *et
al.* (1980). The unit of flux is 10^{10}
cm^{-2} s^{-1}

Source	Los Alamos opacity	Livermore opacity
pp	6.1	6.1
pep	0.015	0.016
^{7}Be	0.41	0.38
^{8}B	0.000585	0.00050
^{13}N	0.046	0.043
^{15}O	0.037	0.033

of-thumb" procedure that was described in Section 7.1, the total
uncertainty is [Bahcall and Ulrich (1988)]

$$\Delta\,(Z/X) = 0.19\,(Z/X)\,. \tag{7.1}$$

The uncertainty given in Eq. (7.1) is only slightly larger than is
obtained from the papers of Grevesse (1984) and Aller (1986) if one
interprets their listed uncertainties as 1σ errors and multiplies by
three in order to approximate an effective 3σ uncertainty.

C Radiative opacity

The basis for the estimate of uncertainties in the radiative opacity has been the comparison of the individual neutrino fluxes that
were calculated using both the Los Alamos Astrophysical Opacity
Library and a preliminary Livermore opacity code. The neutrino
fluxes were evaluated for evolved solar models that were constructed
with opacities from the two independent codes, holding all the other
parameters constant.

Table 7.1 compares the fluxes computed with the two different
opacity codes [see Table XIV of Bahcall *et al.* (1982)]. The computed
fluxes from the pp, pep, ^{7}Be, and ^{13}N reactions agree to better than
10%. The calculated fluxes from ^{8}B decay differ by about 16% and
the fluxes from ^{15}O decay differ by about 11%.

The two opacity codes used in obtaining the fluxes given in Table 7.1 were written independently of each other; they also use different numerical and atomic physics approximations. One is based on a combination of the explicit ion model (detailed configuration accounting) and a mean ion screening constant model, while the other code is based on a mean ion relativistic Hartree–Fock–Slater model.

The fractional differences in the individual fluxes given in Table 7.1 constitute a quantitative estimate of the total theoretical range for state-of-the-art opacity codes with a fixed composition. A detailed description of the physics that underlies these calculations is given in Huebner (1986).

The opacities that were used by Bahcall and Ulrich (1988) include effects not taken into account in the results shown in Table 7.1. One can get a feeling for the significance of these changes by considering the fractional change in the ^{8}B neutrino flux found by constructing detailed standard models with and without each modification. The fractional changes in other neutrino fluxes are generally less important and can be estimated approximately from the partial derivatives given in Section 7.3. The principal improvements in the opacity that were incorporated in 1988 are related to CNO conversion and electron scattering.

CNO conversion. The radiative opacity of the Sun changes as a result of the gradual conversion of the CNO elements to nitrogen via the CNO nuclear burning cycle. Although this process contributes only a small fraction of the solar luminosity, carbon and nitrogen nuclei in the solar core go through the catalytic cycle about a dozen times during the solar lifetime and the oxygen nuclei go partially through the first cycle. Essentially all of the carbon, and about 6% of the oxygen, is converted to ^{14}N. The CNO atoms are important contributors to the opacity in the solar core so that this interconversion alters the mean opacity.

The net result of including the conversion of carbon and oxygen to ^{14}N is an increase of the ^{8}B flux by 8% over what is obtained by ignoring this conversion.

Electron scattering. Collective effects on the electron scattering cross section are included in the Los Alamos opacity tables. In earlier work, Bahcall *et al.* (1982) used the expression that was derived by Diesendorf (1970), which implies a net decrease in the scattering cross section (compared to the Thomson value) of about 35% over most of the Sun. The Los Alamos calculation of collective effects, which is relatively complete, gives a decrease of only 18%, much smaller than implied by the Diesendorf formula [see also Boercker (1987)]. The difference between the Los Alamos and the Diesendorf corrections amounts to a 2.4% difference in the *total* opacity at the solar center.

Between 1982 and 1988, the ^{8}B neutrino flux increased by about 19% as a result of the modifications mentioned above, almost twice the total theoretical range estimated from Table 7.1. The uncertainty in the opacity may have been underestimated by a factor of two. Using the Los Alamos estimate, the calculated ^{8}B neutrino flux increases by a factor of 9% over what was obtained with the Diesendorf correction.

D Solar constant

Recent measurements of the solar constant show agreement with each other at the level of a percent [see, e.g., Willson, Duncan, and Greist (1980), Hickey *et al.* (1980), Willson and Hickey (1977), and Willson *et al.* (1981)]. The systematic errors in some of the experiments appear to be larger than the statistical errors. We adopt a nominal value and an uncertainty that is consistent with all of the recent measurements, that is, an average solar flux at 1 AU of 1372 ± 4 W/m^2 or

$$L_\odot = 3.86 \left(1 \pm 0.005\right) \times 10^{33} \text{ erg s}^{-1}. \tag{7.2}$$

E Solar age

The age of the meteorites is accurately determined and is about 4.55 $\times 10^9$ yr [see, e.g., Wasserburg *et al.* (1977, 1980)]. The time interval between the formation of the meteorites and the formation of the Sun is uncertain, but is expected by most workers to be small on the time scales of interest here. For the standard solar model, we have adopted a solar age of $\approx 4.6 \pm 0.1 \times 10^9$ yr. The precise age of the Sun is not important for solar neutrino calculations as long as the age is in the currently estimated range.

F Equation of state

Contrary to what some physicists intuitively believe, the equation of state of stellar material is well known under the conditions that apply to the solar interior. The uncertainties are small because the solar material is not far from a perfect gas in the region in which nuclear fusion is occurring. The equation of state that is used in the standard solar models described in Chapter 4 includes the thermal pressure, modified by degeneracy effects, as well as radiation pressure. Screening interactions are taken account of according to the Debye–Hückel theory. For the calculations made prior to 1982, we assumed that the solar plasma is completely ionized for all temperatures above 10^5 K (i.e., everywhere except in the surface layers). Ulrich (1982) derived a somewhat improved equation of state that does not make the *ab initio* assumption of full ionization in the solar interior. The differences between the neutrino fluxes calculated with the conventional equation of state and with Ulrich's improved prescription are less than or of order 1.5% for all the neutrino sources [see Bahcall *et al.* (1982)]. These small differences are the only available quantitative measures of how large might be the uncertainties in neutrino fluxes caused by uncertainties in the equation of state.

7.3   Neutrino cross sections

Neutrino absorption cross sections constitute the dominant uncertainties which limit the interpretability of several important solar neutrino experiments (e.g., with detectors of ^{81}Br, ^{98}Mo, ^{115}In, and ^{205}Tl, see Chapters 12 and 14). Because of their importance, the uncertainties for neutrino cross sections are discussed separately in this section.

There are two principal sources of uncertainties in the neutrino absorption cross sections: (1) transitions to excited states and (2) forbidden corrections. For solar neutrino experiments, uncertainties arising from transitions to excited states are usually much larger than uncertainties from forbidden corrections. This section considers some special cases, discusses transitions to excited states, describes the uncertainties from forbidden corrections, and finally indicates how all of the uncertainties can be combined.

A   Some special cases

For the ^{37}Cl detector alone, it is possible to evaluate relatively accurately the allowed matrix elements that connect the ground state of the target nucleus to the excited states of ^{37}Ar. This special calibration is accomplished by using data obtained from the decay of ^{37}Ca, the isotopic analogue of the neutrino capture process. The uncertainties [see Bahcall (1978) and Bahcall and Holstein (1986)] for the high-energy ^{8}B and hep neutrinos are estimated to be about 10%. All of the other sources can only populate the ground state of ^{37}Ar and have a smaller uncertainty, 6%, assigned to their absorption cross sections.

The absorption cross sections for the important deuterium target can be evaluated explicitly using different nuclear models. The estimated uncertainties are 10% [see Nozawa *et al.* (1986), Bahcall and Ulrich (1988), and Bahcall, Kubodera, and Nozawa (1988)].

The uncertainties for neutrino electron scattering cross sections are small and are estimated to be of order 5%. The major source of the uncertainty is radiative corrections [Bahcall (1987)]; the estimated uncertainty could be reduced substantially by a detailed theoretical calculation.

The cross section uncertainties for other specific detectors are discussed in Section IV of Bahcall and Ulrich (1988).

B Calibrations using (p,n) reactions

Transitions, that are not superallowed, to excited states represent the most important source of uncertainties for neutrino absorption cross sections. The transitions at fault are known as **Gamow–Teller transitions** (nuclear spin change of 0 or 1 unit, no parity change). In many important cases, the strength of Gamow–Teller absorption cross sections must be estimated by measuring analogous (but not identical) processes with (p,n) reactions. In **(p,n) reactions**, a portion that is incident on a target replaces a neutron in the initial nucleus, forming a final nucleus with one additional unit of charge plus an outgoing neutron.

How big could the 3σ errors be for Gamow–Teller matrix elements that must be estimated from data on (p,n) interactions? Unfortunately, there is no absolute basis for estimating the errors; reasonable people may assign different plausible values for the 3σ errors.

The existing nuclear models predict mass and configuration dependences that are not observed [see Taddeucci *et al.* (1987)]. Moreover, the quantum mechanical operators describing (p,n) and Gamow–Teller transitions are different, despite some similarities. It therefore seems unlikely that the nuclear theory of (p,n) reactions will be developed in the foreseeable future to the extent required to provide a precise basis for calibrating the relation between (p,n) measurements and Gamow–Teller matrix elements.

The error estimates must be based upon an empirical comparison between matrix elements known *a priori* from β-decay measurements and the same matrix elements inferred from (p,n) experiments. If the errors are Gaussian, then one can calculate a reliable 3σ limit using the results of a large number of accurate calibration experiments for nuclei in the same mass range. However, for the relatively weak Gamow–Teller transitions that are relevant for solar neutrino detectors with large mass numbers A, only of order a half-dozen experiments have been performed for any mass number. More importantly, there is no guarantee that the fluctuations around an average calibration relation will be normally distributed.

The situation is particularly unfavorable for neutrino targets with larger mass numbers such as ^{71}Ga, ^{81}Br, ^{98}Mo, ^{115}In, and ^{205}Tl. No precise calibrations of transitions of the relevant strength have been published for mass numbers in excess of $A = 27$ and no precise calibrations of superallowed or allowed transitions are available for mass numbers in excess of 42 [see Table 3 of Taddeucci *et al.* (1987)].

One reason for the paucity of large A calibrations is that the experiments become more difficult at large mass numbers because the spacing between the levels is smaller and because the Gamow–Teller transitions are often weaker. For weaker transitions, the differences between the quantum mechanical operators for Gamow–Teller matrix elements and for (p,n) processes may be more significant; contributions from orbital angular momentum may be relatively important at large A.

The accurate calibration information that is available is summarized in Table 3 and Figures 20 and 29 of Taddeucci *et al.* (1987). This information is insufficient to allow one to derive a convincing 3σ uncertainty for targets with relatively large mass numbers ($A \gtrsim 70$).

For specificity, I adopt a factor of two uncertainty for the total theoretical range for transitions to excited states whose matrix elements must be inferred from (p,n) measurements. I assume that the lower limit strength is one-half the value obtained from (p,n) measurements by the standard methods and that the upper limit strength is twice the value found by using the (p,n) experiments. Thus, if the cross section inferred from (p,n) measurements is X, I take the upper limit to be $2X$ and the lower limit $0.5X$, which causes a difference in the upper and lower error estimates for some of the detectors. In computing the uncertainties from excited levels, I sum the contributions from individual excited states and neutrino sources and then square the result to obtain the quantity that is added in quadrature to the other estimated uncertainties. I believe this procedure is justified since the most likely large error would result from an inapplicable extrapolation of the calibration of the average relation between (p,n) and neutrino capture cross sections from smaller atomic mass numbers A to a particular large A nucleus. A calibration error would cause the estimated rate of all of the transitions to move up or down together.

The factor of two uncertainty just described represents a guess for the size of systematic errors in using (p,n) experiments to calculate Gamow–Teller matrix elements for solar neutrino experiments; statistical and measurement uncertainties, which are usually smaller than a factor of two, must be considered separately.

More experimental and theoretical work is required on the relation between (p,n) cross sections and weak interaction matrix elements in order to understand the uncertainties. The most urgently required experiments are measurements in the mass range A from ~ 70 to ~ 130 for cases in which the weak interaction matrix elements are known from β-decay experiments. It is especially important to extend to larger A the relation between Gamow–Teller and Fermi matrix elements that has been established at smaller mass numbers. A calibration sample that included of order 10 targets with mass numbers larger than 70 would provide a much needed empirical basis for estimating the errors.

C Forbidden corrections

The uncertainty assigned to forbidden corrections to neutrino absorption cross sections is three times larger than the best estimate value given by Bahcall and Holstein (1986) (which is based upon dimensional analysis since the forbidden matrix elements cannot be reliably calculated). In addition, for ^{8}B neutrinos, there is a further uncertainty that arises from the imprecisely known shape of the ^{8}B neutrino spectrum. The uncertainty adopted for this latter effect, which refers only to ^{8}B neutrino absorption cross sections, is about 3% [see Bahcall and Holstein (1986)].

D   Combining cross section uncertainties

How does one combine the cross section uncertainties for different
neutrino sources that contribute to the capture rate in a given detector? First, divide the sources into two groups, a high-energy group
which contains only ^{8}B and hep neutrinos and a second group that
contains everything else. The most likely errors (calibration errors,
forbidden corrections, uncertain decay rates) will usually affect in
the same way sources within each group but will affect differently
the high-energy and the low-energy sources. Therefore the uncertainties are combined coherently within each group (summed before
squaring) and incoherently between the two groups (square and then
sum). For the ^{115}In experiment, all of the cross section uncertainties are combined coherently since for indium every transition was
calculated using the (p,n) data.

7.4   Power-law dependences

This section shows how to estimate the uncertainties in the calculated neutrino fluxes by assuming that all of the input parameters are
normally distributed with specified best estimates and standard deviations *and* that the probability distribution of each neutrino flux
is Gaussian. This procedure requires only a moderate amount of
computing, that is, constructing a few converged solar models for
each input parameter. Moreover, the partial derivatives can be used
to exhibit explicitly the approximate power-law dependence of the
predicted counting rate upon different parameters and to evaluate
the effects, without constructing a new sequence of standard solar
models, of small changes in the best estimates for parameters. The
following section, 7.4, describes Monte Carlo simulations of the *shape*
of the probability distributions for the neutrino fluxes.

A   Logarithmic derivatives

In order to estimate the uncertainties in the predictions for individual experiments that are caused by the imprecisely known values of
different input parameters, Bahcall and Ulrich have calculated the
logarithmic derivatives of each of the neutrino fluxes with respect to
each of the most significant input parameters. The derivatives were
determined by changing a single parameter, x_i (e.g., the cross section

factor for the pp reaction or the initial ratio of heavy elements to hydrogen), by a small amount (typically by of order 10%) and then recalculating a series of standard solar models until they converged accurately to the assumed present-day luminosity and radius. The differences in neutrino fluxes between the model constructed with the perturbed parameter and the standard solar model were used to form the logarithmic derivatives,

$$\alpha_{i,j} = \frac{\partial \ln\phi_i}{\partial \ln x_j}. \tag{7.3}$$

The partial derivatives calculated here apply to solar models at the present epoch; the derivatives given in this section are different from the partial derivatives that can be derived from a solar model that describes the evolution in time toward the present epoch (see Table 4.6).

In practice the derivatives were computed using the relation

$$\alpha_{i,j} \cong \frac{\ln\left[\phi_i/\phi_i\left(0\right)\right]}{\ln\left[x_j/x_j\left(0\right)\right]}, \tag{7.4}$$

where the quantities denoted by "(0)" refer to the standard solar model. Usually, two converged models (in addition to the standard solar model) were sufficient to estimate the partial derivatives (one model with the parameter on each side of the standard value). The approximation of small changes is equivalent to the statement that the neutrino fluxes have power-law dependences upon the individual parameters, that is,

$$\phi_i = \phi_i\left(0\right)\left[\frac{x_j}{x_j\left(0\right)}\right]^{\alpha_{i,j}}. \tag{7.5}$$

Table 7.2 presents the logarithmic derivatives of the neutrino fluxes with respect to each of the significant parameters. The derivatives with respect to the cross section factor, S_{17}, for the ^{7}Be+p reaction are not listed, since the rarity of this reaction (number 7 of Table 3.1) guarantees that the only derivative that is significantly different from zero is $\partial \ln\phi(^8\text{B})/\partial \ln S_{17} = 1.00$.

The partial derivatives in Table 7.2 can be translated into power-law relations. For example, the basic pp neutrino flux has the following dependencies:

$$\phi\left(\text{pp}\right) \propto S_{11}^{0.14} S_{33}^{0.03} S_{34}^{-0.06} L_\odot^{0.73}\left(Z/X\right)^{-0.08}\left(\text{age}\right)^{-0.07}. \tag{7.6}$$

Table 7.2. Calculated partial derivatives of neutrino fluxes. Each column contains the logarithmic partial derivatives of the neutrino fluxes with respect to the parameter shown at the top of the column. For example, $\partial\ln\phi_{pp}/\partial\ln S_{11} = +0.14$. The low-energy cross section factors S_{11}, S_{33}, and S_{34} refer to reactions 1, 4, and 5, respectively, of Table 3.2; the cross section factor $S_{1,14}$ refers to cross section factor for the ^{14}N + p reaction of Table 3.3.

Source	S_{11}	S_{33}	S_{34}	$S_{1,14}$	$L_\odot$	$R_\odot$	Z/X	Age
pp	+0.14	+0.03	−0.06	−0.02	+0.73	+0.01	−0.08	−0.07
pep	−0.17	+0.05	−0.09	−0.02	+0.87	+0.21	−0.17	+0.00
hep	−0.08	−0.45	−0.08	−0.01	+0.12	−0.09	−0.22	−0.11
^{7}Be	−0.97	−0.43	+0.86	−0.00	+3.40	+0.22	+0.58	+0.69
^{8}B	−2.59	−0.40	+0.81	+0.01	+6.76	+0.48	+1.27	+1.28
^{13}N	−2.53	+0.02	−0.05	+0.85	+5.16	+0.28	+1.86	+1.01
^{15}O	−2.93	+0.02	−0.05	+1.00	+5.94	+0.49	+2.03	+1.27
^{17}F	−2.94	+0.02	−0.05	+0.01	+6.25	+0.37	+2.09	+1.29

The pp neutrino flux is relatively insensitive to all of the input parameters except the total solar luminosity. On the other hand, the experimentally more accessible neutrinos from ^{8}B decay depend sensitively upon a number of the input parameters. Explicitly,

$$\phi\left(^8\text{B}\right) \propto S_{11}^{-2.6} S_{33}^{-0.40} S_{34}^{0.81} S_{17}^{1.0} L_\odot^{6.8} R_\odot^{0.48} \left(Z/X\right)^{1.3} (\text{age})^{1.3}. \quad (7.7)$$

The hep neutrinos are relatively insensitive to all of the input parameters except the rate of the ^{3}He+p and the ^{3}He +^{3}He reactions. Thus

$$\phi\,(\text{hep})$$
$$\propto S_{11}^{-0.08} S_{13}^{1.0} S_{33}^{-0.45} S_{34}^{-0.08} L_\odot^{0.1} R_\odot^{-0.09} \left(Z/X\right)^{-0.2} (\text{age})^{-0.1}. \quad (7.8)$$

The ^{7}Be neutrinos are intermediate between the pp and the ^{8}B neutrinos in the degree of their sensitivity to input parameters. We find

$$\phi\left(^7\text{Be}\right) \propto S_{11}^{-0.97} S_{33}^{-0.43} S_{34}^{0.86} L_\odot^{3.40} R_\odot^{0.22} \left(Z/X\right)^{0.58} (\text{age})^{1.3}. \quad (7.9)$$

In order to interpret geochemical experiments that average the measured ^{8}B neutrino flux over long periods of time (see the discussion of the ^{98}Mo and ^{205}Tl experiments in Chapter 12) a different kind of derivative must be computed. For this application, one needs the rate at which the ^{8}B flux increases with age, holding constant the present model solar age and luminosity. Table 4.6 yields for model ages, t, close to the present age of 4.6×10^9 yr

$$\phi(^8\text{B}, t)\Big|_{L_\odot, t_\text{age}} \propto t^{3.4}. \qquad (7.10)$$

B   Neutrino flux uncertainties

This section describes the calculation of the uncertainties in the production rates of individual solar neutrino fluxes; the results have been summarized previously in Table 6.5 of Chapter 6. The results given here and in Table 6.5 do not include any uncertainties in detector sensitivity.

In general, one can express the total event rate for a given detector in the following form:

$$\text{Event rate} = \sum_i [\phi_i/\phi_i(0)][\phi_i(0)\sigma_i]. \qquad (7.11)$$

The first term on the right-hand side of Eq. (7.11), $[\phi_i/\phi_i(0)]$, can be evaluated using the power-law dependences obtained from Table 7.2 [cf., Eqs. (7.6) to (7.9)]. The second term can be calculated from the neutrino fluxes of the standard solar model, which are given in Table 6.5, and the interaction cross sections, which are given in Chapter 8. The rewriting of Eq. (7.11) in terms of different input parameters takes on an especially simple form for experiments that are sensitive primarily to neutrinos from a single solar nuclear source. For example, the deuterium, neutrino–electron scattering, ^{40}Ar, and ^{98}Mo experiments are sensitive primarily to ^{8}B neutrinos. For these "single-source" experiments, the dependence of the event rate upon input parameters is given by Eq. (7.7).

The total fractional uncertainty in individual neutrino fluxes, ϕ_i, can be computed from the following expression:

$$\frac{\delta\phi_i}{\phi_i} = \left[\sum_i \left[\left(1 + \delta x_j/x_j\right)^{\alpha_{i,j}} - 1\right]^2\right]^{1/2}. \qquad (7.12)$$

Here the ratios $\delta x_j/x_j$ are the fractional uncertainties in each of the input parameters x_j.

Table 6.5 lists the computed total theoretical uncertainties in the individual neutrino fluxes that result from uncertainties in nuclear

Table 7.3. Simultaneous large changes in S_{11}, S_{33}, S_{34}, and S_{17}. Neutrino fluxes are given in the table in units of $cm^{-2}\ s^{-1}$. The fluxes listed in column 2 were calculated from the results of the 1982 standard model; the fluxes given in column 3 were obtained using partial derivatives. The fluxes from the perturbed model are given in column 4; they were calculated by constructing a converged solar model. The fractional difference between the fluxes calculated using partial derivatives and those obtained from an accurate solar model are small and are shown in column 5. This table is adapted from Table XII of Bahcall *et al.* (1982).

Source	'82 standard	Partial derivatives	Perturbed model	Percentage difference
pp	6.07E10	6.27E10	6.25E10	0.5
pep	1.50E08	1.55E08	1.51E08	3
^{7}Be	4.3E09	2.52E09	2.72E09	7
^{8}B	5.6E06	2.22E06	2.05E06	8
^{13}N	5.0E08	3.61E08	3.75E08	4
^{15}O	4.0E08	2.64E08	2.80E08	6

reaction rates, assumed initial heavy element abundances, the radiative opacity, and the solar constant. These results do not include the (sometimes large) uncertainties in the neutrino absorption cross sections, which are discussed in Chapter 8 and in the later sections describing individual experiments.

Are the power-law approximations discussed above seriously in error if derivations from the standard values are large and all in the direction of, for example, decreasing the predicted ^{8}B neutrino flux? The effect of large simultaneous changes was considered by Bahcall *et al.* (1982), who calculated an accurate solar model in which simultaneous changes were made in four nuclear parameters. Each parameter was changed by its estimated 3σ limit in the direction to reduce the flux of ^{8}B neutrinos.

Table 7.3 compares the results calculated with power-law dependences (using the partial derivatives of fluxes on each parameter) with the results obtained from a direct calculation of a solar model constructed using very nonstandard nuclear parameters. The large change in each of the neutrino fluxes for this specific quadruple 3σ variation (1982 estimates of the uncertainties) is given to an accuracy of better than 10% for all of the solar neutrino fluxes of interest. It appears therefore that the effects of specified changes, large or small, can be computed with satisfactory accuracy using the calculated partial derivatives. Nevertheless, the nagging question remains: Is the *distribution* of uncertainties for a given flux or capture rate sufficiently well approximated by a Gaussian with the standard deviation calculated from Eq. (7.11)?

7.5 Monte Carlo simulations

Could enough measured parameters be in error, all in the right direction, so as to reduce the correct standard capture rate to the values observed in the ^{37}Cl and Kamiokande II experiments? The calculations described in Section 7.3 assumed that the errors in the neutrino fluxes are normally distributed and that they can be obtained by calculating partial derivatives of fluxes with respect to the standard model values, extrapolating small changes to large uncertainties. In principle, the assumption of an extrapolated Gaussian could be vitiated by the nonlinear relations that are imposed by the partial differential equations of stellar structure, which are combined with the boundary conditions of matching the observed solar luminosity, effective temperature, and heavy element to hydrogen ratio [cf., Rood (1978)].

Bahcall and Ulrich answered these questions by brute force. They constructed [Bahcall and Ulrich (1988)] 1000 accurate solar models with five input parameters that were chosen each from a normal distribution with the best estimate mean and standard deviation. The five quantities that were allowed to vary randomly were the initial heavy element to hydrogen ratio, Z/X, and the cross section factors for the pp, ^{3}He$+^{3}$He, ^{3}He$+^{4}$He, and ^{7}Be$+$p reactions (reactions 1a, 3, 4, and 7 of Table 3.1). These quantities represent five of the dominant uncertainties in predicting the rate of the ^{37}Cl experiment and most of the other proposed solar neutrino experiments. The inclusion of additional parameters in the simulations would have changed by a small amount the total range of fluxes and capture rates found in the simulations, but would not have affected the general conclusions.

The Monte Carlo simulations agree with the results obtained using partial derivatives of neutrino fluxes and the assumption that the uncertainties are independent and normally distributed. The fractional standard deviations that were calculated from the Monte Carlo simulations (and in parentheses the value obtained using partial derivatives) were, for the pp, pep, hep, ^{7}Be, ^{8}B, ^{13}N, ^{15}O, and ^{17}F fluxes, respectively: 0.006 (0.006), 0.011 (0.011), 0.029 (0.027), 0.050 (0.048), 0.120 (0.120), 0.124 (0.133), 0.137 (0.148), and 0.141 (0.153). The numbers in parentheses were calculated using just the partial derivatives for the five parameters that were allowed to vary in the simulations. In all cases, the standard deviations calculated by the two different methods agree to better than 10%.

What about the very large variations? What is the probability that a random variation of the input parameters would yield a neutrino flux that differs from the best estimate (average of the simulations) by more than some specified number of standard deviations?

Figure 7.1 compares the distribution of fluxes obtained from the simulation with that calculated using the assumption of a Gaussian distribution with a standard deviation evaluated using the partial derivatives given in Table 7.1. These are three separate curves displayed in Figure 7.1: (1) the normal distribution (solid curve); (2) simulations above the average value (short dashes); and (3) simulations below the average value (long dashes).

Figure 7.1a compares the three distributions between zero and four standard deviations. The probability distributions for all three cases are similar, so much so that it is hard to see on this scale how the other two curves deviate from the normal distribution.

For clarity, Figure 7.1b displays the three distributions in the limit of large deviations, more than 2σ differences from the mean. The number of very small values of the ^{8}B flux in the simulated distribution is somewhat lower than for the normal distribution. For the 1000 simulated cases, 6.2 cases that are 2.5σ below the mean would have occurred if the distribution were normal; none were found. The high values are in good agreement with the normal distribution.

The smallest ^{8}B flux found in the simulations was 4.0×10^6 cm^{-2} s^{-1} and the largest value was 7.9×10^6 cm^{-2} s^{-1}. The corresponding 3σ limits that were calculated with the Gaussian partial derivative method are 3.6×10^6 cm^{-2} s^{-1} and 7.7×10^6 cm^{-2} s^{-1}.

The expectations for experiments sensitive only to ^{8}B neutrinos (e.g., neutrino–electron scattering experiments, as well as ^{2}H, ^{40}Ar, and ^{98}Mo detectors) are well described by the power-law dependences discussed in Section 7.3.

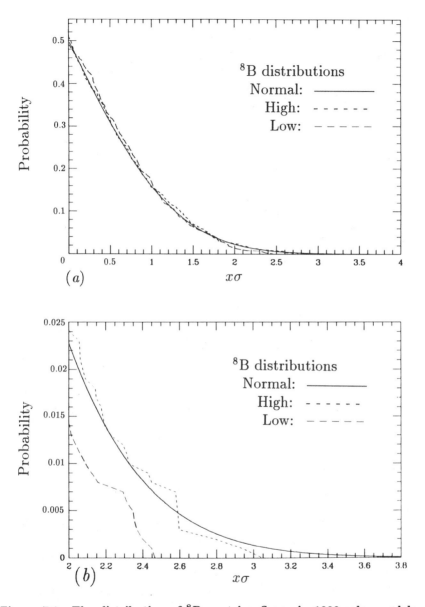

Figure 7.1   The distribution of ^{8}B neutrino fluxes in 1000 solar models
This figure compares the distribution of ^{8}B neutrino fluxes determined
from 1000 accurate solar models with what is expected from a Gaussian
distribution whose mean and standard deviation are calculated from partial
derivatives (Figure 7.1a: 0 to 4σ; 7.1b, 0 to 4σ). The solid curve represents
the normal distribution; short dashes, simulations above the average value;
and long dashes, simulations below the average value.

Table 7.4.   Uncertainties for the neutrino fluxes of the standard solar model.
The table gives in the second column the total fractional uncertainty in the
neutrino fluxes and in the following columns the fractional uncertainties
associated with each important parameter (e.g., pp reaction rate, heavy
element to hydrogen ratio Z/X, or opacity).

Source	Total	pp	^{3}He+^{3}He	^{3}He+^{4}He	^{7}Be+p	^{14}N+p	Z/X	Opacity
pp	0.02	0.01	0.00	0.00	0.00	0.01	0.01	0.00
pep	0.05	0.01	0.00	0.01	0.00	0.01	0.03	0.02
hep	. . .	0.00	0.07	0.00	0.00	0.00	0.03	0.00
^{7}Be	0.15	0.05	0.07	0.05	0.00	0.00	0.11	0.04
*B	0.37	0.12	0.06	0.05	0.22	0.00	0.25	0.08
^{13}N	0.50	0.12	0.00	0.00	0.00	0.30	0.38	0.03
^{15}O	0.58	0.14	0.00	0.00	0.00	0.36	0.42	0.06
^{17}F	0.46	0.14	0.00	0.0	0.00	0.00	0.44	0.06

The agreement between the Monte Carlo simulations and the
Gaussian extrapolation is similarly excellent for the total capture
rates that are predicted for the ^{37}Cl and ^{71}Ga experiments. The
1000 simulations have capture rates varying between 5.8 SNU and
10.5 SNU; the corresponding 3σ limits calculated as described in
Section 7.3 are 5.4 SNU and 10.3 SNU. For the ^{71}Ga detector, the
1000 simulations range from 119 SNU to 144 SNU while the range
calculated via the Gaussian extrapolation is 120 SNU to 143 SNU.

Bahcall and Ulrich could have run more simulations to determine if
the slight differences from a normal distribution are significant. They
decided, however, that more simulations would not reveal anything
of value. The error distributions of the input parameters are not
known sufficiently well to justify detailed study of the tail of the
combined uncertainty distribution.

The question of greatest importance is: Can uncertainties in the
input parameters explain the discrepancy between observation and
calculation for the ^{37}Cl and Kamiokande II experiments? The answer
given in Figure 7.1 and the above cited numerical results is: "No,
not with a significant probability."

Table 7.5. Individual uncertainties in predicted capture rates. The different detectors are listed in the first column. Each subsequent column contains the uncertainty in SNU of the total capture rate caused by the uncertainty in the parameter at the top of the column. This table is adapted from Table XVI of Bahcall and Ulrich (1988).

Detector	pp (SNU)	^{3}He+^{3}He (SNU)	^{3}He+^{4}He (SNU)	^{7}Be+p (SNU)	Z/X (SNU)	Opacity (SNU)	σ_{abs} (SNU)	Total (SNU)
^{2}H	0.7	0.4	0.3	1.3	1.5	0.5	0.6	2.1
^{7}Li	5.1	1.6	1.2	4.9	12.2	2.5	4.2	15
^{37}Cl	0.9	0.5	0.4	1.3	1.8	0.5	0.6	2.4
^{40}Ar	0.2	0.1	0.1	0.4	0.4	0.1	0.1	0.6
^{71}Ga	4.1	2.9	2.2	3.1	10.1	2.8	16	20
^{81}Br	2.6	1.6	1.2	3.4	5.8	1.6	16	17
^{98}Mo	2.1	1.1	0.8	3.8	4.3	1.4	17	18
^{115}In	8.2	6.4	4.9	3.2	22.3	6.9	639	639

7.6 Principal uncertainties

Table 7.4 gives the fractional uncertainty (total theoretical range) in each of the standard model solar neutrino fluxes that is caused by each of the principal sources of uncertainty. The major sources of uncertainty are, in the order in which they appear in Table 7.4, the low-energy cross section factors for the pp, ^{3}He+^{3}He, ^{3}He+^{4}He, ^{7}Be+p, and ^{14}N+p reactions, as well as the heavy element to hydrogen ratio, Z/X, and the radiative opacity. For hep neutrinos, no estimate was made for the total uncertainty because of the large uncertainty in the low-energy cross section factor for reaction 10 of Table 3.1. A measurement of any of the fluxes within the range given in Table 7.5 would be consistent with the standard solar model.

Table 7.4 can be used conveniently for estimating the uncertainty in ν_e-e^- scattering experiments, which are primarily sensitive (at present) to ^{8}B solar neutrinos. The additional uncertainties due to radiative corrections to the scattering cross sections are probably only of order a few percent.

Table 7.5 gives the uncertainty (total theoretical range) in SNU contributed by each of the principal sources of uncertainty to the ^{2}H, ^{7}Li, ^{37}Cl, ^{40}Ar, ^{71}Ga, ^{81}Br, ^{98}Mo, and ^{115}In solar neutrino experiments. The major sources of uncertainty are, in the order in which they appear in Table 7.5, the low-energy cross section factors for the pp, ^{3}He+^{3}He, ^{3}He+^{4}He, and ^{7}Be+p reactions, as well as the heavy element to hydrogen ratio, Z/X, the radiative opacity, and the neutrino cross sections. Table 7.5 is a useful summary when considering uncertainties in individual solar neutrino experiments that are discussed in Chapters 10 to 14.

Bibliographical Notes

Bahcall, J.N., N.A. Bahcall, and R.K. Ulrich (1969) *Ap. J.*, **156**, 559. Presented power-law dependences upon input parameters of the predicted capture rate for the ^{37}Cl experiment. The results continue to be used to make quick estimates of the effect of parameter changes on the calculated neutrino fluxes.

Fowler, W.A. (1958) *Ap. J.*, **127**, 551. Pointed out that the large cross section measured for the ^{3}He+^{4}He reaction (see Chapter 3) might lead, for favorable nuclear and astrophysical parameters, to a ^{8}B neutrino flux in excess of 2×10^{10} cm^{-2} s^{-1}. Fowler's paper was the inspiration for much of the early work on defining the relevant nuclear parameters.

8

Neutrino cross sections

Summary

Neutrino cross sections are as important as stellar models in determining the predicted rates in detectors of solar neutrinos or stellar collapses. The observed rates are products of fluxes times cross sections. For several of the proposed detectors of solar neutrinos, the uncertainties in the neutrino cross sections are much larger than the uncertainties in the standard solar neutrino fluxes.

This chapter summarizes the basic formulae for neutrino cross sections and presents numerical results that apply to the most important neutrino sources and detectors.

Absorption cross sections describe reactions in which an incoming neutrino is absorbed, an electron is created, and a neutron in the target nucleus is transformed into a proton. The cross sections can be written in terms of atomic and kinematic factors times nuclear matrix elements that represent the neutron–proton transformation. In general, all of the factors except the nuclear matrix elements can be calculated accurately. For low-energy neutrinos, the nuclear matrix elements can often be evaluated from laboratory measurements of the radioactive decay between the ground states of the initial and final nuclei involved in the absorption. For the higher-energy ^{8}B and hep neutrinos, transitions to excited states of the daughter nucleus usually introduce large uncertainties. In some important special cases, such as the ^{37}Cl $\rightarrow$ ^{37}Ar reaction, the most important transition probabilities can be calculated accurately using isotopic spin symmetry.

The cross sections for neutrino–electron scattering can be calculated using standard electroweak theory. The results depend upon neutrino flavor: Electron neutrinos are scattered much more strongly, in the energy range of interest, than are muon or tau neutrinos. The energy spectra of the recoil electrons have characteristic shapes that depend upon the energy spectrum of the incident neutrino source but are not sensitive to neutrino flavor. The angular distribution of the recoil electrons is strongly peaked in the forward direction for essentially all practical experimental situations.

Neutral currents are flavor blind. The cross sections for pure neutral-current interactions are independent of neutrino type. Thus neutral-current experiments can distinguish between explanations of the solar neutrino problem that involve "missing neutrinos" (e.g., faulty astrophysics) and explanations in which the total number of neutrinos is constant but the relative numbers with different flavors are altered (e.g., the MSW effect). Neutrino–nucleus elastic scattering can be observed at low temperatures by detecting the recoil of the nucleus. The rate of this reaction is enhanced by a coherence factor equal to the square of the number of neutrons in the nucleus. The cross sections for neutrino excitation of nuclear excited states can be estimated by the traditional methods of calculating nuclear matrix elements. For the special case of neutrino disintegration of the deuteron, the laboratory constraints on the wave functions are sufficient for an accurate calculation of the cross section.

Section 8.1 presents the most important results for neutrino absorption (by charged currents) and Section 8.2 discusses neutrino–electron scattering (by charged and neutral currents). The last section, 8.3, summarizes the nuclear interactions induced by pure neutral currents.

A number of experimentally relevant results are included in this chapter between more extended technical discussions. For the reader's convenience, some of the more directly applicable results are cited explicitly here.

The section on neutrino absorption contains a simple formula, Eq. (8.6), for the angular distribution of the recoil electrons. The use of (p,n) experiments to determine Gamow–Teller matrix elements is described in Section 8.1B. Tables 8.2 and 8.3 give the neutrino absorption cross sections for individual solar neutrino sources incident on the different detectors considered in this book. The cross sections as a function of neutrino energy are summarized in Table 8.4, while the uncertainties in the calculated cross sections are described for six important detectors in Section 8.1D.

The total neutrino–electron scattering cross sections for individual neutrino sources are given in Tables 8.5 to 8.7 for different flavors and experimental thresholds. Figure 8.1 illustrates the strong effect of the energy threshold. The ratio of cross sections for ν_e–e to ν_μ–e scattering is about a factor of six or seven for all of the experimentally relevant cases considered here [see Eq. (8.42)], a key result in considering the influence of the MSW effect on observed event rates (see Section 9.2). The cross sections at specific neutrino energies (1 MeV to 60 MeV) are given in Table 8.8 for ν_e and ν_μ. An approximate analytic formula for the cross sections, valid for ν_e, $\bar{\nu}_e$, ν_μ, and $\bar{\nu}_\mu$, is given in Eq. (8.43). Section 8.2D discusses the energy distributions of the recoil electrons, which are given in Tables 8.9 to 8.13 and are illustrated in Figures 8.2 to 8.4. The angular distributions of recoil electrons from scattering by ν_e and ν_μ are discussed in Section 8.2E and illustrated in Figures 8.6 and 8.7. Pion and muon decay produce neutrinos and antineutrinos whose cross sections are given in Table 8.15.

Section 8.3A describes the advantages and disadvantages of neutrino–nucleus elastic scattering. The deuterium (SNO) and ^{11}B (BOREX) experiments are discussed in Section 8.3B in connection with neutral-current interactions that disintegrate or excite nuclei.

The discussion of absorption cross sections in Section 8.1 is based upon a series of articles by Bahcall (1964a,c, 1978), Bahcall *et al.* (1982), and Bahcall and Ulrich (1988). The material on neutrino–electron scattering in Section 8.2 is adapted from Bahcall (1987) and Bahcall (1964b). The background in weak interaction theory and in nuclear physics that is necessary to derive the results given here is presented in many excellent books; the notation that is used here is adopted from Konopinski (1966) and Okun (1982) [see also, e.g., Bailin (1982), Commins and Bucksbaum (1983), Pietschmann (1983), Halzen and Martin (1984), and Boehm and Vogel (1987)].

8.1   Absorption cross sections

A   General formulae

Solar neutrino experiments have low counting rates because the cross sections for neutrino interactions with matter are small. The transitions that dominate the predicted event rates for all of the projected experiments satisfy the so-called **allowed** selection rules in which the nuclear spin changes by 0 or 1 unit and the parity is unchanged. This section discusses allowed neutrino capture processes; forbidden corrections are of order a few percent and are discussed by Bahcall and Holstein (1986).

The cross section for allowed neutrino capture, symbolized by the process

$$\nu + {}^{Z-1}A \rightarrow e^- + {}^Z A, \tag{8.1}$$

is

$$\sigma = \left(\pi c^3 \hbar^4\right)^{-1} \left[G_V^2 \langle 1 \rangle^2 + G_A^2 \langle \sigma \rangle^2\right] P_e W_e F\left(eZ, W_e\right). \tag{8.2}$$

Here Z is the atomic number of the final nucleus (mass number A), W is the energy of the electron that is produced, P_e is the electron's momentum. The coefficients G_V and G_A are the usual weak interaction coupling constants. The quantity $F(Z, W_e)$ is known in β-decay theory as the Fermi function; it represents the increased (decreased) probability of an electron (positron) being found at the charged nucleus relative to the uniform probability of a plane wave. Fermi functions are discussed in Section 8.1B.

Momentum conservation implies that the recoil energy of the nucleus is small compared to the energies of the leptons (electron and neutrino) for all solar neutrino experiments. The incident neutrino energy, q, and the outgoing electron energy are related by the threshold energy for the reaction, E_{th}, via the relation [cf., Eq. (8.12)]

$$q = E_{th} + W_e. \tag{8.3}$$

In order to be consistent with the most relevant literature, neutrino energy is represented by q in this chapter and by E in Chapter 9. Equation (8.2) neglects the recoil of the nucleus, an excellent approximation for all solar neutrino experiments. The strength of the transition is represented here by the combination of nuclear matrix elements,

$$\zeta = [G_V^2 \langle 1 \rangle^2 + G_A^2 \langle \sigma \rangle^2], \tag{8.4}$$

where $\langle 1 \rangle$ and $\langle \sigma \rangle$ are the standard reduced matrix elements [see, e.g., Rose (1957) or Konopinski (1966)] between the initial and final nuclear states. For example, for neutron decay, $\langle 1 \rangle^2 = 1.0$ and $\langle \sigma \rangle^2 = 3.0$.

For (superallowed) transitions between analogue states having the same total isotopic spin T,

$$\langle 1 \rangle^2_{\mathrm{SA}} = 2T. \tag{8.5}$$

The Fermi matrix element, $\langle 1 \rangle$, is nonvanishing only if the nuclear spin and isospin is unchanged. The Gamow–Teller matrix element is often called $B(\mathrm{GT})$ in the nuclear physics literature.

In direct counting experiments, the angular distribution of the recoil electrons can be measured. For neutrino absorption, the angular distribution is given by the simple formula [Bahcall (1964d)]

$$P(\theta) = \left[1 + \left(\frac{v_e}{c} \right) \alpha \cos \theta \right], \tag{8.6a}$$

where v_e is the recoil velocity of the electron, and the asymmetry parameter α is given by

$$\alpha = \frac{\langle 1 \rangle^2 - \frac{1}{3}(G_A/G_V)^2 \langle \sigma \rangle^2}{\langle 1 \rangle^2 + (G_A/G_V)^2 \langle \sigma \rangle^2}. \tag{8.6b}$$

For neutrino absorption by ^{2}H or ^{115}In, the transformation is pure Gamow–Teller and $\alpha = -1/3$. For the superallowed transition between spin zero the analogue states of ^{40}Ar and ^{40}K, $\alpha = +1$.

It is convenient and customary to replace the nuclear matrix elements in Eq. (8.2) by the product of a phase space factor, f, and a decay lifetime, $t_{1/2}$. This so-called $ft_{1/2}$-**value** is defined by the relation

$$\frac{1}{\left(ft_{1/2} \right)_{I' \to I}} = \frac{\left[G_V^2 \langle 1 \rangle^2 + G_A^2 \langle \sigma \rangle^2 \right]}{[2\pi^3 \ln 2 \, (\hbar/m_e c)^7 \, m_e^2 c^3]}. \tag{8.7}$$

The spins of the initial and final nuclear states are I and I', respectively. For allowed decays, $I - I' = 0$, ± 1. Nuclear physicists use the quantity $ft_{1/2}$ as the standard measure of the strength of a β-decay transition. The dimensionless function $f(Z, W)$ reflects the size of the available phase space and does not depend upon details of the nuclear wave functions. On the other hand, the lifetime of the transition, $t_{1/2}$, is determined by the overlap between the initial and final state vectors. By using the $ft_{1/2}$-value, one replaces the nuclear matrix elements ζ by quantities that can be calculated in principle from measurable parameters.

The phase space factor, f, can be calculated accurately for allowed electron or positron decays or for electron capture (see Section 8.2B below). The magnitude of f depends sensitively upon the mass difference between the initial and final nuclear states. In a number of cases of interest for solar neutrino experiments, one can determine experimentally the $ft_{1/2}$-value that applies for the inverse nuclear process in which the initial and final states are interchanged relative to the nuclear transition shown in Eq. (8.1). Fortunately, there is a symmetry relation, derivable from the Clebsch–Gordon coefficients of the transition, that enables one to obtain the $ft_{1/2}$-value for neutrino absorption from the $ft_{1/2}$-value that is measured in a laboratory β-decay experiment. One finds

$$\frac{(2I+1)}{\left(ft_{1/2}\right)_{I \to I'}} = \frac{(2I'+1)}{\left(ft_{1/2}\right)_{I' \to I}}. \tag{8.8}$$

For computational convenience, the formula for the cross section given in Eq. (8.2) may be written in terms of the $ft_{1/2}$ as

$$\sigma \equiv \sigma_0 \left\langle w_e^2 G\left(Z, w_e\right) \right\rangle, \tag{8.9}$$

where

$$\sigma_0 \equiv \frac{1.206 \times 10^{-42}}{\left(ft_{1/2}\right)_{I \to I'}} \left[\frac{(2I'+1)}{(2I+1)}\right] Z \text{ cm}^2, \tag{8.10}$$

and

$$G\left(Z, w_e\right) \equiv \left(p_e\, F\left(Z, W_e\right) / 2\pi\alpha Z w_e\right). \tag{8.11}$$

Here $w_e \equiv W_e / m_e c^2$ is the energy (in units of the electron's mass) of the electron that is produced and p_e is the electron's momentum (in units of $m_e c$).

The quantity $G(Z, w_e)$ includes only atomic physics effects (except for the finite nuclear size) and has, as written, two simple limits. For electron energies small compared to the electron rest mass, G is approximately equal to unity and for electron energies large compared to $m_e c^2$, G is approximately equal to $(2\pi\alpha Z)^{-1}$.

The following crude approximations can provide useful order of magnitude estimates: $F(Z, W_e) \approx \eta / (1 - \exp -\eta)$ with $\eta = 2\pi\alpha Z w_e / p_e$, $G \sim 1$ and $\sigma \sim \sigma_0 w_e^2$.

All of the nuclear structure effects are included in the dimensional cross section factor σ_0. For many important cases involving transitions among ground states or between isotopic analogue states, the dimensional cross section factor that is given in Eq. (8.10) can be calculated in terms of a measured $ft_{1/2}$-value. Typical $ft_{1/2}$-values for allowed β-decays measured in the laboratory are of order 10^5 s^{-1}. Therefore neutrino absorption cross sections for solar neutrinos are of order 10^{-46} cm^2 for $Z \sim 30$, that is, 10^{-22} barns (the characteristic unit for nuclear physics cross sections).

The usual theory given in the standard textbooks of neutrino capture by atoms neglects two physical effects: (1) the change in the atomic Hamiltonian caused by the increase in nuclear charge by one unit from initial to final atomic state; and (2) the indistinguishability of the created electron from the initially present atomic electrons. Similar atomic physics considerations lead to calculable corrections, known respectively as overlap and exchange effects, to the usual theory of electron and positron emission and to electron capture. Bahcall (1963) showed how to treat weak interaction processes including atomic variables; the predicted corrections are, in some cases, easily measurable and are then in agreement with laboratory experiments [e.g., Bahcall (1965) and Bambynek *et al.* (1977)]. The same formalism can be applied to neutrino capture since electron β-decay and neutrino capture are mathematically very similar. One finds [see Bahcall (1978)] that atomic overlap effects are accounted for to high accuracy if the energy of the continuum electron that is produced in Eq. (8.1) is calculated from an equation that includes the average excitation energy of the final atom $\bar{E}_{\mathrm{ex}}$, that is:

$$W_{\mathrm{e}} = -\bar{E}_{\mathrm{ex}} + q + [M(A, Z - 1) - M(A, Z)] + m_{\mathrm{e}}c^2, \qquad (8.12)$$

where q is the neutrino energy and the term in brackets is the atomic mass difference. Equation (8.12) is the generalization of the usual equation for energy conservation [Eq. (8.3)], when atomic effects are explicitly included. A satisfactory approximation for $\bar{E}_{\mathrm{ex}}$ is [see Eq. (30) of Bahcall (1963)]: $\bar{E}_{\mathrm{ex}} \approx 24.5Z^{1/3}$ eV for $Z < 10$ and $23Z^{2/5}$ eV for $Z > 10$. Exchange effects between the final continuum electron and the electrons bound in the initial atom interfere in a way that reduces the capture rate. Numerical calculations show that atomic overlap and exchange effects change the calculated neutrino absorption cross sections by less than 1% for all the cases considered in this book [see Bahcall (1978)].

A number of groups have performed impressive calculations of the theoretical Gamow–Teller matrix elements to excited states [see the references in Taddeucci *et al.* (1987)]. However there are unresolved questions for each of the investigations as to whether the basis set of wave functions was sufficiently large, whether the Hamiltonian was appropriate, and whether "quenching" was properly included.

For completeness, we need to consider one more process: bound state neutrino capture. In this process, no continuum electrons are produced; a bound electron is created in a previously unoccupied bound state of the final atom. A detailed evaluation [Bahcall (1978)] shows that bound state neutrino capture amounts to less than 1% of the usual continuum process for all the cases considered in this book.

Over the past two and a half decades, the author has developed a convenient computer code that calculates the neutrino absorption cross sections for all the targets of interest in solar neutrino experiments. This code evaluates the weighted average dimensionless phase space factor $\sigma_{av}(Z) \equiv \langle w_e^2 G(Z, w_e) \rangle$ defined implicitly by Eq. (8.2) and more explicitly by

$$\sigma_{av}(Z) = (2\pi\alpha Z)^{-1} \times \int_{w_{min}}^{w^{max}} dw_e w_e p_e F(Z, w_e) \phi(q_\nu). \qquad (8.13)$$

Here q_ν is the dimensionless neutrino energy, $q_\nu = q/(m_e c^2)$, which is related via energy conservation, Eq. (8.12), to the nuclear mass difference and the electron energy. The normalized neutrino energy spectrum is denoted by $\phi(q)$. For radiochemical experiments, all nuclear transformations are equivalent and therefore the minimum electron recoil energy is just $W_{min}/m_e c^2 = 1$. However, for experiments with energy resolution in which individual recoil electrons are counted (e.g., ^{2}H or ^{40}Ar detectors), W_{min} is typically set equal to several MeV in order to reduce background effects to a manageable level.

For most allowed continuum positron emitters, the following approximation is accurate:

$$\phi(q_\nu) \propto q_\nu^2 w_e \left[w_e^2 - 1\right]^{1/2} F(-Z'_{emitter}, w_e). \qquad (8.14)$$

Here $w_e \equiv [1 + q_{\nu,max} - q_\nu]$. For the important special case of ^{8}B decay, in which the predominant final nuclear state is broad, the spectrum of neutrinos must be determined by analyzing laboratory measurements of the resulting α-decay spectrum [Bahcall (1964a) and Bahcall and Holstein (1986)]. Neutrinos produced by electron capture [see Sections 6.2 to 6.4 of Chapter 6] have energy spectra, $\phi(q_\nu)$, that are essentially delta functions, except for the (usually) small effect of thermal broadening.

B Computational details

There are a number of physical effects that must be evaluated in order to obtain accurate neutrino absorption cross sections. These effects manifest themselves in the values of the Fermi functions and in the size of the $ft_{1/2}$-value. We begin with a discussion of the Fermi functions and then consider the phase space factors (f-values).

Fermi functions. Corrections due to special relativity, finite nuclear size, and electron screening in terrestrial atoms must be included in order to compute accurately the Fermi functions $F(Z, w_e)$. The usual definition is [Konopinski (1966)]:

$$F(\pm Z, w_e)$$
$$= 2 \left(1 + \gamma_0\right) \left(2 p_e R\right)^{-2(1-\gamma)} \mid \Gamma\left(\gamma_0 + i\eta\right) \mid^2 / \left(\Gamma\left(2\gamma_0 + 1\right)\right)^2, \quad (8.15)$$

where $\gamma_0 \equiv [1 - (\alpha Z)^2]^{1/2}$ and $\eta \equiv \pm \alpha Z c / v$. The positive sign applies for electron or neutrino capture, as well as electron emission, and the negative sign for positron and neutrino emission. The complex gamma function can be evaluated conveniently by using a numerical approximation due to Lanczos (1964). The nuclear radius R can be represented well by the following expression [Elton (1961)]

$$R = \left[2.908 A^{1/3} + 6.091 A^{-1/3} - 5.361 A^{-1}\right] \times 10^{-3} \left(\hbar/m_e c\right). \quad (8.16)$$

Bahcall (1978) averages the Fermi functions $F(Z, w_e)$ of Eq. (8.15) over a uniform sphere of radius R, which results in a small correction,

$$F(Z, w_e)_{av} = [1 - (2/3)(1 - \gamma_0)]^{-1} F(Z, w_e), \quad (8.17)$$

which reflects the fact that the electron capture can occur anywhere within the nuclear volume.

The prescription described above provides a good approximation to the exact Fermi function. An even more accurate (but much more time consuming) procedure is to solve numerically the Schrödinger equation with a Coulomb potential representing a finite size nucleus. The appropriate numerical methods for obtaining the solutions have been described by Rose (1961) and by Bhalla and Rose (1962). Extensive tables of numerical solutions have been presented by Behrens and Janecke (1969). Comparing Bahcall's results for Fermi functions to theirs yields a correction factor, $(1 + C)$, by which the Fermi functions of Eq. (8.17) should be multiplied in order to bring them into exact agreement with those of Behrens and Janecke (1969). For energies less than or of order 2 MeV (i.e., for all solar neutrino sources except ^{8}B and hep neutrinos)

$$C \approx 0.01 \, \exp\left[2.054 \ln Z - 5.757\right]. \qquad (8.18)$$

The above expression is accurate to a few tenths of a percent from $Z = 4$ to $Z = 80$. For $Z = 18$ (^{37}Cl $\rightarrow$ ^{37}Ar), C is about 1% and for $Z = 32$ (^{71}Ga $\rightarrow$ ^{71}Ga), C is about 4%. The net correction for ^{8}B and hep neutrinos is of the order of a percent for all the targets considered in this book and is small compared to the other uncertainties involved in calculating the cross sections. The quantity C may be set equal to zero for ^{8}B and hep neutrinos.

The correction due to electron screening in terrestrial atoms was made in the same way as in Bahcall (1966a), using the expression for the screening induced potential shift derived by Durand (1964). One can easily show, with the formalism developed by Rose (1936), that the effect of screening is small for the positron continuum wave functions in the solar reactions that produce neutrinos, Eqs. (6.1) to (6.8). The correction to the continuum wave functions can be shown to be of order $V_0/(W - m_e c^2)_{\text{av}}$, where V_0 is the value of the positron potential energy at the nucleus. Numerically,

$$V_0 \sim 90 Z_{\text{nucleus}} \left[64 T_6/p \left(3 + X\right)\right]^{-1/2} \text{ eV} \qquad (8.19)$$

is the value of the Debye–Hückel potential at the nucleus; T_6 is the temperature in units of 10^6 K; X is the mass fraction of hydrogen; and p is the density (in g cm^{-3}). In the solar neutrino emission processes, Eqs. (6.1) to (6.8), the screening correction $V_0/(W - m_e c^2)$ is negligible.

$ft_{1/2}$-*values.*   In the allowed approximation, the $ft_{1/2}$-value can be expressed in terms of the reduced matrix elements by the relation:

$$(1/ft_{1/2}) = K^{-1}[G_V^2 \langle 1 \rangle^2 + G_A^2 \langle \sigma \rangle^2]. \qquad (8.20a)$$

The numerical coefficient is[†] [Wilkinson (1982)]:

$$K^{-1} G_V^2 = (1/6166 \, \text{s}). \qquad (8.20b)$$

[†]The vector coupling constant, G_V, occurs in the description of weak interaction processes involving quark transitions. The relation between G_V and G_F, the Fermi coupling constant for purely weak interactions (see Section 8.2) is $G_V = \cos\theta_c G_F$, where the Cabibbo angle $\theta_c = 13°$.

The above relation can be combined with Eqs. (8.5) and (8.10) to calculate the absorption cross section between isotopic analogue states, provided one can estimate separately $\langle\sigma\rangle^2$. For example, for the transition between the spin zero, isotopic spin two, analogue states of ^{40}Ar and ^{40}K, $\langle 1\rangle^2 = 4$ and $\langle\sigma\rangle^2 = 0$.

If the inverse process to Eq. (8.1), the electron capture process $e^- + {}^Z A \to {}^{Z-1}A + \nu$, can be studied in the laboratory, then the $ft_{1/2}$-value can be determined simply and accurately. This situation applies for example to all the transitions contributing to the ^{37}Cl and ^{71}Ga experiments, except the excited state transitions produced by ^{8}B and hep neutrinos. The value of $ft_{1/2}$ can be calculated directly from laboratory data and tabulated atomic wave functions when the inverse electron capture process has been studied experimentally. The dimensionless phase space factor for allowed electron capture is [Bahcall (1966a)]

$$
\begin{aligned}
&f_{ec} \\
&= 2\pi^2 \left(q_{1s}/m_e c^2\right)^2 \left[1 + \left(q_{2s}^2 L/q_{1s}^2 K\right)\right] \mid \psi_{1s}(R)\mid^2 (\hbar/m_e c)^3, \quad (8.21)
\end{aligned}
$$

where q_{1s} is the neutrino energy when a 1s (i.e., K) electron is captured, L/K is the L to K capture ratio, and $\mid \psi_{1s}(R)\mid^2$ is the value of the square of the modulus of a 1s electron's wave function at the nucleus.

The quantity q_{1s} (or q_{2s}) is to be interpreted [Bahcall (1963), Eq. (70)] as the difference of atomic masses minus the (positive) binding energy of the 1s (2s) electron in the final atomic state that results from electron capture. The K-shell binding energy is important for large Z and small decay energies, q. Bearden and Burr (1967) give a useful table of electron binding energies.

The form in which Eq. (8.21) has been written makes use of the fact that atomic overlap and exchange effects are small for *total* electron capture rates although these effects are easily measurable for capture ratios. There has occasionally been some confusion about this point in the literature because the cancellations that cause the overlap and exchange effects to be small for total decay rates are not obvious with the necessarily imprecise theoretical results (often numerical extrapolations) that are used for many different values of Z. One can show, using Eqs. (71), (80), and (87) of Bahcall (1963), that overlap and exchange effects amount to less than a percent cor-

Table 8.1.   Values of σ_0.

Detector	σ_0 (10^{-46} cm^2)
^{7}Li	22.75
^{7}Li	13.30
^{37}Cl	1.725
^{71}Ga	8.801

rection to the total capture rates for all the cases discussed in this book.

Table 8.1 gives some of the important cross section factors, σ_0, for cases in which the neutrino reaction is the inverse of an electron capture reaction that is observed in the laboratory. The half-lives, $t_{1/2}$, were taken from recent laboratory measurements [see the references in Bahcall (1978) and Bahcall and Ulrich (1988)]. For comparison, the value of σ_0 for the transition between analogue states of ^{40}Ar and ^{40}K is $\sigma_0 = 148.4 \times 10^{-46}$ cm^2, a value which can be computed directly from Eqs. (8.5), (8.7), (8.8), and (8.10). Similarly, the value of σ_0 for the superallowed transition between ^{19}F and ^{19}Ne is $\sigma_0 = 69.9 \times 10^{-46}$ cm^2. The phase space factors f were computed from Eq. (8.21) using self-consistent field Dirac–Hartree wave functions that include relativistic and nuclear size effects as well as electron exchange [Martin and Blickart-Toft (1970), Suslov (1968), and Behrens and Janecke (1969)]. Values of $|\psi_{1s}|^2$ from the three sources listed above agree with each other to within an accuracy of about 1%.

The relativistic and nuclear size effects increase the calculated f_{ec} over their nonrelativistic values by, for example, a factor of 1.13 for the ^{37}Ar decay, and by a factor of 1.4 for the ^{71}Ge decay (the effects are only $\sim 1\%$ for ^{7}Be). The size of these corrections can be estimated by comparing the moduli obtained from published nonrelativistic atomic wave functions with the moduli obtained with wave functions calculated by including relativistic effects. Alternatively, one can estimate the corrections (to an accuracy $\sim 1\%$) by using an approximate formula derived by Racah (1932) [see also Shirley (1964)].

For continuum positron or electron emission, the dimensionless **phase space factor** is:

$$f_{\text{cont.}} = \int_1^{w_{\max}} dw_e \, w_e \, p_e \, F\left(Z, w_e\right) q_\nu^2. \qquad (8.22)$$

The continuum phase space factors, $f_{\text{cont.}}$, can be calculated using many of the subroutines that were developed for evaluating neutrino absorption cross sections. Some numerical values are given in Bahcall (1966a).

Transitions between the ground state of the target nucleus and the ground state of the daughter nucleus dominate the capture rate for all of the solar neutrino sources except ^{8}B and the extremely rare hep neutrinos. However, the ^{8}B and hep neutrinos have sufficient energy to cause transitions to many excited states in the daughter nucleus. The ^{8}B neutrinos in particular induce transitions to excited states that contribute, assuming the standard solar model is correct, the largest amount of any neutrino source to the predicted capture rate for a number of the most important detectors, including ^{2}H, ^{37}Cl, ^{40}Ar, and ^{98}Mo, and are calculated to produce a significant fraction of the capture rate in the ^{71}Ga detector. In addition, neutrinos from electron capture on ^{7}Be can populate in a significant amount the excited states in experiments involving ^{71}Ga, ^{81}Br, ^{115}In, and ^{205}Tl. Since the strength of Gamow–Teller transitions is concentrated at relatively high excitation energies (above a few MeV), phenomenological estimates of the capture rates to excited states that are based upon β-decays between ground states of neighboring nuclei are not very useful.

In some cases, the dominant transitions to excited states are between **isotopic analogue states**, that is, the ground state of the detector and the analogue excited state of the daughter nucleus. This fortunate situation exists for the ^{37}Cl and ^{40}Ar detectors. For these two targets, the capture rate for ^{8}B neutrinos can be estimated accurately using the theoretical matrix elements for the Fermi transition [Eq. (8.5)] which dominate the transition probabilities. For ^{37}Cl, the capture cross section calculated in this way [Bahcall (1964a, 1978)] has been verified by experimental studies [Poskanzer *et al.* (1966) and Sextro, Gough, and Cerny (1974)] of the β-decay of the nucleus ^{37}Ca which is the isotopic analogue of ^{37}Cl and by (p,n) measurements [Rapaport *et al.* (1981)].

(p,n) experiments. Many of the nuclear matrix elements for excited
state transitions have been estimated with the aid of (p,n) reactions
at moderate proton beam energies, of order 100 MeV to 250 MeV [see
Goodman (1985), Krofcheck *et al.* (1985), Rapaport *et al.* (1985),
Krofcheck *et al.* (1987), and Taddeucci *et al.* (1987)]. There is a
significant body of experimental evidence [see, e.g., Taddeucci *et
al.* (1982, 1987)] that the relative (p,n) cross sections for different
transitions between the same pair of nuclei are approximately equal
to the ratio of the squares of the weak interaction matrix elements.
Thus one can normalize the absolute (p,n) cross sections to yield
Gamow–Teller matrix elements by using either the known strength
of the isotopic analogue transition [Eq. (8.5)] or the strength of a
Gamow–Teller matrix element determined by nuclear β-decay. This
technique has been used to estimate the cross sections for excited
state transitions in ^{37}Cl [Rapaport *et al.* (1981)], ^{71}Ga [Krofcheck
et al. (1985)], ^{98}Mo [Rapaport *et al.* (1985)], and ^{115}In [Rapaport
et al. (1985)] although the statistical accuracy of the data – and
the interpretation of the solar neutrino experiments – could be much
improved by refined (p,n) measurements. The uncertainties involved
in using this method are described in Section 7.3B.

Most of the results to date have been obtained by a group that
has worked primarily at the Indiana University Cyclotron Facility
(IUCF). The IUCF team normally presents their results in terms of
the reduced matrix elements $B(\mathrm{GT}) \equiv \langle\sigma\rangle^2$ and $B(F) \equiv \langle 1\rangle^2$. Their
calibrations have been made using the relation

$$B(F) + (1.25)^2 B(\mathrm{GT}) = 6163.4 \ \mathrm{s}/ft_{1/2}, \qquad (8.23)$$

where the quantity $(1.25)^2$ represents the earlier estimate $(G_{\mathrm{A}}/G_{\mathrm{V}})^2$.
In order to remove as much as possible of the dependence on
$(G_{\mathrm{A}}/G_{\mathrm{V}})$ and other uncertain quantities, Bahcall has calculated neutrino absorption cross sections for the IUCF group using only their
inferred *ratios* of $B(\mathrm{GT})$ values. In practice, Bahcall fixed whenever possible the normalizing value of the $B(\mathrm{GT})$ matrix elements
with the aid of independent β-decay measurements, for example, the
measured $ft_{1/2}$-value for the ground state to ground state decay of
^{71}Ge. In some cases, the IUCF has normalized their absolute values to the strength of the (presumably) well-known Fermi transition,
which usually stands out clearly in the experimental results referring
to relatively high excitation energy in the product nucleus.

Table 8.2.   Neutrino absorption cross sections averaged over energy spectra. The units are 10^{-46} cm^2 for all sources except hep and ^{8}B, for which the unit is 10^{-42} cm^2. Contributions from excited states and from forbidden effects are included. Uncertainties are discussed in the text.

Target	pp	pep	hep†	^{7}Be	^{8}B†	^{13}N	^{15}O	^{17}F
^{7}Li	0.0	655	8.4	9.6	3.9	42.4	246	249
^{37}Cl	0.0	16	3.9	2.4	1.06	1.7	6.8	6.9
^{71}Ga	11.8	215	7.3	73.2	2.43	61.8	116	117
^{81}Br	0.0	75	9.0	18.3	2.7	14.5	36.7	37.0
^{98}Mo	0.0	0.0	10	0.0	3.0	0.0	0.0	0.0
^{115}In	78.0	576	6.1	248	2.5	224	355	356

†Unit is 10^{-42} cm^2.

Statistical estimates by Itoh and Kohyama (1981) have been made for the absorption cross sections of the energetic ^{8}B neutrinos for four targets that have also been studied by (p,n) measurements. On the average, the statistical and the (p,n) measurements (see Table 8.2) differ by a factor of two.

C   Absorption cross sections for solar neutrinos

This subsection presents the numerical values for the absorption cross sections of some of the most important solar neutrino detectors.

Table 8.2 presents the neutrino absorption cross sections for individual solar neutrino sources incident on targets of ^{7}Li, ^{37}Cl, ^{71}Ga, ^{81}Br, ^{98}Mo, and ^{115}In. The results of Table 8.2 apply if nothing happens to the neutrinos on their way to the Earth from the Sun, that is, the spectrum of electron neutrinos is unchanged from what it is in the interior of the Sun.

Table 8.3 gives the absorption cross sections for ^{8}B and hep neutrinos incident on ^{2}H and ^{40}Ar as a function of the minimum accepted kinetic energy of the recoil electron. The cross sections refer to transitions from the ground state of ^{40}Ar to the $T = 2$ isobaric analogue state at 4.38 MeV in ^{40}K.

For theoretical calculations of what is implied by different possible solutions of the solar neutrino problem, it is important to know the

Table 8.3.   Absorption cross sections for ^{8}B and hep neutrinos incident on ^{2}H and ^{40}Ar. The absorption cross sections are given for different values of the minimum accepted kinetic energy, T_{min}, of the recoil electrons. The deuterium values were computed using cross sections determined for individual energies by Nozawa (1987) [see Nozawa *et al.* (1986)], averaging over the appropriate neutrino spectra. The cross sections for argon refer to transitions from the ground state of ^{40}Ar to the $T = 2$ isotopic analog state at 4.38 MeV excitation in ^{40}K. The unit for T_{min} is MeV and for cross sections is 10^{-46} cm^2.

T_{min}	$\sigma(^8\text{B})$		$\sigma(\text{hep})$	
	^{2}H	^{40}Ar	^{2}H	^{40}Ar
0.0	1.17E+04	7.85E+03	3.03E+04	3.07E+04
1.0	1.17E+04	7.70E+03	3.03E+04	3.06E+04
2.0	1.17E+04	7.16E+03	3.03E+04	3.03E+04
3.0	1.15E+04	6.11E+03	3.02E+04	2.94E+04
4.0	1.11E+04	4.60E+03	3.01E+04	2.78E+04
4.5	1.08E+04	3.76E+03	2.99E+04	2.67E+04
5.0	1.04E+04	2.92E+03	2.96E+04	2.53E+04
5.5	9.83E+03	2.12E+03	2.93E+04	2.37E+04
6.0	9.17E+03	1.42E+03	2.89E+04	2.20E+04
6.5	8.40E+03	8.57E+03	2.83E+04	2.01E+04
7.0	7.53E+03	4.46E+02	2.76E+04	1.79E+04
7.5	6.59E+03	1.90E+02	2.68E+04	1.57E+04
8.0	5.60E+03	6.20E+01	2.58E+04	1.35E+04
8.5	4.59E+03	1.54E+01	2.46E+04	1.12E+04
9.0	3.62E+03	2.89E+00	2.33E+04	8.98E+03
9.5	2.71E+03	2.91E−01	2.18E+04	6.94E+03
10.0	1.90E+03	0.00E+00	2.02E+04	5.09E+03
11.0	7.14E+02	0.00E+00	1.66E+04	2.09E+03
12.0	1.45E+02	0.00E+00	1.27E+04	4.23E+02
12.5	4.50E+01	0.00E+00	1.07E+04	8.46E+01
13.0	1.07E+01	0.00E+00	8.81E+03	0.00E+00
13.5	1.95E+00	0.00E+00	6.99E+03	0.00E+00
14.0	2.28E−01	0.00E+00	5.31E+03	0.00E+00
14.5	1.10E−02	0.00E+00	3.81E+03	0.00E+00
15.0	0.00E−00	0.00E+00	2.54E+03	0.00E+00

Table 8.4. Absorption cross sections for specific energies. The unit for neutrino energy, q, is MeV and for neutrino cross sections is 10^{-46} cm^2. Contributions from excited states and from forbidden effects are included.

q	$\sigma(^2\mathrm{H})$	$\sigma(^{37}\mathrm{Cl})$	$\sigma(^{40}\mathrm{Ar})$	$\sigma(^{71}\mathrm{Ga})$	$\sigma(^{81}\mathrm{Br})$	$\sigma(^{98}\mathrm{Mo})$
0.130	0.00E+00	0.00E+00	0.00E+00	0.00E+00	0.00E+00	0.00E+00
0.150	0.00E+00	0.00E+00	0.00E+00	0.00E+00	0.00E+00	0.00E+00
0.200	0.00E+00	0.00E+00	0.00E+00	0.00E+00	0.00E+00	0.00E+00
0.250	0.00E+00	0.00E+00	0.00E+00	1.39E+01	0.00E+00	0.00E+00
0.275	0.00E+00	0.00E+00	0.00E+00	1.53E+01	0.00E+00	0.00E+00
0.300	0.00E+00	0.00E+00	0.00E+00	1.70E+01	0.00E+00	0.00E+00
0.325	0.00E+00	0.00E+00	0.00E+00	1.88E+01	0.00E+00	0.00E+00
0.350	0.00E+00	0.00E+00	0.00E+00	2.06E+01	0.00E+00	0.00E+00
0.375	0.00E+00	0.00E+00	0.00E+00	2.26E+01	0.00E+00	0.00E+00
0.400	0.00E+00	0.00E+00	0.00E+00	2.46E+01	0.00E+00	0.00E+00
0.425	0.00E+00	0.00E+00	0.00E+00	2.71E+01	0.00E+00	0.00E+00
1.000	0.00E+00	5.21E+00	0.00E+00	1.01E+02	2.90E+01	0.00E+00
2.000	3.30E+01	3.70E+01	0.00E+00	4.06E+02	1.78E+02	1.16E+02
3.000	4.37E+02	1.15E+02	0.00E+00	1.01E+03	5.20E+02	9.64E+02
4.000	1.46E+03	2.63E+02	0.00E+00	2.18E+03	1.31E+03	2.65E+03
5.000	3.24E+03	5.63E+02	0.00E+00	4.48E+03	3.18E+03	5.67E+03
6.000	5.87E+03	1.52E+03	3.20E+02	8.62E+03	7.04E+03	1.04E+04
7.000	9.41E+03	4.76E+03	2.73E+03	1.56E+04	1.43E+04	1.77E+04
8.000	1.39E+04	1.02E+04	7.09E+03	2.63E+04	2.64E+04	2.91E+04
9.000	1.94E+04	1.79E+04	1.35E+04	4.06E+04	4.41E+04	4.67E+04
10.00	2.59E+04	2.77E+04	2.18E+04	5.84E+04	6.70E+04	7.16E+04
11.00	3.35E+04	3.97E+04	3.21E+04	7.96E+04	9.51E+04	1.03E+05
12.00	4.22E+04	5.38E+04	4.42E+04	1.04E+05	1.28E+05	1.42E+05
13.00	5.21E+04	7.00E+04	5.82E+04	1.32E+05	1.66E+05	1.86E+05
14.00	6.31E+04	8.83E+04	7.40E+04	1.62E+05	2.08E+05	2.36E+05
15.00	7.53E+04	1.09E+05	9.15E+04	1.95E+05	2.54E+05	2.92E+05
16.00	8.87E+04	1.31E+05	1.11E+05	2.32E+05	3.04E+05	3.52E+05
18.00	1.18E+05	1.81E+05	1.54E+05	3.11E+05	4.14E+05	4.84E+05
20.00	1.5 E+05	2.38E+05	2.04E+05	3.99E+05	5.36E+05	6.30E+05
30.00	....	6.11E+05	5.25E+05	8.99E+05	1.21E+06	1.39E+06

individual cross sections as a function of neutrino energy. Different
solutions imply different characteristic modifications of the neutrino
energy spectra.

Table 8.4 presents the absorption cross sections for individual neutrino energies that are relevant for solar neutrino experiments or supernova detections. This table covers the following targets for which
experiments are under way: ^{2}H, ^{37}Cl, ^{40}Ar, ^{71}Ga, ^{81}Br, and ^{98}Mo.
Contributions from excited states and from forbidden effects have
been included. For ^{40}Ar, the cross sections refer to transitions to
the 4.38 MeV isobaric analogue state of ^{40}K. The individual cross
sections given in Table 8.4 are sufficient to reproduce the spectrum
averaged cross sections (given in Tables 8.2 and 8.3) to an accuracy
of order 10%. The results in Table 8.4 are useful for evaluating the
effect of energy-dependent weak interaction processes, such as the
MSW effect, on the event rates in different detectors and in determining the potentialities of the detectors for observing supernovae.
For completeness, $\sigma(^{37}\text{Cl}) = 2 \times 10^{-40}$ cm^2 at a neutrino energy of
50 MeV.

The ^{71}Ga detector is a special case. Somewhat paradoxically, a
majority of the capture rate expected on the basis of the standard
model is from low-energy neutrinos (especially pp neutrinos), while
most of the uncertainty is caused by transitions to excited states.
Therefore the cross sections for transitions between the ground state
of ^{71}Ga and the ground state of ^{71}Ge can be useful in estimating the
capture rates and uncertainties for hypotheses in which the incident
neutrino spectrum is different from the spectrum predicted by the
standard solar model. For the pp, pep, hep, ^{7}Be, ^{8}B, ^{13}N, ^{15}O,
and ^{17}F neutrino sources, respectively, the ground state to ground
state absorption cross sections are (in units of 10^{-46} cm^2): 11.8, 170,
5.44×10^3, 69.1, 2.85×10^3, 58.1, 100.1, and 100.5. Ground state
to ground state transitions are dominant for all except the ^{8}B and
hep neutrino sources.

D   Uncertainties for individual detectors

The uncertainties for the cross sections must be evaluated separately
for each neutrino source and for each target. A detailed discussion
is given in Section IV.B of Bahcall and Ulrich (1988) and in Section 7.3B of this book. This subsection summarizes the final esti-

mated uncertainties of Bahcall and Ulrich, which represent the "total theoretical range" that is defined in Chapter 7.

For the ^{2}H detector, Bahcall and Ulrich assigned an uncertainty of 10%, based upon the comparison of results obtained with different theoretical models.

For the ^{7}Li detector, the nuclear matrix elements are extracted from the superallowed β-decay of ^{7}Be, which is well studied in the laboratory. We estimate the following percentage uncertainties for the different sources: pep (6%), hep (21%), ^{7}Be (5%), ^{8}B (17%), ^{13}N (5.5%), ^{15}O (6%), and ^{17}F (6%).

For the ^{37}Cl detector, only ^{8}B and hep neutrinos have enough energy to populate excited states. All the other sources have cross sections to which Bahcall and Ulrich assigned uncertainties (as defined above) of 6%. For ^{8}B and hep neutrinos, the cross section is uncertain by about 10% [Bahcall and Holstein (1986) and Adelberger and Haxton (1987)].

For the ^{40}Ar detector, forbidden corrections constitute the dominant uncertainty since the superallowed transition to the isobaric analogue state of ^{40}K can be calculated with relatively high precision. For a minimum electron kinetic energy $T_{min} = 0$, 5, or 7 MeV, the 1σ uncertainty for ^{8}B neutrino absorption is 2%, 2.5%, or 3%, respectively, implying an effective 3σ uncertainty $\sim 8\%$ for $T_{min} = 5$ MeV.

For the ^{71}Ga detector, the uncertainties in the cross sections are dominated by transitions to excited states for all but the pp, ^{7}Be, and the ^{13}N sources. For pp neutrinos, the uncertainties are determined almost entirely by the forbidden corrections, whereas for ^{7}Be and ^{13}N neutrinos the estimated uncertainties are comparable from forbidden corrections and excited states. Bahcall and Ulrich (1988) adopt the following upper limit uncertainties in absorption cross sections (uncertainty in parentheses): pp (7%), pep (22%), ^{7}Be (9%), ^{13}N (9%), ^{15}O (19%), and ^{17}F (19%). The corresponding lower limit uncertainties are: 7%, 13%, 8%, 8%, 11%, and 11%. The ^{8}B and hep absorption cross sections are determined almost entirely from the (p,n) measurements. The upper and lower uncertainties for absorption of ^{8}B neutrinos are 88% and 45%, respectively, and for hep neutrinos, 93% and 47%. (About 12% of the total cross section for the ^{8}B neutrinos is due to the ground state to ground state transition; the corresponding number for hep neutrinos is 7.5%.)

All of the allowed transitions for the ^{81}Br detector are to excited states of ^{81}Kr. The total adopted uncertainties for the upper limit are, in the same order of sources, 46%, 33%, 32%, 42%, and 42% [see the discussion of a ^{81}Br detector in Section IV.C of Bahcall and Ulrich (1988)]; the uncertainties for the lower limit are 27%, 26%, 26%, 27%, and 27%. The upper and lower uncertainties for absorption of ^{8}B neutrinos are 97% and 48%, respectively, and for hep neutrinos, 97% and 49%.

For ^{98}Mo, only ^{8}B and hep neutrinos are energetic enough to cause allowed transitions to ^{98}Tc. The experimental 1σ uncertainties in the (p,n) measurements [Rapaport *et al.* (1985)] are by themselves ±30%, so that the total theoretical range in the predicted capture rate for this detector due to the neutrino absorption cross sections is at least a factor of two. For definiteness, a factor of two uncertainty was adopted.

For the ^{115}In detector, all of the relevant transitions are to excited states [see Bahcall (1978) for a discussion of the difficulties of calculating the capture rates for this nucleus]. The best estimate furnished by the (p,n) experiments [Rapaport *et al.* (1985)] for the matrix element to the lowest excited state at 0.61 MeV in ^{115}Sn is in remarkably good agreement with an insightful estimate made by Raghavan (1976) in his original paper on this subject. However, the available (p,n) measurements for all of the relevant transitions have large measuring errors, 1σ uncertainties of order 30%. In addition, the calibration is uncertain for the usual reasons that are discussed in the previous subsection. Thus the total theoretical range in the predicted capture rate for this detector due to the neutrino cross sections is also about a factor of two.

E   Muon decay: neutrinos and antineutrinos

The electron neutrinos (ν_e) from the decay of μ^+'s that are stopped in the beam dump of a meson factory, such as the Los Alamos Meson Factory, can be used in principle to calibrate the overall performance of solar neutrino detectors. Antineutrinos ($\bar{\nu}_e$) from μ^- decay have large cross sections in water solar neutrino detectors (such as the Kamiokande II detector discussed later in Section 13.2) and therefore provide a sensitive indication of the background from muons. The cross sections for neutrinos from muon decay are also important in evaluating the possible detection in solar neutrino experiments of solar flares or other processes involving energetic particle collisions that produce mesons (see Section 10.5).

The total cross sections for neutrinos from stopped muon decay were calculated by Bahcall and Ulrich (1988) for ^{37}Cl using the relatively well-known β-decay matrix elements for this detector and, for ^{71}Ga, using the $B(GT)$ values determined from (p,n) measurements [Krofcheck et al. (1985) and Krofcheck (1987)]. The matrix elements that are most important for absorptions by neutrinos with a stopped muon decay spectrum are the same ones that are dominant for the absorption of ^{8}B neutrinos. The average $B(GT)$ values increase up to the neutrino separation energy. For both the ^{37}Cl and the ^{71}Ga detectors the largest contributions come (for muon and ^{8}B neutrinos) from transitions to excited states at several MeV (typically ~ 5 MeV) above the ground state.

Two calculations of the absorption cross sections were made for each detector: (1) including corrections for forbidden terms according to a plausible prescription [Bahcall and Holstein (1986)]; and (2) ignoring all forbidden corrections. The two calculations show, for both detectors, that the uncertainties are large because of terms that cannot be reliably evaluated. For simplicity, contributions of the opposite sign but the same order of magnitude were omitted; the omitted terms arise from first forbidden corrections to allowed neutrino capture.

For the ^{37}Cl detector,[†]

$$\sigma\left(^{37}\text{Cl}\right)_{\text{incl. forbid.}} = 0.72 \times 10^{-40} \text{ cm}^2, \qquad (8.24)$$

when forbidden corrections are included and

$$\sigma\left(^{37}\text{Cl}\right)_{\text{no forbid.}} = 0.76 \times 10^{-40} \text{ cm}^2, \qquad (8.25)$$

when forbidden corrections are ignored. Similarly, for ^{71}Ga

$$\sigma\left(^{71}\text{Ga}\right)_{\text{incl. forbid.}} = 0.89 \times 10^{-40} \text{ cm}^2, \qquad (8.26)$$

including forbidden corrections and

$$\sigma\left(^{71}\text{Ga}\right)_{\text{no forbid.}} = 1.4 \times 10^{-40} \text{ cm}^2, \qquad (8.27)$$

[†]The numerical values given here differ from those in Bahcall and Ulrich (1988); Fukugita (1988, private communication) pointed out that Bahcall and Ulrich used an incorrect energy spectrum for ν_e from μ^+ decay (see Section 8.2F).

without forbidden corrections. The difference between cross sections calculated with and without forbidden corrections is much larger for neutrinos from μ^+ decay than for ^{8}B neutrinos. The reason is that some of the most important forbidden terms scale as (neutrino energy × nuclear radius)^2. The characteristic neutrino energy for muon decay is about a factor of five larger than for ^{8}B decay.

The calculated forbidden terms include corrections to allowed transitions measured either by the decay of ^{37}Ca (for the case of ^{37}Cl) or by (p,n) measurements (for the ^{71}Ga detector). However, other correction terms that are potentially of the same order are not included. The terms that are omitted are the so-called first forbidden transitions in nuclear β-decay, which would not show up in either the β-decay or the (p,n) measurements; there is no reliable way known to estimate their nuclear matrix elements. Moreover, the prescription [Bahcall and Holstein (1986)] that was adopted for evaluating the forbidden terms provides only a plausible estimate, not a rigorous determination.

The difference between the two ways of calculating the absorption cross sections, with and without forbidden corrections, provides a measure of the theoretical uncertainty. The neutrinos from muon decay can be used to calibrate the overall performance of the solar neutrino detectors to better than a factor of two, but not to high precision.

The cross section for $\bar{\nu}_e$ absorption by protons to produce positrons is

$$\sigma(\bar{\nu}_e + p) = 0.95 \times 10^{-40} \text{ cm}^2 \tag{8.28}$$

for the neutrino spectrum from μ^-'s decaying at rest.

8.2   Neutrino–electron scattering cross sections

Neutrino–electron scattering has four important diagnostic features: (1) the electrons are primarily scattered forward in the direction connecting the Earth and the Sun [Bahcall (1964b)]; (2) the shape of the recoil energy spectrum reflects the solar neutrino spectrum [Reines and Kropp (1964) and Bahcall (1964b)]; (3) the magnitude of the cross section is sensitive to neutrino flavor ['t Hooft (1971)]; and (4) individual events can be counted electronically giving high time resolution. The *angular distribution* (which is strongly peaked in the forward direction of the Earth–Sun axis) allows one to determine experimentally whether the electrons come from the Sun. The measured recoil *energy spectrum* can be used to test for consistency with the theoretical solar neutrino energy spectrum. One can infer the incoming neutrino *flavor* by exploiting the sensitivity of the scattering cross sections to neutrino type. By comparing the results of scattering experiments with absorption rates (involving charge changing interactions only) measured in the ^{37}Cl and other radiochemical experiments, one can determine if neutrino oscillations (resonant or nonresonant) have occurred. The *time resolution* can distinguish between the steady state fluxes expected from the solar core and time-dependent events on the solar surface or elsewhere [see the discussion in Davis (1986) and in Section 10.4]. There may also be a diurnal modulation of the solar neutrino flux caused by resonance scattering in the Earth [see Section 9.2].

Background effects are often the most difficult problem in a solar neutrino experiment, but backgrounds usually decrease as the energy of the detected electron increases. Therefore, most of the discussion of neutrino–electron scattering detectors in the literature concentrates on the neutrinos from ^{8}B decay, which have energies up to 15 MeV.

A number of the planned experiments are expected to have good energy resolution and some angular resolution for the recoil electrons. Therefore detailed spectra are given below for the energy and the angular distributions of the recoil electrons. In order to make the results directly applicable to the proposed solar neutrino experiments, the cross sections for the high-energy ^{8}B and ^{3}He+p neutrinos are presented as a function of the minimum energy, T_{min}, that the recoil electron must have to be counted.

The calculations described in this section assume, except where stated explicitly otherwise, that the spectrum of solar neutrino energies is unchanged as the electron neutrinos travel from the central regions of the Sun to the Earth. This assumption is valid, for most parameters of interest, even if the neutrinos undergo nonresonant oscillations in transit [Gribov and Pontecorvo (1969) and Bahcall and Frautschi (1969)]. For comparative purposes, this section also gives the results if *all* of the electron neutrinos are converted to muon neutrinos within the Sun or in transit from the Sun to the Earth.

The computed differences between the all-ν_e and the all-ν_μ results represent an upper limit to the influence of flavor-changing interactions, such as the MSW effect (see Section 9.2).

The theoretical predictions can be calculated in a similar way if resonant neutrino oscillations, the MSW effect [see Wolfenstein (1978), Mikheyev and Smirnov (1986a), and the discussion in Section 9.2], change some of the electron neutrinos to muon neutrinos. The theoretical expectations that apply if the MSW effect is operative have been described by Bahcall, Gelb, and Rosen (1987). The predicted results depend strongly upon the assumed threshold for the minimum electron recoil energy that is accepted. However, the *minimum* reduction that can be caused by the *complete* conversion of electron neutrinos to muon or tau neutrinos via the MSW effect is almost constant, a reduction factor of about 0.15 for a 5 MeV threshold and 0.14 for a 10 MeV threshold [see Table II of Bahcall (1987)].

Section 8.2A summarizes the basic equations that are used in the numerical calculations; this section can be skipped by readers who are primarily interested in the applications. Section 8.4B gives in Table 8.5 the total neutrino–electron scattering cross sections for all of the important neutrino sources and, for the higher-energy ^{8}B and ^{3}He+p neutrinos, Tables 8.6 and 8.7 tabulate the total cross sections as a function of the minimum accepted kinetic energy of the recoil electrons. The scattering cross sections for specific neutrino energies are given in Table 8.8 and discussed in Section 8.4C. These results are convenient for estimating the expected event rates in different detectors. Because of their experimental importance, Section 8.4D provides the calculated recoil energy spectra that are produced by scattering of electrons by pp, ^{8}B, or ^{3}He+p neutrinos (Tables 8.9 to 8.13), assumed to be either pure ν_e or pure ν_μ. The shapes of the recoil spectra are best appreciated when presented in graphical form. The calculated recoil spectra from the pp, pep, ^{7}Be, ^{8}B, and ^{3}He+p neutrinos are illustrated in Figures 8.2 to 8.6. Section 8.4E describes the expected angular distributions (see especially Table 8.14).

Section 8.2F summarizes the main results and conclusions.

A   Basic relations

The differential cross section for producing a recoil electron with kinetic energy T by scattering with a neutrino of initial energy q is ['t Hooft (1971)]:

$$\frac{d\sigma}{dT} = \sigma_e \left[g_L^2 + g_R^2 (1 - T/q)^2 - g_L g_R (T/q^2) \right] , \qquad (8.29)$$

where

$$g_L = (\pm 1/2 + \sin^2 \Theta_W) , \quad g_R = \sin^2 \Theta_W . \qquad (8.30)$$

The upper sign applies for ν_e–e^- scattering and the lower sign for ν_μ–e^- scattering. Except where indicated otherwise, the unit of energy is the electron's mass, $m_e c^2$, and $\sin^2 \theta_W = 0.23$ [Langacker (1986)]. The dimensional cross section factor is

$$\sigma_e = 2G_F^2 m_e^2 / \pi \hbar^4 = 88.083 \times 10^{-46} \, \mathrm{cm}^2 . \qquad (8.31)$$

The effects of atomic binding of the electrons and of radiative and electroweak corrections are neglected in Eq. (8.29). One can include atomic effects by Fourier analyzing the wave function of the initial bound electron. The calculated corrections [Bahcall (1964b)] are small, of order $(v/c)_{\mathrm{bound}}^2 \sim (\alpha Z_{\mathrm{screened}}/n)^2$, where Z_{screened} is the effective charge of the nucleus seen by the electron and n is the principal quantum number of a hydrogeniclike orbit. The radiative and electroweak corrections are also small, of order a percent in the total cross section for MeV neutrinos [see Sarantakos, Sirlin, and Marciano (1983)].

The total scattering cross section can be computed by integrating Eq. (8.29) analytically between a chosen value of $T_{\min}$ (typically several MeV in most experiments) and the maximum value, $T_{\max}$, of the electron kinetic energy allowed by energy and momentum conservation:

$$T_{\max}(q) = 2q^2 / (1 + 2q) . \qquad (8.32)$$

The result is

$$\sigma_{\mathrm{total},q}(T_{\min}) = \sigma_e \Big[(g_L^2 + g_R^2)(T_{\max} - T_{\min})$$
$$- \left(\frac{g_R^2}{q} + \frac{g_L g_R}{2q^2} \right) (T_{\max}^2 - T_{\min}^2)$$
$$+ \frac{g_R^2}{3q^2} (T_{\max}^3 - T_{\min}^3) \Big]. \qquad (8.33)$$

The final integration over the spectrum of incoming neutrino energies, q, must be performed numerically. Let $\lambda(q)$ be the *normalized* neutrino spectrum incident at Earth. Then

$$\sigma_{\text{total}}(T_{\min}) = \int_0^{q_{\max}} dq \, \lambda(q) \sigma_{\text{total},q}(T_{\min}) . \tag{8.34}$$

Experimentally, the most interesting quantity is often the differential scattering cross section as a function of the recoil energy of the electron. The spectrum averaged differential cross section is

$$\left\langle \frac{d\sigma}{dT} \right\rangle_T \equiv \int_{q_{\min}}^{q_{\max}} dq \, \lambda(q) \frac{d\sigma}{dT} , \tag{8.35}$$

where the minimum neutrino energy that can cause an electron recoil of kinetic energy T is

$$q_{\min}(T) = \left\{ T + [T(T+2)]^{1/2} \right\} / 2 . \tag{8.36}$$

The energy $q_{\min}$ is appropriate when the electron is scattered in the forward direction of the incoming neutrino. As a check on the numerical calculations, the total cross section can also be calculated by integrating $\langle d\sigma/dT \rangle_T$ between $T_{\min}$ and $T_{\max}$.

The angular distribution of the recoil electrons contains important information about the direction of the incoming neutrinos. For solar neutrinos, the angular distribution depends only on the cosine, μ, of the angle between the (instantaneous) Earth–Sun axis and the momentum of the recoil electron. Since the interaction involves two particles, one can specify either the electron's scattering angle (μ) or kinetic energy (T) for a given neutrino energy (q) or, conversely, one can determine the neutrino energy from the recoil kinetic energy and scattering angle. The relations connecting these quantities are:

$$\mu^2 = \frac{T(1+q)^2}{(T+2)q^2} , \tag{8.37}$$

$$T = \frac{2q^2 \mu^2}{\left[(1+q)^2 - q^2 \mu^2\right]} , \tag{8.38}$$

and

$$q(\mu) = \frac{T + |\mu|\sqrt{T(T+2)}}{[(T+2)\mu^2 - T]} . \tag{8.39}$$

The maximum kinetic energy $T_{\max}$ is achieved at $\mu = +1$ [see Eqs. (8.32) and (8.38)]. The minimum value of the cosine of the scattering angle, $\mu_{\min}$ is given by Eq. (8.37) with $T = T_{\min}$. The angular distribution for a fixed neutrino energy is

$$\frac{d\sigma}{d\mu} = \frac{4\left(1 + q\right)^2 q^2 \mu}{\left[\left(1 + q\right)^2 - \mu^2 q^2\right]^2} \left(\frac{d\sigma}{dT}\right)_{T(\mu)}, \qquad (8.40)$$

where for a given μ and q the kinetic energy T is computed from Eq. (8.38) and the differential cross section for a fixed T is given by Eq. (8.29). The spectrum averaged angular distribution is

$$\left\langle \frac{d\sigma}{d\mu} \right\rangle_{T_{\min}} = \int_0^{q_{\max}} dq\, \lambda\left(q\right) \left(\frac{d\sigma}{d\mu}\right)_{T \geq T_{\min}}. \qquad (8.41)$$

As an additional numerical check, one can calculate the total cross section by integrating $\langle d\sigma/d\mu \rangle_{T_{\min}}$ over all directions.

B Total cross sections

The first question to ask about any solar neutrino experiment is: What is the expected rate? To answer this question, one needs the total interaction cross sections for all of the neutrino sources.

Table 8.5 gives the total cross sections for electron neutrino scattering by the solar neutrino sources listed in Chapter 6, as well as two possible calibration sources (^{51}Cr and ^{65}Zn). The neutrinos are assumed to be pure electron neutrinos (ν_e) or pure muon neutrinos (ν_μ) when they reach the target on Earth. (Presumably for the calibration sources only the ν_e cross sections are experimentally relevant.) The values given in the table were calculated for $\sin^2 \theta_W = 0.23$. A 1% change in the square of the sine of the Weinberg angle typically corresponds to $\sim 3\%$ change in the computed total cross sections [see Bahcall (1987)].

Experimentally, the most accessible source of solar neutrinos is from the decay of ^{8}B. How important is it to have a low-energy threshold for counting the recoil electrons? Table 8.6 answers this question, listing the total cross sections for ^{8}B solar neutrinos as a function of the minimum allowed kinetic energy of the recoil electron, $T_{\min}$. The fractional charges in the cross sections for a charge of 1% in $\sin^2 \theta_W$ are $\approx 3\%$ for ν_e–e scattering and $\approx -0.04\%$ to -0.07%

Table 8.5.   Total neutrino–electron scattering cross sections. The neutrinos are assumed to be pure ν_e or pure ν_μ when they reach the detector. The cross sections were calculated for $\sin^2 \Theta_W = 0.23$. The minimum allowed recoil kinetic energy is zero in all cases considered in this table; the maximum recoil energy is given in column 3. The neutrino energy, q, and the maximum electron recoil energy, T_{max}, are given in MeV; the neutrino cross sections, σ_{ν_e}–e and σ_{ν_μ}–e, are given in units of 10^{-46} cm^2.

Source	q	T_{max}	σ_{ν_e}–e	σ_{ν_μ}–e
pp	≤ 0.420	0.261	11.6	3.1
pep	1.442	1.225	112	21.7
hep	≤ 18.773	18.52	884	150
^{7}Be	0.862	0.665	59.3	12.6
^{7}Be	0.384	0.231	19.6	4.98
^{8}B	< 15.0	14.5	608	104
^{13}N	≤ 1.199	0.988	46.5	10.1
^{15}O	≤ 1.732	1.509	71.9	14.7
^{17}F	≤ 1.740	1.517	72.2	14.7
^{51}Cr	0.746	0.556	49.2	10.7
^{51}Cr	0.426	0.266	22.8	5.64
^{65}Zn	0.227	0.107	8.76	2.54
^{65}Zn	1.343	1.128	102.5	20.2
^{65}Zn	≤ 0.330	0.186	6.60	1.95

for ν_μ–e scattering (as T_{min} increases from 0 MeV to 14 MeV) [see Bahcall (1987)].

What is the influence of neutrino flavor on the scattering cross sections? The answer can be obtained by comparing the last two columns of Table 8.5 or columns 2 and 3 of Table 8.6 which give results that were calculated assuming that the neutrinos which reach the Earth are either pure electron neutrinos or pure muon neutrinos. For energy thresholds of interest, the ratio of ν_e–e to ν_μ–e scattering

Table 8.6. ^{8}B **neutrino scattering cross sections.** The scattering cross sections for ^{8}B solar neutrinos incident on electrons are given for different values of the minimum accepted kinetic energy, T_{min}. The neutrinos are assumed to be pure electron neutrinos (ν_e) or muon neutrinos (ν_μ) when they reach the Earth. The cross sections were calculated for $\sin^2\Theta_W = 0.23$. The minimum electron kinetic energy, T_{min}, is given in MeV, and the cross sections, σ_{ν_e}–e and σ_{ν_μ}–e, are expressed in units of 10^{-46} cm^2.

T_{min}	σ_{ν_e}–e	σ_{ν_μ}–e	T_{min}	σ_{ν_e}–e	σ_{ν_μ}–e
0.0	6.08E+02	1.04E+02	8.0	4.64E+01	6.76E+00
1.0	5.09E+02	8.39E+01	9.0	2.44E+01	3.53E+00
2.0	4.15E+02	6.63E+01	10.0	1.10E+01	1.58E+00
3.0	3.27E+02	5.10E+01	11.0	3.93E+00	5.64E−01
4.0	2.48E+02	3.79E+01	12.0	9.88E−01	1.41E−01
5.0	1.80E+02	2.71E+01	13.0	1.36E−01	1.94E−02
6.0	1.23E+02	1.83E+01	13.5	3.60E−02	5.13E−03
7.0	7.90E+01	1.16E+01	14.0	7.4 E−00	1.0 E−03

cross sections is approximately constant,

$$\left(\frac{\sigma_{\nu_e e}}{\sigma_{\nu_\mu e}}\right) \approx 6 \text{ to } 7. \tag{8.42}$$

Because of the slight dependence of recoil spectrum shape upon neutrino flavor and the influence of the threshold energy upon the event rate, the ratio shown in Eq. (8.42) is different from what would be obtained by examining a table of total cross sections at different individual energies (cf., Table 8.8 below).

Figure 8.1 shows that the scattering cross section for ^{8}B neutrinos depends strongly upon the threshold kinetic energy, T_{min}. Below 7 MeV, the effect of an incremental improvement of 1 or 2 MeV in the threshold kinetic energy is relatively moderate ($\lesssim$ a factor of two), but in the region above 7 MeV each additional MeV reduction in the threshold significantly increases the predicted event rate. It is important therefore for solar neutrino experiments to have low backgrounds down to a kinetic energy of order 7 MeV (and even lower if possible).

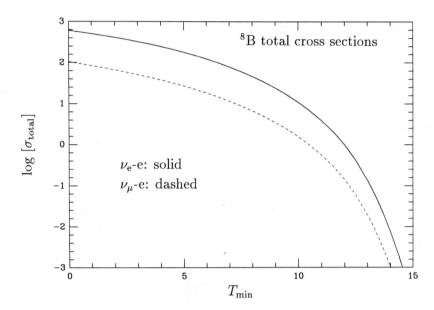

Figure 8.1 **^{8}B neutrino scattering cross sections**   The total cross sections for ν–e^- scattering are shown as a function of the minimum accepted electron kinetic energy, $T_{\min}$. The unit of cross section is 10^{-46} cm^2. The solid curve represents the ν_e–e^- cross sections and the dashed curve represents the ν_μ–e^- cross sections.

The hep neutrinos are of special interest. They would be relatively easy to detect if abundant since the hep neutrinos have the highest end-point energy (18.775 MeV) of any of the appreciable neutrino sources in the standard solar model. Moreover, their abundance may be different in nonstandard solar models in which it is assumed that the temperature gradient is lower (which discriminates against the ^{8}B neutrinos) but fresh ^{3}He is mixed into the core.

Table 8.7 gives the total scattering cross sections that were computed for hep neutrinos on electrons as a function of the threshold kinetic energy, $T_{\min}$. Equation (8.42) again describes reasonably accurately the ratio of ν_e–e to ν_μ–e scattering cross sections.

Hep neutrinos are more important than the ^{8}B neutrinos for producing electron recoil energies in excess of 13.5 MeV if, as estimated [Bahcall and Ulrich (1988)], the ratio of fluxes in the standard solar model is $\approx 10^{2.5}$. Unfortunately, the absolute cross section is

Table 8.7. Hep neutrino scattering cross sections. The scattering cross sections for hep solar neutrinos incident on electrons are given for different values of the minimum accepted kinetic energy, T_{min}. The neutrinos are assumed to be pure electron neutrinos (ν_e) or muon neutrinos (ν_μ) when they reach the Earth. The cross sections were calculated for $\sin^2 \Theta_W = 0.23$. The minimum electron kinetic energy, T_{min}, is given in MeV, and the cross sections, σ_{ν_e}-e and σ_{ν_μ}-e, are expressed in units of 10^{-46} cm^2.

T_{min}	σ_{ν_e}-e	σ_{ν_μ}-e	T_{min}	σ_{ν_e}-e	σ_{ν_μ}-e
0.0	8.84E+02	1.50E+02	11.0	7.45E+01	1.07E+01
1.0	7.84E+02	1.29E+02	12.0	4.71E+01	6.74E+00
2.0	6.87E+02	1.10E+02	13.0	2.73E+01	3.89E+00
3.0	5.94E+02	9.32E+01	13.5	1.99E+01	2.84E+00
4.0	5.04E+02	7.77E+01	14.0	1.41E+01	2.00E+00
5.0	4.20E+02	6.37E+01	15.0	6.11E+00	8.66E−01
6.0	3.42E+02	5.12E+01	16.0	2.01E+00	2.85E−01
7.0	2.71E+02	4.01E+01	17.0	3.88E−01	5.49E−02
8.0	2.09E+02	3.06E+01	17.5	1.11E−01	1.57E−02
9.0	1.55E+02	2.26E+01	18.0	1.5 E−02	2.1 E−03
10.0	1.10E+02	1.59E+01			

small for this large threshold energy so the hep neutrinos will only be observable if either very large detectors are used or the Sun is described by a nonstandard solar model.

C Cross sections at specific energies

Table 8.8 gives the computed cross sections for specific neutrino energies ranging from 1 MeV to 60 MeV. The tabulated values were determined for $T_{min} = 0.0$ MeV, neglecting radiative corrections (which are significant for the larger energies).

The cross sections given in Table 8.8 can be described approximately by the following relation

$$\sigma(q) = \text{const.} \left(\frac{q}{10 \text{ MeV}} \right) 10^{-44} \text{ cm}^2. \qquad (8.43)$$

where the constant is equal to 9.2 for ν_e scattering and 1.6 for ν_μ scattering. For $\bar{\nu}_e$ and $\bar{\nu}_\mu$ scattering, the constants are, respectively,

Table 8.8.   Scattering cross sections for individual
neutrino energies. The tabulated values were computed
for $T_{\min} = 0.0$ MeV. The neutrino energy, q, is given in
MeV and the cross sections, σ_{ν_e}-e and σ_{ν_μ}-e, are given
in units of 10^{-46} cm^2.

q	σ_{ν_e}-e	σ_{ν_μ}-e	q	σ_{ν_e}-e	σ_{ν_μ}-e
1.0	7.15E+01	1.48E+01	16.0	1.49E+03	2.49E+02
2.0	1.63E+02	3.05E+01	18.0	1.68E+03	2.80E+02
3.0	2.57E+02	4.62E+01	20.0	1.87E+03	3.12E+02
4.0	3.51E+02	6.18E+01	25.0	2.34E+03	3.90E+02
5.0	4.45E+02	7.75E+01	30.0	2.82E+03	4.68E+02
7.0	6.35E+02	1.09E+02	40.0	3.77E+03	6.24E+02
10.0	9.19E+02	1.56E+02	50.0	4.71E+03	7.80E+02
12.0	1.11E+03	1.87E+02	60.0	5.66E+03	9.36E+02
14.0	1.30E+03	2.18E+02			

3.9 and 1.3. The recoil spectrum for ν_e-e scattering is relatively flat
[see Eq. (8.29)]. Therefore, for ν_e-e scattering, the values given in
Table 8.8 can be used to estimate reasonably accurately the cross
section for a specified minimum recoil electron energy. One simply
multiplies the tabulated values by $(T_{\max} - T_{\min})/T_{\max}$. This approximation is less appropriate for ν_μ-e scattering, but still will give a
useful first estimate.

D   Energy distributions of recoil electrons

What is the energy spectrum of neutrinos that are incident on the
Earth? Is this spectrum consistent with the theoretical spectrum of
solar neutrinos calculated with the aid of the standard solar model
and the hypothesis of nuclear fusion as the solar energy source? Is
there evidence for resonant conversion of electron neutrinos to some
other flavor? Is the spectrum modified by the decay of the lower-
energy neutrinos? All these questions are fundamental and subject
to experimental tests. This section gives the distributions of recoil
energies that are calculated assuming the incident neutrino spectra
are identical with those calculated with the aid of a standard solar
model and the hypothesis of nuclear fusion. The modifications due
to neutrino decay are described in Section VII of Bahcall (1987).

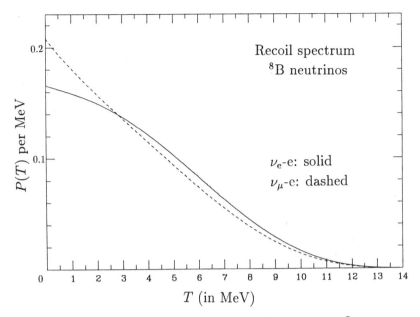

Figure 8.2   Electron recoil spectrum from scattering by ^{8}B neutrinos
The probability distribution, $P(T)$, of producing electrons with different recoil kinetic energies is shown as a function of the kinetic energy, T. The solid curve represents the probability distribution for ν_e-e^- interactions and the dashed curve represents ν_μ-e^- interactions.

The recoil spectra implied by the MSW effect among two flavors are given by Bahcall, Gelb, and Rosen (1987).

Tables 8.9 and 8.10 present the probability distributions, $P(T)$, for the production of recoil electrons with kinetic energies, T, by ν_e-e^- and ν_μ-e^- scattering by ^{8}B solar neutrinos.

Figure 8.2 illustrates the recoil spectra, which were calculated using Eq. (8.35). The general shape of the recoil spectra are similar in the experimentally most accessible region above a few MeV, the main difference is in the magnitude of the cross section [see Eq. (8.42)].

The ^{3}He+p neutrinos are sensitive to different aspects of the solar model than the ^{8}B neutrinos (see Section 6.2). Although the ^{3}He+p neutrinos are expected to be rare, they produce a high-energy tail to the distribution of recoil electrons which may be detected in experiments involving massive amounts of scattering material (see Sections 14.1 and 14.2).

Table 8.9. Recoil spectrum from ν_e–e scattering by ^{8}B neutrinos.
The kinetic energy of the electron is denoted by T (measured in MeV)
and the normalized probability distributions per MeV by $P(T)$.

T	$P(T)$	T	$P(T)$	T	$P(T)$	T	$P(T)$	T	$P(T)$
0.0	0.166	3.0	0.137	6.0	0.083	9.0	2.9E−02	12.0	2.7E−03
0.1	0.165	3.1	0.136	6.1	0.081	9.1	2.7E−02	12.1	2.4E−03
0.2	0.164	3.2	0.135	6.2	0.079	9.2	2.6E−02	12.2	2.1E−03
0.3	0.164	3.3	0.133	6.3	0.077	9.3	2.5E−02	12.3	1.8E−03
0.4	0.163	3.4	0.132	6.4	0.075	9.4	2.3E−02	12.4	1.6E−03
0.5	0.162	3.5	0.130	6.5	0.073	9.5	2.2E−02	12.5	1.3E−03
0.6	0.162	3.6	0.128	6.6	0.071	9.6	2.1E−02	12.6	1.1E−03
0.7	0.161	3.7	0.127	6.7	0.069	9.7	2.0E−02	12.7	9.6E−04
0.8	0.160	3.8	0.125	6.8	0.067	9.8	1.8E−02	12.8	8.1E−04
0.9	0.159	3.9	0.123	6.9	0.065	9.9	1.7E−02	12.9	6.7E−04
1.0	0.159	4.0	0.122	7.0	0.063	10.0	1.6E−02	13.0	5.5E−04
1.1	0.158	4.1	0.120	7.1	0.061	10.1	1.5E−02	13.1	4.5E−04
1.2	0.157	4.2	0.118	7.2	0.059	10.2	1.4E−02	13.2	3.6E−04
1.3	0.156	4.3	0.116	7.3	0.057	10.3	1.3E−02	13.3	2.9E−04
1.4	0.155	4.4	0.114	7.4	0.056	10.4	1.2E−02	13.4	2.3E−04
1.5	0.155	4.5	0.113	7.5	0.054	10.5	1.2E−02	13.5	1.8E−04
1.6	0.154	4.6	0.111	7.6	0.052	10.6	1.1E−02	13.6	1.4E−04
1.7	0.153	4.7	0.109	7.7	0.050	10.7	9.9E−03	13.7	1.1E−04
1.8	0.152	4.8	0.107	7.8	0.048	10.8	9.1E−03	13.8	7.9E−05
1.9	0.151	4.9	0.105	7.9	0.046	10.9	8.4E−03	13.9	5.9E−05
2.0	0.150	5.0	0.103	8.0	0.045	11.0	7.7E−03	14.0	4.4E−05
2.1	0.149	5.1	0.101	8.1	0.043	11.1	7.0E−03	14.1	3.2E−05
2.2	0.148	5.2	0.099	8.2	0.041	11.2	6.4E−03	14.2	2.3E−05
2.3	0.146	5.3	0.097	8.3	0.040	11.3	5.8E−03	14.3	1.7E−05
2.4	0.145	5.4	0.095	8.4	0.038	11.4	5.3E−03	14.4	1.2E−05
2.5	0.144	5.5	0.093	8.5	0.036	11.5	4.8E−03	14.5	8.2E−06
2.6	0.143	5.6	0.091	8.6	0.035	11.6	4.3E−03	14.6	5.7E−06
2.7	0.142	5.7	0.089	8.7	0.033	11.7	3.8E−03	14.7	3.8E−06
2.8	0.140	5.8	0.087	8.8	0.032	11.8	3.4E−03	14.8	2.5E−06
2.9	0.139	5.9	0.085	8.9	0.030	11.9	3.0E−03	14.9	1.6E−06

Figure 8.3 and Tables 8.11 and 8.12 show the recoil spectra for
^{3}He+p neutrino–electron scattering. Again, the general shape of the
recoil spectra are similar for ν_e–e^- and ν_μ–e^- scattering, although
the total cross sections are different. The recoil electrons from ^{3}He+p

Table 8.10. Recoil spectrum from ν_μ–e scattering by ^{8}B neutrinos. The kinetic energy of the electron is denoted by T (measured in MeV) and the normalized probability distributions per MeV by $P(T)$.

T	$P(T)$	T	$P(T)$	T	$P(T)$	T	$P(T)$	T	$P(T)$
0.0	0.208	3.0	0.136	6.0	0.074	9.0	2.4E−02	12.0	2.3E−03
0.1	0.205	3.1	0.134	6.1	0.072	9.1	2.3E−02	12.1	2.0E−03
0.2	0.202	3.2	0.132	6.2	0.070	9.2	2.2E−02	12.2	1.7E−03
0.3	0.199	3.3	0.130	6.3	0.068	9.3	2.1E−02	12.3	1.5E−03
0.4	0.196	3.4	0.128	6.4	0.066	9.4	2.0E−02	12.4	1.3E−03
0.5	0.194	3.5	0.126	6.5	0.064	9.5	1.9E−02	12.5	1.1E−03
0.6	0.191	3.6	0.123	6.6	0.063	9.6	1.8E−02	12.6	9.5E−04
0.7	0.188	3.7	0.121	6.7	0.061	9.7	1.7E−02	12.7	8.0E−04
0.8	0.186	3.8	0.119	6.8	0.059	9.8	1.6E−02	12.8	6.7E−04
0.9	0.183	3.9	0.117	6.9	0.057	9.9	1.5E−02	12.9	5.6E−04
1.0	0.181	4.0	0.115	7.0	0.055	10.0	1.4E−02	13.0	4.6E−04
1.1	0.178	4.1	0.113	7.1	0.053	10.1	1.3E−02	13.1	3.7E−04
1.2	0.176	4.2	0.111	7.2	0.052	10.2	1.2E−02	13.2	3.0E−04
1.3	0.173	4.3	0.109	7.3	0.050	10.3	1.1E−02	13.3	2.4E−04
1.4	0.171	4.4	0.107	7.4	0.048	10.4	1.0E−02	13.4	1.9E−04
1.5	0.169	4.5	0.105	7.5	0.047	10.5	9.7E−03	13.5	1.5E−04
1.6	0.167	4.6	0.102	7.6	0.045	10.6	9.0E−03	13.6	1.1E−04
1.7	0.164	4.7	0.100	7.7	0.043	10.7	8.3E−03	13.7	8.7E−05
1.8	0.162	4.8	0.098	7.8	0.042	10.8	7.7E−03	13.8	6.6E−05
1.9	0.160	4.9	0.096	7.9	0.040	10.9	7.0E−03	13.9	4.9E−05
2.0	0.158	5.0	0.094	8.0	0.038	11.0	6.5E−03	14.0	3.6E−05
2.1	0.155	5.1	0.092	8.1	0.037	11.1	5.9E−03	14.1	2.7E−05
2.2	0.153	5.2	0.090	8.2	0.035	11.2	5.4E−03	14.2	1.9E−05
2.3	0.151	5.3	0.088	8.3	0.034	11.3	4.9E−03	14.3	1.4E−05
2.4	0.149	5.4	0.086	8.4	0.032	11.4	4.4E−03	14.4	9.8E−06
2.5	0.147	5.5	0.084	8.5	0.031	11.5	4.0E−03	14.5	6.8E−06
2.6	0.145	5.6	0.082	8.6	0.030	11.6	3.6E−03	14.6	4.7E−06
2.7	0.142	5.7	0.080	8.7	0.028	11.7	3.2E−03	14.7	3.2E−06
2.8	0.140	5.8	0.078	8.8	0.027	11.8	2.9E−03	14.8	2.1E−06
2.9	0.138	5.9	0.076	8.9	0.026	11.9	2.5E−03	14.9	1.4E−06

neutrino–electron scattering extend about 4 MeV beyond the cutoff for electrons from scattering by ^{8}B neutrinos. The higher-energy ^{3}He+p neutrinos may make possible additional searches for solar neutrinos in some planned detectors.

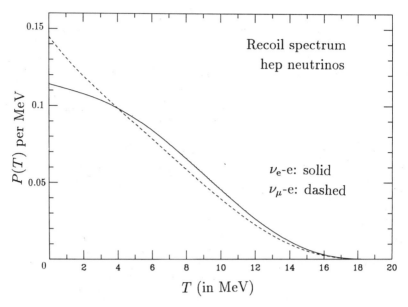

Figure 8.3   Electron recoil spectrum from scattering by hep neutrinos
The probability distribution, $P(T)$, of producing electrons with different recoil kinetic energies is shown as a function of the kinetic energy, T. The solid curve represents the probability distribution for ν_e–e^- interactions and the dashed curve represents ν_μ–e^- interactions.

The neutrinos from the pp reaction are of fundamental importance because they provide a signature of the fusion reaction that is the basis for most of solar energy generation. Table 8.13 gives the probability distributions for both ν_e–e^- and ν_μ–e^- scattering.

Figure 8.4a illustrates the results for pp neutrinos. For comparison, Figure 8.4b shows the probability distributions for the line neutrinos produced by the related pep reaction. For the pp neutrinos, the recoil spectrum falls steeply over the allowed energy range (0.0 MeV to 0.26 MeV), while for the pep neutrinos the probability distribution is relatively flat over most of the allowed energy range (0.0 MeV to 1.2 MeV).

For ^{7}Be neutrinos, Figure 8.5 compares the recoil distributions for the two neutrino lines. Figure 8.5a shows the probability distributions for the production of recoil electrons due to scattering by the 0.862 MeV neutrino line and Figure 8.5b gives the corre-

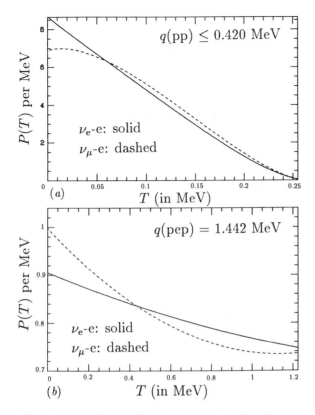

Figure 8.4   Electron recoil spectrum from scattering from pp and pep neutrinos (a) refers to the pp neutrinos and (b) to the pep neutrinos. The probability distribution, $P(T)$, of producing electrons with different recoil kinetic energies is shown as a function of the kinetic energy, T. The solid curve represents the probability distribution for ν_e–e^- interactions and the dashed curve represents ν_μ–e^- interactions.

sponding probabilities for the associated 0.384 MeV line. The recoil spectra from ν_e and ν_μ scattering are qualitatively different for the low-energy line.

For electron recoil energies below 1 MeV, a number of different solar neutrino sources can produce electrons of a given energy. To calculate the expected spectrum from the standard solar model, one must add together the recoil distributions from all of the important neutrino sources with the weights determined by detailed solar model calculations of the individual neutrino fluxes. The recoil dis-

Table 8.11.   Recoil spectrum from ν_e–e scattering by hep neutrinos.
The kinetic energy of the electron is denoted by T (measured in MeV)
and the normalized probability distributions per MeV by $P(T)$.

T	$P(T)$	T	$P(T)$	T	$P(T)$	T	$P(T)$	T	$P(T)$
0.0	0.114	3.6	0.101	7.2	0.074	10.8	3.8E−02	14.4	9.5E−03
0.1	0.114	3.7	0.100	7.3	0.073	10.9	3.7E−02	14.5	9.0E−03
0.2	0.113	3.8	0.100	7.4	0.072	11.0	3.6E−02	14.6	8.5E−03
0.3	0.113	3.9	0.099	7.5	0.071	11.1	3.5E−02	14.7	8.0E−03
0.4	0.113	4.0	0.098	7.6	0.070	11.2	3.4E−02	14.8	7.5E−03
0.5	0.113	4.1	0.098	7.7	0.069	11.3	3.3E−02	14.9	7.1E−03
0.6	0.112	4.2	0.097	7.8	0.068	11.4	3.2E−02	15.0	6.6E−03
0.7	0.112	4.3	0.097	7.9	0.067	11.5	3.1E−02	15.1	6.2E−03
0.8	0.112	4.4	0.096	8.0	0.066	11.6	3.0E−02	15.2	5.8E−03
0.9	0.111	4.5	0.095	8.1	0.065	11.7	2.9E−02	15.3	5.4E−03
1.0	0.111	4.6	0.095	8.2	0.064	11.8	2.8E−02	15.4	5.0E−03
1.1	0.111	4.7	0.094	8.3	0.063	11.9	2.8E−02	15.5	4.6E−03
1.2	0.110	4.8	0.093	8.4	0.062	12.0	2.7E−02	15.6	4.3E−03
1.3	0.110	4.9	0.093	8.5	0.061	12.1	2.6E−02	15.7	4.0E−03
1.4	0.110	5.0	0.092	8.6	0.060	12.2	2.5E−02	15.8	3.6E−03
1.5	0.109	5.1	0.091	8.7	0.059	12.3	2.4E−02	15.9	3.3E−03
1.6	0.109	5.2	0.091	8.8	0.058	12.4	2.3E−02	16.0	3.0E−03
1.7	0.109	5.3	0.090	8.9	0.057	12.5	2.2E−02	16.1	2.8E−03
1.8	0.108	5.4	0.089	9.0	0.056	12.6	2.2E−02	16.2	2.5E−03
1.9	0.108	5.5	0.088	9.1	0.055	12.7	2.1E−02	16.3	2.3E−03
2.0	0.108	5.6	0.087	9.2	0.054	12.8	2.0E−02	16.4	2.0E−03
2.1	0.107	5.7	0.087	9.3	0.053	12.9	1.9E−02	16.5	1.8E−03
2.2	0.107	5.8	0.086	9.4	0.052	13.0	1.9E−02	16.6	1.6E−03
2.3	0.107	5.9	0.085	9.5	0.051	13.1	1.8E−02	16.7	1.4E−03
2.4	0.106	6.0	0.084	9.6	0.050	13.2	1.7E−02	16.8	1.3E−03
2.5	0.106	6.1	0.083	9.7	0.049	13.3	1.6E−02	16.9	1.1E−03
2.6	0.105	6.2	0.083	9.8	0.048	13.4	1.6E−02	17.0	9.4E−04
2.7	0.105	6.3	0.082	9.9	0.047	13.5	1.5E−02	17.1	8.1E−04
2.8	0.105	6.4	0.081	10.0	0.046	13.6	1.4E−02	17.2	6.9E−04
2.9	0.104	6.5	0.080	10.1	0.045	13.7	1.4E−02	17.3	5.8E−04
3.0	0.104	6.6	0.079	10.2	0.044	13.8	1.3E−02	17.4	4.8E−04
3.1	0.103	6.7	0.078	10.3	0.043	13.9	1.2E−02	17.5	3.9E−04
3.2	0.103	6.8	0.077	10.4	0.042	14.0	1.2E−02	17.6	3.2E−04
3.3	0.102	6.9	0.076	10.5	0.041	14.1	1.1E−02	17.7	2.5E−04
3.4	0.102	7.0	0.075	10.6	0.040	14.2	1.1E−02	17.8	1.9E−04
3.5	0.101	7.1	0.074	10.7	0.039	14.3	1.0E−02	17.9	1.4E−04

Table 8.12.   Recoil spectrum from ν_μ–e scattering by hep neutrinos. The
kinetic energy of the electron is denoted by T (measured in MeV) and the
normalized probability distributions per MeV by $P(T)$.

T	$P(T)$	T	$P(T)$	T	$P(T)$	T	$P(T)$	T	$P(T)$
0.0	0.145	3.6	0.103	7.2	0.067	10.8	3.2E−02	14.4	8.0E−03
0.1	0.143	3.7	0.102	7.3	0.066	10.9	3.2E−02	14.5	7.6E−03
0.2	0.142	3.8	0.100	7.4	0.065	11.0	3.1E−02	14.6	7.2E−03
0.3	0.140	3.9	0.100	7.5	0.064	11.1	3.0E−02	14.7	6.7E−03
0.4	0.139	4.0	0.099	7.6	0.063	11.2	2.9E−02	14.8	6.3E−03
0.5	0.138	4.1	0.097	7.7	0.062	11.3	2.8E−02	14.9	5.9E−03
0.6	0.136	4.2	0.097	7.8	0.061	11.4	2.7E−02	15.0	5.6E−03
0.7	0.135	4.3	0.096	7.9	0.060	11.5	2.7E−02	15.1	5.2E−03
0.8	0.134	4.4	0.095	8.0	0.059	11.6	2.6E−02	15.2	4.8E−03
0.9	0.132	4.5	0.094	8.1	0.058	11.7	2.5E−02	15.3	4.5E−03
1.0	0.131	4.6	0.093	8.2	0.057	11.8	2.4E−02	15.4	4.2E−03
1.1	0.130	4.7	0.092	8.3	0.056	11.9	2.3E−02	15.5	3.9E−03
1.2	0.129	4.8	0.091	8.4	0.055	12.0	2.3E−02	15.6	3.6E−03
1.3	0.127	4.9	0.090	8.5	0.054	12.1	2.2E−02	15.7	3.3E−03
1.4	0.126	5.0	0.089	8.6	0.053	12.2	2.1E−02	15.8	3.0E−03
1.5	0.125	5.1	0.088	8.7	0.052	12.3	2.0E−02	15.9	2.8E−03
1.6	0.124	5.2	0.087	8.8	0.051	12.4	2.0E−02	16.0	2.5E−03
1.7	0.123	5.3	0.086	8.9	0.050	12.5	1.9E−02	16.1	2.3E−03
1.8	0.122	5.4	0.085	9.0	0.049	12.6	1.8E−02	16.2	2.1E−03
1.9	0.120	5.5	0.084	9.1	0.048	12.7	1.8E−02	16.3	1.9E−03
2.0	0.119	5.6	0.083	9.2	0.047	12.8	1.7E−02	16.4	1.7E−03
2.1	0.118	5.7	0.082	9.3	0.046	12.9	1.6E−02	16.5	1.5E−03
2.2	0.117	5.8	0.081	9.4	0.045	13.0	1.6E−02	16.6	1.3E−03
2.3	0.116	5.9	0.080	9.5	0.044	13.1	1.5E−02	16.7	1.2E−03
2.4	0.115	6.0	0.079	9.6	0.043	13.2	1.4E−02	16.8	1.0E−03
2.5	0.114	6.1	0.078	9.7	0.042	13.3	1.4E−02	16.9	9.2E−04
2.6	0.113	6.2	0.077	9.8	0.041	13.4	1.3E−02	17.0	7.8E−04
2.7	0.112	6.3	0.076	9.9	0.040	13.5	1.3E−02	17.1	6.8E−04
2.8	0.111	6.4	0.075	10.0	0.040	13.6	1.2E−02	17.2	5.8E−04
2.9	0.110	6.5	0.074	10.1	0.039	13.7	1.2E−02	17.3	4.9E−04
3.0	0.109	6.6	0.073	10.2	0.038	13.8	1.1E−02	17.4	4.0E−04
3.1	0.108	6.7	0.072	10.3	0.037	13.9	1.0E−02	17.5	3.3E−04
3.2	0.107	6.8	0.071	10.4	0.036	14.0	9.9E−03	17.6	2.7E−04
3.3	0.106	6.9	0.070	10.5	0.035	14.1	9.5E−03	17.7	2.0E−04
3.4	0.105	7.0	0.069	10.6	0.034	14.2	9.0E−03	17.8	1.6E−04
3.5	0.104	7.1	0.068	10.7	0.033	14.3	8.5E−03	17.9	1.2E−04

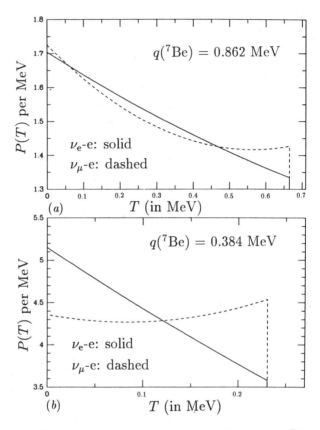

Figure 8.5   Electron recoil spectrum from scattering by ^{7}Be neutrinos (*a*) refers to the neutrino line with energy 0.862 MeV and (*b*) to the line of energy 0.384 MeV. The probability distribution, $P(T)$, of producing electrons with different recoil kinetic energies is shown as a function of the kinetic energy, T. The solid curve represents the probability distribution for ν_e–e^- interactions and the dashed curve represents ν_μ–e^- interactions.

tributions that are appropriate to ^{13}N, ^{15}O, and ^{17}F solar neutrinos are shown in Figures 7*a* to 7*c* of Bahcall (1987).

E   Angular distributions of recoil electrons

The angular distribution of the recoil electrons is strongly peaked in the forward direction of the Sun–Earth axis. A convenient analytic formula representing this forward-peaking is given by Bahcall

Table 8.13. Electron recoil spectrum from scattering by pp neutrinos. The kinetic energy of the electron is denoted by T (measured in MeV) and the normalized probability distributions per MeV by $P(T)$.

T	$P(T)_{\nu_e}$	$P(T)_{\nu_\mu}$	T	$P(T)_{\nu_e}$	$P(T)_{\nu_\mu}$
0.000	8.6882	6.8750	0.130	3.6546	4.0034
0.010	8.2602	6.9883	0.141	3.2869	3.6254
0.020	7.8558	6.9768	0.151	2.9213	3.2403
0.030	7.4608	6.8862	0.161	2.5669	2.8618
0.040	7.0742	6.7426	0.171	2.2196	2.4850
0.050	6.6891	6.5476	0.181	1.8812	2.1134
0.060	6.3058	6.3132	0.191	1.5545	1.7511
0.070	5.9252	6.0487	0.201	1.2487	1.4103
0.080	5.5442	5.7544	0.211	0.9604	1.0869
0.090	5.1641	5.4371	0.221	0.6938	0.7864
0.100	4.7861	5.1022	0.231	0.4541	0.5152
0.110	4.4063	4.7466	0.241	0.2482	0.2819
0.120	4.0309	4.3822	0.251	0.0903	0.1026

(1964b). This section presents the detailed angular distributions that are calculated using Eq. (8.40) and the standard solar neutrino spectra.

It is convenient to define a **critical scattering angle**, Θ_c, such that 90% of the recoil electrons are scattered within a cone of half opening-angle Θ_c about the forward direction of the Sun–Earth axis. Table 8.14 gives the values of Θ_c for both ^{8}B and hep solar neutrino spectra. The computed values of Θ_c depend sensitively upon the minimum electron energy, T_{min}, that is accepted in a given experimental arrangement. If *all* of the recoil electrons are counted ($T_{min} = 0.0$ MeV), the angular distributions are rather broad (opening half-angle in excess of 45°). However, for the larger values of T_{min}, the angular distributions are strongly forward peaked. Typical angles are $\Theta_c \approx 15°$ for a 5 MeV threshold and $\Theta_c \approx$ a few degrees for the highest-energy electrons.

Table 8.14 shows that the values of Θ_c are similar for scattering by ν_e and ν_μ neutrinos, except when low-energy thresholds are used.

Table 8.14. Angle within which 90% of electrons are scattered. The minimum kinetic energy of the electron is denoted by T_{min} (measured in MeV) and the angle within which 90% of the electrons are scattered is represented by Θ_c (in degrees). The second and third columns were computed using a ^{8}B neutrino spectrum and the fourth and fifth columns were calculated with a hep neutrino spectrum.

T_{min}	Θ_{c,ν_e} ^{8}B	Θ_{c,ν_μ} ^{8}B	Θ_{c,ν_e} hep	Θ_{c,ν_μ} hep
0.0	48.26	51.44	43.62	47.05
1.0	33.00	33.98	32.18	33.45
2.0	25.38	25.89	25.76	26.45
3.0	20.60	20.90	21.53	21.96
4.0	17.20	17.40	18.44	18.75
5.0	14.60	14.74	16.07	16.29
6.0	12.51	12.60	14.16	14.32
7.0	10.76	10.82	12.56	12.68
8.0	9.25	9.29	11.18	11.27
9.0	7.92	7.94	9.98	10.04
10.0	6.71	6.72	8.90	8.94
11.0	5.61	5.61	7.93	7.95
12.0	4.58	4.58	7.03	7.04
13.0	3.69	3.69	6.18	6.19
14.0	2.97	2.97	5.36	5.37

Figure 8.6 shows the angular distributions for both ^{8}B and hep electron neutrinos for two typical values of T_{min}, namely, 5 MeV and 9 MeV. The corresponding integral probability distributions for electrons to be scattered within different angles are shown in Figure 9 of Bahcall (1987).

Figure 8.7 illustrates the broad angular distributions that are expected from two low-energy sources of neutrinos, the pp neutrinos and the ^{7}Be (0.862 MeV) neutrinos. This figure shows both the differential and the integral probability distributions. Note that the

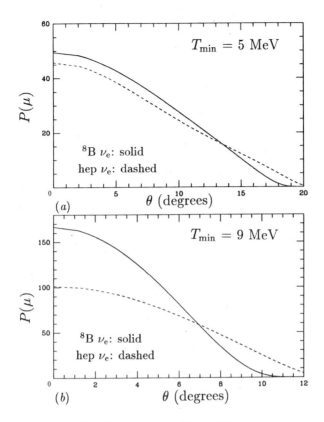

Figure 8.6   Angular distributions for ^{8}B and hep neutrinos   Here
$P(\mu)\,d\mu$ is the differential probability for an electron to be scattered to
an angle θ, $\mu = \cos\theta$. The solid curves represent the probability distributions for ^{8}B neutrinos and the dashed curves represent the scattering
probabilities for hep neutrinos. Results are shown for two typical values
of T_{min}, 5 MeV and 9 MeV.

differences between ν_e and ν_μ are somewhat more pronounced for
these low-energy sources.

F   Neutrinos from muon and pion decay

Neutrinos from pion and muon decays can cause neutrino–electron
scattering events in solar neutrino detectors. The decaying mesons
could originate from cosmic rays that are incident on the Earth or
from nuclear interactions on the surface of the Sun.

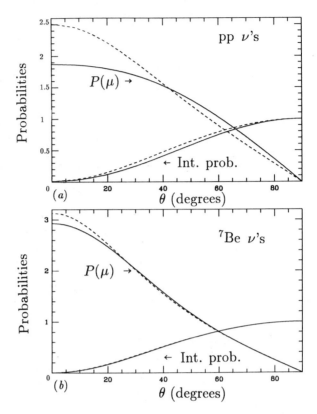

Figure 8.7   Angular distributions for pp and ^{7}Be neutrinos (a) refers to the pp neutrinos and (b) pertains to neutrinos with an energy of 0.862 MeV   (^{7}Be neutrinos). Here $P(\mu)\,d\mu$ is the differential probability for an electron to be scattered to an angle θ, $\mu = \cos\theta$, and Int. Prob. is the integrated probability distribution. The solid curves represent the probability distributions for electron neutrinos and the dashed curves represent the scattering probabilities for muon neutrinos.

Table 8.15 gives the relevant cross sections for neutrinos and antineutrinos from pion and muon decays. The energy spectrum of ν_e from μ^+ decay (or $\bar\nu_e$ from μ^- decay) has the form

$$P\left(x\right) \;=\; x^2\left(1 - 2x\right), \qquad x \;=\; \frac{q}{m_\mu}, \qquad (8.44a)$$

Table 8.15. Neutrino scattering cross sections from muon and pion decays at rest.

ν type	Decay particle	σ (10^{-46} cm^2)
ν_e	μ^+	2.98×10^3
$\bar{\nu}_e$	μ^-	1.25×10^3
ν_μ	μ^-	5.77×10^2
$\bar{\nu}_\mu$	μ^+	4.93×10^2
ν_μ	π^+	4.65×10^2
$\bar{\nu}_\mu$	π^-	3.97×10^2

and the energy spectrum of $\bar{\nu}_\mu$ from μ^+ decay (or ν_μ from μ^- decay) has the form

$$P(x) = x^2 \left(1 - \frac{4}{3}x\right), \qquad x = \frac{q}{m_\mu}, \qquad (8.44b)$$

where the probability that the neutrino energy divided by the rest mass of a muon, m_μ, lies in the interval dx is proportional to $P(x)$.

G   Discussion and conclusions

This section summarizes the main results for neutrino–electron scattering that are described in the previous subsections and makes some comments on the implications of these results for future solar neutrino experiments.

(1) *Flavor dependence.* Electron neutrinos have larger cross sections than muon neutrinos by a typical factor of order six or seven [see Eq. (8.42)] for the higher-energy ^{8}B and hep neutrino spectra, if the minimum accepted electron recoil energy, $T_{\min}$, is above several MeV. Thus the event rate for an electron scattering detector is sensitive to neutrino flavor. Electron neutrino scattering experiments, when combined with absorption experiments that measure the incident beam of ν_e's, can test effectively whether or not electron neutrinos have been resonantly converted to neutrinos of a different flavor (the MSW effect).

(2) *Total cross sections.* The total cross sections depend strongly upon the minimum recoil energy, T_{min} (see Figure 8.1). For ${}^8B$ neutrinos, this is especially true for thresholds larger than 7 MeV. For each MeV that the threshold is lowered, the expected counting rate increases significantly. It is important therefore that solar neutrino experiments have low backgrounds down to an electron kinetic energy of order 7 MeV (even lower values are desirable) so that T_{min} may be set as low as possible.

(3) *hep neutrinos.* Electron recoil energies above ~ 13.5 MeV are produced almost entirely by neutrinos from the rare hep reaction. Therefore, the energy range between 13.5 MeV and 18.5 MeV can be used to study the solar reaction, ${}^3He+p$ [Eq. (6.3)], which is broadly distributed in the Sun [see Bahcall and Ulrich (1988) and Section 6.2, Figure 6.1]. In contrast, the ${}^8B$ neutrinos are produced mainly in the innermost few percent of the solar mass. The event rate for hep neutrinos will be small if the standard solar model is correct, for example, only of order 0.15 events with ≥ 13.5 MeV per kiloton of argon per year. Fortunately, however, the hep reaction is sensitive to the assumptions of some of the nonstandard solar models (especially the mixing of ${}^3He$) and could be larger in some of the models. Therefore, hep neutrinos can provide a useful diagnostic of the stellar physics, once the hep flux is calculated in the various nonstandard solar models. The fact that a ${}^3He+p$ neutrino spectrum produces some recoil electrons with higher energies than is possible with ${}^8B$ neutrinos may enable additional underground experiments to be used as solar neutrino detectors.

(4) *Electron recoil spectra.* How sensitive is the observed recoil spectrum to the incident neutrino energy spectrum? Figure 8.8 provides an answer.

Figure 8.8 shows the recoil electron spectra that were calculated with two different assumed neutrino spectra: (1) the theoretical ${}^8B$ neutrino spectrum (the solid curve in Figure 8.8); and (2) a fictitious linear neutrino spectrum,

$$\lambda(q) = \frac{2 \times q}{q_{max}^2}. \tag{8.45}$$

For simplicity, the flavor of both spectra was assumed to be pure ν_e. The recoil spectrum calculated for the fictitious linear spectrum is shown as a dashed line in Figure 8.8 and is very different from the

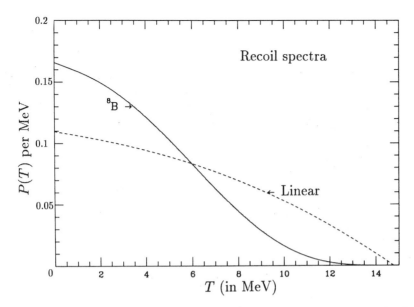

Figure 8.8   Sensitivity of scattered electron energy spectrum to the incident neutrino spectrum   This figure compares the recoil distributions of electrons scattered by linear and ^{8}B neutrino spectra. The solid curve represents the probability distribution for scattering by a ^{8}B neutrino spectrum and the dashed curve represents the probability distribution for scattering by a fictitious linear spectrum. The probability distribution, $P(T)$, of producing electrons with different recoil kinetic energies is shown as a function of the kinetic energy, T.

recoil distribution that is predicted for a ^{8}B neutrino spectrum (the solid line in Figure 8.8).

Measurements of the recoil electron energy spectrum can provide an important consistency test for the predictions of solar models and of neutrino physics.

(5) *Angular distributions.* The higher-energy electrons are strongly forward peaked in the direction of the Sun–Earth axis. For electrons with energies above 5 MeV, 90% of the electrons are scattered into a cone with an opening angle of 15° about the forward direction and this critical scattering angle decreases to just a few degrees for the highest-energy electrons. If lower-energy electrons are accepted, the angular distribution is rather broad.

8.3   Neutral currents

Neutral currents distinguish between explanations of the solar neutrino problem that involve revised astrophysics and explanations that involve neutrino oscillations. For astrophysical explanations, (i.e., nonstandard solar models) the flux of neutrinos is reduced relative to the standard solar model while the energy spectrum remains the same. For explanations involving oscillations, the total flux of neutrinos is the same as in the standard model although their flavor distribution is radically different.

Neutral currents are flavor blind. In the standard electroweak theory, neutrinos of different flavors have the same neutral-current interactions. Therefore neutral-current experiments can distinguish between "missing" neutrinos (e.g., faulty astrophysics) or new physics.

This section discusses coherent neutrino–nucleus scattering (in Section 8.3A), quasielastic neutrino–nucleus scattering (in 8.3B), and neutrino–electron scattering (in 8.3C). The discussion is based upon work by Freedman (1974), Drukier and Stodolsky (1984), Cabrera, Krauss, and Wilczek (1985), Freedman, Schramm, and Tubbs (1977), Chen (1985), Raghavan, Pakvasa, and Brown (1986), and Weinberg (1987).

A   Neutrino–nucleus elastic scattering

The first pure neutral-current process to be discussed in connection with solar neutrinos [Drukier and Stodolsky (1984)] was neutrino–nucleus elastic scattering. This process can be represented symbolically by the reaction

$$\nu + A \rightarrow \nu + A. \tag{8.46}$$

Here A is a nucleus with N neutrons and Z protons. At the low energies relevant for solar neutrinos, the nucleus can be treated as a point scatterer all of whose nucleons respond coherently. The differential cross section is then given to a good approximation [Freedman (1974)] by the expression

$$\frac{\mathrm{d}\sigma}{\mathrm{d}\mu} = \frac{\sin^2 \theta_W}{2\pi} G_V^2 (N + Z)^2 q^2 (1 + \mu), \tag{8.47}$$

where μ is the cosine of the scattering angle. Integrating over all directions, the total cross section is

$$\sigma \approx 4 \times 10^{-43} \text{ cm}^2 \ N^2 \left(q/10 \text{ MeV}\right)^2. \tag{8.48}$$

The above expressions for the cross sections are strictly valid only for spin zero nuclei with $N = Z$, although they are a reasonable approximation for more general cases [Freedman (1974), Tubbs and Schramm (1975), and Drukier and Stodolsky (1984)].

The coherence factor ($\propto N^2$) that occurs in Eq. (8.48) represents one of the main advantages of using neutrino–nucleus elastic scattering. For practical cases, the cross section can be a factor of 10^3 larger than for the corresponding neutrino absorption and neutrino–electron scattering processes (see Sections 8.1 and 8.2).

The main difficulty in studying neutrino–nucleus scattering is that the only observable product is the recoil nucleus. The average energy of the nuclear recoil is [Drukier and Stodolsky (1984)]

$$\langle E_A \rangle = \frac{2}{3A} \left(\frac{q}{1 \text{ MeV}}\right)^2 \text{keV}, \tag{8.49}$$

where A is the mass number of the nucleus. The recoil energy is small because the neutrino energy is much less than the mass of the nucleus.

In designing an experiment with a large mass thermal bolometer, one must compromise between the desirable feature that the cross section increases as N^2 and the undesirable fact that the recoil energy decreases inversely as the mass number A. For many purposes, silicon is a reasonable compromise [Drukier and Stodolsky (1984), Cabrera, Krauss, and Wilczek (1985), but see also Martoff (1987) and Section 13.8].

B Quasielastic neutrino–nucleus scattering

Reactions in which the final state of the nucleus is altered have a practical advantage over purely elastic scattering [Freedman (1974), Donnelly et al. (1974), Raghavan, Pakvasa, and Brown (1986), and Aardsma et al. (1987)]. One can detect the characteristic decay products of the excited nucleus (which may be, e.g., gamma rays of specific energies or protons, neutrons, or α-particles if the excited nucleus is unstable to emission of strongly interacting particles). The general reaction can be represented by

$$\nu + A \rightarrow \nu' + A', \tag{8.50}$$

without specifying the decay product(s) of the excited nucleus, A'.

Raghavan and his collaborators have emphasized the special advantages of a ^{11}B detector which can observe both the neutral-current reactions via

$$\nu + {}^{11}\text{B} \rightarrow \nu' + {}^{11}\text{B}^* \tag{8.51}$$

and the charged-current reactions via the corresponding ^{11}B–^{11}C transitions. The cross sections for most of the transitions have been estimated by using shell model calculations or by assuming that the magnetic dipole spin excitation strength derived from electromagnetic decays can be used to calculate the dominant matrix elements. It is difficult to assign a well-determined uncertainty to the estimated cross sections that have been derived for the ^{11}B transitions, except for the ground state to ground state charged-current reaction.

Weinberg (1987) has proved an unfortunate theorem which states that for nuclear excitation reactions like that shown above for ^{11}B there is no angular correlation between the signal and the direction of the incoming neutrino. This means that it will not be possible to establish from a directional correlation that the neutral-current reactions involving ^{11}B are caused by neutrinos from the Sun.

The Sudbury solar neutrino collaboration has proposed observing the neutral-current disintegration of the deuterium by detecting the gamma rays emitted when the produced neutrons are captured in a doped impurity [Chen (1985) and Aardsma et al. (1987)]. The basic reaction is

$$\nu + {}^2\text{H} \rightarrow \nu' + \text{n} + \text{p}. \tag{8.52}$$

The cross sections for these reactions can be calculated with satisfactory accuracy [Bahcall, Kubodera, and Nozawa (1988)]:

$$\langle \sigma \left({}^8\text{B} \right) \rangle \; = \; 0.41 \left(1 \pm 0.1 \right) 10^{-42} \text{ cm}^2, \tag{8.53}$$

and

$$\langle \sigma \left(\text{hep} \right) \rangle \; = \; 1.15 \left(1 \pm 0.1 \right) \times 10^{-42} \text{ cm}^2. \tag{8.54}$$

In addition, the ratio of neutral-current to charged-current reaction cross sections is determined by the calculations to an accuracy of $\pm 1/2\%$ [Bahcall, Kubodera, and Nozawa (1988)].

C Neutrino–electron scattering

Neutrino–electron scattering,

$$\nu + e \rightarrow \nu' + e', \qquad (8.55)$$

which was discussed in Section 8.2, is sensitive to neutrinos of all types and has the advantage that it is highly directional. The main disadvantage of using this method is that the scattering by nonelectron type neutrinos is reduced by a factor of order of six or seven for the energies of greatest interest (see Table 8.8).

Bibliographical Notes

Bahcall, J.N. (1964) *Phys. Rev. Lett.*, **B136**, 1164. Discussion of neutrino–electron scattering in astrophysical contexts, using CVC theory.

Bahcall, J.N. (1964) *Phys. Rev. Lett.*, **12**, 300. First calculation of enhanced neutrino absorption by ^{37}Cl leading to the isotopic analogue state in ^{37}Ar.

Drukier, A. and L. Stodolsky (1984) *Phys. Rev.*, **30**, 2295. Original paper on using neutral-current elastic scattering on nuclei to detect solar neutrinos, including superconductivity grains as detectors.

Freedman, D.Z. (1974) *Phys. Rev. D*, **9**, 1389. Original analysis of coherent effects in elastic neutrino-nucleus scattering and their possible importance in stellar collapse and neutron star cooling.

Weinberg, S. (1987) *International Journal of Modern Physics* **A**, 2301. Cogent discussion of the importance of neutral-current experiments.

9

Beyond the standard model
of electroweak interactions

Summary

The **standard electroweak model** describes the electromagnetic and weak interactions of quarks and leptons. This quantum field theory is in excellent agreement with an enormous body of accurate data. In its simplest form, the model conserves lepton number, has zero masses for all of the neutrinos, negligible neutrino-magnetic moments, and no massive weakly interacting particles (WIMPs).

Is there physics beyond the standard model? Yes. Nearly all theoretical physicists believe that there is new physics to be discovered, that the standard model is incomplete. One reason for this belief is that the standard model contains a large number of parameters whose values are not explained and which must be determined from experiment. Also, the symmetries of the model are postulated, not derived, from an underlying principle.

This chapter describes several ideas related to particle physics that represent extensions of the standard electroweak model. The discussion begins in Section 9.1 with oscillations in vacuum of neutrinos. Oscillations between different neutrino states can occur if at least one neutrino eigenstate that is coupled to the electron neutrino has a nonzero mass and if the neutrino states that are created in weak-interaction decays are not states of definite mass, that is, not stationary states of the free Hamiltonian. The neutrino states, $|\nu_e\rangle$, $|\nu_\mu\rangle$, $|\nu_\tau\rangle$, that are produced in weak-interaction decays in association with particular charged leptons are called **flavor or current eigenstates**. The flavor eigenstates are linear combinations of the **mass**

eigenstates, the states that do diagonalize the free Hamiltonian. A flavor eigenstate is not a normal mode of the system. The wave function of an electron neutrino can be expanded in terms of the wave functions of mass eigenstates. Different mass eigenstates move with different speeds causing the relative phases of the expansion coefficients to vary with time. Depending upon the time of observation (i.e., the relative phase of the expansion coefficients), the original flavor eigenstate has more or less probability of being detected as an electron–neutrino eigenstate. Vacuum oscillations are an unlikely solution to the solar neutrino problem because the amount of mixing required is large. Fine tuning of a relation between neutrino masses, the average neutrino energy, and the Earth–Sun distance is necessary in order to obtain a big effect of vacuum oscillations on solar neutrino fluxes. In addition, the mixing angles must be much larger than the known quark mixing angles, a possible but *a priori* unattractive situation.

Neutrino oscillations in matter are considered in Section 9.2. The effect of electron neutrinos scattering off electrons in the Sun can cause the almost complete conversion of solar neutrinos of the electron type to neutrinos of a different flavor; this process is known as the **MSW effect**. Fine tuning is not required: The MSW effect can occur for a wide range of possible neutrino parameters. The mixing angles and mass differences can each vary by orders of magnitude and still remain within the parameter domain that solves the solar neutrino problem.

Flavor oscillations require that individual lepton number is not conserved and that some neutrinos have mass. These requirements are not revolutionary; they are satisfied by some well-studied examples of Grand Unified Theories (GUTs). The extension of the electroweak model to include neutrino oscillations that solve the solar neutrino problem implies that some neutrinos have masses between 10^1 eV and 10^{-6} eV.

Was the Sun designed to provide a convenient laboratory for studying neutrino masses via oscillations? It almost seems that way.

The ideas discussed in the latter part of this chapter – large magnetic moments, WIMPs with very special properties, and fast neutrino decay – are "way beyond the standard model." They represent significant changes in electroweak theory.

The published literature for this subject is enormous. Fortunately, there are systematic review articles that describe the history of the subject and cite many of the most important papers. The reader may wish to consult some of these reviews for additional material [see, e.g., Bilenky and Pontecorvo (1978), Mikheyev and Smirnov (1986d), Wolfenstein (1986), Bilenky and Petcov (1987), Harari and Nir (1987), Rosen and Gelb (1987), Weinberg (1987), Bahcall, Davis, and Wolfenstein (1988), and Boehm and Vogel (1988)].

9.1   Vacuum oscillations

Neutrino oscillations can explain the solar neutrino problem if the flavor eigenstate ν_e that is produced by weak-interaction processes in the Sun is largely converted to neutrinos of different flavors in transit to the Earth. Neutrino absorption experiments, which use for example ^{37}Cl or ^{71}Ga detectors, are sensitive only to ν_e (and not ν_μ or ν_τ) at the energies that are relevant for solar neutrinos.

How much can oscillations in vacuum decrease the terrestrial flux of ν_e's? The answer is: On the average not very much, but a lot for special choices of the parameters. The rate predicted by the standard model for the ^{37}Cl experiment, 7.9 SNU (see Chapter 10), is a factor of 3.8 times the observed rate of 2.1 SNU. The full reduction of 3.8 cannot be achieved by vacuum oscillations, unless the parameters are fine tuned and some unexpectedly large mixing angles exist. Of course, the solution of the solar neutrino problem could be a combination of two unlikely possibilities, namely, that the theoretical rate is near its extreme lower limit of 5.3 SNU (see Chapter 10) and the reduction from vacuum oscillations is a factor of 2.2.

There is some confusion in the literature about the magnitude of the probable reductions from vacuum oscillations. The next subsection is an attempt to clarify this situation by exhibiting the steps that are involved in deriving the minimum ν_e flux and presenting the probability distribution of observing a specific decrease in ν_e flux if the mixing angles were chosen randomly. The following subsection discusses the explicit solutions for oscillation among two neutrino states and shows what can be achieved by fine tuning the parameters. The last subsection expresses the problem of vacuum oscillations in a form that is convenient for discussing matter oscillations. In most equations, units are adopted in which $\hbar = c = 1$; the factors of $\hbar$ and c are shown for equations in which specific dimensional numbers are given.

A   Minimum flux

The flavor eigenstates $(\nu_e, \nu_\mu, \nu_\tau$, which are denoted here by Greek
subscripts $|\nu_\alpha\rangle, |\nu_\beta\rangle, ...)$ can be represented as linear combinations of
the mass eigenstates (denoted by numerical subscripts $|\nu_1\rangle, |\nu_2\rangle, ...$).
The relation between a state $|\nu_\alpha\rangle_t$ that was initially a flavor eigen-
state and the mass eigenstates $|j\rangle$ can be written

$$|\nu_\alpha\rangle_t = \sum_j U_{\alpha j} \exp(-iE_j t)|j\rangle, \qquad (9.1)$$

where U is a unitary matrix that can be chosen to be real if CP is
conserved. For simplicity in presentation, each state $|j\rangle$ is assumed
to be a momentum eigenstate (same momentum p); hence each state
$|j\rangle$ has a slightly different energy $E_j(p, m_j)$, where m_j is the (small)
neutrino mass. The amplitude for observing an initially created state
$|\nu_\alpha\rangle$ as the (different or same) flavor eigenstate $|\nu_\beta\rangle$ at some future
time t is

$$\langle \nu_\beta|\nu_\alpha\rangle_t = \sum_j U_{\alpha j} U_{\beta j}^* \exp(-iE_j t). \qquad (9.2)$$

Thus the probability for a transition from the state $|\nu_\alpha\rangle$ to the state
$|\nu_\beta\rangle$ is

$$|\langle \nu_\beta|\nu_\alpha\rangle_t|^2 = \sum_{j,k} U_{\alpha j} U_{\beta j}^* U_{\alpha k}^* U_{\beta k} \exp\left[-i\left(E_j - E_k\right)t\right]. \qquad (9.3)$$

For typical solar neutrino applications, the average over neutrino
energies causes the oscillating terms to cancel out for $j \neq k$ [see
Bahcall and Frautschi (1969)].

The smallest possible neutrino flux, ignoring oscillatory terms, is
found by minimizing the probability for an electron neutrino to remain an electron neutrino subject to the condition that the total
probability for all transitions is unity. The probability of an electron neutrino remaining an electron neutrino after a time t, or after
traveling a distance $d \cong ct$, is

$$|\langle \nu_e|\nu_e\rangle_t|^2 \rightarrow \sum_i |U_{ei}|^4 = \sum_i x_i^2, \qquad (9.4)$$

where

$$x_i \equiv |U_{ei}|^2. \qquad (9.5)$$

The average probability is, by design, independent of energy.

The condition that the total probability is conserved is enforced by introducing a Lagrangian multiplier, λ. With total probability conserved, the smallest probability for a ν_e remaining a ν_e is found by minimizing the function F, where

$$F = \sum_i x_i^2 + \lambda \left[\left(\sum_i x_i \right) - 1 \right]. \qquad (9.6)$$

Differentiating F, one sees that the minimum flux is achieved when all of the x_i are equal,

$$x_i = -\frac{\lambda}{2} = \frac{1}{N}. \qquad (9.7)$$

In this special case, each of the transition probabilities is equal to $1/N$, where N is the total number of neutrino flavors (usually assumed to be three). Therefore the probability of an electron neutrino remaining an electron neutrino satisfies

$$|\langle \nu_e | \nu_e \rangle_t|^2 \geq \frac{1}{N}. \qquad (9.8)$$

Thus the maximum diminution in flux is a factor of three if there are three neutrino flavors. This is sufficient to reduce the predicted neutrino event rate in the ^{37}Cl and Kamiokande II experiments to the observed rates (see Chapters 10 and 13) but only barely.

Figure 9.1 shows the probability for $N = 3$ that the flux is reduced by a factor of less than N_{reduce} if the matrix elements U_{ei} are drawn from a random distribution that conserves probability; the case $N = 6$ was studied by Nussinov (1976). There is a 50% chance that the reduction is less than a factor of 1.8 and a 74% chance that the reduction does not exceed a factor of 2.0. To achieve the minimum required reduction rate to explain the low event rate in the ^{37}Cl experiment, a factor of 2.2 reduction is required; there is only about one chance in five that the reduction factor is this large. There is a 12% chance that the reduction is as large or larger than a factor of 2.5 and only a 5% chance of exceeding 2.75.

Large reductions are unlikely for vacuum oscillations. But, of course, the matrix elements are not drawn from a random distribution. The actual matrix elements have a *post facto* probability of unity, independent of their magnitudes.

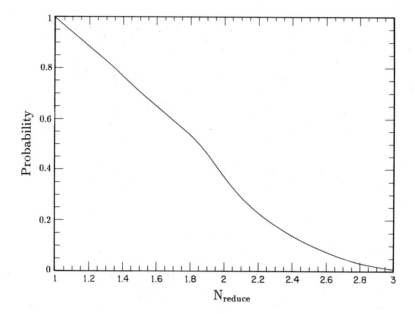

Figure 9.1   Vacuum oscillation probabilities for three neutrino flavors
The probability that the flux of electron-neutrinos is reduced by a factor
of more than N_{reduce} by vacuum oscillations. The probability distribution was calculated by assuming that all transition matrix elements U_{ei}
are equally likely.

B   Two flavors

Basic relations.   The most important aspects of neutrino oscillations can be understood by studying the explicit solution for two
interacting neutrinos. For this two-flavor problem, the flavor eigen-
states, $|f\rangle$, can be expressed in terms of the mass eigenstates, $|m\rangle$,
with the aid of a two-dimensional orthogonal matrix, $|f\rangle = U_{\mathrm{V}}|m\rangle$,
where,

$$U_{\mathrm{V}} = \begin{pmatrix} \cos\theta_{\mathrm{V}} & \sin\theta_{\mathrm{V}} \\ -\sin\theta_{\mathrm{V}} & \cos\theta_{\mathrm{V}} \end{pmatrix}. \tag{9.9}$$

Here θ_{V} denotes the **vacuum mixing angle**. Without loss of gener-
ality, we can choose $0 \leq \theta_{\mathrm{V}} < \pi/4$ so that $|\nu_e\rangle$ is "mostly"
$|\nu_1\rangle$. The second flavor eigenstate, $|\nu_x\rangle$, may be identified in practice
with ν_μ, ν_τ, with a linear combination of ν_μ and ν_τ, or even with a
fourth-generation neutrino.

The time evolution of an electron–neutrino state may be expressed

$$|\nu_e\rangle_t = \cos\theta_V \exp\left(-iE_1 t\right)|\nu_1\rangle + \sin\theta_V \exp\left(-iE_2 t\right)|\nu_2\rangle, \quad (9.10)$$

where E_1 and E_2 are the energies of the two mass eigenstates with the same momentum p. The probability amplitude for an electron neutrino remaining an electron neutrino after traveling for a time t is

$$\langle\nu_e|\nu_e\rangle_t = \cos^2\theta_V \exp\left(-iE_1 t\right) + \sin^2\theta_V \exp\left(-iE_2 t\right), \quad (9.11)$$

which corresponds to a probability of a ν_e remaining a ν_e of

$$|\langle\nu_e|\nu_e\rangle_t|^2 = 1 - \sin^2 2\theta_V \sin^2\left[\tfrac{1}{2}\left(E_2 - E_1\right)t\right]. \quad (9.12)$$

The two mass eigenstates are assumed to have the same momentum [for a discussion of this point see Kayser (1981)], which implies that they have slightly different energies if they have finite masses. The energy difference for relativistic neutrinos is

$$E_2 - E_1 = \frac{m_2^2 - m_1^2}{2E} \equiv \pm\frac{\Delta m^2}{2E}, \quad (9.13)$$

where the appearance of the plus and minus signs reflects the introduction of a positive-definite quantity, Δm^2,

$$\Delta m^2 \equiv |m_1^2 - m_2^2|. \quad (9.14)$$

The plus sign applies if $m_2 > m_1$ and the minus sign in the opposite case.

The probability that an electron neutrino remains an electron neutrino may be written therefore in the following form that appears frequently in the literature:

$$|\langle\nu_e|\nu_e\rangle_t|^2 = 1 - \sin^2 2\theta_V \sin^2\left(\frac{\pi R}{L_V}\right), \quad (9.15)$$

where R is the distance traveled in a time t and $L_V \equiv 4\pi E/\Delta m^2$. The probability of observing a neutrino of a different flavor, $|\nu_x\rangle$, is

$$|\langle\nu_x|\nu_e\rangle_t|^2 = \sin^2 2\theta_V \sin^2\left(\frac{\pi R}{L_V}\right). \quad (9.16)$$

Also, the probability of a ν_x neutrino being observed as an electron neutrino is the same as the reverse process

$$|\langle\nu_e|\nu_x\rangle_t|^2 = |\langle\nu_x|\nu_e\rangle_t|^2. \quad (9.17)$$

The magnitude of vacuum mixing is proportional to $\sin^2\theta_V$, which is expected to be small because the possibly analogous quark mixing angles are small. The sine squared of the Cabibbo angle, which measures the admixture of down quarks with strange quarks, is $(0.22)^2 = 0.05$. If the Cabibbo angle is inserted for θ_V in Eq. (9.15), then the average survival probability of an electron neutrino is about 90%. Prior to the discovery of the MSW effect, most theoretical physicists were skeptical of the possibility that neutrino oscillations played an important role in the solar neutrino problem. The analogy with the known small quark mixing angles made the required large mixing angles seem unattractive.

The **vacuum oscillation length**, L_V can be written in the form

$$L_V \equiv \frac{4\pi E \hbar}{\Delta m^2 c^3} = 2.48 \left(\frac{E}{\text{MeV}}\right) \left(\frac{\text{eV}^2}{\Delta m^2}\right) \text{ m}, \qquad (9.18)$$

which is often used in discussing terrestrial oscillation experiments that employ beams from reactors or accelerators [see Bilenky and Pontecorvo (1978) and Boehm and Vogel (1987)].

To observe variations with distance or time due to neutrino oscillations, a neutrino source must be localized within a region that is much less than L_V. By the uncertainty principle, the neutrino wave function must contain a spread of momenta, $\Delta p \gg \hbar/L_V$. Thus a rigorous treatment of neutrino oscillations requires the construction of wave packets, but for practical situations (nearly) always yields the same results as the simpler analysis with a fixed momentum that is presented here and elsewhere in the literature. For a delightfully clear pedagogical description of the quantum mechanics of neutrino oscillations, see Kayser (1981).

For discussing solar neutrinos, it is more convenient to rewrite the distance-dependent dimensionless argument of the sine function as follows:

$$\frac{\pi R}{L_V} = 1.9 \times 10^{11} \left(\frac{1 \text{ MeV}}{E}\right) \left(\frac{\Delta m^2}{\text{eV}^2}\right) \left(\frac{R}{1 \text{ AU}}\right), \qquad (9.19)$$

where 1 AU is the average distance between the Earth and the Sun.

Solar neutrino experiments are sensitive to mass differences for which the argument $\pi R/L_V$ can significantly affect the probability of observing an electron neutrino at Earth. The minimum mass difference squared, $\Delta m^2_{\text{solar}}$, that can be studied with solar neutrinos is

$$\Delta m^2_{\text{solar}} = 1.6 \times 10^{-12} \text{ eV}^2 \left(\frac{E}{1 \text{ MeV}}\right), \qquad \left(\frac{\pi R}{L_V}\right) = 0.3, \quad (9.20)$$

Table 9.1.   Values of R/E
accessible in different
experiments. The values of
the distance R are measured
in m and the energies E in
MeV.

Experiment	R/E
Accelerator	$10^{-2} - 10^{1}$
Reactor	$10^{0} - 10^{2}$
Atmospheric	$10^{2} - 10^{4}$
Solar	$10^{10} - 10^{11}$
Supernova	$10^{19} - 10^{20}$

which is many orders of magnitude smaller than can be achieved in terrestrial experiments.

Table 9.1 compares the relative sensitivity of different types of experiments to neutrino mass differences. The figure of merit is R/E, which determines the magnitude of the mass-dependent term in the expression for $|\langle \nu_e | \nu_e \rangle_t|^2$.

Averages. The event rate from each neutrino source must be averaged over the region of emission in the Sun, the region of absorption in the Earth (the distance from the Sun to the Earth depends upon the time of year), and the energy spectrum of the neutrino source. This triple average may be written symbolically as

$$\langle \phi_i \sigma_{i,\alpha} \rangle = \left\langle \int dE \, \phi_{\text{model},i}(E) \, \sigma_i(E) \, |\langle \nu_e | \nu_e \rangle_t|^2 \right\rangle_{\text{Sun, Earth}} \tag{9.21}$$

In their seminal paper, Gribov and Pontecorvo (1969) [see also Pontecorvo (1968); both papers are "must" reading, see Bibliographical Notes] discussed the first two averages, which affect the value of the distance R in the oscillatory term for the survival probability [see Eq. (9.15)]. For vacuum oscillations, the different positions of production in the core of the Sun change the argument of the oscillatory term by a fractional amount $\sim 10^{-4}$ and the Earth–Sun distance changes the argument by $\sim 10^{-2}$.

The average over energies is more important since solar neutrino sources generally have broad energy spectra. The phase ϕ that occurs in the equation for the survival probability is

$$\phi = \frac{\pi R}{L_{\mathrm{V}}} = 2 \times 10^{10} \left[\left(\frac{\Delta m^2}{1 \ \mathrm{eV}^2} \right) \left(\frac{10 \ \mathrm{MeV}}{E} \right) \left(\frac{R}{10^{13} \ \mathrm{cm}} \right) \right]. \quad (9.22)$$

A small change in the neutrino energy E or the distance R causes ϕ to change by many factors of 2π. Most of the important sources, for example, the ^{8}B and pp neutrinos, have continuous energy spectra (see Section 6.3), causing the argument of the oscillatory function, ϕ, to change by many orders of magnitude ($\sim 10^{10}$) as the energy varies from its smallest to its largest significant value.

For line neutrino spectra, the thermal broadening and the variation of the distance R cause a fractional phase change of order

$$\frac{\delta\phi}{\phi} \sim \left[\left(\frac{kT}{E} \right) + \left(\frac{\delta R}{R} \right) \right]. \quad (9.23)$$

The thermal energy broadening is of order $kT/E \sim 10^{-3}$ for pep and ^{7}Be line neutrinos. The thermal motion of the ions causes large phase changes, $\delta\phi \sim \pi$, for $\Delta m^2/E > 10^{-8}$ eV^2/MeV. The variation of the positions at which the neutrino is created in the Sun causes a fractional phase change of almost the same size, $\delta R/R \sim 5 \times 10^{-3}$. The periodic change in $\delta R/R$ due to the ellipticity of the Earth's orbit is about an order of magnitude larger.

Performing the average over energies, Bahcall and Frautschi (1969) showed that in most cases of interest for vacuum oscillations of solar neutrinos

$$|\langle \nu_{\mathrm{e}} | \nu_{\mathrm{e}} \rangle_t|^2_{\mathrm{average}} = 1 - \tfrac{1}{2} \sin^2 2\theta_{\mathrm{V}}. \quad (9.24)$$

The average over energies removes, for typical parameters, the resonances that were discussed by Gribov and Pontecorvo.

Fine tuning. How much can one do by fine tuning the parameters? A lot, but the tuning must be precise.

Figure 9.2 shows the effective cross section for absorption of ν_{e} by ^{37}Cl, averaged over the ^{8}B neutrino spectrum and weighted by the probability of an electron neutrino remaining an electron neutrino, $|\langle \nu_{\mathrm{e}} | \nu_{\mathrm{e}} \rangle_t|^2$. The ordinate of Figure 9.2 is the ratio, $\sigma(D)/\sigma(D=0)$, of

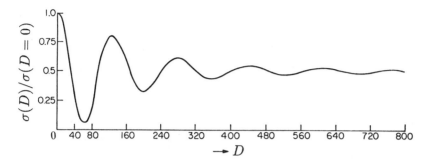

Figure 9.2   The effect of averaging over energy   The figure shows the dependence of the average ^{8}B neutrino absorption cross section versus a quantity D, defined in Eq. (9.26), which is proportional to $|\Delta m^2|/R$. A large decrease is seen at $D \sim 60$, corresponding to $\Delta m^2 \sim 5 \times 10^{-11}$ eV^2. For large values of D, the reduction in capture rate is always a factor of two [from Bahcall and Frautschi (1969)].

the effective ^{8}B neutrino absorption cross section for ^{37}Cl computed with and without vacuum oscillations. Thus

$$\frac{\sigma(D)}{\sigma(0)} = \frac{\int dE\, \phi(E)\sigma(E)|\langle \nu_e|\nu_e\rangle_t|^2}{\int dE\, \phi(E)\sigma(E)}. \tag{9.25}$$

The abscissa of Figure 9.2 is

$$D = \frac{|\Delta m^2|}{(\mathrm{eV}^2)} \left(\frac{R}{1\ \mathrm{AU}}\right) \times 10^{12}. \tag{9.26}$$

The argument of the oscillatory term, $\pi R/L_V$, is $(m_e/E_\nu)D$, where m_e is the electron's mass. For specificity, Figure 9.2 was plotted assuming that the diagonal components of the mass matrix were equal (i.e., $\theta_V = 45°$).

The reduction in effective cross section caused by the factor of $|\langle \nu_e|\nu_e\rangle_t|^2$ is not seen in Figure 9.2 until $m_e D/\langle E_\nu\rangle \sim \pi/2$ or $D \sim 30$. A large decrease occurs at $D \sim 60$, corresponding to $m_e D/\langle E_\nu\rangle \sim \pi$, or

$$\Delta m^2 \sim 5 \times 10^{-11}\ \mathrm{eV}^2. \tag{9.27}$$

For large values of D, the effect of vacuum oscillations is a factor of two independent of the precise values of the neutrino masses, the region of production, or the Earth–Sun distance.

Oscillations with the value of Δm^2 given in Eq. (9.27) have been called "just-so" oscillations by Glashow and Krauss (1987) [see also

Barger *et al.* (1981)]. If the mass difference has the just-so value, then the gallium signal will be suppressed by a factor of between 0.40 and 0.65. Glashow and Krauss also show that there is a characteristic semiannual periodicity in both the event rate and the shape of the spectrum due to the varying Earth–Sun distance in Eq. (9.19) [see also Gribov and Pontecorvo (1969)].

Coherence. Are solar neutrino eigenstates with different masses coherent when they arrive at the Earth? The answer is not obvious. Nussinov (1976) pointed out that nondegenerate mass eigenstates propagate at different velocities and therefore must have been emitted at slightly different times (or in different places) if they arrive at the same time at Earth. Interference between the different eigenstates will not occur unless the source is coherent over a time interval comparable to the difference in arrival times. The condition for quantum mechanical coherence is

$$(v_1 - v_2)t \lesssim c\tau_{\mathrm{coh}}, \tag{9.28}$$

where τ_{coh} is the coherence time of the source. The original value of τ_{coh} estimated by Nussinov was too short; Krauss and Wilczek (1985a) estimate $\tau_{\mathrm{coh}} \sim 10^{-15}$ s. An improved estimate is given by Loeb (1989). The distance, d, over which different mass eigenstates are coherent is, when expressed in astronomical units (1 AU = 1.5 $\times 10^{15}$ cm, the Earth–Sun distance),

$$d_{\mathrm{coh}} \lesssim 4\ \mathrm{AU} \left[\left(\frac{E}{10\ \mathrm{MeV}} \right)^2 \left(\frac{10^{-4}\ \mathrm{eV}^2}{\Delta m^2} \right) \left(\frac{\tau_{\mathrm{coh}}}{10^{-15}\ \mathrm{s}} \right) \right]. \tag{9.29}$$

Solar neutrinos of typical energies 10 MeV and squared mass differences less than $10^{-3.5}$ eV^2 can in principle be coherent mixtures of two or more mass states when they arrive at the Earth. However, the phase-averaging caused by the finite-energy width of the neutrino spectrum or by the source extension [see Eqs. (9.22) and (9.23)] is more important in practice, except for neutrino lines from pep or ^{7}Be.

C   The mass matrix

The equations describing vacuum oscillations can be rewritten in the form of a Schrödinger equation with a mass matrix playing the role of the Hamiltonian. This formulation is particularly convenient in discussing the MSW effect, which is considered in the next section. For simplicity, the equations are often written as if we knew that only two neutrino flavors are coupled to each other. The generalization to three or more flavors is straightforward although sometimes cumbersome.

An arbitrary neutrino state can be written in the flavor basis as

$$|\nu\rangle_t = c_e(t)|\nu_e\rangle + c_x(t)|\nu_x\rangle. \qquad (9.30)$$

In particular, one can write for $|\nu_e\rangle_t$,

$$|\nu_e\rangle_t = \cos\theta_V \exp(-iE_1 t)|\nu_1\rangle + \sin\theta_V \exp(-iE_2 t)|\nu_2\rangle, \qquad (9.31)$$

with an analogous relation for $|\nu_x\rangle$.

The matrix describing the time dependence of an arbitrary state expressed in the flavor basis can be derived by differentiating the equations for $|\nu_e\rangle_t$ and $|\nu_\mu\rangle_t$, reexpressing $|\nu_1\rangle$ and $|\nu_2\rangle$ in terms of the flavor basis eigenstates, and combining the results. Denoting the mass matrix in the flavor representation by

$$M_{0,\text{ flavor}} = \begin{pmatrix} a_1 & b \\ b & a_2 \end{pmatrix}, \qquad (9.32)$$

one finds

$$a_1 = \frac{m_1^2 \cos^2\theta_V + m_2^2 \sin^2\theta_V}{2E}, \qquad (9.33)$$

$$a_2 = \frac{m_1^2 \sin^2\theta_V + m_2^2 \cos^2\theta_V}{2E}, \qquad (9.34)$$

and

$$b = \frac{\Delta m^2 \sin 2\theta_V}{4E}. \qquad (9.35)$$

The most symmetric form of the mass matrix is obtained by subtracting an appropriate constant times the unit matrix, which only changes the overall state vector by a phase and does not affect probability amplitudes. The final result is

$$i\frac{d}{dt}\begin{pmatrix} c_e(t) \\ c_x(t) \end{pmatrix} = \pm\frac{\Delta_V}{2}\begin{pmatrix} -\cos 2\theta_V & \sin 2\theta_V \\ \sin 2\theta_V & \cos 2\theta_V \end{pmatrix}\begin{pmatrix} c_e(t) \\ c_x(t) \end{pmatrix}, \qquad (9.36)$$

where the plus sign applies if $m_2 > m_1$ and the minus sign if $m_2' < m_1$ [see Eqs. (9.13) and (9.14)]. The positive-definite quantity Δ_V is the energy difference between the (relativistic) states $|\nu_e\rangle$ and $|\nu_x\rangle$,

$$\Delta_V = |\Delta m^2/2E|. \tag{9.37}$$

The mass eigenvalues obtained by diagonalizing the mass matrix are

$$E_{1,2} = \text{const.} \pm 1/2\Delta_V, \tag{9.38}$$

where the constant term is included to remind the reader that an arbitrary constant multiple of the unit matrix was subtracted from the mass matrix. Only energy differences are significant for the problem of interest.

The **vacuum oscillation length** is defined by

$$L_V = \frac{2\pi}{|E_1 - E_2|} = \frac{2\pi}{\Delta_V}, \tag{9.39}$$

which is equivalent to Eq. (9.18).

The Hamiltonian formulation of vacuum neutrino oscillations given above is the starting point for investigating matter oscillations.

9.2   Matter oscillations

The published literature on the MSW effect is rich and beautiful, reflecting the nature of the subject which began with the fundamental papers of Wolfenstein (1978, 1979) and Mikheyev and Smirnov (1986a,b,c). Readers can spend profitable and enjoyable hours studying original papers which treat fully aspects of the subject that are not discussed in detail here. For convenience, references to a representative set of papers on this subject are collected together in Section 9.2G.

A   Time (or spatial) evolution

An arbitrary state vector can again be written in terms of the flavor eigenstates as

$$|\nu\rangle_t = c_e(t)|\nu_e\rangle + c_\mu(t)|\nu_\mu\rangle + c_\tau(t)|\nu_\tau\rangle. \tag{9.40}$$

The time evolution of the state vector is determined by the equation

$$i\frac{\mathrm{d}}{\mathrm{d}t}\begin{pmatrix} c_e(t) \\ c_\mu(t) \\ c_\tau(t) \end{pmatrix} = M \begin{pmatrix} c_e(t) \\ c_\mu(t) \\ c_\tau(t) \end{pmatrix}, \qquad (9.41)$$

where the mass matrix consists of the vacuum mass matrix, M_0, that was discussed in Section 9.1D and a contribution due to the presence of matter, M_{matter}:

$$M = M_0 + M_{\text{matter}}. \qquad (9.42)$$

The form of M_{matter} was derived first by Wolfenstein (1978, 1979), who pointed out that neutrino–electron scattering contributes a term to the mass matrix that is not present in vacuum. The neutral-current scattering is the same to high accuracy, in the standard electroweak model, for all flavors of neutrinos and therefore only adds to the mass matrix an unimportant constant times the unit matrix.

The term introduced by Wolfenstein arises from the Feynman diagram shown in Figure 9.3, which can only occur via charged-current scattering. An electron neutrino emits an electron and a positively charged intermediate boson, which is then absorbed by an ambient electron to become an electron neutrino. There is no corresponding diagram for neutrinos of other flavors.

The key point is that the charged-current neutrino–electron scattering introduces a term in the mass matrix that singles out electron neutrinos. This term can be written

$$M_{\text{matter}} = \sqrt{2}\,G_{\mathrm{F}}\,n_e\,P_e = \sqrt{2}\,G_{\mathrm{F}}\,n_e\,|\nu_e\rangle\langle\nu_e|, \qquad (9.43)$$

where P_e is the projection operator for electron neutrinos and n_e is the electron number density. For antineutrinos of the electron flavor, $\bar{\nu}_e$, the coefficient of the projection operator has the same magnitude but the opposite sign. Here G_{F} is the usual Fermi coupling constant.

The **Wolfenstein term** is the coefficient of P_e in Eq. (9.43) and represents the contribution of neutrino–electron scattering to the index of refraction of neutrinos in matter. The classical expression for the index of refraction is $2\pi N f(0)/p^2$, were N is the number density of scatterers, p the momentum, and $f(0)$ the forward scattering ampli-

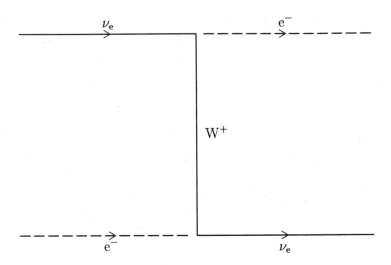

Figure 9.3   Feynman diagram responsible for the MSW effect   This process can occur for ν_e, but not for ν_μ or ν_τ.

tude. The reason for the appearance of the factors G_F and n_e in the Wolfenstein term is obvious from the classical result.[†]

The total mass matrix M can be expressed, following the argument given in Section 9.1D, in terms of the vacuum mixing angle, θ_V, the mass difference squared, Δm^2, and the energy via the term, Δ_V, and the product $G_F n_e$.

[†] The Fermi constant appears linearly since the index of refraction depends upon the forward scattering amplitude, not on a transition probability. There is a saga associated with the coefficient of the Wolfenstein term; the factor of $\sqrt{2}$ was first given correctly by Lewis (1980) and the minus sign was derived by Langacker *et al.* (1983). Mikheyev and Smirnov (1986a) used the incorrect sign in their first paper, which was based upon the original papers of Wolfenstein (1978, 1979). The faulty sign imposed the implausible requirement that the mass of ν_e exceed the mass of ν_μ or ν_τ. For this reason, the discovery of Mikheyev and Smirnov that matter oscillations could be resonant did not initially attract the attention it deserved. The situation was publicly clarified, and the MSW effect gained greatly in apparent plausibility, when Langacker pointed out to Bethe (1986) the correct sign. A simple heuristic derivation of the Wolfenstein term is given in Section 5.8 of Boehm and Vogel (1988). The equation for the time evolution in terms of a mass matrix has been derived from relativistic theory by Halperin (1986), Mannheim (1988), and Baltz and Weneser (1988).

Two neutrino flavors are sufficient for understanding most of the physical results of the MSW effect. In this case, the flavor eigenstates are related to the mass eigenstates by a unitary transformation U_M that is analogous to the vacuum transformation introduced in Section 9.1B. In matter, the mixing angle is denoted by θ_M.

The time development of the flavor eigenstates in matter is described by an equation that is analogous to the vacuum Schrödinger equation:

$$i\frac{d}{dt}\begin{pmatrix} c_e(t) \\ c_x(t) \end{pmatrix} = \pm\frac{\Delta_M}{2}\begin{pmatrix} -\cos 2\theta_M & \sin 2\theta_M \\ \sin 2\theta_M & \cos 2\theta_M \end{pmatrix}\begin{pmatrix} c_e(t) \\ c_x(t) \end{pmatrix}. \quad (9.44)$$

The plus sign applies for $m_2 > m_1$ and the minus sign for $m_1 > m_2$. The expressions relating Δ_M and θ_M to their vacuum counterparts are obtained by equating the elements of M given above to the elements obtained in terms of the vacuum expressions using the explicit formulae for M_0 and M_{matter}.

The energy eigenvalues are

$$E_{1,2} = \text{const.} \pm 1/2\Delta_M, \quad (9.45)$$

where, as in Eq. (9.38), the constant is included because an arbitrary multiple of unity has been subtracted from the mass matrix. The explicit form for Δ_M is

$$\Delta_M = \left[\left(\pm\Delta_V \cos 2\theta_V - \sqrt{2}G_F n_e\right)^2 + (\Delta_V \sin 2\theta_V)^2\right]^{1/2}. \quad (9.46)$$

The plus sign is again to be used for $m_2 > m_1$ and the minus sign for the opposite case.

The **mixing angle in matter** is determined by the relation

$$\tan 2\theta_M = \frac{\tan 2\theta_V}{[1 \pm (L_V/L_e) \sec 2\theta_V]}, \quad (9.47)$$

where in Eq. (9.47) the plus sign applies for $m_2 < m_1$ and the minus sign for $m_2 > m_1$. The **neutrino–electron interaction length**, L_e which appears in the formula for $\tan 2\theta_M$ is defined by

$$L_e = \frac{\sqrt{2}\pi\hbar c}{G_F n_e} = 1.64 \times 10^5 \text{ m} \left(100 \text{ g cm}^{-3}/\mu_e\rho\right). \quad (9.48)$$

The electron mean molecular weight, μ_e, can be calculated with the aid of the hydrogen mass fraction, X, using

$$\mu_e = \frac{(1 + X)}{2}. \quad (9.49)$$

The **Fermi constant** $G_F = 1.436 \times 10^{-49}$ erg cm^3 [or $10^{-5}/m_p^2$ in units ($\hbar = c = 1$) preferred by particle physicists]. A useful relation for numerical calculations is

$$\frac{G_F N_A R_\odot}{\hbar c} = 1.94 \times 10^2, \quad \frac{n_e(0)}{N_A} = 98.6 \text{ cm}^{-3}, \tag{9.50}$$

where $n_e(0)$ is the electron number density at the center of the standard solar model and N_A is Avogadro's number ($N_A = 6.02 \times 10^{23}$).

The interaction length, L_e, is independent of energy, unlike the oscillation lengths in vacuum or in matter.

The resonant character of matter oscillations is shown by Eq. (9.47) for $\tan\theta_M$. If $m_2 \geq m_1$, the mixing angle in matter is maximal ($\theta_M = \pi/4$) for an electron density determined by

$$\left(\frac{L_V}{L_e}\right)_{\text{res}} = \cos 2\theta_V, \tag{9.51}$$

which is often referred to as the "resonance condition."

There is no resonance for electron neutrinos if $m_2 < m_1$. (In this case there would be a resonance for $\bar{\nu}_e$.) The subsequent discussion in this section assumes $m_2 > m_1$.

Even for a very small vacuum mixing angle, the matter mixing angle is $\theta_M = \pi/4$ if the electron density satisfies the resonance condition. At the resonant density, the two diagonal matrix elements of the mass matrix are equal. Mixing between ν_e and ν_x can become very large at resonance and the smallness of the vacuum mixing angle is no longer important.

The **MSW resonance density** is

$$n_{e,\text{res}} = \frac{|\Delta m^2| \cos 2\theta_V}{2\sqrt{2} G_F E}, \tag{9.52}$$

or

$$\frac{n_{e,\text{res}}}{N_A} \simeq 66 \cos 2\theta_V \left(\frac{|\Delta m^2|}{10^{-4} \text{ eV}^2}\right)\left(\frac{10 \text{ MeV}}{E}\right). \tag{9.53}$$

Numerical values of the electron density, n_e, in the standard solar model are given in Table 4.5; the radial dependence of n_e is illustrated in Figure 4.1d and discussed in Section 4.3.

An electron neutrino will always pass through resonance if its energy exceeds the resonance value for an electron density corresponding to the solar center. Therefore, the minimum neutrino energy that can be resonant in the Sun is

$$E_{\min} = 6.6 \cos 2\theta_{\rm V} \left(\frac{\Delta m^2}{10^{-4}\ {\rm eV}^2}\right) {\rm MeV}. \qquad (9.54)$$

A significant fraction of the ^{8}B neutrinos created in the solar core will pass through a resonance density for mass differences of order 10^{-4} eV^2 or smaller.

B Constant density case

Many of the features of the full MSW effect can be understood by considering the explicit solution of the two-flavor problem for a constant density. The results for this case were given first by Wolfenstein (1978, 1979).

The difference between the energy eigenvalues for the two mass eigenstates is

$$E_1 - E_2 \equiv \Delta_{\rm M} = \Delta_{\rm V} \times D_{\rm M}, \qquad (9.55)$$

where the ubiquitous quantity $\Delta_{\rm M}$ has been written in terms of $\Delta_{\rm V}$ and the conveniently expressed matter correction,

$$D_{\rm M} = \left[1 - 2\left(\frac{L_{\rm V}}{L_{\rm e}}\right)\cos 2\theta_{\rm V} + \left(\frac{L_{\rm V}}{L_{\rm e}}\right)^2\right]^{1/2}. \qquad (9.56)$$

At the resonance density, $L_{\rm V}/L_{\rm e} = \cos 2\theta_{\rm V}$,

$$D_{\rm M}({\rm resonance}) = \sin 2\theta_{\rm V}, \qquad (9.57)$$

which simplifies some of the algebraic manipulations.

The probability that an electron neutrino remains an electron neutrino has the same algebraic form as for vacuum oscillations, that is,

$$|\langle \nu_x | \nu_e \rangle_t|^2 = 1 - \sin^2 2\theta_{\rm M} \sin^2\left[\frac{\pi R}{L_{\rm M}(E)}\right]. \qquad (9.58)$$

The probability of being observed as a neutrino of a different flavor is

$$|\langle \nu_x | \nu_e \rangle_t|^2 = \sin^2 2\theta_{\rm M} \sin^2\left[\frac{\pi R}{L_{\rm M}(E)}\right]. \qquad (9.59)$$

The **matter oscillation length** is defined in analogy with the vacuum oscillation length [see Eq. (9.39)],

$$L_{\rm M} \equiv \frac{2\pi}{|E_1 - E_2|} = \frac{L_{\rm V}}{D_{\rm M}}. \qquad (9.60)$$

The mixing angle in matter is given by the relation

$$\sin 2\theta_M = \frac{\sin 2\theta_V}{D_M},$$ (9.61)

where, for any nonzero vacuum mixing angle,

$$\sin 2\theta_M\Big|_{\text{resonance}} = 1$$ (9.62)

or $\theta_M = \pi/4$.

At very high electron densities, the matter mixing angle approaches $\pi/2$,

$$\theta_M \to \pi/2, \quad n_e \to \infty,$$ (9.63)

provided $m_2 > m_1$. This result is of special importance in understanding the analytic formulae for the variable density case that are discussed in the next section.

The probability of an electron neutrino making a transition to a different flavor state can be written more simply in the three limiting cases in which the vacuum oscillation length is much larger than, much smaller than, or equal to $\cos 2\theta_V$ times the neutrino–electron interaction length. The simplified results are

$$|\langle \nu_x | \nu_e \rangle_t|^2$$
$$= \begin{cases} \sin^2 2\theta_V \sin^2 (\pi R/L_V), & L_V/L_e \ll 1 \\ (L_e/L_V)^2 \sin^2 2\theta_V \sin^2 (\pi R/L_e), & L_V/L_e \gg 1 \\ \sin^2 (\pi R \sin 2\theta_V/L_V), & L_V/L_e = \cos 2\theta_V. \end{cases}$$ (9.64)

If the vacuum oscillation length is much less than the neutrino–electron interaction length (very few ambient electrons), then matter oscillations reduce to vacuum oscillations. The formula for this case is identical to the result obtained in Section 9.1 [see Eq. (9.16)]. If the vacuum oscillation length is much greater than the neutrino–electron interaction length [many electrons present, an example that was discussed by Wolfenstein (1978) in connection with the solar neutrino problem], then mixing is *inhibited* by a factor of $(L_e/L_V)^2$. In the resonance case, the third limit in the equation displayed above, the amount of mixing is equal to the square of an energy-dependent oscillatory function, which will typically yield $1/2$ when averaged over energy.

For a constant density, the maximum amount of mixing is a factor of two after averaging over a continuous energy spectrum. The possibility of large flavor conversion, which is the essence of the MSW effect, exists only for a slowly varying electron density. This case is discussed next.

C Variable density

The full power of matter oscillations shows up only when the electron density is variable. In this case, neutrinos of one flavor can be almost completely converted to neutrinos of another flavor.

Physical picture. The two mass eigenvalues, $\pm\Delta_M/2$, that were evaluated in the previous section are as close together as possible when the electron density satisfies the resonance condition,

$$\sqrt{2}G_F n_e = \Delta_V \cos 2\theta_V. \quad (9.65)$$

The process of neutrino flavor conversion in matter, the MSW effect, can be visualized most easily in terms of the near crossing of almost degenerate energy levels [Bethe (1986)]. At the resonant density, the two propagation eigenstates are separated by a tiny energy difference δE, which is

$$\delta E = 5 \times 10^{-12} \sin 2\theta_V \text{ eV} \left[\left(\frac{\Delta m^2}{10^{-4} \text{ eV}^2} \right) \left(\frac{10 \text{ MeV}}{E} \right) \right]. \quad (9.66)$$

Figure 9.4 illustrates how a neutrino beam can change completely its flavor when the electron density varies slowly. At very small electron densities, like those available in normal laboratory experiments, the electron neutrino almost coincides, for small vacuum mixing angles, with the lower mass eigenstate $|\nu_1\rangle$ and the second flavor eigenstate approximately coincides with the mass eigenstate $|\nu_2\rangle$. The mass of the heavier mass eigenstate increases monotonically with ambient electron density and becomes asymptotically proportional to the electron density. If $m_2 \gtrsim m_1$, then at large electron densities the heavier mass eigenstate, $|\nu_2\rangle$, becomes close to $|\nu_e\rangle$ and the lower mass eigenstate, $|\nu_1\rangle$, remains essentially unchanged as a function of electron density (neglecting neutral-current interactions that are the same for all neutrinos). An electron neutrino that is created in the solar interior moves from high densities to lower densities. If the electron density changes at a sufficiently slow rate (a requirement known as the adiabatic condition), no transitions occur between the mass eigenstates; the electron neutrino remains close to the mass eigenstate $|\nu_2\rangle$ and ex-

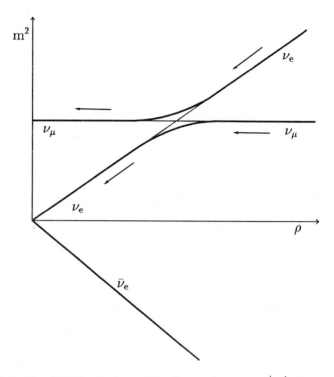

Figure 9.4   The MSW effect   The flavor eigenstate $|\nu_e\rangle$ is created at
high densities, ρ, in the solar interior where it is approximately the same
as the heavier mass eigenstate $|\nu_2\rangle$.   As the electron density decreases
slowly, the flavor eigenstate remains close to the mass eigenstate $|\nu_2\rangle$. Upon
emerging from the Sun to an essentially vacuum environment, the original
$|\nu_e\rangle$ is close to the vacuum flavor eigenstate $|\nu_x\rangle$ and is not observable
in the ^{37}Cl or ^{71}Ga detectors, which (for the relevant neutrino energies)
are sensitive only to ν_e. The sign of the Wolfenstein term is opposite for
neutrinos and antineutrinos, which explains why the mass of $\bar{\nu}_e$ decreases
with increasing density. (For simplicity, the effects of neutral currents are
omitted.)

its from the Sun as essentially a different flavor eigenstate, $|\nu_x\rangle$.
The path of a flavor eigenstate created in the Sun is illustrated
in the figure by the arrows pointed toward lower electron densities.

The process of flavor conversion is reversible. If one shines into
the Sun a beam of muon neutrinos of the right energies to undergo
the MSW effect, the muon neutrinos will become electron neutrinos

when they reach the solar center. The evolution of muon neutrinos will be in the opposite direction to the arrows in Figure 9.4.

Matter oscillations are similar to phenomena in a number of different fields, including the coherent regeneration of K-mesons [Wolfenstein (1978)] and spin precession in a changing magnetic field [Bouchez *et al.* (1986)]. A particularly simple classical analogy was described independently by Mikheyev and Smirnov (1986d) and Weinberg (1987), who likened the MSW effect to a system of weakly coupled oscillators (for definiteness, taken to be two pendulums). Oscillations of one pendulum correspond to the propagation of ν_e; oscillations of the second pendulum to the propagation of ν_x. The creation of a ν_e is analogous to the excitation of the "ν_e-pendulum." The coupling causes oscillations to be transferred from one pendulum to the other; the periodic exchange of pendulum oscillations corresponds to neutrino oscillations. The presence of matter changes the frequencies of the oscillators, affecting differently the ν_e and the ν_x pendulums. The coincidence of the oscillation frequencies of the pendulums corresponds to the near equality of the eigenvalues of the neutrino mass eigenstates. In both systems, oscillating pendulums and neutrinos, the exchange of excitations may be complete. In an experimental *tour de force*, Weinberg (1987) performed a public demonstration of the equivalence of the MSW effect to phenomena observed with a pair of mechanical oscillators.

Figure 9.5 illustrates the rotation of the state vector $|\nu_2\rangle$ in a plane defined by the flavor eigenstates, $|\nu_e\rangle$ and $|\nu_x\rangle$. At the center of the Sun, a ν_e is, because of the large ambient electron density, close to ν_2. As the density decreases toward the MSW resonance density, the ν_2 gradually becomes a mixture of equal amplitudes of the vacuum flavor eigenstates. At very low electron densities, ν_2 is very close to ν_x. If the change in electron density is sufficiently slow (the adiabatic approximation), then the original ν_e will remain close to ν_2 and therefore be transformed into an almost pure ν_x state on exiting from the Sun.

Analytic results. How can this physical picture be represented mathematically[†]? Parke (1986a,b) gives a simple analytic treatment

---

[†]The first analytic results were obtained in the adiabatic approximation, which will be discussed later in this section. In the adiabatic approximation, the con-

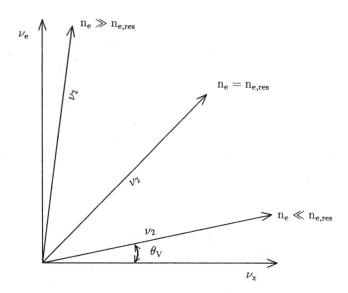

Figure 9.5   Matter rotation of the heavier neutrino eigenstate   This figure shows the rotation of the heavier mass eigenstate in a coordinate system in which the flavor eigenstates are the orthogonal axes. At densities much above the resonance density, the eigenstates $|\nu_e\rangle$ and $|\nu_2\rangle$ are very similar. As the electron density decreases, the mass eigenstate $|\nu_2\rangle$ becomes more similar to $|\nu_x\rangle$. If the density changes sufficiently slowly, then a $|\nu_e\rangle$ that is created at the center of the Sun can remain close to $|\nu_2\rangle$, emerging at low densities as essentially a $|\nu_x\rangle$.

of the variable density problem that contains the essential physical results. The following discussion is based upon his papers. The limits of validity of the approximations used here are evaluated carefully in papers that are referred to in the introduction to this section, including Haxton (1986), Petcov (1987), Toshev (1987a,b), and Baltz and Weneser (1988). The differences between the analytical results and the precise numerical results are usually small but are perceptible in some limited regions of parameter space.

The flavor eigenstates $|\nu_e\rangle$ and $|\nu_x\rangle$ can be represented at any point in the Sun as linear combinations of the local mass eigenstates, the eigenstates that correspond to the ambient electron density. The

---

version probability was calculated analytically for any density distribution by Mikheyev and Smirnov (1986b,c), Messiah (1986), and Barger et al. (1986).

coefficients of the expansion of $|\nu_e\rangle$ and $|\nu_x\rangle$ in terms of the mass eigenstates can again be represented by a unitary transformation with an instantaneous mixing angle θ_M. The value of θ_M can be found from the relations given in the previous subsection.

For a slowly varying electron density, the matter mass eigenstates evolve independently in time. Consider an electron neutrino that is created at a higher density than the resonant density. In the region before the resonant density is reached by a ν_e on its way out of the Sun, it is convenient to use as basis states (for describing an arbitrary neutrino state) the adiabatic states

$$\exp\left[-i\int_t^{t_r} E_1(t')\,\mathrm{d}t'\right]|\nu_1 n_e(t)$$

and

$$\exp\left[-i\int_t^{t_r} E_2(t')\,\mathrm{d}t'\right]|\nu_2 n_e(t)$$

where $|\nu n_e(t)$ represents a neutrino state created at time t in the presence of an electron density n_e and the exponential factor describes the subsequent adiabatic time evolution.

As the ν_e goes through resonance, these basis states may be mixed, but on the other side of resonance the neutrino state can again be written as a linear combination of adiabatic eigenstates that depend upon the local electron density. Therefore the evolution of a basis state that is produced at a time t, goes through resonance at a time t_r, and is detected at a time t_f can be represented schematically by

$$\exp\left[-i\int_t^{t_r} E_1(t')\,\mathrm{d}t'\right]|\nu_1 n_e(t)\rangle$$
$$\longrightarrow a_1 \exp\left[-i\int_{t_r}^{t_f} E_1(t')\,\mathrm{d}t'\right]|\nu_1 n_e(t_f)\rangle$$
$$+ a_2 \exp\left[-i\int_{t_r}^{t_f} E_2(t')\,\mathrm{d}t'\right]|\nu_2 n_e(t_f)\rangle, \qquad (9.67)$$

where $|a_1|^2 + |a_2|^2 = 1$. There is a corresponding relation for the second adiabatic state, with coefficients that are simply derived from a_1 and a_2.

The probability for an electron neutrino that is created at a density larger than the resonance density to remain a ν_e can be evaluated as in the similar calculations for vacuum and for constant density oscillations. The result is given by Parke (1986a,b) and is

$$|\langle \nu_e | \nu_e \rangle_t|^2 = \tfrac{1}{2} + \left(\tfrac{1}{2} - P_{\text{jump}} \right) \cos 2\theta_M \cos 2\theta_V, \qquad (9.68)$$

where the value of θ_M is determined from the constant-density relation, Eqs. (9.47) and (9.61), using the density at which the ν_e was formed. Equation (9.68) is often referred to as **Parke's formula**.

The form of Eq. (9.68) is similar to the expression for the average survival probability in the case of vacuum oscillations, Eq. (9.24). However, there is a crucial difference. Equation (9.68) contains an extra factor of $\cos 2\theta_M$, which is negative (~ -1) if the neutrino is created at high densities [see Eq. (9.47) or Eq. (9.61)].

The negative factor of $\cos 2\theta_M$ in Eq. (9.68) can cause almost complete cancellation in the survival probability for matter oscillations, provided that the quantity P_{jump} is not large. Since θ_M approaches $\pi/2$ at large densities, the two factors of $1/2$ in the above equation can cancel each other to high accuracy for a wide range of θ_V, leaving only a small electron–neutrino flux.

The quantity P_{jump} represents the probability of jumping from one adiabatic mass eigenstate to another, $P_{\text{jump}} = |a_2|^2$, that is, a departure from the smooth adiabatic evolution described in Figure 9.3. The probability of level jumping was calculated by Haxton (1986), Parke (1986a), and Dar et al. (1987), using an approximation developed by Landau (1932) and Zener (1932) in which the electron density varies linearly in the region near the resonant electron density. The jumping probability is, in this approximation,

$$P_{\text{jump}} = \exp \left[\frac{-\pi \Delta m^2 \sin^2 2\theta_V}{4E \cos 2\theta_V} \left(\frac{n_e}{|dn_e/dr|} \right)_{\text{res}} \right], \quad n_e > n_{\text{res}}.$$
$$(9.69)$$

The formula in Eq. (9.69) is accurate if the density depends linearly upon radius. Precise expressions have been derived also assuming that the density depends exponentially upon radius [see, e.g., Toshev (1987a), Petcov (1988), and Krastev and Petcov (1988)]; an efficient and accurate numerical algorithm has been described by Haxton (1987a).

Table 9.2 gives the values of the oscillation parameters in matter for different electron densities, ranging from vacuum densities to very large densities. This table is adapted from Parke (1986b) and

Table 9.2.   Parameters for matter oscillations.

n_e	0	$n_{e,res}$	$2n_{e,res}$	$\rightarrow \infty$
$E_2 - E_1$	$\Delta m^2/2E$	$\Delta m^2/2E \sin 2\theta_V$	$\Delta m^2/2E$	$\rightarrow \infty$
θ_M	θ_V	$\pi/4$	$\pi/2 - \theta_V$	$\rightarrow \pi/2$
L	L_V	$L_V/\sin 2\theta_V$	L_V	$\rightarrow 0$

provides a convenient summary of some of the important character-istics of the analytic solutions.

The accuracy of the **adiabatic approximation** and the magnitude of P_{jump} are determined by the same dimensionless quantity, the ratio of the total width, Δr, of the resonance in physical space to the matter oscillation length, L_M, at resonance. The width of the resonance in energy space is $\Delta_V \sin 2\theta_V$. The width in physical space is defined by the region in which the change in the electron density causes the eigenvalues to separate by an amount equal to the intrinsic energy difference, that is, $\sqrt{2}G_F \delta n_e = \pm\Delta_V \sin 2\theta_V$. The ratio of the total spatial width to the matter oscillation length is therefore

$$\frac{\Delta r}{L_{M,res}} = \frac{\Delta m^2 c^3 \sin^2 2\theta_V/\cos 2\theta_V}{2\pi\hbar E|n_e^{-1}\,dn_e/dr|}. \tag{9.70}$$

The jump probability can be rewritten in an illuminating fashion in terms of this ratio,

$$P_{jump} = \exp\left[-\frac{\pi^2(\Delta r)}{2L_{M,res}}\right]. \tag{9.71}$$

When $\Delta r \gtrsim L_M$, the propagating neutrino state has sufficient time to adjust itself to the slowly changing matter eigenstate as the res-onance is traversed. Thus, large values of $\Delta r/L_M$ correspond to the adiabatic approximation in which the probability of jumping from one adiabatic state to another is small. In the opposite limit, $\Delta r \ll L_M$, the electron density changes so rapidly near resonance that the flavor state cannot adiabatically keep up with the mass eigenstate. In this nonadiabatic limit, the probability P_{jump} of level crossing is large.

For ν_e produced at high densities, $n_{e,prod} \gg n_{e,res}$,

$$|\langle\nu_e|\nu_e\rangle_t|^2 \approx \sin^2\theta_V + P_{jump}(E)\cos 2\theta_V. \tag{9.72}$$

For small vacuum mixing angles, the survival probability is approximately equal to P_{jump}.

The adiabatic approximation of no level jumping is more accurate the larger the mixing angle. How large does θ_V have to be in order for the adiabatic approximation to apply? Very small angles are sufficient.

The critical angle at which the resonance width is equal to the matter mixing length is given by the relation

$$\left.\frac{\sin^2 2\theta_V}{\cos 2\theta_V}\right|_{\text{crit}} = 4 \times 10^{-4} \left(\frac{E}{10 \text{ MeV}}\right)\left(\frac{10^{-4} \text{ eV}^2}{\Delta m^2}\right), \qquad (9.73)$$

or

$$\theta_V (\text{crit}) = 0.01 \text{ rad} \left[\left(\frac{E}{10 \text{ MeV}}\right)\left(\frac{10^{-4} \text{ eV}^2}{\Delta m^2}\right)\right]^{1/2}. \qquad (9.74)$$

For a 10 MeV neutrino and a mass difference squared of 10^{-4} eV^2, the adiabatic condition is satisfied if θ_V is larger than 0.6°. At smaller masses, the adiabatic approximation becomes progressively worse and the jump probability becomes larger. In deriving the numerical relations for θ_V, the logarithmic derivative of the electron density (see Section 4.3) was evaluated at $0.1R_\odot$, the region within which most of the ^{8}B neutrinos are produced (see Figure 6.1). The logarithmic derivative is about a factor of two smaller at the peak of the ^{8}B neutrino production ($0.05R_\odot$) and about a factor of five larger beyond $0.2R_\odot$ (see discussion of Figure 4.1d in Section 4.3).

The adiabatic condition can also be written in terms of a critical neutrino energy, E_{crit}. For energies below E_{crit}, the adiabatic criterion is satisfied and for energies larger than E_{crit}, the probability of level crossing is appreciable. Numerically,

$$E_{\text{crit}} = 5 \times 10^4 \text{ MeV} \left(\frac{\sin^2 2\theta_V}{\cos 2\theta_V}\right)\left(\frac{\Delta m^2}{10^{-4} \text{ eV}^2}\right). \qquad (9.75)$$

The equations given in this section determine the probability that an electron neutrino remains an electron neutrino as a function of the electron density at the point at which the neutrino is created. Neutrinos created below their resonance density on the far side of the Sun can pass through two resonances. The formula for the average probability of observing an electron neutrino outside the Sun has the same form as before except that

$$P_{\text{jump}} \Longrightarrow P_{\text{jump}_1}(1 - P_{\text{jump}_2}) + P_{\text{jump}_2}(1 - P_{\text{jump}_1}), \qquad (9.76)$$

that is, by $2P_{\text{jump}}(1 - P_{\text{jump}})$. Neutrinos created below their resonance density propagate essentially as in vacuum.

The results for $|\langle \nu_e | \nu_e \rangle_t|^2$ must be averaged over the mass distribution in the Sun (and the location and mass traversed in reaching the detector, which will be discussed later), using the probability as a function of position of the production of a neutrino of each nuclear source, i (e.g., ^{8}B, ^{7}Be, or pp). These probabilities are specified for the standard solar model in Table 4.5. The neutrino flux $\phi_i(E)$ reaching a terrestrial detector may therefore be written symbolically

$$\phi_i(E) = \phi_{i,0}(E) \int_{\text{Sun,det.}} dM(r) \left(\frac{d\phi_i(E)}{dM(M)} \right) |\langle \nu_e | \nu_e \rangle_t|^2, \qquad (9.77)$$

where $\phi_{i,0}$ is the spectrum of electron neutrinos that is created in the Sun (assumed unaffected by oscillations).

Matter oscillations change both the magnitude of the flux and the shape of the neutrino energy spectrum from a given nuclear source, since the survival probability depends upon energy.

Illustrative MS diagram.   In their original paper, Mikheyev and Smirnov (1986) introduced a convenient representation of the effect of matter oscillations on the observed rates in solar neutrino experiments. Mikheyev and Smirnov plotted the contours of fixed experimental rates in a plane determined by Δm^2 and θ_V. Nearly all subsequent workers have adopted some form of the Mikheyev–Smirnov representation which may be appropriately called an **MS diagram**.

Figure 9.6 shows an illustrative MS diagram plotted for the ^{37}Cl experiment. The orthogonal coordinates in this case are $\sin^2 2\theta_V / \cos 2\theta_V$ and Δm^2. Most of the early papers on the MSW effect used $\sin^2 2\theta_V$ instead of $\sin^2 2\theta_V / \cos 2\theta_V$. The quantity $\sin^2 2\theta_V / \cos 2\theta_V$ has two advantages: It expands the crucial region near $\theta_V = \pi/4$ and it makes the expression for P_{jump} a straight line on a logarithmic plot [see Eq. (9.69)]. The lines in Figure 9.6 represent contours of equivalent rates in the experiment, in this case, about 2 SNU in the ^{37}Cl detector.

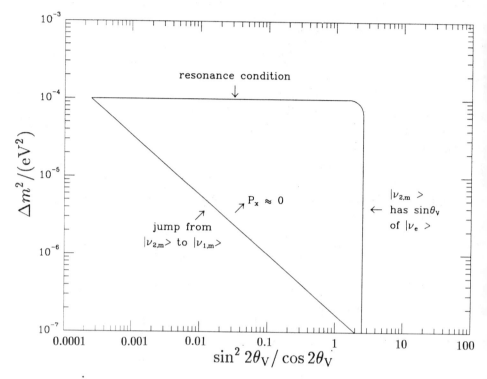

Figure 9.6   Illustrative MS diagram   The lines shown are equal rate contours ($\sim$ 2 SNU) for the ^{37}Cl experiment. The horizontal and vertical lines represent solutions which satisfy the adiabatic criterion that the width of the resonance is large compared to the matter oscillation length. The diagonal line corresponds to a solution in which the level-crossing probability P_{jump} is large.

The shape and location of the isorate contours in an MS diagram can be understood using the analytic formulae derived previously. The horizontal and vertical lines represent a reduction of about a factor of four that is achieved in the adiabatic approximation.

Consider first the horizontal line. Neutrinos of energy above $\sim$ 7 MeV can pass through a resonance and be converted to a different flavor for values of $\Delta m^2 \sim 10^{-4}$ eV^2 [see Eq. (9.54)]. A larger reduction in the predicted event rate would be achieved by lowering the cutoff E above which flavor conversion is nearly complete. To reach a lower cutoff, the value of Δm^2 must be decreased by a corresponding factor, since $n_{\mathrm{e,res}}$ is proportional to $\Delta m^2/E$. The horizontal line covers a wide range in vacuum oscillation angles, $0.02 \lesssim \sin 2\theta_{\mathrm{V}} \lesssim 0.85$. The lower limit on $\sin 2\theta_{\mathrm{V}}$ is determined by the validity of the adiabatic condition. The upper boundary is fixed by the term $\sin \theta_{\mathrm{V}}$ in the formula for the survival probability. At large values of θ_{V}, the heavier mass eigenstate, $|\nu_2\rangle \sim \sin \theta_{\mathrm{V}} |\nu_{\mathrm{e}}\rangle$, to which the electron neutrino is adiabatically converted, contains a relatively large component of the flavor eigenstate $|\nu_{\mathrm{e}}\rangle$.

The vertical solution represents a survival probability equal to $\sin^2 \theta_{\mathrm{V}} \sim 0.25$ [see Eq. (9.72)]. The mass differences are sufficiently small for this solution that all neutrinos which contribute significantly to the observed event rate pass through a resonant density and are converted to a different flavor. Nearly all of the contribution of the dominant ^{8}B neutrinos comes from the higher energies in the spectrum (see Tables 6.3 and 8.4). Once Δm^2 becomes sufficiently low that neutrinos with energies $E \sim 5$ MeV are converted, no further significant reduction occurs in the ^{8}B-induced event rate. The strong increase of the ^{8}B neutrino absorption cross section with increasing energy explains the essentially vertical character of the solution at large θ_{V}.

The diagonal line represents the nonadiabatic contribution of P_{jump}. The equation of the line is $\Delta r = \mathrm{const.} \times L_{\mathrm{M}}$, where the value of the constant determines the fraction of the converted ν_{e} flux that is restored by level jumping. Smaller values of Δm^2 are associated with the diagonal (nonadiabatic) solution than with the horizontal (adiabatic) solution because the resonance width is proportional to Δm^2.

The energy spectra for arriving ν_{e} are different along the three portions of the triangle, which constitutes an important experimental diagnostic. Along the horizontal (sometimes called "adiabatic") line, the higher-energy neutrinos are mainly absent; very few of the ^{8}B neutrinos with energies above 7 MeV survive. The lower-energy neutrinos are essentially unaffected since the resonance density for neutrinos of energies less than 5 MeV is larger than the central solar density if $\Delta m^2 \sim 10^{-4}$ eV^2. Along the vertical (large $\sin \theta_{\mathrm{V}}$) solution, the reduction in flux is practically independent of energy just as it is for vacuum neutrino oscillations. The lower-energy neutrinos are mainly depleted along the diagonal (nonadiabatic) solution. The pp and ^{7}Be neutrinos are preferentially converted in flavor for this solution, whereas the higher-energy ^{8}B neutrinos survive with greater probability. The energy spectrum of the nonadiabatic solution is the same as is expected from neutrino decay.

Since the survival probability, $|\langle \nu_e | \nu_e \rangle_t|^2$, depends upon energy in a characteristic way for each of the branches, the measurement of the neutrino energy spectrum can distinguish between solutions with different oscillation parameters.

D   MS diagrams for some experiments

Figure 9.7 shows an accurate MS diagram computed by Parke (1988) using the approximations described in Parke and Walker (1986) and in Section 9.2C. The generic triangular structure discussed in the previous subsection is apparent in Figure 9.7. The flaring up of the contours at large mixing angles and at large SNU values can be understood from the formula for the minimum neutrino energy that passes through resonance, Eq. (9.54). Because the hypothetical SNU rate increases toward the standard value of 7.9 SNU, the value of Δm^2 must be increased at large θ_V in order that only a few ^{8}B neutrinos pass through a resonance density.

Figure 9.8 shows an MS diagram for ^{71}Ga, again computed by Parke (1988). The band formed by the 1σ uncertainties about the ^{37}Cl experiment result, 2 ± 0.3 SNU, are shown as dotted contours in this figure.

Even in the limited two neutrino space considered here, there are two solutions, with either a small or a large mixing angle, that satisfy both the ^{37}Cl results and many of the possible outcomes of the ^{71}Ga experiment. This ambiguity in mixing angle must be resolved by future experiments that are sensitive to the energy spectrum of the incident neutrinos.

E   Regeneration in the Earth

All astronomers know that the stars are seen most clearly at night. Why should the Sun be different? When observing electron neutrinos, the Sun may indeed be brightest at night. Electron neutrinos that are converted to ν_x while passing through the Sun may be reconverted to ν_e on their way to a solar neutrino detector on the opposite side of the Earth from the Sun. If the MSW explanation is correct, at least half of the ^{8}B neutrinos created in the Sun have had their flavor changed by matter interactions. Therefore, regeneration

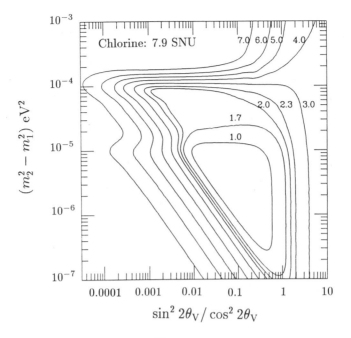

Figure 9.7   MS diagram for the ^{37}Cl experiment   Each contour is labeled by the corresponding SNU value. This figure is similar to Figure 3 of Parke and Walker (1986), but the current version is based upon the standard solar model described in Chapter 4.

in the Earth will on the average produce more ν_e from ν_x rather than vice versa and may make the Sun appear to shine brighter in electron neutrinos at night.

The main features of neutrino regeneration can be inferred from the equations for propagation in a constant or slowly varying electron density. The value of energy over mass difference squared that is relevant is [see Eq. (9.53)]

$$\frac{E}{\Delta m^2} \approx 10^5 \cos 2\theta_V \left[\left(\frac{66 \text{ cm}^{-3}}{(n_e/N_A)} \right) \right], \tag{9.78}$$

where E is measured in MeV and Δm^2 in eV^2. The electron density in the Earth varies from $n_e \sim 1.6 N_A$ for the mantle to $n_e \sim 6 N_A$ in the core [see de Braemaecker (1985) or Stacey (1985)]. For neutrinos with energies of order 10 MeV, the relevant mass range is

$$\Delta m^2_{\text{regen}} \sim 10^{-6} \text{ eV}^2. \tag{9.79}$$

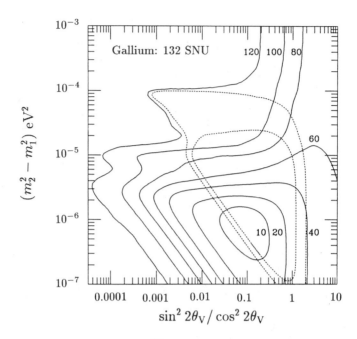

Figure 9.8   MS diagram for the ^{71}Ga experiment   The solid curves are iso-SNU contours for the ^{71}Ga experiment. The band formed by the 1σ uncertainties about the ^{37}Cl experiment result, 2.1 ± 0.3 SNU, is indicated by dotted contours. This figure is similar to Figure 4 of Parke and Walker (1986) but was recomputed by Parke using the standard solar model described in Chapter 4.

The vacuum mixing angle must also be reasonably large in order that the regeneration probability [see Eq. (9.64)] not be a smaller number. For representative mass differences, this requires

$$\sin^2 2\theta_V \geq 0.01, \tag{9.80}$$

i.e., θ_V must be larger than $3°$. For smaller values of θ_V, the matter oscillation length becomes so large that no oscillations occur within the Earth.

The mixing parameters given in Eqs. (9.79) and (9.80) are within the range found earlier for MSW solutions without taking account of the effect of the Earth. Therefore, regeneration of two neutrino species does not introduce major new parameter domains as solutions for the solar neutrino problem. Instead, regeneration shuffles slightly the relation between observed rates and mixing parameters. The labels of the average iso-SNU plots passing through a given point in the MS diagram are changed somewhat, but the overall solution space remains essentially unaffected.

The best thing that could happen to solar neutrino astronomy is the observation of terrestrial regeneration. The enhanced event rate at night caused by the reconversion of ν_x to ν_e could provide a characteristic signal indicating the origin of the solar neutrino problem. The observed rate might be seen to vary in a calculable and striking way depending upon the amount of terrestrial material that is traversed on the way to the detector, that is, upon the time of day or upon the season of the year. The difference between the daytime and nighttime rates is generally known as the **day–night effect**.

A number of authors have carried out detailed calculations evaluating the day–night and season effects [see especially Mikheyev and Smirnov (1986d), Carlson (1986), Bouchez et al. (1986), Dar and Mann (1987), Dar et al. (1987), Hiroi et al. (1987b), and Baltz and Weneser (1987, 1988)]. For precise results, the equations for the time evolution [Eqs. (9.41) and (9.42)] must be solved numerically including the effects of passage through solar matter, travel in the vacuum between the Sun and the Earth, and propagation to the laboratory site. Fortunately, the analytic approximations developed in the previous subsections are accurate for most of the relevant parameter space so that the net amount of computation is not excessive.

For the ^{37}Cl experiment, the effect of transmission through the Earth distorts somewhat the iso-SNU contours in a typical MS diagram. This is particularly noticeable on a precise MS diagram in the region of $\Delta m^2 \sim 3 \times 10^{-6}$ eV^2. The effect of Earth transmission moves the iso-SNU contours toward the left in an MS diagram from $\sin 2\theta_V \approx 0.85$ to $\sin 2\theta_V \approx 0.6$. The net effect of the distortion of the iso-SNU contours by Earth transmission is comparable to the spread in the iso-SNU contours introduced by the uncertainties in the predictions of the standard solar model (see Table 6.5).

The nighttime counting rate might be a factor of two or three larger than the daytime value for experiments that are mainly sensitive to ^{8}B neutrinos. However, if agreement is required with the average ^{37}Cl result, then big effects are only expected for large mixing angles, $\theta_V \approx 25°$. The seasonal dependence is significant, but somewhat smaller, typically $\lesssim 50\%$. Davis and his colleagues are beginning preparations to extract samples twice a day from the

^{37}Cl detector (see Chapter 10). For the ^{71}Ga experiments described in Chapter 11, the relative day–night and seasonal effects are expected to be significant only if the absolute counting rate is small. Therefore, one would not expect to be able to measure a statistically significant variation with this detector.

Real-time experiments, which use neutrino–electron scattering (see Chapter 13) or direct counting of neutrino absorption events (see Chapter 14) will automatically be sensitive to day–night and seasonal effects. As in the case of ^{37}Cl, large effects are predicted only for relatively large mixing angles. However, direct counting experiments provide important additional data that is not available for radiochemical measurements, namely, the spectrum of the recoil electrons (which reflects the incident energy spectrum of the neutrinos) and the precise time a neutrino event occurs (which determines the terrestrial chord that was traversed). If we are fabulously lucky, then not only will the total rate be found to vary systematically with the amount of Earth that is traversed, but also the observed shape of the incident neutrino spectrum may depend upon the path of the neutrinos through the Earth.

F   What can be learned experimentally?

There are three experimental signatures of the MSW explanation of the solar neutrino problem: the shape of the detected ν_e energy spectrum, the total number of incident neutrinos of all flavors, and, for specific parameters, periodic variations.

Over the next half a decade, the number of solar neutrino experiments will increase to at least five. There will be one or more experiments that use neutrino–electron scattering in water (see Chapter 13, especially the discussion of the ongoing Kamiokande II experiment in Section 13.2), ^{71}Ga (see the discussion of the GALLEX and SAGE detectors in Chapter 11), and ^{98}Mo (see Section 12.1), as well as the ongoing ^{37}Cl detector (see Chapter 10).

Suppose the MSW effect is the correct solution of the solar neutrino problem. Will the existing and planned experiments be able to demonstrate empirically that matter oscillations are causing the problem? Let us examine what can be learned from each of the three MSW signatures.

Energy spectrum. The horizontal (adiabatic) solution in an MS diagram corresponds to a depletion primarily of the high-energy neutrinos and the diagonal (nonadiabatic) solution is especially short of low-energy ν_e's. The vertical (large $\sin\theta_V$) solution reduces neutrinos of all energies by approximately the same factor.

If the horizontal solution is correct, then the high-energy neutrinos will be largely missing from the standard solar neutrino spectrum (which is described in Section 6.3). This solution will produce counting rates below the total theoretical range for neutrino–electron scattering, ^{37}Cl, and ^{98}Mo detectors, but will reduce the rate for the ^{71}Ga experiment by only a small amount [10 SNU out of a total of 132 SNU, see Eq. (11.2) and Table 11.1]. The same results would be expected if one of the nonstandard solar models discussed in Chapter 5 were correct.

A similar ambiguity will be present if the vertical adiabatic solution is correct. Both vacuum neutrino oscillations and the vertical adiabatic solution can cause the reduction in ν_e flux to be independent of energy.

If the diagonal (nonadiabatic) solution is correct, the fundamental pp neutrinos will be largely flavor converted in the Sun. The ^{71}Ga experiment will yield a very low event rate (see Figure 9.8) while the neutrino–electron scattering and ^{98}Mo detectors will register only a moderate reduction. This result would be striking and, after calibrations of the ^{71}Ga detectors were made with a ^{51}Cr source, probably would lead to a consensus that the MSW effect was operating on solar neutrinos.

Total number of neutrinos. No existing experiment can measure the total number of incident solar neutrinos independent of flavor. However, there are proposed experiments, for example, the SNO D_2O experiment (see Section 14.1) and the BOREX ^{11}B detector (see Section 14.4) that could measure neutral-current interactions (see also the discussions of coherent scattering of nuclei in Section 14.5). The advantage of neutral-current interactions is that they are flavor blind and therefore indicate the total number of neutrinos that are incident upon the Earth, independent of whether they have suffered a flavor change. Realistically, we will probably have to wait a decade before obtaining definitive results from one of these experiments.

Temporal variations. If the MSW parameters lie in certain limited domains (see Section 9.2E), then striking day–night and seasonal variations may be observed with the ^{37}Cl detector or with the Kamiokande II experiment. Other real-time solar neutrino experiments, such as those described in Chapters 13 and 14, would be sensitive to temporal variations. The *a priori* probability of observing temporal variations is not large since observable variations occur only for a small fraction of the possible solution space for the MSW parameters.

Attempts to observe the temporal variations should have high priority since they would provide a unique signature of neutrino matter oscillations. The product of *likelihood* × *importance* is large.

Synopsis. Experiments in progress or planned for the next decade may demonstrate that the MSW effect operates powerfully upon solar neutrinos and that the electron neutrino interacts with at least one heavier neutrino via a mass difference squared $\Delta m^2 \sim (10^{-3\pm1}$ eV^2) and a mixing angle $\theta_V \gtrsim 1°$. The existing and planned experiments may not be able to establish with certainty the identity of the ν_e's interaction partner, ν_x, nor even whether the interaction occurs with only one other neutrino flavor. Theoretical prejudice may produce a consensus, but the number of experimental facts will still be too small to constrain rigorously the MSW solutions to a small parameter domain.

Will the limited experimental clarification be significant for particle physics? The answer is: "Yes."

The first indication of physics at the mass scale at which the electroweak and strong interactions are unified could come from the observation of the MSW effect. No reason is known why neutrinos must have exactly zero masses, but there is an attractive explanation for why neutrinos may have very small masses. This explanation involves the grand unification mass scale. A typical value for the mass at which the strong and electromagnetic interactions are unified is $M_{\mathrm{GUT}} \sim 10^{15}$ GeV. The mass scale at which the weak and electromagnetic interactions are unified, $M_{\mathrm{EW}} \sim 10^2$ GeV, is much smaller than M_{GUT}. Thus one can form dimensionally a small mass by considering $M_{\mathrm{EW}}^2/M_{\mathrm{GUT}} \sim 10^{-2}$ eV. An expression of exactly this form is embodied in the famous **see-saw formula** of Gell-Mann, Ramond, and Slansky (1979) and Yanagida (1978) [see also Langacker *et al.* (1987) and Weinberg (1987)], which expresses the mass of a neutrino in terms of the square of the mass of its associated quark or lepton,

$$m(\nu_i) \cong \frac{m^2(q_i)}{M_{\text{GUT}}} \sim \frac{m^2(l_i)}{M_{\text{GUT}}}. \tag{9.81}$$

The see-saw formula can be obtained by diagonalizing a neutrino mass matrix having off-diagonal (Dirac) matrix elements of the order of the observed lepton or quark masses and one nonzero diagonal element representing a Majorana mass resulting from physics that occurs at the grand unified mass scale. The mass of the heaviest (top) quark may be of order the mass of the intermediate W boson. Inserting $m(q_i) \approx 100$ GeV in Eq. (9.81) gives $m(\nu_i) \sim 10^{-2}$ eV, about what is needed for the MSW effect to work effectively in the Sun. This argument suggests that ν_x, the neutrino into which ν_e is converted by matter oscillations, is most likely the neutrino associated with the heaviest quark or lepton, that is, ν_τ. In some theories, the large mass appearing in the denominator of the see-saw formula corresponds to an intermediate scale of order 10^{11} GeV. If these models are correct, then ν_e is most likely to be converted to ν_μ.

Neutrino masses of the order of magnitude required for the MSW effect to explain the solar neutrino problem are obtained naturally from the see-saw mechanism using a grand unification mass derived previously. The estimates of M_{GUT} are obtained from analyses of the energy scale at which the strong, weak, and electromagnetic couplings become equal.

Solar neutrino studies constitute the low-energy frontier of particle physics in which energy differences as small as 10^{-20} GeV [see Eq. (9.66)] can give us information about energies on the scale of the grand unification mass, 10^{15} GeV.

G   Rich and beautiful

There are a number of interesting and important questions which arise in solving numerically the time-evolution equations. Illuminating discussions are given by, among others, Barger *et al.* (1980, 1986), Bethe (1986), Rosen and Gelb (1986), Kolb, Turner, and Walker (1986), Messiah (1986), Parke and Walker (1986), Bouchez *et al.* (1986), Haxton (1986, 1987a), Dar *et al.* (1987), and Baltz and Weneser (1988). The Landau–Zener method has been discussed by Haxton (1986), Parke (1986a,b), Dar *et al.* (1987), and Petcov

(1987). Some authors have provided analytic solutions for specific assumed functional forms for the solar electron number density. Treatments that are particularly instructive include Notzold (1987), Pizzochero (1987), Toshev (1987a,b), and Petcov (1987, 1988). The physical interpretation of the MSW effect is given special attention by Messiah (1986), Bethe (1986), Kim, Nussinov, and Sze (1987), Langacker *et al.* (1987), Mikheyev and Smirnov (1987), Kim, Nussinov, and Sze (1987), Weinberg (1987), and Kim, Kim, and Sze (1988). Three-flavor oscillations have been analyzed by Barger *et al.* (1980, 1986), Langacker *et al.* (1987), Kuo and Pantaleone (1987), Petcov and Toshev (1987), and Mikheyev and Smirnov (1988). Day-night and seasonal effects have been examined by Carlson (1986), Bouchez *et al.* (1986), Dar and Mann (1987), Dar *et al.* (1987), Hiroi *et al.* (1987b), and Baltz and Weneser (1987, 1988). Neutrino-electron scattering has been discussed by Bahcall, Gelb, and Rosen (1987) and by Hiroi, Sakuma, Yanagida, and Yoshimura (1987a).

9.3 Neutrino magnetic moment

If the electron neutrino has a large magnetic moment (and/or a large electric dipole moment),[†] then traversal of the solar magnetic field may flip the neutrino's spin. The spin flip would change the left-handed ν_e that is detectable in solar neutrino experiments (with, e.g., ^{37}Cl or ^{71}Ga detectors) into a right-handed neutrino that would not be detectable in these experiments. This idea was originally considered by Cisneros (1971). More recently, Okun, Voloshin, Vysotskii, and their collaborators have included the possibility of flavor oscillations [see Voloshin and Vysotskii (1986), Okun (1986), and Voloshin, Vysotskii, and Okun (1986a,b)] and have drawn attention to the important question of time dependences. The Soviet scientists suggested that the effects of neutrino magnetic moment precession may explain the previously suggested correlation between sunspots and event rate in the ^{37}Cl experiment (see Section 10.5A) and also lead to a distinctive half-year period in the intensity of the flux of

---

[†]Okun (1986) pointed out that the manifestations of an electric dipole moment and a magnetic dipole moment are practically indistinguishable for a highly relativistic (light) neutrino. For simplicity, the term "magnetic moment" will be used in the present section to represent the combined interaction of an electric dipole and a magnetic moment.

left-handed ^{8}B ν_e's that reach the Earth. The key ideas are that the magnetic field in the solar convective zone may be time dependent (as reflected in the changing sunspot intensity) and that the region of the convective zone through which ^{8}B neutrinos pass on their way to the Earth will depend upon the season of the year (since the plane of the Earth's orbit is slightly inclined to the equator of the Sun).

More recently Akhmedov and Khlopov (1988) and Lim and Marciano (1988) have considered the combined effects of flavor mixing, magnetic spin flip, and matter interactions. They have shown that resonances may occur in the spin flip process that are analogous to the MSW resonance.

A Spin flip in the convective zone

To estimate the orders of magnitude required for the relevant quantities so that the spin flip mechanism may work, consider first the case of no flavor mixing. A flux of left-handed electron neutrinos, $\phi_0(E)$, that is produced in the Sun will be partially converted into unobservable right-handed neutrinos. The fraction that survives, $\phi_B(E)$, will depend upon the neutrino magnetic moment, the field strength (and configuration) that is traversed, and the spatial extent of the field. A neutrino[†] with a diagonal electromagnetic moment μ that is moving in the presence of a magnetic field B will oscillate between left- and right-handed components, ν_{left} and ν_{right},

$$\phi_B(E) = \phi_0(E) \cos^2 \left(\mu \int B_\perp \mathrm{d}l \right), \qquad (9.82)$$

where $B_\perp$ is the magnitude of the magnetic field perpendicular to the direction of propagation.

[†]**Dirac neutrinos** are described by a conventional four-component field, ψ; a neutrino satisfies the same special relativistic equation as does an electron or a positron (in the absence of an electromagnetic field). However, a neutrino has zero charge and its wavefunction may be, therefore, a linear combination, with equal amplitudes, of ψ and $C\psi$, where C is the charge conjugation operator. In this case, the neutrino would be its own antiparticle (like the photon) and is called a **Majorana neutrino** [see, e.g., Commins and Bucksbaum (1983)]. The Majorana neutrino has a zero (diagonal) magnetic moment but could have a nonzero off-diagonal transition moment, representing transitions between different flavors. The orthogonal linear combination of the observed Majorana neutrino might not have a a physical realization or might be much heavier than the ordinary neutrino, as envisioned in the see-saw mechanism.

The reduction in flux implied by Eq. (9.82) is independent of energy, a characteristic that is not preserved when interactions with matter are included.

The spin precession described by Eq. (9.82) occurs in an arbitrarily weak transverse field if the left- and right-handed neutrinos are degenerate in energy (as for Dirac masses). Oscillations in spin direction between nondegenerate mass eigenstates of different handedness are appreciable only if the phase difference caused by the magnetic mixing moment is at least as large as the difference caused by unequal masses,

$$|\mu_{ij} B_\perp| \gtrsim \Delta_V, \tag{9.83}$$

where μ_{ij} is the magnetic moment matrix in the flavor basis and Δ_V is defined by Eq. (9.37). In other words, the magnetic interaction energy must exceed the energy splitting due to unequal masses. Numerically, this requires a somewhat smaller mass difference than is relevant for most of the MSW solutions discussed in Section 9.2. The mass difference must satisfy

$$\Delta m^2 \gtrsim 10^{-8} \text{ eV}^2 \left[\left(\frac{10 \text{ MeV}}{E} \right) \left(\frac{10^3 \text{ gauss}}{B} \right) \left(\frac{10^{-10} \mu_B}{\mu} \right) \right]. \tag{9.84}$$

Okun, Voloshin, and Vysotskii suggest that magnetic fields of order $B_\perp \sim (1\text{--}5) \times 10^3$ gauss may exist in the convective zone of the Sun, which has an extent (see Table 4.1) of order $L \sim 2 \times 10^{10}$ cm. The order of magnitude that is required for the magnetic moment to cause a significant decrease in the left-handed flux of ν_e can be inferred from Eq. (9.82). The probability of a spin flip depends upon the dimensionless quantity

$$\frac{\mu B L}{\hbar c} \sim 2 \left(\frac{\mu}{2 \times 10^{-10} \mu_B} \right) \left(\frac{B}{10^3 \text{ gauss}} \right) \left(\frac{L}{R_\odot} \right). \tag{9.85}$$

Thus a magnetic moment of the order of

$$\mu \approx (0.3 - 1) \times 10^{-10} \mu_B, \quad \mu_B = e\hbar/2m_e c \tag{9.86}$$

is required in order to produce a significant probability of spin flip. Neutrino spin precession cannot be observed under laboratory conditions because the required product of magnetic field and path length is too large.

The required magnetic moment, Eq. (9.86), is many orders of magnitude greater than typical theoretical values obtained from calculations made with relatively conventional models [see Fujikawa and Schrock (1980), Liu (1987), Lim and Marciano (1988), and references quoted therein]. Conventional theoretical calculations suggest a typical moment of the order

$$\mu_{\nu_e,\text{typical}} \sim 10^{-19} \left(\frac{m_{\nu_e}}{1 \text{ eV}} \right) \mu_{\text{B}}. \tag{9.87}$$

The ratio of the typical "expected" moment, $10^{-19}\mu_{\text{B}}$, to the required moment, $10^{-10}\mu_{\text{B}}$, is not a fair estimate of the probability of this hypothesis to be correct, but is some indication of its lack of inevitability. Voloshin and Vysotskii (1986), Voloshin, Vysotskii, and Okun (1986a,b), Fukugita and Yanagida (1987), and Stefanov (1988) have described electroweak models in which electromagnetic moments $\mu \sim 10^{-11}$ to $10^{-10}\mu_{\text{B}}$ are possible, although somewhat contrived.

The existing experimental limits permit a much larger magnetic moment than is implied by the typical theoretical calculations. The analysis of $\bar{\nu}_e$-scattering experiments made with reactors can be used to argue that [Kyuldjev (1984) and Lim and Marciano (1988)]

$$\mu_{\text{lab}} \lesssim 2 \times 10^{-10}\mu_{\text{B}}, \tag{9.88}$$

90% confidence limit, although some treatments give a less stringent limit.

The astrophysical limits are less certain but are ostensibly more restrictive. Results in the literature from the analysis of ordinary stars suggest upper limits of the order [see Bernstein, Ruderman, and Feinberg (1963), Domogatski and Nadezhin (1971), Sutherland *et al.* (1976), and Marciano and Parsa (1986)]

$$\mu_{\text{astroph.}} \lesssim 10^{-10}\mu_{\text{B}}. \tag{9.89}$$

Much more stringent limits have been proposed by some authors who argue that the conditions inside Supernova 1987A, at the time of collapse and neutrino production, are sufficiently well understood to rule out much smaller magnetic moments. These arguments suggest that [see, e.g., Goldman *et al.* (1988), Lattimer and Cooperstein (1988), and Barbieri and Mohapatra (1988)]

$$\mu_{\text{astroph.}} \lesssim 10^{-12}\mu_B. \tag{9.90}$$

Until the solar neutrino problem is solved, conclusions like Eq. (9.90) that are based upon an assumed understanding of the dynamics of stellar collapse will be treated with caution by some workers. Solar evolution (on time scales of 10^{10} yr and densities of 10^2 g cm^{-3}) is mild compared to stellar collapse (on time scales of milliseconds and nuclear densities). If we do not understand main sequence solar evolution, we may not understand nonequilibrium stellar collapses.

Adding a flavor off-diagonal magnetic moment interaction to the vacuum flavor oscillations discussed in Section 9.1C, one obtains

$$
\left|\langle \nu_x | \nu_e \rangle_t \right|^2 = \cos^4(\theta_V) \cos^2 \left(\mu_{\nu_1} \int B_\perp dl \right)
$$
$$
+ \sin^4(\theta_V) \cos^2 \left(\mu_{\nu_2} \int B_\perp dl \right). \qquad (9.91)
$$

For small magnetic fields (presumably corresponding to sunspot minimum), the minimum observed flux is $1/2$ (at $\theta_V = \pi/4$), in agreement with the earlier results for mixing without matter or magnetic fields (see Section 9.1). The parameters in Eq. (9.91) can be chosen so that the average over the 11 year cycle of the solar sunspots can be equal to the observed ratio ($\sim 1/4$, see Chapter 10).

The origin of the semiannual dependence might be delightfully simple astronomy. Voloshin, Vysotskii, and Okun (1986a) pointed out that the ^{8}B flux should be modulated by a half-year cycle caused by the combined effects of the inclination ($7°$) of the plane of the ecliptic to the solar equator and the weakening of the toroidal magnetic field near the equator. The toroidal field has opposite signs on different sides of the equator and therefore the field must vanish at the equator. The size of the characteristic transition region is about 7×10^9 cm. This region is sufficiently large to shade well the ^{8}B neutrinos (which are produced in a small region in the solar center, $\sim 0.05R_\odot$, see Table 4.5 and especially Figure 6.1) but only partially the pp neutrinos (which are produced in a much more extended region, $\sim 0.1R_\odot$).

When the Earth is near one of the intersection points of the Earth's orbit with the equatorial plane of the Sun (at the beginning of June and the beginning of December), the production region of the ^{8}B neutrinos is seen from the Earth through the equatorial slit in the magnetic field and the source of left-handed neutrinos is not weakened as a result of the leakage into the right-handed helicity state. On the other hand, when the Earth is maximally distant from the plane of the solar equator (at the beginning of September and the beginning of March), the neutrinos arriving at Earth traverse a strong toroidal magnetic field and the detected ^{8}B flux should be minimal.

An accurate prediction for the amount of shading as a function of time of year could be calculated using the results given in Table 4.5 for the production of different neutrino sources as a function of distance from the solar center. This has not yet been done.

B Matter enhanced spin flip

For higher electron densities than exist in the solar convective zone, there is a spin flip analogue of the MSW resonance that is represented by Eq. (9.51), (9.52), and (9.53). Akhmedov and Khlopov (1988) and Lim and Marciano (1988) have described the spin flip resonances in beautiful papers. They begin by considering a mass matrix that contains, as before [see Eq. (9.42)], a contribution from a vacuum mass matrix, M_0, as well as contributions from electromagnetic interactions and from matter interactions.

For simplicity, consider first the case of two flavors with four-component Dirac neutrinos. In the convenient basis in which the neutrinos have a definite handedness (i.e., spin projection along the direction of propagation), there are four basis states $\nu_{e,L}$, $\nu_{e,R}$, $\nu_{x,L}$, and $\nu_{x,R}$. The matrix of electromagnetic moments can be written symbolically

$$M_{em} = \begin{pmatrix} \mu_{ee} & \mu_{ex} \\ \mu_{xe} & \mu_{xx} \end{pmatrix}. \tag{9.92}$$

The matter contribution for the left-handed components has the form

$$M_{matter} = \begin{pmatrix} a(\nu_e) & 0 \\ 0 & a(\nu_x) \end{pmatrix}. \tag{9.93}$$

The right-handed components, $\nu_{e,R}$ and $\nu_{x,R}$, do not interact electroweakly with matter and may be considered mass eigenstates (with an associated diagonal mass matrix). In the present case, the Wolfenstein terms, $a(\nu_e)$ and $a(\nu_x)$, include contributions from neutral-current scattering off neutrons; these neutral-current contributions do not cancel out when considering oscillations between left-handed and right-handed components. The appropriate elements of the mass matrix are for neutral matter [see Voloshin, Vysotskii, and Okun (1986)]

$$a(\nu_e) = \frac{G_F}{\sqrt{2}}(2n_e - n_n), \tag{9.94}$$

and

$$a(\nu_x) = \frac{G_F}{\sqrt{2}}(-n_n), \tag{9.95}$$

where n_n is the ambient density of neutrons. The difference between $a(\nu_e)$ and $a(\nu_x)$ is independent of the neutron density and is equal to the familiar Wolfenstein term, given in Eq. (9.43).

There is a resonant conversion of $\nu_{e,L} \to \nu_{x,R}$ at an electron density for which the diagonal matrix elements are equal, namely,

$$\frac{n_{e,res}(B)}{n_{e,res}(\text{MSW})} = \frac{\cos^2 \theta_V}{\cos 2\theta_V [1 - n_n/2n_e]}, \quad \nu_{e,L} \to \nu_{x,R}. \tag{9.96}$$

The ratio of electron number density to neutron number density can be calculated from the composition of the standard solar model that is given in Table 4.4. The limiting values are:

$$\frac{n_e}{n_n} = \begin{cases} 5.8, & r \geq 0.3R_\odot, \\ 2.05, & r = 0.0R_\odot. \end{cases} \tag{9.97}$$

The ratio can be represented approximately by the formula $n_e/n_n \approx 2.04 + 140(R/R_\odot)^2$ within the central region of the Sun $(R < 0.1R_\odot)$.

The MSW resonance and the spin flip resonance of $\nu_{e,L} \to \nu_{x,R}$ are well separated in space for some values of the parameters and can be treated separately in some illustrative cases. The spin flip conversion is then governed by the 2×2 submatrix

$$M_0 + M_B = \begin{pmatrix} \Delta_V \sin^2 \theta_V + a(\nu_e) & \mu_{ex}B \\ \mu_{xe}B & \Delta_V \end{pmatrix}. \tag{9.98}$$

The mathematical problem is the same as for pure MSW conversion in a variable density. The analytic technique of Parke (1986a,b) and others that was discussed in Section 9.2C can be applied to spin conversion by expanding the state vectors, as before, in terms of adiabatic states [see Eqs. (9.67) and (9.67)]. The results have been given explicitly by Lim and Marciano (1988).

Lim and Marciano find that for some possible vacuum mixing parameters, large interior magnetic fields ($B_\perp > 10^4$ gauss) can produce sufficient spin conversion to explain the solar neutrino deficit without additional help from the MSW effect. At smaller fields, the problem reduces to the previously discussed pure MSW resonance.

For Majorana neutrinos, the rule that the Wolfenstein term changes sign for antineutrinos is important. The mass matrix can again be written explicitly [see Akhmedov and Khlopov (1988), Lim and Marciano (1988), or Voloshin *et al.* (1986b)]. Spin flavor conversion leads to processes such as $\nu_e \to \bar{\nu}_x$ or, for very large mixing, $\nu_e \to \bar{\nu}_e$. The resonance for the $\nu_e \to \bar{\nu}_x$ conversion occurs under approximately the same conditions as the MSW conversion,

$$\frac{n_{e,res}(B)}{n_{e,res}(\text{MSW})} = \frac{1}{[1 - n_n/n_e]}, \quad \nu_e \to \bar{\nu}_x. \tag{9.99}$$

As in the previous example, illustrative cases can be found in which magnetic spin flip is by itself sufficient to reduce the predicted neutrino event rate to the observed value.

The conversion products (antineutrinos) are detectable in the case of Majorana neutrinos. If $\bar{\nu}_e$ are produced in any appreciable numbers, they should be visible in water Cerenkov detectors because of their large absorption cross sections on protons.

C   My cup runneth over

Resonant spin conversion in the Sun introduces four additional astrophysical parameters into the analysis of solar neutrino oscillations, namely, the magnitude and extent of the magnetic field in the convective zone and in the radiative zone. These astronomical parameters cannot be determined with confidence using available data. The neutrino electromagnetic moments add further complications.

If significant spin conversion of ^{8}B neutrinos occurs in the convective zone, then a characteristic semiannual periodicity should be observed in the intensity of the higher-energy neutrino flux. The suggested 11 year periodicity should also be verified by future observations. Otherwise, it will be difficult to disentangle the competing effects of matter conversion and spin conversion.

The solar acoustic oscillations (p-modes) can be used to determine if extremely strong magnetic fields are present in the Sun (see the discussion of p-modes in Section 4.4). Existing observations set an upper limit of 10^7 gauss on the interior field [see Brown and Murrow (1987)], within three orders of magnitude of the value required for spin flip.

9.4 Neutrino decay

Neutrinos are not expected to decay on the time scale of 500 s, the transit time for the Sun–Earth distance. The original reason for considering the possibility was a false alarm, presumably due to the statistics of small numbers. One of Davis' early ^{37}Cl counting runs yielded zero events, a result that could be consistent with neutrinos decaying on the way to the Earth from the Sun [Bahcall, Cabibbo, and Yahil (1972)]. The finite rate of 2 SNU that was later established with good statistics (see Chapter 10) requires fine tuning if neutrino decay is important, namely a mean decay distance comparable to 1 AU (the Sun–Earth distance) for ^{8}B solar neutrinos. The observation of $\bar{\nu}_e$'s from SN1987A [see Hirata et al. (1987) and Bionta et al. (1987), and Section 15.5] shows that significant neutrino decay on the way to the Earth from the Sun is extremely unlikely (but not impossible).

Why still consider neutrino decay? Because it is a useful test of the discriminatory powers of solar neutrino experiments; simple predictions of what should be seen in other experiments follow if one assumes that the ^{37}Cl results are explained by ν_e decay. Also, neutrino decay is (barely) consistent with the observations of SN1987A and with the solar results, provided one takes account of vacuum neutrino flavor mixing [Frieman et al. (1988)] as well as matter enhancement of the mixing [Raghavan et al. (1988)].

The flux of ν_e's with energy E that reach the Earth without decaying, assuming no flavor mixing, is

$$\phi_{i,\mathrm{decay}}(E) \; = \; \phi_{i,\mathrm{stable}}(E) \exp\left[-(500 \text{ s})/\tau(E)\right]. \tag{9.100}$$

The lifetime, $\tau(E)$, for decay depends inversely upon energy because of time dilation,

$$\tau(E) \propto \gamma^{-1} \propto E^{-1}. \tag{9.101}$$

Therefore, the undecayed spectrum has a single parameter that can be adjusted to fit the results of the ^{37}Cl experiment. The result is [Bahcall, Cabibbo, and Yahil (1972) and Bahcall, Petcov, Toshev, and Valle (1986)]

$$\phi_{i,\mathrm{decay}}(E) \; = \; \phi_{i,\mathrm{stable}}(E) \exp\left[-(10 \text{ MeV}/E)\right]. \tag{9.102}$$

The most dramatic result of Eq. (9.102) is an approximately 20-fold decrease in the event rate predicted for the ^{71}Ga detectors. The predicted rate is reduced drastically because all of the low-energy pp and ^{7}Be neutrinos are removed by decay and only the highest-energy ^{8}B neutrinos survive.

The no-mixing decay hypothesis is wrong. The observations of SN1987A show that, if ν_e is a mass eigenstate, 10 MeV ν_e are stable over distances of order 50 kpc ($\sim 10^{10}$ AU).

What happens to decaying neutrinos and antineutrinos from SN1987A if the flavor eigenstate are not mass eigenstates? Frieman, Haber, and Freese (1988) addressed this question. The flavor eigenstates $|\nu_e\rangle$, $|\nu_\mu\rangle$, $|\nu_\tau\rangle$ will be mixtures of the mass eigenstates $|\nu_1\rangle$, $|\nu_2\rangle$, $|\nu_3\rangle$ when leaving the neutron star. In the two component models considered by Frieman $et\ al.$ (1988), the heavier mass eigenstate, ν_2, will decay to the lighter eigenstate, ν_1. The most plausible decay mode is [Bahcall $et\ al.$ (1972)]

$$\nu_2 \rightarrow \bar{\nu}_1 + \phi, \tag{9.103}$$

where ϕ is a massless scalar (or pseudoscalar) boson. Although the component of $\bar{\nu}_e$ that is $\bar{\nu}_2$ ($\sin\theta_V$) decays on the way to Earth, the $\bar{\nu}_e$ beam is replenished by decays of ν_e and ν_x. For $\theta_V \gtrsim 20°$, Frieman $et\ al.$ find that the $\bar{\nu}_e$ signal is actually increased over what is expected if there is no decay.

Table 9.3 is reproduced from the paper of Frieman $et\ al.$ (1988). Since the energy release in neutrinos in Type II supernovae is of order $(3 \pm 1.5) \times 10^{53}$ ergs [see Eq. (15.1)], the argument of Frieman $et\ al.$ (1988) excludes very large mixing angles,

$$\theta_V \lesssim 60°. \tag{9.104}$$

If the heavier mass eigenstate has a decay lifetime of the form Eq. (9.100), then the survival probability is (using similar arguments to those developed in Section 9.1)

$$|\langle\nu_e|\nu_e\rangle_t|^2 = \cos^4\theta_V + \sin^4\theta_V \exp\left[-(1000\ \text{s})/\tau(E)\right]. \tag{9.105}$$

In order to satisfy the observational constraints for the ^{37}Cl and Kamiokande II detectors (see Sections 10.3 and 13.2),

$$\theta_V \gtrsim 45°, \quad \tau(10\ \text{MeV}) < 500\ \text{s}. \tag{9.106}$$

There is therefore a small range of mixing angles,

$$45° \lesssim \theta_V \lesssim 60°, \tag{9.107}$$

for which the available data on neutrino observations of the Sun and SN1987A do not yet exclude neutrino decay as the solution of the

Table 9.3. Supernova energy versus neutrino mixing angle. The table gives the total neutrino energy estimated for SN1987A versus the neutrino mixing angle [from Frieman, Haber, and Freese (1988)].

Mixing angle (degrees)	$E_{SN}/10^{53}$ erg	
	no decay	decay[†]
0	2.5±0.8	2.2±0.7
10	2.4±0.7	2.3±0.7
30	2.0±0.6	3.0±0.9
50	1.6±0.5	5.4±1.6
60	1.4±0.4	9.0±2.7
70	1.3±0.4	19.1±5.7
90	1.3±0.4	— —

[†]$\nu_2 \to \bar{\nu}_1 \phi$ decay assumed.

solar neutrino problem. Frieman *et al.* (1988) mention the possibility that astrophysical arguments based upon supernova dynamics, an often controversial subject, may further limit the acceptable parameter range if the decay is assumed to occur via majorons.

The analysis of the SN1987A observations should be repeated taking into account the possible effect of enhancement of the oscillations by matter.

Matter can greatly enhance the decay probability of an essentially stable electron neutrino [Raghavan *et al.* (1988)]. Small mixing angles can lead to a large decay probability via the MSW effect. The electron neutrino is assumed to be, in vacuum, mainly composed of the lightest mass eigenstate, ν_1. In passing through solar material, ν_e's are resonantly converted to the heavier mass eigenstate, ν_2, which can then decay rapidly. The basic scheme is

$$\nu_e \to \text{MSW} \to \nu_2 \to \text{decay.} \tag{9.108}$$

Neutrino decay that is mediated by the MSW effect, Eq. (9.108), will produce an experimental signal that is different from the pure MSW effect in some cases and not in others. In measurements of neutrino absorption with radiochemical detectors (such as ^{37}Cl or

^{71}Ga) or in real-time absorption detectors (such as D_2O or ^{40}Ar), the surviving electron neutrino flux will be the same as for a pure MSW effect. Detectors that are sensitive to neutral currents (via electron–neutrino scattering or by special modes, see discussion of SNO and BOREX in Chapter 14) might be able to observe the spectrum change that results when the "lost" ν_e's decay via the indirect route shown in Eq. (9.108). The typical energy of the neutrino produced by the decay is, because the energy is shared between two particles, less than the energy of the ν_e that is converted by the MSW effect [see Raghavan *et al.* (1988)].

Raghavan *et al.* (1988) pointed out that Eq. (9.103) implies that solar ν_e's will decay, through the intermediary ν_2, into $\bar{\nu}_e$'s. These solar $\bar{\nu}_e$'s can in principle be observed in water Čerenkov detectors (such as Kamiokande II, SNO, SUNLAB, or Super Kamiokande II, see Chapter 13) as well as in liquid scintillation detectors (such as LVD, or LSD, see Sections 13.4 and 15.5A). The cross section for the absorption reaction

$$\bar{\nu}_e + p \rightarrow n + e^+ \qquad (9.109)$$

is about 10^2 times larger than, for the same incident energy, the neutrino–electron scattering cross sections which produce solar neutrino events from ν_e's in these detectors. Unfortunately, the $\bar{\nu}_e$'s from solar ν_e decay will have, on the average, lower energies than the parent ν_e's and, more importantly, will not produce highly directional e^+'s (as in neutrino–electron scattering, see Section 8.2E). However, in the solar neutrino experiments BOREX and SNO that are discussed in Chapter 14, the low-energy $\bar{\nu}_e$ events may be identified by detecting in delayed coincidence the characteristic gamma ray that is produced when the accompanying neutron [see Eq. (9.109)] is captured. Searching for antineutrinos from the Sun constitutes a sensitive test of the neutrino decay mode described in Eq. (9.103), provided nature has arranged for the MSW effect to operate in the Sun. The search will also set interesting limits on the low-energy $\bar{\nu}_e$ cosmic background radiation.

9.5 WIMPs

Weakly interacting massive particles (WIMPs) can in principle solve both the solar neutrino problem and the dark matter problem.[†] Heavy neutral particles that are present in the massive halo of the Galaxy, and which contribute a lot of mass without much light, could be captured gravitationally by the Sun [Press and Spergel (1985)]. Energy transport by captured WIMPs could be sufficiently efficient to lower slightly the central temperature gradient of the Sun and thereby decrease the ^{8}B solar neutrino flux [see Section 5.13 and Faulkner and Gilliland (1985), Spergel and Press (1985), and Gilliland *et al.* (1986)]. It is at least an extraordinary coincidence, and conceivably a beautiful synthesis, that for WIMPs the required particle physics characteristics are the same (to order of magnitude) for solving the dark matter and solar neutrino problems.

WIMP characteristics. The basic principle of WIMP physics can be stated succinctly: If you are going to invent a new particle, you might as well invent a useful one.

WIMPs facilitate energy transport and therefore decrease the temperature gradient that is required to transport energy in the solar core. Because they have only weak interactions at solar thermal energies, WIMP cross sections are small compared to typical photon cross sections.

Stellar evolution theory has been successful in explaining astronomical observations made with photons (see Chapter 2). Therefore, energy transport in addition to the usual photon and convective transport should be efficient only in a small region of the Sun, namely, in the region in which the ^{8}B neutrinos are produced. If the WIMPs are mainly confined to this region, but are effective where present, they will have the maximum impact on the solar neutrino flux without affecting classical astronomical observables.

Therefore, the scale height, h, of a WIMP is ideal for solving the solar neutrino problem if $h \approx R(^8\mathrm{B})$, where $R(^8\mathrm{B})$ is the radial extent of the region in which ^{8}B neutrinos are produced. In the standard solar model, $R(^8\mathrm{B}) \approx R_\odot/20$ (see Table 4.5 and Figure 6.1). This constrains the WIMP mass to

$$2 \text{ GeV} \lesssim M \lesssim 10 \text{ GeV}. \tag{9.110}$$

[†]The fact that much of the matter in galaxies and between galaxies is not observed to emit photons is known as the **dark matter problem**.

Unless the scattering cross sections is exceptionally large (see Section 5.13), the WIMP mass must satisfy an additional constraint that follows from the condition that WIMPs do not evaporate from the solar core, namely [see Gould (1987a)]

$$M \gtrsim 4 \text{ GeV}. \tag{9.111}$$

For maximum efficiency in heat transport, WIMPs should interact with the ambient matter once or twice as they travel a scale height. Hence, the most desirable mean free path, λ, is

$$\lambda = \frac{1}{n\sigma_{sc}} \sim h \sim R(^8\text{B}) \sim \frac{R_\odot}{20}, \tag{9.112}$$

where n is the number density of particles that scatter WIMPs with a cross section σ_{sc}.

The average scattering cross section per baryon should be, for maximum effectiveness,

$$\sigma_{sc} \sim 5 \times 10^{-36} \text{ cm}^2. \tag{9.113}$$

This cross section is very large for a weakly interacting massive particle, $\sim 10^3$ times larger than the cross section due to ordinary Z^0 exchange [see Goodman and Witten (1985) and Raby and West (1988)].

A cross section of the magnitude given in Eq. (9.113) would make WIMPs much more efficient at transporting energy than photons, which have typical interaction cross sections that are $\gtrsim 10^{-24}$ cm^2.

The ratio of the energy transport efficiency of WIMPs to the efficiency of photons is of order $\sigma_{sc}(\text{photons})/\sigma_{sc}(\text{WIMPs})$.

The number density of WIMPs, n_W, required to match approximately the photon energy transport in the solar interior is, in terms of the proton number density, n_p,

$$\frac{n_W}{n_p} \sim \frac{\sigma_{sc}(\text{WIMPs})}{\sigma_{sc}(\text{photons})} \sim 10^{-11}. \tag{9.114}$$

The annihilation rate that follows naturally from Eqs. (9.113) and (9.114) is too large [see Krauss et al. (1985)]; the WIMPs and anti-WIMPs self-destruct before they have a chance to carry energy out of the solar core. In order to keep the abundance high enough to affect the solar neutrino flux, the annihilation cross section must be suppressed relative to the scattering cross section or there must be many fewer anti-WIMPs than WIMPs [cf. Krauss et al. (1985)].

The annihilation time may be written

$$t_{\text{ann}} = \frac{1}{n_{\text{anti}-\text{W}}\sigma_{\text{ann}}v(T)} \, , \tag{9.115}$$

or using Eq. (9.114) and the density of protons in the solar core,

$$t_{\text{ann}} \approx 3 \times 10^5 \text{ yr} \left[\left(\frac{n_{\text{W}}}{n_{\text{anti}-\text{W}}} \right) \left(\frac{\sigma_{\text{sc}}}{\sigma_{\text{ann}}} \right) \right] . \tag{9.116}$$

Two different particle physics scenarios, aimed at the two separate factors in the brackets of Eq. (9.116), have been constructed in order to make the annihilation time longer than the age of the Sun. The first scenario assumes a cosmic asymmetry between anti-WIMPs and WIMPs left over from the Big Bang [see, e.g., Gelmini, Hall, and Lin (1987)]. In this scenario, there are not enough anti-WIMPs in the solar core to cause a problem. The required large asymmetry in the current solar interior [$(n_{\text{W}}/n_{\text{anti}-\text{W}}) \gtrsim 10^4$] could be caused by a modest asymmetry in the galactic halo or in the primordial Big Bang, which is amplified by the subsequent almost complete annihilation of the minority anti-WIMPs. The second class of scenarios arranges an unusually small ratio of annihilation cross section to scattering cross section [see, e.g., Gelmini *et al.* (1987), Greist and Seckel (1987), and Ross and Segré (1987)]. One requires, [$(\sigma_{\text{sc}}/\sigma_{\text{ann}}) \gtrsim 10^4$]. In some cases, the small annihilation rate is "achieved" by assuming a conserved particle number.

Particle physics scenarios. Several ingenious scenarios have been proposed to meet the requirements that are described in the previous subsection. In most of the models, the WIMP is taken to be a not-yet-discovered electrically neutral spin 1/2 fermion. In each case, a new interaction (beyond ordinary Z^0 exchange) is introduced in order to achieve a sufficiently large scattering cross section [see Eq. (9.113)]. Gelmini *et al.* (1987) assumed large Yukawa interactions with a color triplet scalar that has a mass of order 100 GeV. Raby and West (1987) hypothesized a massive neutrino with an anomalously large magnetic moment that allows it to interact electromagnetically. Ross and Segré (1987) postulate a stable heavy neutrino that interacts with quarks via a neutral gauge boson whose effective strength is 50 times stronger than a conventional Z^0, but which does not couple to ordinary leptons. This last scenario can be implemented in models inspired by superstring theory.

9.6 Ockham's razor

Knowing what we do now, the MSW effect is, in my opinion, the most natural particle physics explanation of the solar neutrino problem. All other explanations postulate something that is surprising, in some cases at least as surprising as the solar neutrino problem itself. Vacuum neutrino oscillations require large mixing angles and fine tuning in order to work. The spin flip mechanism needs a large magnetic moment. The WIMP hypothesis requires a new particle with a large scattering cross section and a relatively small annihilation cross section. Neutrino decay is effective only if the electron neutrino has a large coupling constant for the decay mode.

The MSW explanation can work provided the masses and mixing angles lie within a wide parameter domain that is made plausible by the ideas of grand unification.

Bibliographical Notes

Barger, V., K. Whisnant, S. Pakvasa, and R.J.N. Phillips (1980) *Phys. Rev. D*, **22**, 2718. Recognized that matter oscillations are maximal at the resonance density [see their Eq. (19)], but did not apply their results to the solar neutrino problems. First discussed analytic solutions for three neutrino flavors.

Bethe, H.A. (1986) *Phys. Rev. Lett.*, **56**, 1305. A beautiful physical explanation of the MSW effect that ignited widespread interest in matter oscillators among western scientists.

Boehm, F.A. and P. Vogel (1987) *Physics of Massive Neutrinos* (Cambridge: Cambridge University Press). An excellent summary of neutrino properties with special emphasis on experiments that provide information about masses and charge conjugation.

Glashow, S. (1961) *Nucl. Phys.*, **22**, 579; Weinberg, S. (1967) *Phys. Rev. Lett.*, **19**, 1264; and Salam, A. (1968) in *Elementary Particle Theory*, edited by N. Svartholm (Stockholm: Almqvist and Wiksells) p. 367. The prophetic papers that laid the foundations for the successful synthesis of electromagnetic and weak theories that is known as the standard electroweak model. The standard model has been so persistently correct in predicting the results of new experiments that the model has become a source of frustration for particle physicists seeking new (i.e., not understood) worlds to conquer.

Kayser, B. (1981) *Phys. Rev. D*, **24**, 110. Informative wave packet treatment of oscillations. Serious students must read this article.

Maki, Z., M. Nakagawa, and S. Sector (1962) *Prog. Theor. Phys.*, **28**, 870. First discussion of neutrino flavor mixing.

Mikheyev, S.P. and A.Yu. Smirnov (1986) *Sov. J. Nucl. Phys.*, **42**, 913; (1986) *Nuovo Cimento*, **9C**, 17; (1986) *Sov. JETP*, **64**, 4; (1986) in *Proceedings of 12th Intl. Conf. on Neutrino Physics and Astrophysics*, edited by T. Kitagaki and H. Yuta (Singapore: World Scientific) p. 177. Epochal papers, exciting to read. Mikheyev and Smirnov obtained by numerical integration the principal results for matter oscillations in the Sun and presented them succinctly, together with a clear physical explanation.

Parke, S.J. (1986a) *Phys. Rev. Lett.*, **57**, 1275. Derived analytic results for the MSW effect that account for the general behavior of the solutions.

Pontecorvo, B. (1968) *Sov. JETP*, **26**, 984; Gribov, V. and B. Pontecorvo (1969) *Phys. Lett.*, **B28**, 493. The original papers suggesting that neturino flavor oscillations explain the solar neutrino problem. Founded a subject. Revolutionary ideas presented with clarity and brevity. The importance of averaging over energies was pointed out almost immediately by Bahcall and Frautschi (1969).

Press, W.H. and D.N. Spergel (1988) in *Dark Matter in the Universe*, edited by J. Bahcall, T. Piran, and S. Weinberg (Singapore: World Scientific) p. 206. An excellent summary of the status of WIMP theory and of experimental detection schemes.

Rosen, S.P. and J.M. Gelb (1986) *Phys. Rev. D*, **34**, 969. One of the earliest numerical explorations and interpretations of MSW solutions. Emphasized the importance of the energy spectrum as a diagnostic test.

Voloshin, M.B., M.I. Vysotskii, and L.B. Okun (1986) *Sov. JETP*, **64**, 446. Systematic discussion of nonresonant neutrino spin flip by the initiators of the reexamination of magnetic field effects on neutrino propagation. Stresses observable time dependencies if spin flip occurs in the convective zone.

Wolfenstein, L. (1978) *Phys. Rev. D*, **17**, 2369; (1979) *Phys. Rev. D*, **20**, 2634. Presented the fundamental equations for neutrino propagation in matter, the basis for the MSW effect. It took seven years for the physics community to recognize the significance of Wolfenstein's brilliant insight.

10

The ^{37}Cl experiment

Summary

The unexpected difference between the observed and the calculated capture rate in the ^{37}Cl detector created the solar neutrino problem. This chapter describes the ^{37}Cl experiment and its most direct implications.

The reaction that is used to detect solar neutrinos is the inverse of the laboratory decay of ^{37}Ar. The neutrino absorption reaction is

$$\nu_e + {}^{37}\text{Cl} \rightarrow e^- + {}^{37}\text{Ar}, \quad E_{\text{th}} = 0.814 \text{ MeV}. \quad (10.1)$$

The 0.8 MeV threshold energy permits the detection of all the major solar neutrino sources except the basic pp neutrinos. Some of the characteristics of the detector are summarized in Table 10.1.

The event rate predicted by the standard model for a ^{37}Cl detector is 7.9(1 ± 0.33) SNU, where the indicated uncertainty represents the total theoretical range. The total theoretical range is calculated by evaluating the 3σ uncertainties for all measured input parameters and using the full spread in calculated values for input quantities that cannot be measured; the uncertainties from different quantities are combined quadratically. The observed rate is (2.1 ± 0.9) SNU, where the observational error is three standard deviations. This discrepancy between calculation and observation has existed for two decades. The solution of this problem may involve new physics or new astronomy; future experiments will decide which.

The most intriguing and important observational question is: Does the event rate in the ^{37}Cl detector vary with time? Careful monitoring of the signal over the next few years will answer this question.

Table 10.1.   Characteristics of the ^{37}Cl detector.

Location	Homestake Mine, Lead, South Dakota
Depth	4850 ft
Tank	20 ft diameter × 48 ft long
Detector fluid	C_2Cl_4
Total weight of fluid	615 tons
Volume	3.8×10^5 liters
Threshold	0.814 MeV
^{37}Cl atoms	2.16×10^{30}
Half-life ^{37}Ar	35.0 days
Neutrino sensitivity	^{8}B, ^{7}Be

The first section in this chapter, 10.1, summarizes the theoretical expectations and the next section, 10.2, describes the experiment including the original reasons for choosing ^{37}Cl, the location, chemical processing, low-level counting, and backgrounds. Section 10.3 presents the observational results. The following section, 10.4, uses the experimental data to derive conservative limits on the ^{8}B, ^{7}Be, and CNO neutrino fluxes, quantitative implications for the theory of stellar evolution, and an upper bound to the cosmological energy density in neutrinos. Section 10.5 discusses the controversial question of whether or not the observed fluxes vary with time. The most recent data have generated great interest; Section 10.6 presents a summary of these data. Section 10.7 describes some of the most important future goals of the ^{37}Cl experiment.

Ray Davis has described the development of the ^{37}Cl experiment in the Fowler Festschrift [see the Appendix in this book reprinted from Bahcall and Davis (1982)]. Davis and his collaborators have also presented the experimental details and the observational results in a number of clear articles, including especially Davis (1964), Davis (1968), Davis, Harmer, and Hoffman (1968), Davis, Harmer, and Neeley (1969), Davis, Evans, Radeka, and Rogers (1972), Bahcall and Davis (1976), Davis, Evans, and Cleveland (1978), Davis (1978), Cleveland, Davis, and Rowley (1984), Rowley, Cleveland, and Davis (1985), and Davis (1987).

Table 10.2. Capture rates
predicted by the standard model
for a ^{37}Cl detector.

Neutrino source	Capture rate (SNU)
pp	0.0
pep	0.2
hep	0.03
^{7}Be	1.1
^{8}B	6.1
^{13}N	0.1
^{15}O	0.3
^{17}F	0.003
Total	7.9 SNU

10.1 Theoretical expectations

The capture rate predicted for a ^{37}Cl detector by the standard model is [Bahcall and Ulrich (1988)]:

$$\sum_i \phi_i \sigma_i = 7.9\,(1 \pm 0.33)\ \text{SNU}, \qquad (10.2)$$

where the indicated uncertainty represents the total theoretical range defined in Chapter 7. The ^{37}Cl detector is sensitive to neutrinos with energies above 0.814 MeV, which excludes the abundant pp neutrinos. However, all of the other neutrinos discussed in Chapter 6 are above threshold and contribute in various amounts to the total expected event rate.

Table 10.2 shows the contribution to the predicted capture rate of each of the neutrino sources discussed in Chapter 6. The ^{37}Cl experiment is primarily sensitive to neutrinos from ^{8}B decay, because of the great sensitivity of the detector to neutrinos with sufficiently high energy to excite the state in ^{37}Ar that is the analogue of the ground state of ^{37}Cl.

Approximately 77% (6.1 SNU) of the predicted event rate is contributed by ^{8}B neutrinos; the next largest contribution is the 14% (1.1 SNU) from ^{7}Be neutrinos. Most of the calculated uncertainty is caused by the ^{8}B neutrino flux. About 87%, or 2.3 SNU out of the total estimated uncertainty of 2.6 SNU, is associated with the ^{8}B flux.

Most of the predicted capture rate depends upon the rare mode for the completion of the pp chain, reaction 9 of Table 3.1, which occurs only about twice in every 10^4 terminations of the chain. Moreover, the ^{8}B flux is sensitive to the physical conditions in the solar interior [see, e.g., Figure 6.2a and Eq. (6.13a)]. The predicted flux is also sensitive to a number of input parameters [see Eq. (7.7)]. This sensitivity of the ^{8}B flux to input parameters has motivated many of the theoretical and experimental investigations over the past two years that have refined the astrophysical and nuclear physics data and have made possible the definite error estimate given in Eq. (10.2).

The total uncertainty is made up of contributions of comparable size from several different input parameters (see row 3 of Table 7.4): 1.8 SNU from the initial ratio of heavy elements to hydrogen, Z/X; 1.3 SNU from the low-energy cross section factor (S-value, see Chapter 3) for the ^{7}Be–p reaction; 0.9 SNU from the pp reaction; and 0.6 SNU from the neutrino absorption cross sections. Since several different parameters contribute appreciably to the total uncertainty, it seems unlikely that the total theoretical uncertainty will be reduced much below 2 SNU, or of order 25% of the predicted capture rate, in the foreseeable future.

How reliable is the theoretical prediction? Chapter 7 describes the formal logic that was used in determining the total theoretical uncertainty and also specific individual uncertainties that were assigned for different quantities. This formal procedure was the basis for the error estimates cited in the above paragraph. There is also an independent but more informal way of evaluating the total uncertainty. This informal method is based upon the time dependence of the published values and was discussed briefly in Section 7.1. A skeptical but not unreasonable position is that the total uncertainty is equal to the total range of the published values.

Figure 10.1 shows, as a function of the date of publication, all of the predicted capture rates, and their quoted errors, for every paper in which I (as an author or as a co-author) have published theoretical capture rates. This figure also shows the quoted errors whenever I was rash enough to publish estimated uncertainties.

All of the 14 values published since 1968 are consistent with the range given in Eq. (10.2). The first calculation in 1963 was made

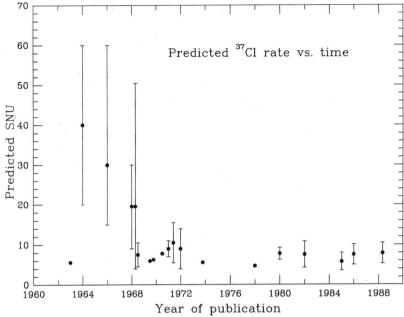

Figure 10.1 **Predicted capture rates as a function of time** The published predictions of neutrino capture rates in the ^{37}Cl experiment are shown as a function of the date of publication. The values and their error bars are from Bahcall, Fowler, Iben, and Sears (1963), Bahcall (1964a), Bahcall (1966), Bahcall and Shaviv (1968), Bahcall, Bahcall, Fowler, and Shaviv (1968), Bahcall, Bahcall, and Shaviv (1968), Bahcall (1969), Bahcall and Ulrich (1970), Bahcall and Ulrich (1971), Bahcall *et al.* (1980), Bahcall *et al.* (1982), Bahcall *et al.* (1985), Bahcall (1986), and Bahcall and Ulrich (1988). Similar results have been obtained by many authors, see, for example, Abraham and Iben (1971), Chitre, Ezer, and Stothers (1973), Wheeler and Cameron (1975), Rood (1978), Filippone and Schramm (1982), Cassè, Cahen, and Doom (1986), and Cahen, Doom, and Cassè (1986).

before the realization [Bahcall (1964a)] that the capture rate for ^{8}B neutrinos is enhanced by a factor of about 17 by transitions to excited states, especially to the analogue state.

Figure 10.2 shows the dramatic increase in detailed knowledge of the nuclear properties of ^{37}Cl and ^{37}Ar in a quarter century of experimental work on these two isotopes. The first part of the figure shows the meager information that was available in 1964 [Bahcall (1964b)]. None of the spins and parities of the nuclear levels of ^{37}Ar were known (nuclear properties in parentheses were inferred

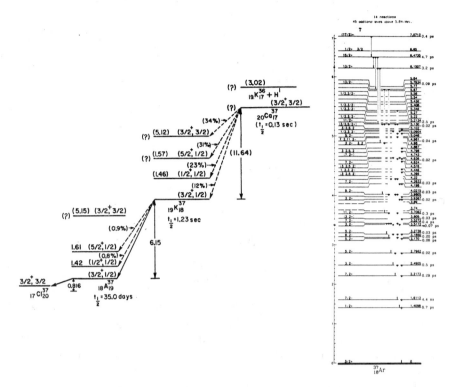

Figure 10.2   Before and after: the mass 37 system   This figure illustrates the increase in detailed knowledge of the mass 37 system that can be
used to calculate solar neutrino capture cross sections for ^{37}Cl. Figure 10.2a
shows the meager information that was available to Bahcall (1964b); nearly
all of the important characteristics were calculated on the basis of a nuclear
model (properties inferred from the model are shown in parentheses). The
β-decay of the isotope ^{37}Ca was predicted, not measured, since ^{37}Ca had
not been observed. Figure 10.2b indicates the abundant information that
is now available for ^{37}Ar, as the result of many precise measurements of
spins, parities, and energy levels [taken from the *Table of Isotopes, Seventh
Edition* (1978) (New York: Wiley)]. The position of the isotopic analogue
state in ^{37}Ar and the β-decay of ^{37}Ca have been observed experimentally,
confirming and refining the original ideas of the strength of the capture
cross section for ^{8}B neutrinos.

on the basis of a model, not measured) and the crucial analogue level corresponding to the ground state of ^{37}Cl was drawn in with a question mark to indicate that it was calculated, not observed. The stability of the isotope ^{37}Ca to the emission of neutrons and protons was suggested and it was proposed that ^{37}Ca β-decayed with a measurable lifetime to ^{37}K. The second part of Figure 10.2 shows a partial summary of the information about ^{37}Ar that is available in 1988. As a result of beautiful nuclear physics experiments, the spins, parities, and excitation energies of many states of ^{37}Ar are now well known, including the crucial analogue state. Also, the isotope ^{37}Ca was detected and found to have properties in good agreement with the predicted characteristics. The many accurate experiments on the mass 37 system have confirmed the ideas used originally to calculate the capture cross section of ^{8}B neutrinos. The measurement of the β-decay of ^{37}Ca makes possible an accurate empirical determination of the capture cross sections for ^{8}B and hep neutrinos.

10.2   The experiment

A   Why ^{37}Cl?

Chlorine was chosen for the first solar neutrino experiment because of its unique combination of physical and chemical characteristics, which were favorable for building relatively cheaply a large scale neutrino detector. The absorption reaction to the ground state of ^{37}Ar that is shown in Eq. (10.1) has a relatively low threshold (0.814 MeV) and a favorable cross section (relatively small $ft_{1/2}$-value, log $ft =$ 5.), nuclear properties that facilitate the observation of all of the relatively low-energy neutrinos except those from the pp reaction. In addition, the key step that made the experiment appear feasible in 1964 was the realization that ^{8}B could excite transitions to the analogue state of ^{37}Ar, increasing the expected rate from this source by more than an order of magnitude.

Neutrino absorption by ^{37}Cl, Eq. (10.1), is also favorable from a chemical point of view. Chlorine is abundant and inexpensive enough that one can afford the hundreds of tons that are needed to observe solar neutrinos. In addition, ^{37}Cl is a significant fraction, 24.23% by number, of all the chlorine atoms. The most suitable chemical compound is perchloroethylene, C_2Cl_4, a pure liquid, which is manufactured on a large scale for cleaning clothes. Neutrino capture produces a tiny amount of ^{37}Ar, a noble gas, as dissolved atoms in the liquid. The capture process creates an ^{37}Ar atom with sufficient recoil energy to break free of the parent perchloroethylene molecule and penetrate the surrounding liquid, where it reaches thermal equilibrium. The dissolved argon atoms can be removed easily from the liquid by purging with helium gas. The simplicity of the chemical process is crucial both for carrying out the experiment and for convincing skeptics of the validity of the results. The lifetime for ^{37}Ar, 35 days, is convenient; a much shorter lifetime would require chemical extractions too frequently and a much longer lifetime would severely limit the number of exposures that could be obtained.

B   Location and specifications

The ^{37}Cl detector was built deep underground in order to avoid the production of ^{37}Ar in the detector by cosmic rays. The construction was carried out with the cooperation of the Homestake Gold Mining Company (Lead, South Dakota), who excavated a large cavity in their mine and who have provided an efficient and pleasant environment for the experiment for the more than two decades of its operation.

The neutrino target is 2.2×10^{30} atoms (133 tons) of ^{37}Cl in the form of 3.8×10^5 liters, 615 tons, of liquid perchloroethylene, C_2Cl_4. The tank containing this material is located at the 4850 ft level underground in the Homestake Gold Mine. The resulting effective depth of the chlorine detector is 4100 m of water equivalent. Figure 10.3 shows the ^{37}Cl detector in the Homestake Mine.

The number of ^{37}Cl atoms in the detector provides the basis for converting between the theoretical unit of SNU (10^{-36} captures per target atom per second) and the observational unit of atoms of ^{37}Ar produced per day in the tank. One finds:

$$\text{One } ^{37}\text{Ar atom per day} = 5.35 \text{ SNU}. \tag{10.3}$$

C   Chemical processing

The chemical processing is relatively simple, especially when one considers that the sample weighs 615 tons. The first step is to place a small amount (~ 0.1 cm^3 STP) of isotopically pure ^{36}Ar (or ^{38}Ar) carrier gas in the tank containing the perchloroethylene and to stir

Figure 10.3 The ^{37}Cl neutrino detector The figure shows the tank containing 100,000 gallons of perchloroethylene in the cavity 4850 ft below ground in the Homestake Mine in Lead, South Dakota. R. Davis, Jr., principal scientist of the ^{37}Cl experiment, is leaning on the catwalk above the tank and J. Galvin, expert technician, is standing below. The photograph is by courtesy of the Brookhaven National Laboratory (circa 1967).

the carrier gas into the liquid to ensure that it dissolves. The amount of inert carrier gas that is extracted at the end of the run is a direct measure of the efficiency of recovery for ^{37}Ar produced by neutrinos. None of the chemical or macroscopic procedures used to remove argon from the tank and to place it in a small counter distinguishes between different isotopes of argon.

The liquid in the tank is allowed to stand exposed to solar neutrinos for some time, usually of the order one to three months. The longer exposure permits the neutrino induced ^{37}Ar activity to grow to nearly saturation value.

After exposure, the argon in the tank is removed by circulating about 4×10^5 liters (1 tank volume) of He through the gas and liquid phases in the tank, then through a condenser at $-32\ °$C and a molecular sieve at room temperature, and finally through a charcoal trap cooled to the temperature of liquid nitrogen where the argon is absorbed. Gas circulation is accomplished and the tank is stirred by two large pumps, each connected to an eductor system of 40 nozzles. The nozzles draw helium from the top of the tank and force it through the liquid in the form of fine bubbles. About 95% of the argon is removed (and retained on the charcoal trap) by the helium circulation.

Figure 10.4 shows Davis inspecting the newly installed gas circulation pumps in 1967.

The large charcoal trap in which the argon has been condensed is heated in order to transfer the argon to a separate line where it is purified. Next, the volume of the recovered sample of argon is measured. Finally, the inert carrier and radioactive ^{37}Ar are loaded into a small (0.3 to 0.5 cc) proportional counter along with tritium-free methane, which serves as a counting gas. In processing the sample, special precautions are taken to remove krypton, radon, and tritium in order that they not contribute radioactive background to the counting of ^{37}Ar (see below).

The recovery efficiency is determined for each exposure using the measured volume and isotopic composition (determined by mass spectrometry at a later time) of the extracted sample, which is compared with the amount of isotopically pure carrier gas that was introduced at the beginning of each run. If the standard solar model were correct, one would expect about 50 ^{37}Ar atoms in the 380,000 liters of liquid at the time it was purged. These few atoms of ^{37}Ar behave chemically in the same way as the $\sim 10^{19}$ atoms of ^{36}Ar or

Figure 10.4 Inspecting the pumps   R. Davis, Jr. (in the doorway) enters the room to inspect the newly installed gas circulation pumps; G. Friedlander (kneeling) is shown holding a wrench. The photograph is by courtesy of Brookhaven National Laboratory (circa 1967).

^{38}Ar that were introduced as a carrier. Therefore, the recovery efficiency for ^{37}Ar atoms can be inferred directly from a determination of the efficiency with which the carrier is recovered.

Two additional tests have been performed to ensure that ^{37}Ar produced in the large tank is indeed extracted efficiently. In one, a small neutron source was placed in the center of the tank through a reentrant tube. Neutrons produce ^{37}Ar in the liquid by a series of nuclear reactions. Davis and his collaborators verified that the ^{37}Ar produced in the tank by the neutron source is recovered along with the carrier gas. The second test was to introduce a measured number of ^{37}Ar atoms (500) into the detector and then remove them, measuring the overall recovery and counting efficiency. Both of these tests show that ^{37}Ar is recovered with high efficiency.

A special test was performed in order to show that ^{37}Ar, which is produced in the form of a recoiling ion, becomes a neutral argon atom. The calibration of the recovery efficiency rests upon the assumption that the Ar$^+$ ion that is produced does not form a molecule but instead moves around as a free neutral atom which is recovered together with the carrier gas of inert argon isotopes. Tetrachloroethylene labeled with ^{36}Cl was added to a tank of perchloroethylene. The ^{36}Cl decays by β-minus emission to ^{36}Ar:

$$C_2Cl_3{}^{36}Cl \rightarrow C_2Cl_3(\text{fragments}) + {}^{36}Ar^+ + e^- + \bar{\nu}_e. \qquad (10.4)$$

This process is essentially equivalent to the capture process ^{37}Cl$(\nu_e, e^-)^{37}$Ar. After standing a definite length of time, the quantity of neutral ^{36}Ar produced by ^{36}Cl decay [see Eq. (10.4)] was determined by activation analysis after separation using a helium purge. In this experiment (performed by J. Evans, H. Vera Ruiz, and R. Davis) the ^{36}Ar was recovered with $100 \pm 3\%$ yield [see Rowley et al. (1985)].

The ultimate test of the ^{37}Ar recovery efficiency would be to place a source of known neutrino intensity and energy in or adjacent to the chlorine detector. Alvarez (1973) first studied, in a careful investigation, the possibility of carrying out a direct calibration using ^{65}Zn [see Eq. (6.10) for the decay scheme]. A source of 1.0 megacurie of ^{65}Zn would be required to obtain an ^{37}Ar production rate of four ^{37}Ar atoms day^{-1}, a rate 10 times higher than that presently observed in the detector (from the combination of solar neutrinos and background events). In 1981 and 1982 a trial irradiation was performed in the HFIR reactor at Oak Ridge to measure the ^{65}Zn production in zinc. This test irradiation indicated that only 0.5 megacurie could be prepared at HFIR and further tests at HFIR were abandoned [Rowley et al. (1985)].

D Counting

The entire gas sample from the tank is placed in a small proportional counter with an internal volume of less than a cubic centimeter, about 2.5 cm long and 0.4 cm inside diameter. The counter is filled with the argon recovered from the tank along with 5 to 10% methane. In some experiments, a small amount of neon or argon gas is added to bring the total gas pressure in the counter from 1.0 to 1.5 atmospheres.

The ^{37}Ar counting depends upon observing the 2.82 keV Auger electrons from the electron capture decay of ^{37}Ar. The counters have a resolution of about 25% (full width of the peak at half the peak height, FWHM) for electrons of this energy.

In a typical experimental run, only of order six events are observed that have the proper characteristics to correspond to an ^{37}Ar decay. The decay of ^{37}Ar is characterized in part by the **energy** deposited in the counter by the Auger electrons that are produced following the electron capture process. The decay is also characterized by the short **rise time** of the electronic pulses associated with the detection of their Auger electrons. These low-energy electrons have a short range in the gas, only about 0.1 mm. As a result, the hundred or so ion pairs produced by the event are contained initially in an extremely small volume. A fast rise time pulse results from the collection of these electrons on the center wire of the proportional counter. Hence a measurement of the rise time can distinguish the fast ^{37}Ar pulses from the slower pulses that result from β rays, cosmic rays, and Compton electrons. The charge pulse from the counter is measured during a short period of time, usually about 10 ns. The charge collected during this period from a fast rising pulse, like one from an ^{37}Ar decay, will be greater than that collected from a slower rising pulse with the same energy. On the other hand, noise pulses typically have extremely fast rise times, shorter than those associated with the Auger electrons from ^{37}Ar decay.

A combined measurement of energy and pulse rise time is a powerful discriminant against background processes and has been used to characterize the ^{37}Ar decays in the most sensitive experimental runs. Each pulse is recorded in a two-dimensional plane whose axes are the measured electron energy and the pulse rise time. Pulses from background sources and from electronic noise generally occupy a different region of the energy rise time plane than do the neutrino-induced ^{37}Ar decays. The calibration of the energy rise-time plane is achieved with the aid of an ^{55}Fe source which produces X-rays of an energy of 5.9 keV and a short rise time that is similar to what is observed from ^{37}Ar Auger electrons. Although the counter background is greatly reduced by using a two-dimensional characterization of the desired pulses, a small number of the counts within the accepted area are estimated to be from background sources (see below).

The sample is counted for a period of approximately eight months or longer in order to determine accurately the (low) background characteristic of the particular counter that was used. The counters containing experimental samples are tested approximately every two months with an external ^{55}Fe source to see if any gains or drifts have occurred in the amplifiers. Essentially all of the ^{37}Ar decays in the first few months.

Over the more than two decades of operation of this experiment many improvements in the counting have been made. Reductions in background have been achieved by using low radioactivity material in the counters and other counting components, by using a large well-type sodium iodide crystal as an anticoincidence counter, and by moving the counting apparatus to the underground laboratory at Homestake, where the counters are surrounded by a heavy shield (30 cm Pb + 5 cm Hg).

The counts obtained in a given run within the acceptable region of pulse rise time and energy are analyzed by the method of maximum likelihood; the detailed procedure is described by Cleveland (1983). For each run, the counts are assumed to arise from a constant background and a constant production rate with a single decaying component. Consistency tests have been performed in which the half-life of the decaying component was determined empirically to be equal to that of ^{37}Ar by using the raw data and the maximum likelihood analysis [see Eq. (10.9) below]. However, for the most accurate determination of the production rate of ^{37}Ar in the tank, the half-life of the decaying component was assumed to be equal to 35 days, the half-life of ^{37}Ar. A likelihood function is calculated using the recorded times of all counts within the accepted range of pulse height and rise time. This likelihood function includes fluctuations in ^{37}Ar production, extraction and processing, as well as fluctuations during counting. The most likely values of the production rate and of the background rate are those that maximize the likelihood function. The production rate is required to be zero or positive; negative values are excluded. This requirement introduces an asymmetry for small counting rates that must be evaluated by Monte Carlo simulations. Results from many runs are combined by multiplying individual likelihood functions to give a combined likelihood function.

The maximum likelihood method of analyzing the data was developed in 1977; all the data obtained since 1970 have been analyzed using this method. It has been extensively tested by Monte Carlo simulations with varying input production rates and background counting rates. The results of these simulations show that the most likely values of the ^{37}Ar production rate and of the counter background agree well with the input values.

The background rates in the counters (from sources other than ^{37}Ar) have been derived from the maximum likelihood analysis of the times of the recorded events. The average counter background rate over 61 runs from 1970 to 1984 was 0.033 day^{-1}. The total range was from 0 to 0.137 day^{-1}. *For runs made since 1984, the average background counting rate is about 0.010 day^{-1}, that is, only 3.6 counts per year!* Experimental runs were carried out on a rather irregular schedule during the period 1969 to 1975. During 1976, an attempt was made to perform measurements more frequently, every 35 to 50 days. After 1976, measurements were made on a nearly regular basis of six runs per year until 1985 when observations were terminated temporarily because of the failure of the two liquid circulation pumps (in July 1984 and May 1985). In 1986, the University of Pennsylvania assumed responsibility for the experiment from Brookhaven National Laboratory and regular observations were resumed in October 1986 with one new pump. A second new pump was added in April 1988. Occasional special runs have been performed from time to time to look for increased neutrino fluxes from solar flares and from unusual astronomical events.

E Backgrounds

Background events constitute the most difficult obstacle in most solar neutrino experiments. One cannot, as in laboratory experiments, turn off the source and assume that the background is the rate observed when the Sun is not shining in neutrinos. Moreover, the event rate is small, so that rare background processes can sometimes be confused with the solar signal.

There are four major recognized background sources for the ^{37}Cl experiment: (1) cosmic ray muons and products of their interaction ($\pi^{\pm}$, energetic protons and neutrons, and evaporation protons); (2) fast neutrons from the rock wall via (α,n) reactions and from spontaneous fission of ^{238}U; (3) α-particle interactions from uranium and thorium in the perchloroethylene; and (4) cosmic ray neutrinos.

The main source of background is believed to be caused by cascades initiated by deeply penetrating muons that are themselves produced by cosmic rays in the upper atmosphere. The muon cascades contain energetic pions, protons, and neutrons; these cascade particles ultimately produce ^{37}Ar via (p,n) reactions on ^{37}Cl.

The background from high-energy muons has been estimated in two independent ways: (1) by exposing 600 gallon tanks of C_2Cl_4 at higher levels in the mine and then extrapolating the production rates of ^{37}Ar to the lower level at which the chlorine detector is located; and (2) by multiplying an *in situ* measurement of the vertical muon flux by a theoretical estimate [Zatsepin, Kopylov, and Shirakova (1981) and Wolfendale, Young, and Davis (1972)] of the ^{37}Ar yield per muon at the average muon energy at the depth of the ^{37}Cl chamber. The two estimates of the background are in good agreement and yield:

$$\text{Background} = (0.08 \pm 0.03) \text{ day}^{-1} \ [0.4 \pm 0.16 \text{ SNU}]. \quad (10.5)$$

The background can also be measured by a method that depends upon the fast muon process $^{39}K(\mu^+; \mu^+ + n + p)^{37}$Ar developed by Fireman *et al.* (1985). Preliminary results from the measurement carried out at the full depth of the chlorine detector agree with the value used above. Improved background estimates can and should be made with the aid of ^{39}K.

10.3   The results

Rowley, Cleveland, and Davis (1985) have presented the results of 61 runs completed during the period from 1970 to 1984. In these 61 runs, there were 774 counts having both the correct energy and the correct rise time. These counts were divided by the maximum likelihood procedure of Cleveland (1983) (described in Section 10.2D) into 435 background counts and 339 counts resulting from ^{37}Ar decay.

Table 10.3 shows the results of the 61 experimental runs. Listed in the table for each run are the time at the start of the exposure, the time at the end of the exposure, and the mean time of exposure, defined as

$$t_{\text{mean}} = t_{\text{start}} + (1/\lambda) \ln [1/2 + 1/2 \exp (\lambda t_{\text{end}} - \lambda t_{\text{start}})], \quad (10.6)$$

together with the maximum likelihood production rate and the lower and upper 1σ error limits. The quantity, t_{mean}, is the time at which half of the observed atoms of ^{37}Ar were accumulated. This time is useful in analyzing the data for time variations, especially for experiments with a long exposure period.

Table 10.3.   Exposure times and ^{37}Ar production rates
from individual runs using the chlorine detector.

	Exposure dates, years			Atoms per day		
Run	Start	End	Mean	^{37}Ar production rate	Lower limit	Upper limit
18	70.279	70.874	70.780	0.214	0.0	0.498
19	70.874	71.180	71.098	0.490	0.150	0.830
20	71.180	71.462	71.383	0.349	0.067	0.630
21	71.462	71.755	71.675	0.0	0.0	0.555
22	71.755	71.951	71.885	0.289	0.0	0.779
24	72.168	72.380	72.311	0.497	0.226	0.768
27	72.517	72.848	72.765	1.226	0.820	1.633
28	72.848	73.073	73.002	0.0	0.0	1.165
29	73.073	73.287	73.218	0.608	0.211	1.006
30	73.287	73.668	73.581	0.147	0.0	0.365
31	73.668	73.952	73.873	0.505	0.0	1.080
32	73.952	74.070	74.023	0.277	0.0	0.928
33	74.070	74.487	74.398	0.302	0.066	0.539
35	74.500	74.591	74.553	0.0	0.0	0.509
36	74.591	75.121	75.028	0.671	0.355	0.987
37	75.121	75.454	75.370	0.877	0.455	1.298
38	75.454	75.733	75.654	0.279	0.0	0.755
39	75.733	76.062	75.978	0.580	0.252	0.909
40	76.065	76.180	76.134	0.419	0.078	0.760
41	76.180	76.270	76.232	0.569	0.152	0.987
42	76.270	76.386	76.340	0.605	0.0	1.534
43	76.386	76.542	76.485	0.058	0.0	0.260
44	76.542	76.676	76.625	0.047	0.0	0.371
45	76.676	76.772	76.732	0.337	0.025	0.648
46	76.722	76.924	76.868	0.491	0.137	0.844
47	76.924	77.076	77.020	0.979	0.583	1.376
48	77.076	77.290	77.221	0.407	0.138	0.676
49	77.290	77.385	77.345	1.075	0.593	1.558
50	77.385	77.594	77.526	0.910	0.588	1.232
51	77.594	77.824	77.754	0.849	0.525	1.173
52	77.824	78.054	77.982	0.588	0.308	0.867
53	78.054	78.361	78.279	0.200	0.0	0.404

Table 10.3. (*continued*)

Run	Start	End	Mean	^{37}Ar production rate	Lower limit	Upper limit
	Exposure dates, years			Atoms per day		
54	78.361	78.595	78.523	0.626	0.365	0.887
55	78.595	78.827	78.755	0.468	0.162	0.774
56	78.827	79.051	78.980	0.853	0.546	1.160
57	79.051	79.150	79.110	0.215	0.0	0.668
58	79.150	79.375	79.304	0.853	0.295	1.410
59	79.375	79.586	79.517	0.237	0.034	0.439
60	79.586	79.818	79.746	0.0	0.0	0.158
61	79.818	80.065	79.991	0.090	0.0	0.387
62	80.065	80.281	80.211	0.023	0.0	0.254
63	80.281	80.451	80.391	0.0	0.0	0.325
64	80.451	80.604	80.548	0.488	0.222	0.754
65	80.604	80.739	80.687	0.224	0.0	0.649
66	80.739	80.892	80.836	0.361	0.0	1.614
67	80.892	81.059	80.999	0.319	0.051	0.588
68	81.059	81.290	81.218	0.359	0.175	0.544
69	81.290	81.519	81.448	0.477	0.166	0.788
70	81.519	81.673	81.616	0.081	0.0	0.301
71	81.673	81.826	81.770	1.209	0.844	1.574
72	81.826	81.966	81.913	0.636	0.337	0.935
73	81.966	82.210	82.136	0.077	0.0	0.228
74	82.210	82.361	82.305	0.478	0.237	0.720
75	82.361	82.810	82.719	0.503	0.176	0.830
76	82.810	83.040	82.968	0.475	0.144	0.806
77	83.040	83.194	83.137	0.461	0.237	0.684
78	83.194	83.366	83.305	0.752	0.465	1.040
79	83.366	83.531	83.471	0.604	0.332	0.875
80	83.531	83.654	83.606	0.824	0.299	1.348
81	83.654	83.884	83.812	0.330	0.089	0.571
82	83.884	84.095	84.026	0.545	0.257	0.832
All sixty-one runs combined				0.462	0.421	0.502

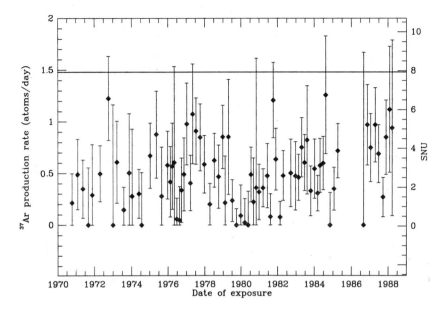

Figure 10.5   Observational results from the chlorine solar neutrino experiment   For details see Davis (1968), Davis, Harmer, and Hoffman (1968), Davis, Harmer, and Neeley (1969), Davis, Evans, Radeka, and Rogers (1972), Bahcall and Davis (1976), Davis, Evans, and Cleveland (1978), Davis (1978), Cleveland, Davis, and Rowley (1984), Rowley, Cleveland, and Davis (1985), Davis (1987), and Section 10.6. This figure contains data on more recent runs, 83-99, discussed in Section 10.6 and generously made available by R. Davis, Jr. and B. Cleveland. The line at 7.9 SNU across the top of the figure represents the prediction of the standard model.

Figure 10.5 shows the results for each of the runs. The vertical error bars indicate 1σ errors. The combined production rate for the 61 runs that is inferred with the aid of the maximum likelihood analysis is

$$\text{Production rate} = 0.462 \pm 0.040 \ ^{37}\text{Ar atoms day}^{-1}. \quad (10.7)$$

Subtracting the small background rate given in Eq. (10.5) and converting to solar neutrino units using Eq. (10.3), the capture rate in the tank is estimated to be

$$\text{Capture rate} = (2.05 \pm 0.3) \text{ SNU}. \quad (10.8)$$

In a separate analysis, the half-life of the decaying component and the production rate were *both* estimated from the data. This treatment yielded a most likely half-life of

$$\tau_{\text{best estimate}} = 37 \pm 6 \text{ days} \qquad (10.9)$$

and a production rate of 0.45 ± 0.05 ^{37}Ar atoms day^{-1}. The agreement of the inferred half-life of the decaying component with that of ^{37}Ar (35 days) provides independent evidence that ^{37}Ar is being detected.

The observations were interrupted in 1985 due to successive failures of the circulating pumps. The first pump failed in July 1984, during run 85. The second pump failed in May 1985 at the end of run 89. Observations began again in October 1986 (run 90) with one pump. A second pump was added in April 1988 (for run 99).

A number of runs have been performed since the resumption of operations in late 1986. These measurements provide additional information about the possible time dependence of the capture rate (see Sections 10.5 and 10.6).

10.4   Theory versus experiment: the implications

Figure 10.5 shows all of the observing runs from 1970 to 1988 and compares them with the theoretical value of Eq. (10.2).

A   The ^{8}B flux

The ^{37}Cl experiment yields an allowed range for the flux of electron neutrinos with a ^{8}B spectrum that reaches the Earth from the Sun. The ratio of the observed to the calculated ^{8}B flux is the crucial quantity that must be compared with models of weak interactions to determine if something happens to neutrinos on the way to the Earth from the center of the Sun. However, the conversion of the experimental limit to the desired ratio of observed to predicted fluxes depends upon what one is willing to assume, as illustrated in the following discussion.

The experimental result given in Eq. (10.8) can be written as a limit on the ^{8}B neutrino flux by dividing the event rate in SNUs by the absorption cross section given in Table 8.2. Ignoring for the moment all of the estimated uncertainties, one finds

$$\phi(^8\text{B})_{\text{no errors}} \leq 2.0 \times 10^6 \text{ cm}^{-2} \text{ s}^{-1}, \qquad (10.10)$$

which is about one-third the value of 5.8×10^6 cm^{-2} s^{-1} obtained from the standard model.

If one assumes instead that the 3σ upper limit to the *entire* observed capture rate, that is, SNU, is due to ^{8}B neutrinos, then one obtains a *conservative* upper limit to the ^{8}B neutrino flux. Making this conservative assumption, one finds

$$\phi(^8\text{B})_{\text{observed}} \leq 2.7 \times 10^6 \, \text{cm}^{-2} \, \text{s}^{-1}. \qquad (10.11)$$

The limit given in Eq. (10.11) is a factor of two smaller than the best estimate flux given in Chapter 6.

Since the theoretical uncertainties are much larger than the observational errors, one can set a conservative limit to the "missing flux" by using, instead of the best estimate provided by the standard solar model, the value obtained by subtracting the estimated total theoretical uncertainty (see Table 6.5 and Sections 8.3 and 8.4) from the best estimate. The "theoretical lower limit" is

$$\phi_{\text{theoretical}} \geq 3.7 \times 10^6 \, \text{cm}^{-2} \, \text{s}^{-1}, \qquad (10.12)$$

which is about 1.9 times larger than the experimental upper limit given in Eq. (10.10).

There is another extreme possibility. The experimental result cited in Eq. (10.10) is consistent with *no* observed ^{8}B neutrino flux. For the standard solar model, the contribution of all but ^{8}B neutrinos is 1.8 SNU (see Table 10.1), which is consistent with the observed capture rate.

The observed event rate could, in principle, be due partly or wholly to interactions caused by something other than neutrinos from the solar interior. If, as has been suggested [e.g., Davis (1987)], there are physical variations in the event rate associated with, for example, the sunspot cycle or solar flares, then it is likely that the observed rate is an overestimate of the rate from neutrinos produced by nuclear fusion in the solar core. If this hypothesis is correct, then the ^{37}Cl experiment does not provide any evidence that the ^{8}B neutrino flux has been detected.

Therefore the experimental limit on the ratio of observed to predicted ^{8}B neutrino flux is in the range

$$0 \leq \left(\frac{\phi(^8\text{B})_{\text{observed}}}{\phi(^8\text{B})_{\text{predicted}}} \right) \leq 0.5. \qquad (10.13)$$

The value of $\phi(^8B)_{\text{predicted}}$ that is used in Eq. (10.13) is the lowest calculated value that is consistent with the total theoretical range described in Chapters 6 and 7, namely, the value given in Eq. (10.12).

B Implications for stellar evolution

Something is wrong with the standard theory of stellar evolution *if* nothing happens to solar neutrinos on the way to the Earth from the interior of the Sun. The discrepancy between calculated and observed rate for the ^{37}Cl experiment has existed for two decades. Many hundreds of papers have been published describing improved calculations of input parameters, a better treatment of some aspect of the standard model, or a new experimental determination of a quantity that is used in deriving the standard solar model. Despite the intensive effort of many people, the discrepancy has persisted. The quantitative estimate of the uncertainty has become better defined [see Eq. (10.1)], but the experimental and theoretical values continue to disagree.

It seems likely to me that the explanation of the solar neutrino problem will not be trivial if it lies in the astrophysical realm. Whatever is wrong must be reasonably subtle and may well affect other aspects of the theory of stellar evolution.

Solar neutrino experiments are crucial tests of the theory of stellar evolution. We know more about the Sun than about any other star. Moreover, the Sun is in the simplest (main sequence) stage of stellar evolution, quietly converting hydrogen to helium in a quasistatic fashion. If we do not understand how the Sun shines, then there is a good chance that we have not understood correctly the more complicated (and fashionable) stages of stellar evolution, for example, the formation of giant stars, the evolution of normal stars into degenerate stars (white dwarfs or neutron stars) or black holes, and the production of heavy elements during stellar explosions. Much of what we take for granted in interpreting astronomical observations will be under a cloud of suspicion until we know for sure whether or not the explanation of the solar neutrino problem requires new astrophysics.

All of the previous discussion in this chapter assumes the correctness of one fundamental prediction of the standard solar model, namely, that nearly all of the Sun's luminosity is derived from the pp chain. This result is not obvious, since stars only slightly more massive than the Sun are predicted to burn primarily via the CNO cycle. In fact, Bethe (1939) in his epochal paper on nuclear energy generation suggested that the CNO cycle was dominant in the Sun. In order to derive accurately the demarcation in mass between the pp chain and the CNO cycle, one must construct accurate and detailed numerical models for the evolution of main sequence stars.

If one hypothesizes, contrary to prediction of the standard solar model, that the CNO cycle is the primary mode of solar energy generation, then one can derive from this assumption alone a predicted rate for the ^{37}Cl experiment. One finds [Bahcall (1978)] that

$$\text{Rate if CNO cycle dominant} = 28 \text{ SNU.} \qquad (10.14)$$

Stellar evolution theorists may take some comfort in the fact that this large rate, 28 SNU, implied by the dominance of the CNO cycle, is *not* observed. Although something may be wrong, we may also have gotten something right.

The observed capture rate, Eq. (10.8), and the neutrino absorption cross sections, Table 8.2, can be combined to set an upper limit on the CNO neutrino fluxes. Making the good approximation that the ^{13}N and ^{15}O neutrino fluxes are equal, one obtains

$$\phi(\text{CNO}) \leq 3.5 \times 10^9 \text{ cm}^{-2} \text{ s}^{-1}. \qquad (10.15)$$

The upper limit given in Eq. (10.15) is within a factor of six of the value predicted by the standard model (see Table 6.5). The corresponding upper limit on the flux of ^{7}Be neutrinos is

$$\phi(^7\text{Be}) \leq 1.5 \times 10^{10} \text{ cm}^{-2} \text{ s}^{-1}. \qquad (10.16)$$

The observational upper limit on the flux of ^{7}Be neutrinos is only a factor of three larger than the flux predicted by the standard model.

The upper limits on ^{7}Be and CNO neutrino fluxes given in Eqs. (10.15) and (10.16) provide valuable constraints on models of the solar interior.

The basic pp reaction is below threshold in the ^{37}Cl experiment. However, the flux of the relatively rare pep neutrinos [reaction 2 of Table 3.1, see also Eq. (6.2)] is closely related to the flux of pp neutrinos. Unfortunately, the capture rate from this fundamental source is too low to stand out above the background. In the standard model (see Table 10.1),

$$\text{pep rate} = 0.2 \text{ SNU.} \qquad (10.17)$$

The observed rate of 2 SNU can be achieved in some nonstandard solar models that are described in Chapter 5. All of the nonstandard models require the acceptance of at least one implausible assumption. One cannot *prove* that all nonstandard models are incorrect, only that no one has yet come up with an attractive alternative to the standard solar model that is consistent with what we know from laboratory physics and astronomical observations. The radical nature, from the point of view of stellar evolution theory, of some of the proposed alternatives to the standard solar model (see Chapter 5) is a measure of the importance of the solar neutrino problem to astrophysical theory.

C Bounds on cosmological energy density

The observed capture rate can be used to set an upper limit on the background energy density, $\rho(q)$, in a hypothetical background of cosmological neutrinos. One has [Bahcall (1977)]

$$\rho(q) \leq \frac{(3 \text{ SNU }) q}{c\sigma(q)}, \tag{10.18}$$

using the 3σ upper limit on the observed rate from Eq. (10.8). The closure density, the fraction Ω of the closure density of the universe that could be in the form of neutrinos of energy q, is therefore

$$\Omega(q) \leq 0.084 \left[\left(\frac{q}{10 \text{ MeV}} \right) \left(\frac{10^{-42} \text{ cm}^2}{\sigma(q)} \right) h^2 \right], \tag{10.19}$$

where h is the Hubble constant in the units of 100 km s^{-1} Mpc^{-1}. Inserting the appropriate cross sections $\sigma(q)$ from column 3 of Table 8.4, we find, for example, $\Omega(10 \text{ MeV}) \leq 0.03h^2$ and $\Omega(30 \text{ MeV}) \leq 0.004h^2$. Thus the universe cannot be closed by high-energy neutrinos: not a surprising fact – but a fact nonetheless.

D Implications for particle physics

The Sun provides a well-collimated beam of low-energy neutrinos that can be used to perform exquisitely sensitive tests of fundamental neutrino properties. In many suggested particle physics models, the electron neutrino ν_e that is created, in laboratory β-decay or in nuclear fusion reactions in the solar interior, is not a mass eigenstate. Instead, ν_e is a linear combination of two (or more) mass eigenstates

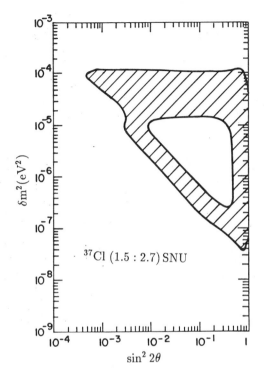

Figure 10.6   MSW effect   The figure shows the 95% confidence region in the δm^2–$\sin^2 2\Theta_V$ plane, where Θ_V is the vacuum mixing angle defined in Section 9.1. The cross hatched region was determined using 1000 standard solar models created in Monte Carlo simulations (see Section 7.5). If one assumes the standard solar model is correct (and the uncertainties estimated in Chapter 7 are valid) and the MSW effect is the only relevant new physics beyond the standard electroweak model, then the parameter region outside the cross hatched area in Figure 10.6 is excluded. The calculations for this figure were performed by Haxton [see Bahcall and Haxton (1989)].

[Gribov and Pontecorvo (1969)]. There is in these models a phase difference, or oscillation angle, which represents the mixing between the two eigenstates as ν_e propagates from the Sun to the Earth. The phase angle, ϕ, can be written (see Section 9.2B) as

$$\phi = \frac{\pi R}{L_V} = 1.9 \times 10^{11} \left(\frac{1 \text{ MeV}}{E}\right) \left(\frac{\delta m^2}{\text{eV}^2}\right) \left(\frac{R}{1 \text{ AU}}\right), \quad (10.20)$$

where 1 AU is the distance between the Earth and the Sun and $\delta m^2 = m_1^2 - m_2^2$. Thus low-energy solar neutrino experiments are in principle sensitive to mass squared differences as small as 10^{-12} eV^2.

Many suggestions have been made of how neutrinos are affected by the long propagation path through the solar material and interplanetary space. The suggestions that are most often discussed are reviewed in Chapter 9; they involve neutrino oscillations (in vacuum or enhanced by matter), a large neutrino dipole moment, and neutrino decay. In order to decide which, if any, of these ideas are correct, we must perform diagnostic experiments that test both the shape of the energy spectrum and the flavor content (neutrino type) of the particles that reach us. At present, we cannot say with confidence whether or not the solar neutrino problem requires new physics.

The most currently popular explanation of the solar neutrino problem is the MSW effect, which is discussed in Section 9.2. Figure 10.6 shows for the MSW effect the 95% confidence region in the $\delta m^2 - \sin^2 2\Theta_V$ plane, where Θ_V is the vacuum mixing angle defined in Section 9.1. The cross hatched region was determined using 1000 standard solar models created in Monte Carlo simulations (see Section 7.5). If one assumes the standard solar model is correct (and the uncertainties estimated in Chapter 7 are valid) and the MSW effect is the only relevant new physics beyond the standard electroweak model, then the parameter region outside the cross hatched area in Figure 10.6 is excluded.

10.5   Do solar neutrino fluxes vary with time?

Is the rate of neutrino events observed in the ^{37}Cl experiment consistent with statistical fluctuations? Or is there a significant time dependence? Perhaps a correlation with the season of the year or the intensity of solar cosmic radiation? Or does the time sequence of observed capture events imply that the neutrino has a finite magnetic moment and that what we observe is influenced by the solar magnetic field configuration along the line of sight [as suggested by Voloshin, Vysotskii, and Okun (1986)]?

The answers to the above questions are important. If the observed capture rate is time dependent, then the detected event rate does not reflect the steady neutrino luminosity of the solar interior. The typical time scales for changes in the physical conditions of the Sun exceed the Kelvin–Helmholtz time, which is of order 10^7 yr [see discussion of Eq. (2.1) in Chapter 2]. The important branches of the pp fusion chain have time scales of order 10^4 yr or more (see Table 3.1). Hence, if the capture rate varies on any time scale that can be measured by human experimentalists, then most likely the results of the ^{37}Cl experiment are not telling us a direct or simple fact about the solar interior. Something other than astrophysics is involved. In this case, the arguments for some particle physics explanations would be greatly strengthened.

The statistical questions are complicated and controversial because the signal is small (a few counts per observing run) and because the observations have been carried out at irregular times. Also, some of the variations that have been discussed in the literature have periods that are comparable to the total time span of the data. More data must be obtained and confirming observations should be attempted with new detectors.

The discussion in this section reflects two viewpoints which emphasize different aspects of the question. These two viewpoints, which might be labeled, respectively, as "stimulating" and "conservative," are expressed in papers by Davis (1987) [see also Davis, Cleveland, and Rowley (1987)] and by Bahcall, Field, and Press (1987). The following discussion explains how the same statistical facts have been interpreted in seemingly contradictory ways in the two viewpoints.

The time dependence of the neutrino capture rate has been investigated using a number of different techniques. The interested reader should consult the papers by Basu (1982), Ehrlich (1982), Gavrin, Kopylov, and Makeev (1982), Haubold and Gerth (1983), Kopylov and Gavrin (1984), Lanzerotti and Raghavan (1981), Raychaudhuri (1986), and Subramanian (1983).

A   SNUs versus spots

Bazilevskaya, Stozhkov, and Charakhch'yan (1982) and Davis (1987) have described an apparently remarkable correlation between sunspot number and the capture rate of solar neutrinos. This correlation is illustrated in Figure 10.7. Following Davis, this figure is based upon the 61 experimental runs that are tabulated in Table 4 of Rowley, Cleveland, and Davis (1985). Monthly average sunspot numbers are taken from *Solar Geophysical Data for 1985,1986*. See Davis (1987) and Gavrin, Kopylov, and Makeev (1982) for references to earlier discussions of a possible dependence of SNUs versus spots. Section 9.3

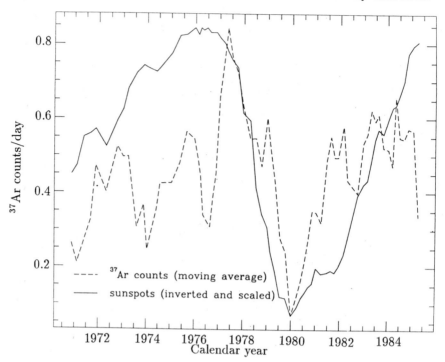

Figure 10.7   SNUs versus spots   The figure shows the average monthly sunspot number (solid curve) versus the moving averaged SNU rate (dashed curve) as a function of calendar year. The scale of the ordinate is arbitrary for sunspots, which are scaled to the same peak to peak range, and inverted (small sunspot number at the top).

discusses a possible theoretical interpretation of the suggested correlation in terms of a large neutrino magnetic moment.

Figure 10.7 shows both the observed sunspot number and the five point moving average of the neutrino capture rate [in captures per day, which is linearly related to SNUs via Eq. (10.3)]. The neutrino capture rates were smoothed, following Davis (1987), because the individual measurements constitute a rather noisy data set. The scale of the ordinate for sunspots is arbitrary. For convenience in visualizing the results, Figure 10.7 has the same peak to peak range for spots and ^{37}Ar captures per day. Also for convenience, the sunspot number is plotted so that it decreases from the bottom to the top of Figure 10.7, while the capture rate is plotted in the more conventional way with larger values at the top of the figure.

An anticorrelation between SNUs and sunspot number is apparent in Figure 10.7. Davis (1987) pointed out that in the period from 1977 (corresponding to the beginning of sunspot cycle 21) through 1983 the correlation coefficient between yearly average ^{37}Ar production rate and sunspot numbers is 0.94. The change in ^{37}Ar production rate in the three year period 1977 to 1980 corresponds to a change in the rate from 0.84 atoms day^{-1} (4.4 SNU) to 0.1 atoms day^{-1} (0.5 SNU), essentially equal to the cosmic ray background rate (0.4 SNU). After the drop, the ^{37}Ar production rate gradually recovers as the solar activity declines.

The most remarkable feature of the apparent correlation is that the capture rate is very close to the background level of 0.1 capture per day [see Eq. (10.5)] for the four measured points near 1980.0.

Is this apparent correlation real? Davis (1987) suggests that the effect that is apparent in Figure 10.7 has a statistical significance of about 5 standard deviations. Moreover, he stresses that the largest changes are seen at a characteristic time, when sunspot activity is at a minimum. On the other hand, Bahcall, Field, and Press (1987) have concluded on the basis of an analysis of the same data that the correlation is not significant at a definitive level.

How can one draw such different conclusions on the basis of the same facts? The following discussion shows that there are different plausible ways of looking at the data.

Consider a hypothetical linear relation between neutrino capture rate and sunspot number,

$$SNU = a + b \times \text{ sunspots.} \tag{10.21}$$

The simplest statistic to investigate is whether the χ^2 distribution for the relation of Eq. (10.21) is significant or not. The numerical values of the pairs of variables (SNU, sunspots) can be taken from the paper of Rowley et al. (1985). Remarkably enough, the significance of the correlation depends sensitively on exactly how one estimates the standard deviations associated with each measured SNU value. The upper and lower error estimates are different in Cleveland's (1983) application of the maximum likelihood method. The reason for this asymmetry is that Cleveland does not allow negative values for the neutrino capture rate.

Bahcall et al. (1987) defined two different estimates of the individual standard deviations: (1) average error, which is equal to one-half the difference between the upper and lower limits quoted by Rowley *et al.* (1985); and (2) *upper error*, which is equal to the difference between the upper limit and the best estimate value. For most experimental situations in which the counting rate is large, observers quote symmetric estimates of the uncertainties, in which case the average error is equal to the upper error.

The null hypothesis is that the capture rate is constant in time ($b = 0$). This hypothesis can be marginally rejected (2% significance level) for average errors. However, the null hypothesis is *not* rejected for upper errors (61% significance level).

The reader will immediately see how both points of view are consistent with the above cited result. The "stimulating" view of Davis interprets the 2% significance level of rejection for the null hypothesis as quantitative confirmation of the apparent correlation. In Davis' view, 2% is a highly suggestive result. On the other hand, I am impressed by the fact that the results depend sensitively upon how the errors are estimated. Even with the most favorable interpretation of the errors, the significance level for rejection of a constant flux is as *high* as 2%. Since there was no *a priori* physical reason for expecting an anticorrelation with sunspot number, many different variables could be (and perhaps were) considered in a search for a possible correlation. This makes, in my view, the 2% significance level seem inconclusive.

Since χ^2 probabilities depend sensitively on assumed Gaussian errors, Bahcall, Field, and Press (1987) also computed a different, more model-independent test for statistical significance. For the null hypothesis that the SNU flux is constant in time, the data (whatever its error distribution) should be statistically invariant under scrambling in time, as long as the pairing of each measurement with its own standard deviation is preserved. Therefore Bahcall *et al.* (1987) randomly permuted the data 1000 times. For each permutation, they fit to sunspot number as indicated in Eq. (10.21) and compared the value of χ^2 for the scrambled data with the value for the observed data. This test determines the acceptability of the null hypothesis: The neutrino capture rates are independent of time.

The fit of Eq. (10.21) to the real data is better than is obtained with the scrambled data in all but 5% of the permitted cases. This result confirms the conclusion of Davis (1987) (although not at the suggested extremely high significance level) that there is an apparent correlation between SNUs and sunspot number. However, in my view the significance level is not sufficiently small to justify rejecting definitively the null hypothesis of no time variation.

How robust, under editing of the data set, is the conclusion that there is a correlation? Not very. If the four low neutrino capture rates around 1980 are removed and an identical statistical analysis (scrambling on the data) is performed, the best fit SNU rate is practically independent of sunspot number. In fact, 86% of the scrambled data sets yield a better fit to the sunspot data than do the actual data [see Bahcall *et al.* (1987)].

The suggested correlation between neutrino capture rate and sunspot number depends almost entirely upon the four low points near the beginning of 1980. I believe that this sensitivity to a small number of points suggests that the correlation is less likely to be physically real. On the other hand, the "stimulating" point of view rejoins that the four points that were removed are valid data and are especially significant because they occur near maximum sunspot activity.

In any event, the reader can see that there is some analogy between the current controversy and the traditional source of academic polemics, namely: "Is it more correct to describe the coffee cup as half empty or half full?"

B   Correlations with solar flares

Bazilevskaya *et al.* (1982) and Davis (1987) have suggested that solar flares may have produced neutrino pulses that showed up in his ^{37}Cl detector as unusually high observations. This result is of great potential importance since conventional theoretical calculations indicate that, while flare dynamics is not understood in detail, there is not enough energy available in neutrinos produced by flares to cause a significant signal in the ^{37}Cl detector [Lingenfelter *et al.* (1985) and Bahcall, Field, and Press (1987)]. Moreover, many of the suggested theoretical interpretations of the ^{37}Cl experiment would have to be revised if it turned out that part of the observed event rate of 2 SNU was due to neutrinos produced by solar flares on the surface of the Sun, rather than by neutrinos produced by nuclear fusion in the solar interior.

The following discussion summarizes the observational evidence on the relation between solar flares and the ^{37}Cl neutrino capture rate [see Davis (1987) for a fuller discussion of the observations]. Run

27, the highest experimental run, corresponded in time with the great solar flares of August 1972. Unfortunately, the flares occurred early in the exposure interval and therefore most of the ^{37}Ar atoms that might have been produced by flares would have decayed before the extraction was completed. However, Bazilevskaya *et al.* (1981) pointed out that run 71 might also be high because a burst of high-energy protons was produced by the Sun in October 1981. The extraction for this period indeed indicated an enhanced capture rate. In his Table 2, Davis (1987) lists several other solar flares for which there was an excess of observed ^{37}Ar atoms in the tank. However, not all major solar flares yielded an excess of ^{37}Ar and the uncertainties in the number of atoms is large in all cases. If flare events were responsible for the increase in the capture rate above the average rate, the approximate numbers of ^{37}Ar atoms produced in the detector in runs 27, 51, and 71 were 250 ± 130, 20 ± 15, and 56 ± 30, respectively. According to Davis, Monte Carlo simulations of the data indicate that approximately 1 or 2 runs as high as the runs that might be correlated with flares are expected in 60 experiments.

A simple calculation is sufficient to confirm the result of Lingenfelter *et al.* (1985) that neutrinos from the solar flares are not expected to produce a significant number of events in the ^{37}Cl detector. The total energy in the great flares of August 1972 was about 10^{33} ergs [Lin and Hudson (1976)]. If this amount of energy is assumed, as an extreme upper limit, to be present all in neutrinos from π and μ decay, then approximately $10^{9.5}$ cm^{-2} ν_e would have been incident on the Earth. This fluence corresponds to only ~ 1 ^{37}Ar atom produced in the ^{37}Cl tank, down by more than two orders of magnitude form the number estimated by Davis to be required to explain the high result of run 27. A more detailed calculation based upon the current understanding of solar flares gives a much smaller predicted fluence of flare neutrinos [see Lingenfelter *et al.* (1985)].

Could the chlorine experiment be responding to the production of mesons or energetic positron emitters in the Earth's atmosphere that are produced by flare particles? A conceivable mechanism is that the cosmic rays are modulated by the solar wind encountered on the way to the Earth. However, quantitative estimates of the implications of this mechanism show that it is inadequate to explain high runs in the chlorine experiment. Gaisser and Stanev (1985) showed that cosmic ray secondaries that decay in flight to produce high-energy neutrinos are insufficient to account for the suggested effect by at least three orders of magnitude. de la Zerda-Lerner and O'Brien (1987) calculated the neutrino emission from positrons emitted by radioactive isotopes produced by cosmic rays as they strike the Earth's atmosphere and showed that the positron neutrino emission falls short of explaining the correlation by nine orders of magnitude.

C   Tests of the flaring hypothesis

Is some unknown mechanism operating in the Sun or in the Earth's atmosphere? Further observations are necessary to clarify whether the suggested enhancement in neutrino rates is real and, if so, how it depends upon the characteristics of the flares that are observed by photon observations. Fortunately, there are operating neutrino detectors that will have a greater sensitivity to flare neutrinos than the ^{37}Cl experiment. Observations with these other experiments, when combined with continuing studies using the ^{37}Cl detector, will settle the question of whether flares contribute to the observed event rate in solar neutrino detectors.

The basic assumption used in the following discussion, adapted from Bahcall (1988), is that the source of flare neutrinos is from collisions that produce pions and muons whose decays produce the neutrinos of interest. For specificity, the meson decays are assumed to occur at rest. This assumption minimizes the calculated ratio of predicted counts in nonradiochemical experiments (like Kamiokande II, IMB, Baksan, and LVD) compared to the ^{37}Cl experiment, since the ^{37}Ar nucleus is torn apart by very high-energy neutrinos. Independent of the source of the neutrinos, other detectors must observe at least as many events as calculated here if the high ^{37}Cl runs are explained by neutrinos from π and μ decay.

Table 10.4 summarizes the sensitivity of several existing neutrino detectors to neutrinos from pion and muon decay. All of the detectors are discussed in this book (see Sections 10.2, 11.2, 13.2, 13.4, and 15.4A). The cross sections are taken from Chapter 8. The absorption cross sections for ν_e are taken from Section 8.1E. The effective scattering cross sections given in Table 10.4 are the sum of the scattering cross sections given in Table 8.15 for the three neutrinos that are produced in the $\pi^+-\mu^+-e^+$ decay chain. This prescription yields the appropriate cross section to use in comparing expected rates of scattering events with the number of absorption events that may be observed by radiochemical detectors, provided that the neutrinos are created by relatively low-energy pp collisions which mainly produce π^+'s and μ^+'s. If equal numbers of mesons of both charges are produced, then the effective scattering cross section per ν_e is increased by about 55%, which is the reason for the inequality signs in Table 10.4. The largest cross section given in the table is for the reaction $\bar{\nu}_e + p \rightarrow e^+ + n$. The $\bar{\nu}_e$'s are produced by μ^- decay and occur in significant numbers if the colliding protons (or other baryons) that produce the decaying mesons have energies in excess of a few GeV.

Solar neutrinos provide the main background in the radiochemical experiments that attempt to detect solar flares with ^{37}Cl and ^{71}Ga. The saturation (i.e., maximum) number of radioactive atoms produced in equilibrium with the solar neutrino flux is about 20 for both the ^{37}Cl detector and the GALLEX ^{71}Ga detector. A large flare would produce of order $10^{2.0\pm0.5}$ radioactive atoms, many of which might decay before the tanks were purged.

The direct counting experiments (Kamiokande II, IMB, and LVD) are more sensitive to flares than are the radiochemical experiments. The total signal is expected to be larger in the direct counting experiments, 10^2 to 10^5 events, because of their large masses. For comparison, the 1987A supernova produced only 11 events in the Kamiokande II detector [Hirata et al. (1987)] and 8 events in the IMB detector [Bionta et al. (1987)]. Also, the direct counting experiments provide accurate time measurements, which could be correlated with electromagnetic observations of flares.

Table 10.4 shows that an extraordinary signal should be recorded in the Kamiokande II, IMB, Baksan, and LVD detectors if a ^{37}Cl run was high because of a flarelike event. The last column of the table shows the expected number of neutrino events that should occur in each of the detectors assuming that, as suggested by Davis (1987) for the great flare of August 1972, 250 ± 130 ^{37}Ar atoms are produced in the ^{37}Cl detector. The number of events ranges from a minimum of 35 in the SAGE ^{71}Ga detector (approximately 17 events would be expected in the GALLEX detector) to a maximum of about 10^5 events in the IMB detector presuming that $\bar{\nu}_e$'s are also produced in abundance.

The Kamiokande II collaboration [Zhang et al. (1988)] has reported on a relevant search for relic antineutrinos from past supernovae. The search with Kamiokande II data is based upon 357.4

Table 10.4. Sensitivity of detectors to neutrinos from pion and muon decay. Here N is the number of target particles (electrons or nuclei), σ is the cross section, and N_{events} is the number of events expected assuming 250 events occurred in Davis' ^{37}Cl tank.

Detector	N (10^{30})	σ $(10^{-40}$ cm$^2)$	N_{events}
^{37}Cl (Davis)	2.16	0.7	250 $\pm$130
^{71}Ga (SAGE)	0.21	1	35
Kamiokande II (2.1 kilotons H$_2$O)			
ν–e	7.1$\times10^2$	$\geq$ 0.0039	4.5$\times10^4$
$\bar{\nu}_e$ + p	1.4 $\times10^2$	0.95	2 $\times10^4$
IMB (6.8 kilotons H$_2$O)			
ν–e	2.3$\times10^3$	$\geq$ 0.0037	1$\times10^3$
$\bar{\nu}_e$ + p	4.5 $\times10^2$	0.95	7 $\times10^4$
Baksan (0.2 kilotons scintillator)			
ν–e	7.0$\times10^1$	$\geq$ 0.0039	45
$\bar{\nu}_e$ + p	1.9 $\times10^1$	0.95	3 $\times10^3$
LVD (1.8 kilotons scintillator)			
ν–e	6$\times10^2$	$\geq$ 0.0039	4 $\times 10^2$
$\bar{\nu}_e$ + p	1.6 $\times10^2$	0.95	2 $\times10^4$

detector days between the period of January 6, 1986 and December 31, 1987. For this study, a reduced fiducial volume of 0.6 kgm was used and recoil positron energies between 19 MeV and 35 MeV were considered. In order to take account of the smaller fiducial volume and the reduced energy range that was searched, the number of counts given in Table 10.4 should be multiplied by 0.062.

If a flarelike process had produced during this period 250 events in the ^{37}Cl tank, then it would have produced in the Kamiokande

II detector between 28 events (from scattering of neutrinos from μ^+ and π^+ decay) and 1300 events (if $\bar{\nu}_e$ from μ^- decay were incident in equal number with ν_e from μ^+ decay). Only two widely separated events in the energy range between 19 MeV and 35 MeV were observed in the entire (~ 1 yr) period of observation.[†] Unfortunately, no strong solar flares were observed during the period of observation since the epoch under consideration was near the minimum in the solar cycle. However, this observation shows that a continued search with Kamiokande II during the more active phases of the solar cycle should provide information on flares of between 10^{-1} and 10^{-3} the strength of the August 1972 flares, depending upon what one assumes about the production of $\bar{\nu}_e$'s, supposing always that the 1972 flares produced 250 ^{37}Ar atoms in Davis' ^{37}Cl tank. The other detectors shown in Table 10.4 can also provide interesting limits on flarelike events.

D   Galactic flares

Can flarelike events occurring on other galactic stars be detected with the existing ^{37}Cl detector? This question arises from time to time as a result of some suggestive temporal correlation between a spectacular galactic event (such as a large gamma ray burst) and an unusually high capture rate in an individual run with the ^{37}Cl detector. The burst energy, E_{burst}, that is required to produce an observable signal from a galactic object (such as a gamma ray burster) is[‡]

$$E_{\text{burst}} = 0.2 M_\odot c^2 \left[\left(\frac{0.01}{\epsilon} \right) \left(\frac{100}{N_{\text{cap}}} \right) \left(\frac{d}{1 \text{ kpc}} \right)^2 \right], \qquad (10.22)$$

where ϵ is the fraction of the total burst energy that is converted to ν_e that result from μ^+ decay, N_{cap} is the number of ^{37}Ar atoms produced

---

[†]Hirata *et al.* (1988b) have also examined the data on high energy neutrino events observed in Kamiokande I and II between July 1983 and July 1988. These data do not show a correlation of Kamiokande events with solar flares. However, the large energy thresholds used in the Hirata *et al.* (1988b) analysis, 100 MeV (and 50 MeV), exclude nearly all of the neutrinos from the decays of low energy muons and are in the region in which neutrinos often disrupt the nucleus rather than produce bound ^{37}Ar nuclei.

[‡] The detection of galactic supernova with the ^{37}Cl detector is considered in Section 15.5.

by neutrino capture ($N_{cap} \sim 10^2$ corresponds to of order 10 SNU for a typical chlorine run), and d is the distance of the flare star. As in the preceding discussion, the neutrinos are assumed to be generated by the decays of mesons produced by collisions among high-energy protons. Eq. (10.22) shows that a burst energy of the order of the total binding energy of a neutron star (0.1 $M_{\odot}c^2$) would be required to give an observable signal from a galactic flare. Therefore, the ^{37}Cl detector is not expected to be able to detect flares from other stars in the Galaxy.

10.6   Recent data

Table 10.5 lists recent data obtained from the ^{37}Cl experiment in the same format as was used in Table 10.3. These data were generously supplied by R. Davis, Jr. and B. Cleveland for inclusion in this chapter and are shown together with the earlier data in Figure 10.5.

The analysis is complete for runs 83 through 89. The average of the neutrino capture rates for these runs is

$$\text{Average rate} \; = \; 0.46 \; ^{37}\text{Ar atoms day}^{-1}, \tag{10.23}$$

that is, 2.4 SNU, which is in good agreement with the production rate computed for all the runs through number 82 [see Eq. (10.7)].

The 10 most recent runs, numbers 90 through 99, have been analyzed since the failure of the two original pumps and the resuscitation of the ^{37}Cl experiment in September 1986.[†]

Some explanations are required for individual runs. Run 90 was a long run of more than one and a half years. Because of pump failure, the ^{37}Cl tank received a gulp of air that caused the extracted sample for run 90 to be very large. Two separate event rates are given for run 90. The first entry for this run was calculated using the same maximum likelihood analysis as for all other runs that are listed in Tables 10.3 and 10.5; the value given in this row for the production rate, 0.00 ^{37}Ar per day, is preferred for statistical investigations since no special assumptions are made. However, the counter background

---

[†]Mass spectrometry has not yet been performed on runs 90 to 99 in order to verify that the chemical yield was as high as in previous runs. However, Davis believes that the mass spectrometric measurements will confirm that the efficiency was as high as usual. If the chemical yield were lower than normal, the production rates will be increased over what is given in Table 10.5.

Table 10.5.   Recent data for the ^{37}Cl detector. Exposure times and ^{37}Ar production rates from individual runs using the chlorine detector. Runs that are marked with a dagger (90, 93, and 96 to 98) are special and are commented on in the text.

	Exposure dates, years			Atoms per day		
Run	Start	End	Mean	^{37}Ar production rate	Lower limit	Upper limit
83	1984.095	1984.251	1984.193	0.308	0.136	0.481
84	1984.251	1984.360	1984.316	0.576	0.315	0.837
85	1984.360	1984.553	1984.487	0.599	0.341	0.857
86	1984.553	1984.671	1984.624	1.260	0.691	1.829
87	1984.671	1984.922	1984.847	0.0	0.0	0.313
88	1984.922	1985.128	1985.060	0.351	0.142	0.559
89	1985.128	1985.337	1985.268	0.718	0.456	0.981
90†	1985.337	1986.773	1986.671	0.00	0.00	1.67
(90†	1985.337	1986.773	1986.671	1.21	0.75	1.67)
91	1986.773	1986.993	1986.875	0.97	0.58	1.36
92	1986.993	1987.123	1987.058	0.75	0.42	1.08
93†	1987.123	1987.171	1987.149			
94	1987.171	1987.372	1987.305	0.97	0.61	1.33
95	1987.372	1987.539	1987.479	0.69	0.41	0.97
96†	1987.539	1987.796	1987.720	0.27	0.08	0.46
97†	1987.796	1987.944	1987.889	0.85	0.49	1.21
98†	1987.944	1988.155	1988.086	1.12	0.51	1.73
99	1988.155	1988.268	1988.222	0.94	0.09	1.79

for run 90 appears to have been low; the number of events was small outside the ^{37}Ar region in the pulse height rise-time plane. The second entry for run 90 (enclosed in parentheses) was calculated assuming that the counter background was zero (which is different from the result found with the maximum likelihood analysis). Run 93 was extracted at the time of Supernova 1987A. Only one ^{37}Ar event was observed for this run; between one and two events were expected from solar neutrinos. Analysis of this run is continuing because of the unusual behavior of the counter during extended measurement of the background. The final uncertainty has not been computed yet for this run but will be larger than for the other runs listed in Table 10.5

because of the shortness of the exposure and because of difficulties in measuring the counter background. Runs 96 to 98 are still counting at the time of this writing; the production rates for these runs are preliminary but are unlikely to be affected significantly by further measurements of the counter backgrounds.

The results for these new experiments sent shock waves through the community of people interested in solar neutrino experiments. The mean capture rate for the 10 recent experiments, runs 90 through 99, computed using the maximum likelihood method with no special assumptions is [Cleveland and Davis (1988), private communication]

$$\text{Average capture rate} = (3.6 \pm 0.7) \text{ SNU}, \qquad (10.24)$$

which is 2 standard deviations larger than the average event rate prior to 1984 [see Eq. (10.8)]. Moreover, the most recent experiments have been performed at a time when the number of sunspots was small. The average capture rate computed for all of the runs, including the most recent data, is

$$\text{Capture rate} = (2.2 \pm 0.3) \text{ SNU}, \qquad (10.25)$$

which is not significantly different from the average rate that was computed for the data of 1970 to 1984 [see Eq. (10.8)].

What is the meaning of the new results? Has the experiment been changed significantly by the installation of the new pumps? Is the higher event rate due to a very unlikely statistical fluctuation during the period of low sunspot activity? Or is there a physical correlation between the capture rate in the ^{37}Cl experiment and the phase of the sunspot cycle?

New experimental data that will be obtained over the next few years with the ^{37}Cl and Kamiokande II detectors are required to answer these questions. It is worth noting, however, that Davis and his associates are convinced that the introduction of new pumps did not alter the experimental setup.

10.7   The future of the ^{37}Cl experiment

The ^{37}Cl experiment has been operating for more than two decades. Is there something more that should be done with the existing detector? Or have essentially all of the important things already been done?

There are several crucial things that must be done with the existing ^{37}Cl detector.

Further background measurements, especially with ^{39}K, are essential. The existing background determination, cited in Eq. (10.5), requires a significant extrapolation (see Section 10.2E). If correct, the background contribution to the total capture rate is relatively small, about 17% [cf. Eqs. (10.5) and (10.7)]. Suppose, for illustrative purposes, that the background has been underestimated by a factor of three, then the inferred solar neutrino capture rate would be reduced from 2.1 SNU to 1.2 SNU. This three-times larger estimate for the background would reduce the observed solar neutrino rate to a value below that expected from pep and ^{7}Be neutrinos (cf. Table 10.1). This hypothetical revision of the background rate would change qualitatively the astrophysical interpretation of the experimental results. In the imaginary case being considered, most of the so-far suggested nonstandard solar models would be in disagreement with the experimental results. Also, all of the inferences about neutrino characteristics, for example, neutrino masses and mixing angles, as well as the neutrino magnetic moment, would require revision.

Data must be obtained through another sunspot cycle to test whether the suggested correlation between SNUs and spots is physically real. If this correlation is verified, for example, by obtaining data which show a much reduced capture rate in late 1990 or early 1991 (see Figure 10.5), then the interpretation of the experiment will be changed. If a large amplitude correlation is established, then most or all of the observed capture rate is not due to neutrinos from the solar interior. The implications of this possibility would be enormous. All of the particle physics explanations that explain a constant reduction factor of order three or four would have to be revised or abandoned. The only suggested way of imposing an 11 year periodicity on solar neutrinos requires a large neutrino magnetic moment (see Section 9.3).

The detector is sensitive to ν_e from time-dependent events in the Galaxy, for example, from galactic supernovae [see Bahcall (1978)]. A Type II supernova like the one that exploded recently in the LMC, SN1987A, would produce of order 5 to 50 ^{37}Ar atoms (see Table 15.6), a signal that could be detected easily. A number of authors have also suggested [see Davis (1987)] that large solar flares will produce a large excess of ^{37}Ar atoms. Solar flares are not expected, on the basis of conventional theoretical models, to produce a measurable enhancement in the ^{37}Cl capture rate. If an enhancement due to solar flares were confirmed, this would have great consequences for our understanding of phenomena that occur near the solar surface. A calibration of the entire experiment with a radioactive source, probably ^{65}Zn, would be valuable [see Alvarez (1973)]. This would test directly the throughput efficiency of the ^{37}Cl detection system. It is important to keep the ^{37}Cl experiment running through the period when other solar neutrino experiments are producing data in order to be able to compare unusual runs. There may be high runs caused by neutrino bright, but optically dark, astronomical events. The discovery of "neutrino fluxes" could be confirmed only if more than one neutrino detector saw an unusual event.

Bibliographical Notes

Alvarez, L.W. (1949) University of California Radiation Laboratory Report UCRL-328. Thorough investigation of the possibility of ^{37}Cl detector near a reactor as a test of Fermi's theory of weak interactions.

Alvarez, L.W. (1973) Physics Notes, Memo No. 767, Lawrence Radiation Lab. (March 23). Proposal for calibrating the ^{37}Cl experiment with a ^{65}Zn source.

Bahcall, J.N. (1964) *Phys. Rev. Lett.*, **12**, 300; R. Davis, Jr. (1964) *Phys. Rev. Lett.*, **12**, 303. Back-to-back papers proposing a ^{37}Cl solar neutrino experiment.

Bahcall, J.N. and R. Davis, Jr. (1976) *Science*, **191**, 264. The first decade of solar neutrinos summarized.

Bahcall, J.N. and R. Davis, Jr. (1982) in*Essays in Nuclear Astrophysics*, edited by C.A. Barnes, D.D. Clayton, and D. Schramm (Cambridge: Cambridge University Press) p. 243. Our joint account of the early history of the solar neutrino problem, which appears as the Appendix to this book.

Davis, R., Jr., *et al.* (1978) in *Solar Neutrinos and Neutrino Astronomy*, edited by M.L. Cherry, W.A. Fowler, and K. Lande (New York: AIP Conf. Proc. No. 126.) Vol. 1, p. 1. Classic description of the experiment by the master.

Davis, R., Jr., D.S. Harmer, and K.C. Hoffman (1968) *Phys. Rev. Lett.*, **20**, 1205; Bahcall, J.N., N.A. Bahcall, and G. Shaviv (1968) *Phys. Rev. Lett.*, **20**, 1209. Back-to-back papers presenting and discussing the first experimental results. Despite an enormous refinement in the results, the general picture did not change for the next two decades.

Davis, R., Jr. (1987) in *Proceedings of Seventh Workshop on Grand Unification, ICOBAN'86, Toyama, Japan,* edited by J. Arafune (Singapore: World Scientific) p. 237. An excellent summary from a contemporary viewpoint.

Pontecorvo, B. (1946) Chalk River Laboratory Report PD-205. Breathtakingly beautiful proposal to use neutrino absorption near a reactor as a test of weak interaction theory.

11

The ^{71}Ga experiments

Summary

Two ^{71}Ga experiments that are in progress in the Soviet Union and in Italy will provide the first observational information about the basic pp neutrinos, the only funded experiments that can detect these very low-energy neutrinos ($E_{pp} \leq 0.42\,\mathrm{MeV}$). The flux of pp neutrinos that reach the Earth can be calculated precisely if the standard solar and electroweak models are correct. Therefore, the observations with gallium detectors will constitute a critical test of explanations of the solar neutrino problems.

The reaction that is used in these radiochemical experiments is

$$\nu_e + {}^{71}\mathrm{Ga} \rightarrow e^- + {}^{71}\mathrm{Ge}, \quad E_{th} = 0.2332 \text{ MeV}. \qquad (11.1)$$

The low threshold energy makes possible the detection of pp neutrinos. The radioactive ^{71}Ge decays by electron capture, the inverse of reaction (11.1), with a lifetime of $\tau_{1/2} = 11.43$ days.

The capture rate predicted by the standard model for a ^{71}Ga experiment is relatively well determined, since the fraction ($\sim$ half) of the capture rate that is contributed by pp neutrinos is insensitive to input parameters and the ground state to ground state cross sections can be calculated accurately. About one-quarter of the event rate is calculated to come from ^{7}Be neutrinos, whose contribution in the standard model can be evaluated with good accuracy. The total event rate derived from the standard model is $\sum_i \phi_i \sigma_i = 132^{+20}_{-17}$ SNU. The minimum expected event rate is 79 SNU, provided only that nothing happens to neutrinos inside the Sun or on their way to the Earth.

The GALLEX collaboration, consisting primarily of European scientists with U.S. and Israeli participation, will use 30 tons of gallium in an aqueous solution of gallium chloride and hydrochloric acid. The experiment will be carried out in an underground facility in the Gran Sasso Laboratory in Italy. The SAGE collaboration, primarily Soviet scientists with U.S. participation, will use 60 tons of gallium metal in an underground laboratory dug into the side of a mountain in the Baksan Valley, in the Caucasus region of the Soviet Union. The chemical processing is simple (at least in the chloride solution) and the counting of the radioactive ^{71}Ge can be accomplished with low background counters that are similar to those used in the ^{37}Cl experiment. Both the GALLEX and the SAGE collaborations have completed pilot experiments that establish the feasibility and efficiency of the detection procedures. A throughput test can be made of the overall efficiency for observing low-energy neutrinos in the final full scale detectors by using an intense laboratory source of ^{51}Cr neutrinos.

The first section of this chapter, 11.1, presents the theoretical expectations and their known uncertainties. The next section, 11.2, discusses and compares the two ^{71}Ga experiments that are in progress. The final section, 11.3, formulates some of the questions that can be answered by the ^{71}Ga detectors.

Detailed descriptions of the two gallium experiments can be found in a number of clear and informative articles, for example, Dostrovsky (1978), Bahcall et al. (1978), Hampel (1981), Zatsepin (1982), Kirsten (1984), Hampel (1985), Barabanov et al. (1985), Hampel (1986), Kirsten (1986), and Cribier et al. (1988). The earliest proposal for a gallium experiment was made by Kuzmin (1966), at a time when the required amount of gallium exceeded the annual world production by an order of magnitude.

11.1   Predictions of the standard model

The capture rate predicted for a ^{71}Ga detector by the standard model is [Bahcall and Ulrich (1988)]

$$\sum_i \phi_i \sigma_i = 132^{+20}_{-17} \text{ SNU},\qquad(11.2)$$

where the indicated uncertainty represents the total theoretical range

Table 11.1. Capture rates predicted by the standard model for a ^{71}Ga detector.

Neutrino source	Capture rate (SNU)
pp	70.8
pep	3.0
hep	0.06
^{7}Be	34.3
^{8}B	14.0
^{13}N	3.8
^{15}O	6.1
^{17}F	0.06
Total	132 SNU

defined in Chapter 7. This rate corresponds to 1.17 events per day in a 30 ton target of gallium.

The energy threshold for absorption of ν_e by ^{71}Ga is 0.233 MeV; therefore, a gallium detector is sensitive to, in different proportions, all the sources of solar neutrinos discussed in Chapter 6.

Table 11.1 shows the contribution to predicted capture rate of each of the neutrino sources discussed in Chapter 6. Neutrinos from the basic pp reaction produce approximately half, 54% or 71 SNU, of the computed total capture rate. The other main contributors are ^{7}Be neutrinos, 26% or 34 SNU, and ^{8}B neutrinos, 11% or 14.0 SNU.

The principal uncertainties in the calculated rate are shown in the fifth row of Table 7.5. The dominant uncertainty is caused by the transitions to excited states whose matrix elements are inferred from (p,n) measurements which have a significant uncertainty [see the discussion in Section IV of Bahcall and Ulrich (1988)]. According to the prescription for calculating errors that is described in Chapter 7, the maximum increase that could be caused by excited state transitions is +16 SNU and the corresponding extreme decrease is −8 SNU. Excited state transitions contribute 88%, or 12 SNU, of the total ^{8}B contribution. The curve of ^{8}B absorption cross sections versus excitation energy has a broad peak in the range 3 MeV to 5 MeV, which contains about one-third of the total calculated transition strength. If we were to ignore all of the uncertainties associated with the excited state transitions, then the remaining total calculated uncertainty would be only 12 SNU (or 9% of the total capture rate).

Excited state transitions cause a significant uncertainty – of order 10% in the total capture rate – for the standard solar model predictions, an uncertainty which seems unavoidable unless an unanticipated major improvement is achieved in the accuracy with which GT matrix elements to excited states can be determined.

The transition from the ground state of ^{71}Ga to the isobaric analogue state in ^{71}Ge does not contribute significantly to the expected capture rate because the analogue state decays mostly by particle emission. The cross section for absorption of ^{8}B neutrinos is large, 3.14×10^{-43} cm^2, but the upper limit for the γ-decay of this state is less than 10% [Champagne et al. (1987)]. Thus transitions to the isobaric analogue state account for less than 0.2 SNU. This result is different from the situation with ^{37}Cl where the transition to the analogue state alone accounts for almost half (46%) of the total predicted capture rate.

The cross section for absorption of neutrinos from a calibrating source of ^{51}Cr is [Bahcall and Ulrich (1988)]:

$$\sigma \left(^{51}\text{Cr on} \, ^{71}\text{Ga} \right) \; = \; 59 \left(1 \pm 0.1 \right) \times 10^{-46} \text{ cm}^2. \qquad (11.3)$$

Excited states contribute only 6% to the cross section; states above 0.5 MeV require too much energy to be populated by ^{51}Cr neutrinos. Therefore the calibration of the ^{71}Ga detector with a ^{51}Cr source [see Cribier et al. (1988)] cannot remove the significant uncertainty in the sensitivity caused by transitions to highly excited states. The ^{51}Cr calibration can be thought of as a throughput test for the *overall* efficiency of the detecting system.

The ^{71}Ga detectors could also be calibrated with a source of ^{37}Ar neutrinos, which Haxton (1988b) has suggested may have some practical advantages over a ^{51}Cr source. The cross section for absorption of ^{37}Ar neutrinos is

$$\sigma \left(^{37}\text{Ar on} \, ^{71}\text{Ga} \right) \; = \; 72 \left(1 \pm 0.1 \right) \times 10^{-46} \text{ cm}^2. \qquad (11.4)$$

The numerical value given in Eq. (11.4) was obtained using the (p,n) data of Krofcheck et al. (1985) and Krofcheck (1987) to estimate the $\sim 5.5\%$ contribution of excited states, which arises mostly from the state at 0.5 MeV excitation energy in ^{71}Ge.

How large are the uncertainties if, as appears likely, the standard predictions are incorrect? The answer depends upon what is wrong with the standard predictions. The fractional uncertainties would be increased if, as is possible for one MSW solution, lower-energy neutrinos are preferentially affected by resonant matter oscillations and only the higher-energy electron neutrinos reach the Earth without having their flavor changed. In this case, nearly all of the small expected event rate (~ 10 SNU) could come from transitions whose strengths are determined by (p,n) interactions and are therefore uncertain. On the other hand, most nonstandard solar models would give rise to smaller uncertainties in the predicted event rate since they primarily suppress the higher-energy neutrinos that populate excited states. The most extreme of these nonstandard models is the so-called "No ^{8}B" model (see Chapter 5), in which all of the higher-energy neutrinos are artificially removed. This "model" predicts a capture rate of 1.8 SNU for the ^{37}Cl experiment and is therefore in satisfactory agreement with the neutrino observations (although it is inconsistent with a number of excellent laboratory experiments on the nuclear reaction cross section). The "No ^{8}B" hypothesis effectively minimizes the uncertainty in the predicted event rate since the capture rates for high-energy neutrinos are less well known, due to uncertainties in both the production rates and the absorption cross sections. For this particularly favorable case (with regard to uncertainties), the calculated rate is $\sum_i \phi_i \sigma_i = 118^{+13}_{-12}$ SNU. The total theoretical uncertainty is, also for this case, approximately 10%.

11.2   The experiments

A   Overview

Two major solar neutrino experiments using ^{71}Ga are under way, one by a primarily European collaboration with U.S. and Israeli participation [GALLEX, see, e.g., Kirsten (1986) and Hampel (1985, 1986)] and the second by a group in the Soviet Union with some American collaborators [see Barabanov et al. (1985)].

The GALLEX collaboration will use 30 tons of gallium (1.03 × 10^{29} atoms of ^{71}Ga[†]) in an aqueous solution of gallium chloride and

---

[†]The isotopic abundance of ^{71}Ga is 39.6%, Gramlich and Machlan (1985).

hydrochloric acid; the detector will be located in the Gran Sasso Laboratory in Italy. Measurements are expected to begin sometime in 1990. The Soviet experiment will use 60 tons of gallium metal as a detector in a solar neutrino laboratory constructed in the Baksan Valley. The Soviet experimentalists expect to have a detector that is operating in early 1989. The initial chemical extraction is different in the GALLEX and the Soviet experiments, but the final chemical procedures and the methods of low-level counting will be similar for both collaborations. Background effects are expected to be small for this experiment. A direct test for some of the backgrounds is possible by measuring the amount of ^{69}Ge that is produced, solar neutrino absorption by ^{69}Ga being negligible [see Dostrovsky (1978) and Bahcall *et al.* (1978)]. Both the GALLEX and the Soviet collaborations have successfully completed pilot experiments with large quantities of gallium.

The comparison of the results from the two experiments will be a valuable check on any possible systematic errors. Since the GALLEX group is using a $GaCl_3$–HCl solution while the Soviet group is using metallic gallium, agreement between the results of the two experiments will confirm the conclusion that the chemical extraction can be performed with high efficiency. The two methods have complementary advantages and disadvantages.

A calibration of the overall efficiency of a gallium detector for low-energy neutrinos is possible using a radioactive source of ^{51}Cr [see Cribier *et al.* (1988)]. The difficulty and expense of performing such a calibration is comparable to what is required in carrying out some of the other solar neutrino experiments discussed in Chapters 12 to 14. Nevertheless, a calibration experiment is a wise use of valuable people-time and other resources because gallium solar neutrino experiments are, for the present, unique in providing information about the fundamental pp neutrinos. Also, the signal in a radiochemical experiment does not provide many diagnostic handles. A calibration experiment with ^{51}Cr would show directly that a gallium experiment measures what it is supposed to measure.

B   The GALLEX experiment

The GALLEX experiment developed from a collaboration begun at Brookhaven National Laboratory [Bahcall *et al.* (1978) and Hampel

(1981)]. Many of the techniques are similar to those used by Davis and his collaborators in the ^{37}Cl experiment.

The GALLEX detector will be placed in the Gran Sasso Underground Laboratory, where an excavated laboratory facility has already been prepared. Figure 11.1 shows the GALLEX underground facility. The detector consists of 30 tons of gallium in the form of a concentrated $GaCl_3$–HCl solution. The neutrino-induced ^{71}Ge atoms form the volatile molecule $GeCl_4$. A measured amount of inactive Ge carrier atoms, which also forms $GeCl_4$, is added at the beginning of a run to provide a sufficiently large sample for extraction so that the efficiency can be determined experimentally after each run. At the end of an exposure, the $GeCl_4$ is swept out of the solution by bubbling air or nitrogen gas through the tank. The gas stream is then passed through two gas scrubbers where the $GeCl_4$ is absorbed in water. The $GeCl_4$ is then extracted into CCl_4, back-extracted into tritium-free water, and finally reduced to ~ 0.1 to 1 cm^3 of the gas germane, GeH_4, by means of $NaBH_4$. The GeH_4, together with xenon, is introduced into a small proportional counter, where the number of ^{71}Ge atoms is determined by observing their radioactive decay.

A pilot experiment with 4.6 tons of $GaCl_3$ solution (equivalent to 1.26 tons of gallium) was performed by an international collaboration (Brookhaven National Laboratory, MPI Kernphysik Heidelberg, WIS Rehovot, IAS Princeton). This pilot experiment was successfully completed in 1983. In a typical experimental run, more than 99% of the Ge that was added to the $GaCl_3$ solution was extracted in 28 hours. It was shown that the entire chemical procedure (extraction, purification, and subsequent conversion to GeH_4) can be carried out with high efficiency.

The ^{71}Ge decay rate expected for a full scale solar neutrino experiment is about 1 per day. A counting system capable of measuring such a low decay rate has been developed by the group working at Heidelberg. Figures 11.2 and 11.3 depict the GALLEX counting system. The germanium is counted in a miniaturized proportional counter. The energy deposition from Auger electrons and from X-rays emitted in the ^{71}Ge electron capture decay result in an energy spectrum with two peaks: an L peak at 1.2 keV and a K peak at 10.4 keV. All provisions typical for low-level counting have to be applied in order to reach the low background rates required: use

**Figure 11.1 GALLEX facility in the Gran Sasso Underground Labora-
tory**   The figure shows a view of the GALLEX main building and (be-
hind it) the low-level counting building. The GALLEX main building
(12m × 10m × 9m) will house the process tank containing the 105 tons
of gallium chloride solution, a spare tank, and the equipment for the
germanium extraction. The counting of ^{71}Ge will be performed in the
low-level counting laboratory (10m × 10m × 6m). (Photograph courtesy of
GALLEX collaboration.)

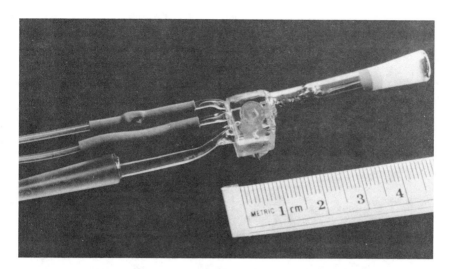

Figure 11.2   Miniaturized proportional counter for the GALLEX experiment   The figure shows a miniaturized proportional counter for counting small numbers of ^{71}Ge atoms; the counter is made from suprasile quartz. The active volume of the counter (to the right in the picture) is 0.5 cm^3 . The two wire connections to the preamplifier and the tube for the counter filling can be seen on the left hand side of the figure. (Photograph courtesy of the GALLEX collaboration.)

of ultra pure materials for counter construction; an anticoincidence shield with NaI and plastic scintillation detectors; and heavy passive shielding with lead and iron. The residual background in the L and K peak energy windows is of the order of 1 count per day. This remaining background is from β-particles that are produced by natural radioactivity in the construction materials, from Compton electrons caused by external gamma rays, and from electronic noise pulses. Fortunately, these background events produce, in most cases, pulse shapes different from those of ^{71}Ge decays. The combination of pulse energy and rise time can therefore be used to count the ^{71}Ge with a very small background rate (see the discussion of counting in the ^{37}Cl experiment in Section 11.2D). The computer-controlled counting systems are therefore designed to record the whole shape of each proportional counter pulse by means of a transient digitizer.

The most serious background reaction is ^{71}Ga(p,n), the protons being generated in the GaCl$_3$ solution as secondaries from $(\alpha$,p) and (n,p) reactions and from cosmic ray muon interactions. In the course

**Figure 11.3   Computer-controlled counting system for GALLEX propor-
tional counters**   The heart of the counting system is a transient digitizer
which records the shape (pulse height versus time) of each proportional
counter pulse. Analysis of the pulse shape distinguishes ^{71}Ge decay events
from counter background events. (Photograph courtesy of GALLEX collaboration.)

of the pilot experiment, the tolerable levels of uranium, thorium, and
radium impurities in the $GaCl_3$ solution were determined; the $GaCl_3$
solution that is being purchased by the GALLEX collaboration is
being carefully tested to make sure that it meets the background
specifications. The ^{71}Ge production by fast neutrons entering the
solution from outside has been studied with a Pu–Be neutron source.
This effect sets limits on the fast neutron flux at the detector site,
which (if necessary) can be met by use of a water shield around the
detector. Measurements *in situ* suggest that the background from
fast neutrons is small ($< 2\%$).

The depth dependence of the ^{71}Ge production in $GaCl_3$ by cosmic ray muons has been derived from measurements and calculations on the same effect for the Cl detector and from the measured cross section ratio (^{71}Ge from $GaCl_3$)/(^{37}Ar from C_2Cl_4) for 225 GeV muons. For the shielding depth of the Gran Sasso Underground Laboratory, the ^{71}Ge production rate is estimated to be of order 0.01 atoms per day in 30 tons of gallium, corresponding to a muon background of a few SNU or less.

C The USSR experiment

There are two main differences between the Soviet and the GALLEX experiments: (1) the Soviet group has chosen to use a metal gallium target, rather than $GaCl_3$ solution; and (2) the Soviet target will contain 60 tons of gallium, instead of 30 tons. After the initial extraction steps (see below), the further experimental procedures including the chemical processing and the counting are similar in the two experiments.

The USSR experiment will be carried out in an underground chamber that was excavated underneath a mountain. The excavation was carried out in the Andyrchi mountain massif of the North Caucus region in the Baksan river valley; the facility is known as the Baksan Neutrino Observatory [see Pomansky (1988) for a description of the laboratory]. Figures 11.4 and 11.5 depict the SAGE facility. The large Soviet group is led by V.N. Gavrin and G.T. Zatsepin and has American collaborators from several universities and from Los Alamos National Laboratory; the collaboration is sometimes referred to as SAGE [Soviet-American-Gallium-Experiment]. The details of the Soviet experiment have been described by Zatsepin (1982), Barabanov et al. (1985), and Pomansky (1986).

Since the USSR and the GALLEX experiments are similar, it is only necessary to summarize the main advantages and disadvantages of a metallic target and then describe how the initial stages of the extractions differ.

The main advantages are: (1) the metal target is less sensitive to the background reactions produced by radioactive impurities; (2) a metal target has a smaller volume (which reduces some other backgrounds) because of its much larger density. The smaller volume also enhances the production rate in a laboratory calibration experiment with, for example, ^{51}Cr. The main disadvantage of the metal target

Figure 11.4   Underground chamber for the SAGE experiment   The figure shows the large underground chamber that will be used for the Soviet-
American gallium experiment. The ten reactor vessels with motors on their
tops are installed in one section of the chamber. The four vessels on the
right were filled in 1988 with thirty tons of metallic gallium and will be
used in the first stage of the experiment. Part of the chemical extraction system for the thirty tons is seen along the right wall. (Photograph
courtesy of T. Bowles)

is that each time the germanium is separated it is necessary to add
fresh chemical reagents. One must therefore control strictly the germanium impurities. The greater complexity of the initial extraction
procedure makes it more difficult to demonstrate that the chemical
processing is free of unknown systematic effects.

Gallium metal melts at about 30° C, permitting the liquid (which
has a density of 6.0 g cm^{-3}) to be mixed with dilute hydrochloric
acid. To remove germanium, hydrogen peroxide is added to the
dilute acid and the entire mass is mixed vigorously. Upon mixing,

Figure 11.5   Another view of the SAGE underground chamber   The four reactor vessels with attached stirring motors at the left of the chamber contain thirty tons of gallium for use in the first stage of the experiment. The extraction equipment for the full sixty ton experiment is being installed in the area in the background of the figure. Chemistry laboratories and the low-level counting systems are housed in rooms at the far end of the chamber. Note in the foreground the two scientists hiding from the camera. (Photograph courtesy of T. Bowles.)

the metallic gallium is dispersed in the acid in the form of small globules each coated with an oxide layer. The mass appears as a black mud. If the concentration of HCl is chosen correctly, a few minutes of vigorous stirring causes the oxide coating of the globules of metal to dissolve and a clean metallic layer to form with the acid on top. The reaction is carried out in a Teflon vat provided with a mechanical mixer. In the full scale experiment, 10 identical reactor vessels will contain the gallium. The dilute acid from the 10 units will be combined and, after the solution is concentrated by evaporation, concentrated acid will be added. The germanium, which is in the concentrated acid, can be removed by purging with gas and collected in a water scrubber using a procedure similar to that described above for the GALLEX experiment. The germanium chloride collected is subsequently converted to germane by reduction with sodium borohydride. Since very large volumes of hydrochloric acid are used in this method, a procedure was developed to recover the hydrochloric acid and reuse it.

Figures 11.1 and 11.2 show the underground chamber for the SAGE experiment, including the reactor vessels that will contain the gallium. The chamber also contains the chemical extraction system; analytic chemistry laboratories and low level counting areas are located in separate rooms at the far end of the chamber.

The Soviet group has carried out successful pilot experiments with a 7 ton gallium detector, demonstrating that the chemical extraction is reliable and efficient. Both the Los Alamos and the Moscow collaborations have constructed counters with low levels of background events and with high counting efficiencies.

11.3   What can be learned?

This section describes some of the theoretical background that will be relevant for interpreting the ^{71}Ga experiments.

A   The minimum stellar rate

Using only energetic considerations, Bahcall, Cleveland, Davis, and Rowley (1985) calculated a **minimum astronomical rate** provided only that nothing happens to the neutrinos on the way to the Earth from the solar interior. This minimum rate is achieved if only pp and pep neutrinos are produced by the Sun (reactions 1 and 2 of Table 3.1). The reaction rate expected in this case is

$$\sum (\phi\sigma)_{\text{min. astronomical}} = 80 \text{ SNU}. \qquad (11.5)$$

The corresponding minimum neutrino fluxes are given in Table 6.6.

The rate given in Eq. (11.4) is slightly larger than is obtained if one uses the pp and pep fluxes from the standard solar model that are given in Table 6.5. The reason is that in the minimum model *all* of the pp fusion reactions are assumed to terminate via the ^{3}He+^{3}He reaction (reaction 4 of Table 3.1), which produces two pp or pep neutrinos for each termination of the chain. In the standard model, about 15% of the terminations proceed through the ^{3}He+^{4}He reaction (reaction 5 of Table 3.1), which produces only one pp or pep neutrino per termination.

If the observed capture rate is less than 80 SNU, then (with only one proviso) something must have happened to the neutrinos on the way to the Earth from the solar interior. This inference is independent of details of the solar model. The only alternative to the conclusion that new neutrino physics is required would be that the Sun is not currently burning nuclear fuel at a rate sufficient to balance the loss of energy via photons radiated from the solar surface. The hypothesis of a main sequence Sun that is not in quasistatic energy equilibrium is, in the eyes of many physicists and astrophysicists, a more radical hypothesis than, for example, neutrino mixing.

B Dependence of standard predictions on solar models

The flux of neutrinos from the pp reaction is the most accurately calculated of all the solar neutrino fluxes. The total estimated uncertainty is only 2%. It has often been stated mistakenly in the literature that the flux of this reaction is fixed by the observed solar luminosity. In fact, the computed flux of pp neutrinos would be about one-half as large as obtained for the standard solar model if ^{3}He in the solar interior were burned primarily by interactions with an α-particle rather than by interactions with other ^{3}He nuclei. The reason for the difference is that two pp reactions are required to terminate the chain via reaction 5 of Table 3.1 (which predominates in the standard solar model) whereas only one pp reaction is necessary if reaction 6 predominates [see the comment above concerning the minimum astronomical rate given in Eq. (11.4)].

If the model dependence of the predicted pp flux is really a factor of two, why is the estimated uncertainty so small (2%)? To answer this question, one must distinguish between uncertainties within the framework of the standard solar model and dependence upon more general ideas of how main sequence stars shine. Within the context of standard models, the uncertainty is indeed tiny. The cross sections for the reactions that burn ^{3}He are relatively well known (see data for reactions 5 and 6 in Table 3.1); they indicate that reaction 5 occurs about six times as often as reaction 6 under average solar interior conditions. For the calculated solar interior conditions, this ratio is rather stable with only a small uncertainty. However, if one does not accept the numerical results of standard solar models, then the in-principle uncertainty of the pp flux within the pp chain is a factor of two. The dependence of the pp flux upon the central temperature of the model is approximately [see Eq. (6.13b)]: Flux $\propto T^{-1.2}$.

C   Nonstandard models

If the observed capture rate satisfies

$$\sum (\phi\sigma)_{\text{stellar}} \gtrsim 80 \text{ SNU}, \tag{11.6}$$

then the rate could in principle be caused by one or more combinations of nonstandard stellar physics and new neutrino physics (e.g., the MSW effect or other explanations discussed in Chapter 9). This domain of ambiguity can be reduced significantly if one is willing to *assume* that the solar neutrino problem is due *either* to nonstandard astrophysics *or* new neutrino physics, but not both operating together.

The largest event rates are predicted by the Q-nuclei model, 200 to 400 SNU [see Sur and Boyd (1985) and Section 5.14] and by the pure CNO model, 610 SNU (see Section 5.17). Neither of these models is very attractive; they are both based upon *ad hoc* hypotheses about microscopic physics.

If we assume the correctness of the standard electroweak model (with no neutrino oscillations) and require consistency with the upper limit deduced from the ^{37}Cl experiment (see Chapter 10), then the range corresponding to so-far calculated nonstandard models is between about 80 SNU and 120 SNU.

D   The MSW effect

The gallium experiments could provide strong support for the MSW explanation of the solar neutrino problem. If the observed counting rate is much below the minimum astronomical rate of 80 SNU, this would be strong evidence in favor of the nonadiabatic MSW solution or the large mixing angle solution (see Figure 9.8 and related discussion in Sections 9.2D and 9.2F).

E Electric charge nonconservation

The conservation of electric charge is widely believed to be an *absolute* conservation law, valid to arbitrary accuracy. The validity of this law, like all laws of physics, ultimately rests upon experiment. Feinberg and Goldhaber (1959) and Goldhaber (1975) have discussed a variety of possible experimental tests of charge conservation. The most stringent lower limits on the lifetime of the electron (conceivable decays are, e.g., e $\rightarrow \gamma + \nu_e$, e $\rightarrow 2\nu_e + \bar{\nu}_e$, etc.) are of order 10^{25} yr for electrons in germanium atoms [Avignone *et al.* (1986), also earlier limits by Steinberg *et al.* (1975) and Moe and Reines (1965), but see also Nussinov (1987)].

The validity of charge conservation in reactions involving nucleons is not guaranteed by experiments that demonstrate the stability of the electron. In principle, there could be a process which permitted, at a certain level, n $\rightarrow$ p + γ or n $\rightarrow$ p + ν_e + $\bar{\nu}_e$ while forbidding, to the same order of small quantities, all decays for the electron. The lifetimes for nucleon decays (n $\rightarrow$ p + anything) that are accessible to solar neutrino experiments are very long, primarily because of the large amounts of material and the long counting times that are required to detect solar neutrinos.

Solar neutrino radiochemical experiments that involve targets for which the neutrino capture threshold is less than the mass of the electron provide sensitive tests of charge conservation with no extra effort [Bahcall (1978)]. Some examples of the kinds of processes that are forbidden by charge conservation but are allowed by energy conservation and all the other known laws of physics are: ^{71}Ga $\rightarrow$ ^{71}Ge + ν_e + $\bar{\nu}_e$ and ^{71}Ga $\rightarrow$ ^{71}Ge + γ (E_γ $\equiv$ 275 keV). Targets composed of ^{55}Mn, ^{81}Br, ^{87}Rb, and ^{205}Tl are also suitable, in principle, for tests of charge conservation.

The great sensitivity of solar neutrino experiments to charge nonconservation can be seen immediately from the definition of a SNU, that is, 10^{-36} daughter atoms produced per target atom per second. Translated into lifetimes measurable in a radiochemical experiment designed to detect ν_e + $^Z A \rightarrow$ e$^-$ +$^{(Z+1)} A$, this gives:

$$\tau \left(^Z A \rightarrow {}^{(Z+1)} A + \text{anything} \right) \geq \frac{2.2 \times 10^{28} \text{yr}}{P \left(\text{SNU} \right)}. \qquad (11.7)$$

Here $P(\text{SNU})$ is the measured production rate in SNUs of daughter atoms $^{(Z+1)}A$ which, in principle, could be due either to solar neutrino captures, to background processes, or to charge nonconservation. Note that radiochemical experiments are sensitive to any mode by which the charge nonconservation is effected; this is indicated by the appearance of the word "anything" in Eq. (11.6). If pp solar neutrinos reach the Earth, then the ultimate sensitivity to charge nonconservation obtainable with a ^{71}Ga experiment alone, $P \approx 10^2$, will be a lifetime $\approx 10^{26.5}$ yr [see Steinberg (1976)].

The interpretation of charge nonconservation experiments requires some theoretical assumptions. A reasonably general, and perhaps plausible, assumption is that the interaction matrix element factors into a nuclear part times something else. The nuclear part itself can then be factored (for the relevant low-momentum transfers) into an ordinary nuclear β-decay matrix element (obtainable, e.g., from the electron capture rate of ^{71}Ge for a ^{71}Ga target) times a charge nonconserving nuclear matrix element, $\langle n|H_0 Q|p\rangle$. Using Fermi's golden rule with the above assumption, one can write

$$| \langle n|H_Q|p\rangle |^2 \rho \; = \; (\ln 2\hbar/2\pi) \left[\tau_{1/2}(Q)\right]^{-1} \left[ft_{1/2}/6 \times 10^3 \text{ s}\right], \quad (11.8)$$

where ρ is the phase space for the decay (which depends on the decay mode and theory assumed), and $\tau_{1/2}(Q)$ is the lifetime, or upper limit to the lifetime, for charge-nonconserving decays [cf., Eq. (11.6)]. The nuclear $ft_{1/2}$-factor that appears in Eq. (11.7) must be corrected for the statistical factor, $[(2I+1)/(2I'+1)]$, that takes account of the difference in spin between initial and final nuclear states. For ^{71}Ga $\rightarrow$ ^{71}Ge, $ft_{1/2} \approx 5 \times 10^4$ s. In the form given in Eq. (11.7), the results of a solar neutrino experiment can be interpreted in terms of a limit on a nucleon noncharge-conserving matrix element in a manner that is entirely analogous to, but independent of, the matrix element $\langle e^-|H_Q|0\rangle$ that is determined by electron decay experiments.

If one assumes that the weak interactions include a small charge-nonconserving part that has the usual form except for a neutrino replacing the electron in the lepton current, $H_Q \equiv \epsilon H_{\text{usual form}}$, then one can obtain an interesting limit on ϵ. The result may be expressed in terms of the ratio of branching ratios for the elementary neutron decays:

$$\epsilon^2 \; = \; \Gamma\left(n \rightarrow p + \nu_e + \bar{\nu}_e\right)/\Gamma\left(n \rightarrow p + e^- + \bar{\nu}_e\right). \quad (11.9)$$

Table 11.2.   Upper limits to the charge-nonconserving weak interaction.

Decay	$\epsilon^2_{\text{upper limit}}$	Reference
^{87}Rb $\rightarrow$ ^{87}Sr^m	3×10^{-19}	Norman and Seamster (1978)
^{71}Ga $\rightarrow$ ^{71}Ge^m	9×10^{-24}	Barabanov et al. (1980)
^{87}Rb $\rightarrow$ ^{87}Sr^m	8×10^{-21}	Vaidya et al. (1983)
^{113}Cd $\rightarrow$ ^{113}In^m	10^{-17}	Roy et al. (1983)
^{127}I $\rightarrow$ ^{127}I^{excited}	10^{-23}	Holjevic et al. (1987)

Bahcall (1978) finds:

$$\epsilon^2 = \left[\frac{P(\text{SNU})t_{1/2}(\text{n})}{2.2 \times 10^{28} \text{ yr}} \right] \left[\left(\frac{W(\text{n})}{W(^Z A)} \right)^5 \frac{(ft)_{ZA}}{(ft)_{\text{n}}} \right]. \qquad (11.10)$$

Here $W(\text{n})$ is the mass difference (1.29 MeV) between a neutron and a proton; $W(^Z A)$ is the *nuclear* mass difference between the isotopes $^Z A$ and $^{(Z+1)}A$; $(ft)_{ZA}$ includes the statistical spin factor; and $(ft)_{\text{n}} = 1.1 \times 10^3$ s, $t_{1/2}(\text{n}) = 6.5 \times 10^2$ s. For ^{71}Ga one finds

$$\epsilon^2 = 6 \times 10^{-29} P \left(^{71}\text{Ga}; \text{SNU} \right). \qquad (11.11)$$

Thus a ^{71}Ga solar neutrino experiment will be sensitive to a noncharge-conserving part of the weak interactions that is more than 26 orders of magnitude smaller than the main part of the weak interactions.

There are several beautiful laboratory experiments that place stringent limits on ϵ^2. The results are summarized in Table 11.2.

There is, in principle, some ambiguity in interpreting radiochemical solar neutrino experiments in which the neutrino capture threshold is less than the mass of the electron. How can one be sure that a measured counting rate in, for example, the ^{71}Ga experiment is caused by solar neutrinos instead of charge nonconservation? There are two answers to this question: one theoretical, one experimental. First, all current theoretical considerations suggest that it is much more likely that the solar-induced rate is of order 10 SNU to 100 SNU (consistent with stellar evolution predictions, perhaps modified by electroweak effects beyond the standard model) than that electric charge is not conserved at a level that accidentally is of the right order of magnitude to be detectable in a solar neutrino experiment. To see an effect dominated by charge nonconservation in the ^{71}Ga experiment, either the event rate must exceed significantly 10^2 SNU or both charge-nonconservation and strong suppression of solar neutrinos must be occurring. Second, the event rates in the ^{71}Ga and other solar neutrino experiments can be compared. If the counting rates in the ^{71}Ga and the other solar neutrino experiments are consistent with the same solar neutrino fluxes, then the ^{71}Ga experiment may be interpreted additionally in terms of an upper limit on charge-nonconserving reactions.

Bibliographical Notes

Bahcall, J.N., B.T. Cleveland, R. Davis, I. Dostrovsky, J.C. Evans, W. Frati, G. Friedlander, K. Lande, J.K. Rowley, R.W. Stoenner, and J. Weneser (1978) *Phys. Rev. Lett.*, **40**, 1351. The U.S. proposal which contained many of the ingredients that are being used in the European and Soviet experiments.

Barabanov, I.R., *et al.* (1985) in *Solar Neutrinos and Neutrino Astronomy*, edited by M.L. Cherry, W.A. Fowler, and K. Lande (New York: AIP Conf. Proc. No. 126.) p. 175. A clear description of the Soviet experiment.

Cribier, M., *et al.* (1988) *Nucl. Inst. Meth.*, **A265**, 574. An excellent discussion of the experimental possibilities for using ^{51}Cr as a calibration source for the GaCl$_3$ detector.

Dostrovsky, I. (1978) in *Proceedings of Informal Conference on Status and Future of Solar Neutrino Research*, edited by G. Friedlander (Brookhaven National Laboratory) Report No. 50879, Vol. 1, p. 231. A beautiful description of how the gallium experiment works. This was an electrifying lecture: Anyone who was present will remember the test tube demonstration that Dostrovsky presented and circulated among the listeners.

Kirsten, T. (1986) in *Massive Neutrinos in Astrophysics and in Particle Physics*, edited by O. Fackler and J. Tran Thanh Van (Editions Frontieres) p. 119. The GALLEX experiment described authoritatively by the senior spokesman.

Kuzmin, V.A. (1966) *Sov. JETP*, **22**, 1051. The visionary suggestion of a ^{71}Ga solar neutrino experiment at a time when the required amount of gallium greatly exceeded the annual world production.

12

Geochemical and radiochemical detectors: Mo, Tl, Li, Br, and I

Summary

This chapter discusses the five radiochemical or geochemical experiments that have received the most attention in the literature and which appear to have the best chances of being performed: ^{98}Mo, ^{205}Tl, ^{7}Li, ^{81}Br, and ^{127}I.

The ^{98}Mo and ^{205}Tl experiments (Sections 12.1 and 12.2, respectively) utilize long-lived daughter isotopes that are produced by neutrino absorption and that are extracted from ores buried deep underground for millions of years. The reactions by which the neutrinos are detected are: $\nu_e + {}^{98}\text{Mo} \rightarrow e^- + {}^{98}\text{Tc}$ and $\nu_e + {}^{205}\text{Tl} \rightarrow e^- + {}^{205}\text{Pb}$. The buried ores contain ^{98}Tc and ^{205}Pb. The geological stability of the ore deposits over the past 10^7 yr, which influences the shielding from cosmic rays of the neutrino-absorbing isotopes, is an important consideration in interpreting observational results.

The ^{98}Mo and ^{205}Tl experiments are sensitive to neutrino fluxes averaged over millions of years. If the standard solar model is correct, the average neutrino fluxes to which the buried ores are exposed should be the same, to within 1%, as the contemporary fluxes that are measured, for example, by the ^{37}Cl, ^{71}Ga, and neutrino–electron scattering experiments. For some speculative solutions of the solar neutrino problem, there could be a significant difference between the average and the contemporary neutrino fluxes.

The proposed ^{7}Li, ^{81}Br, and ^{127}I experiments, which are discussed in Sections 12.3, 12.4, and 12.5, respectively, use radiochemical techniques to measure contemporary neutrino fluxes. Like the ^{37}Cl de-

363

tector, the proposed ^{7}Li and ^{81}Br experiments would be sensitive to several different neutrino sources. In addition to ^{8}B neutrinos, a ^{7}Li detector is especially sensitive to CNO neutrinos and a ^{81}Br detector to ^{7}Be neutrinos. If the ^{8}B neutrino flux is much less than the standard model value and the ^{7}Be and CNO neutrinos are not depleted by as large a fraction, then the ^{7}Li and ^{81}Br detectors could furnish unique information about the ^{7}Be and CNO neutrinos. The iodine detector is sensitive to ^{8}B and ^{7}Be neutrinos, the relative sensitivity to these two sources is not known.

The neutrino absorption cross sections can be calculated accurately for ^{7}Li, but not for the other four targets discussed in this chapter.

12.1 ^{98}Mo

A heroic geochemical experiment using ^{98}Mo is being performed by a team of workers at the Los Alamos National Laboratory [see Cowan and Haxton (1982) and Wolfsberg et al. (1985)]. The reaction by which the neutrinos are detected is:

$$\nu_e + {}^{98}\text{Mo} \rightarrow e^- + {}^{98}\text{Tc}. \tag{12.1}$$

The ^{98}Mo experiment is of special interest since the ^{98}Tc that is created is itself unstable, β-decaying to ^{98}Ru with a mean life, τ_m, of 6×10^6 yr. Thus the ^{98}Mo experiment provides some information about the ^{8}B neutrino flux averaged over the past several million years (see below for a quantitative description of the temporal sensitivity).

The ^{98}Mo experiment is sensitive only to ^{8}B (and hep) neutrinos. The ground state to ground state transition, which has a threshold energy of 1.68 MeV, is forbidden. All of the strong transitions are to excited states of ^{98}Tc and require more energy than is available for any of the neutrino sources except ^{8}B (and hep).

The capture rate predicted by the standard model is [Bahcall and Ulrich (1988)]:

$$\sum_i \phi_i \sigma_i = 17.4^{+18.5}_{-11} \text{ SNU}, \tag{12.2}$$

of which all but 0.08 SNU comes from ^{8}B neutrinos. The dominant uncertainty in the predicted capture rate is from the neutrino capture cross sections; the total uncertainty from all other sources is only 6

Figure 12.1   Mass spectrograph used for ^{98}Mo experiment   The figure shows the mass spectrometer used to detect technetium. The spectrograph has a detection limit of about 10^6 atoms of ^{98}Tc; it is a tandem magnetic-sector ($90°$, 30 cm radius) mass spectrometer with both faraday and pulse counting ion detection.

SNU. The matrix elements for the excited state transitions must all be estimated from (p,n) experiments. The 1σ uncertainties in the (p,n) experiments amount by themselves to uncertainties of about $\pm$ 30% [see Rapaport *et al.* (1985)]. In addition, there are systematic uncertainties in calibrating the (p,n) measurements for the relevant mass range. The total theoretical uncertainty in the absorption cross sections is therefore at least a factor of two [see Section 8.2F and Bahcall and Ulrich (1988)].

A geochemical experiment has been developed at Los Alamos National Laboratory to extract ^{98}Tc produced by solar neutrino capture on ^{98}Mo. In order to obtain a usable signal, about 2600 tons of ore from a deeply buried deposit must be processed, yielding about 13 tons of molybdenite. The original ore is extracted from the Henderson molybdenum mine in Clear Creek County, Colorado. AMAX corporation uses the ore in the commercial production of molybdenum oxide. Flotation separation at the Henderson molybdenum mine produces a concentrate that is more than 90% molybdenite, reducing the original sample of ore to about 13 tons of molybdenum. The concentrate is then shipped to the AMAX roasting plant at Fort Madison, Iowa, for conversion of the MoS_2 to MoO_3. The required 13 tons of molybdenite amounts to only a fraction of a day's production of the Fort Madison plant. The fate of technetium in the chemical processing is inferred from measurements on its chemical homologue rhenium, which has very similar properties.

An ultrasensitive mass spectrometer, shown in Figure 12.1, will be used [see Wolfsberg *et al.* (1985)] to detect the expected 10^7 atoms of ^{98}Tc. Also, ^{99}Tc will be present at several orders of magnitude greater concentration than ^{98}Tc, since ^{99}Tc is a spontaneous fission product of ^{238}U in the molybdenite. Mass spectrometry will be performed on the sample to measure the ratios of ^{97}Tc, ^{98}Tc, and ^{99}Tc. The absolute abundances will be fixed by measuring the ratio ^{97}Tc/^{99}Tc in a small part of the sample to which a known amount of ^{97}Tc has been added. The concentrations of ^{99}Tc (and ^{97}Tc) can then be used to determine the amount of neutrino-produced ^{98}Tc that is present, somewhat analogous to the way ^{36}Ar and ^{38}Ar are used in the ^{37}Cl experiment (cf., Section 10.2C).

The most difficult task in any geochemical experiment is to demonstrate that backgrounds from natural radioactivity and from cosmic rays are tolerable. The backgrounds for this experiment have been discussed carefully and extensively by Cowan and Haxton (1982) in their classic paper, and the reader is referred to this work for details. Cowan and Haxton present strong arguments which suggest that the backgrounds will not interfere with a meaningful measurement of the ^{8}B neutrino flux. They also discuss how a determination of the ^{99}Tc concentration can be used to show experimentally that the radioactive backgrounds are not significant. A further experimental test is to demonstrate that ^{99}Tc is in approximate secular equilibrium with uranium, which would show that technetium is not lost from the ore over geological times.

In order to provide convincing proof of the validity of the measurement, the experiment should be performed with samples from different locations in the mine or from other mines. If one can demonstrate that the answer is independent of the local environment, then we are on a firmer basis in ascribing the observed amount of ^{98}Tc to the action of solar neutrinos.

What does the ^{98}Mo experiment measure? Is the ^{8}B flux to which the buried molybdenum has been subject over many eons expected to be different from the contemporary flux that is registered by all other detectors? The experiment is sensitive to processes that occur on a Kelvin–Helmholtz time scale, $\sim 10^7$ yr (see Section 2.1), but there is no know mechanism for changing the flux of ^{8}B neutrinos on a time scale this short (see Chapter 5). The more obviously relevant time scale is the nuclear burning time scale, which is of order the current age of the Sun (see Section 2.1).

The sensitivity of the ^{98}Mo experiment to a time-dependent ^{8}B neutrino flux can be explored by examining the theoretical ratio of the number of ^{98}Tc atoms in the ore, N_{Tc}, to the number of molybdenum atoms in the ore, N_{Mo}.

Let us define the time dependence, $f(t)$, of the ^{8}B neutrino flux by the relation,

$$\phi(t) = \phi_{\mathrm{SM}} \times f(t)f, \qquad (12.3)$$

where ϕ_{SM} is the contemporary value of the ^{8}B flux predicted by the standard solar model (5.8×10^6 cm^{-2} s^{-1}). Then the ratio of ^{98}Tc to ^{98}Mo atoms is

$$\frac{N_{\mathrm{Tc}}(t)}{N_{\mathrm{Mo}}} = (\phi_{\mathrm{SM}} \, \sigma \, \tau_{\mathrm{m}}) \, I(t), \qquad (12.4)$$

where τ_{m} is the mean life of the ^{98}Tc that is produced by neutrino absorption [see Eq. (12.1)] and $I(t)$ is a simple integral whose explicit form is given below. Of course, for a constant neutrino flux, $I(t) = 1.0$.

The equilibrium fraction of ^{98}Tc atoms in the ore is infinitesimal. Numerically, the coefficient of $I(t)$ in Eq. (12.4) is

$$\epsilon = (\phi_{\mathrm{SM}} \, \sigma \, \tau_{\mathrm{m}}) = 3.3 \times 10^{-21}. \qquad (12.5)$$

The time dependence of the measured ratio is given by the expression
[cf., Eq. (12.4)]

$$I(t) = \int_{t_0/\tau_{\mathrm{m}}}^{t/\tau_{\mathrm{m}}} \mathrm{d}x \, f(\tau_{\mathrm{m}}x) \exp - (t/\tau_{\mathrm{m}} - x), \qquad (12.6)$$

where t_0 is a time before solar neutrino capture produces any ^{98}Tc.

How different is the ^{8}B flux on the time scales to which the ^{98}Mo
experiment is sensitive? The time dependence of the ^{8}B flux in the
standard solar model is (see Table 4.6)

$$\phi(t) \propto t^{3.4}. \qquad (12.7)$$

Let t_{pres} represent the present epoch. Then $\phi(t_{\mathrm{pres}} - \tau_{\mathrm{m}}) \simeq \phi_{\mathrm{SM}} [1 - \tau_{\mathrm{m}}/t_{\mathrm{pres}}]^{3.4} \approx 0.995\phi_{\mathrm{SM}}$. Therefore, in the standard model the ^{8}B
neutrino flux is effectively constant on the time scales to which the
^{98}Mo experiment is sensitive.

We can understand the implications of these results [Eqs. (12.4)
to (12.7)] by considering a simple case. Suppose that at some time
earlier in the Sun's history, t_{event}, the ^{8}B neutrino flux changed from
the constant value, ϕ_{SM}, to a different value, $\beta \times \phi_{\mathrm{SM}}$, which is the
contemporary value being measured by the ^{37}Cl and the electron
scattering experiments. Then

$$f(x) = \begin{cases} \beta, \ t > t_{\mathrm{event}}, \\ 1, \ t < t_{\mathrm{event}}. \end{cases} \qquad (12.8)$$

The integral giving the time dependence is then

$$I(t) = [\beta + (1 - \beta) \exp - (t - t_{\mathrm{event}})/\tau_{\mathrm{m}}]. \qquad (12.9)$$

As anticipated, $I(t) \equiv 1.0$ if the suppression factor $\beta = 1.0$.

If the "event" occurred a long time ago, then the measured value
will be equal to the contemporary value. Thus

$$I(t) \to \beta, \quad (t - t_{\mathrm{event}})/\tau_{\mathrm{m}} \gg 1. \qquad (12.10)$$

The accuracy with which the ^{8}B flux can be inferred from the
abundance ratio of ^{98}Tc and ^{98}Mo is about a factor of two [see Bahcall
and Ulrich (1988) and Section 7.3] because of existing uncertainties
in the neutrino absorption cross sections. In order for the average
^{8}B neutrino flux that is measured in the ^{98}Mo experiment to differ
by more than a factor of two from the contemporary value, the solar

Table 12.1.   Maximum
time ^{98}Mo is sensitive to
neutrino variations. The
table gives the maximum
time, Δt ago, that an event
in the solar core could
have occurred and still be
revealed by a ^{98}Mo solar
neutrino experiment. Here
τ_{m} is the mean life of ^{98}Tc,
$\tau_{\mathrm{m}} = 6 \times 10^{6}$ yr.

β	$\Delta t / \tau_{\mathrm{m}}$
1/2	0
1/4	1
1/8	2
1/16	3

event must have been more recent than a critical time $\Delta t = (t-t_{\mathrm{event}})$ ago, where

$$\Delta t = \tau_{\mathrm{m}} \times \ln \left[(1 - \beta) / \beta \right]. \tag{12.11}$$

Table 12.1 shows the ratio of Δt to τ_{m} for a large range of values of the contemporary suppression factor, β, of the ^{8}B neutrino flux. For β ranging from 1/2 to 1/16, Δt varies from $0 \times \tau_{\mathrm{m}}$ to $3 \times \tau_{\mathrm{m}}$. Even if the suppression were an enormous factor of 16, the effect would be observable only if the event occurred as recently as 2×10^{7} yr ago.

What should we expect for the ratio $\Delta t / \tau_{\mathrm{m}}$? What is the characteristic time over which major "events" or changes might occur in the solar interior? Cowan and Haxton (1982) in their original paper suggest that time scales of the order of a few million years may be relevant, based upon a suggestion by Dilke and Gough (1972) that certain low-order g-modes may be unstable in the Sun (see the discussion of this possibility in Section 5.10). More conventional ideas about solar evolution based upon the standard solar model suggest that the characteristic time for significant change may be the calculated main sequence lifetime of the Sun (10^{10} yr), the average lifetime of a proton in the solar interior (also 10^{10} yr), or the current age of the Sun (5×10^{9} yr). Thus on the basis of standard ideas of stellar evolution, we would expect a very large ratio

$$\Delta t / \tau_{\mathrm{m}} \simeq 10^3, \qquad (12.12a)$$

and therefore

$$I(t) = \beta \qquad (12.12b)$$

to high accuracy. The average ^{8}B neutrino flux to which the ^{98}Mo experiment is sensitive is expected to be the same to an accuracy of 1 % or better as the contemporary flux measured by the ^{37}Cl, Kamiokande II, and other experiments.

Thus standard ideas about the time scale for solar evolution suggest that a molybdenum experiment should yield a value for the ^{8}B ν_{e} flux that is consistent with the value found with contemporary measurements. This conclusion is independent of the reason, faulty astrophysics or new physics, that the neutrino flux is low.

12.2 ^{205}Tl

A ^{205}Tl experiment would have the lowest effective threshold for neutrino absorption, 0.05 MeV, of any of the solar neutrino detectors currently being actively investigated. The detection reaction is

$$\nu_{\mathrm{e}} + {}^{205}\mathrm{Tl} \rightarrow \mathrm{e}^- + {}^{205}\mathrm{Pb}, \qquad E_{\mathrm{th}} = 0.054 \text{ MeV}. \qquad (12.13)$$

A ^{205}Tl detector would provide important information since it is sensitive to pp neutrinos emitted over the past 10^7 yr [see Freedman et al. (1976), Freedman (1978), and Henning et al. (1985)]. However, the neutrino absorption cross sections for the strongest transitions between low-lying states cannot be calculated accurately [see Bahcall (1978), Section IV.I] and it is difficult to give a rigorous upper limit to the uncertainties [see Braun and Talmi (1986)]. This situation is frustrating because the time dependence of the pp neutrinos reflects accurately the time dependence of the solar luminosity, a quantity of fundamental importance in the theory of stellar evolution (see Table 4.6). Kienle (1988) and Freedman (1988) have suggested that it may be possible to determine the Gamow–Teller strength of the most important low-energy transition by detecting the β-decay into bound states [see Bahcall (1961)] of ^{205}Tl that is completely ionized in a heavy ion synchrotron. However, a measurement of the bound-state decay rate, while of great importance for determining the capture probability of pp neutrinos, will not affect the (factor of two) uncertainties for transitions to the more highly excited states that dominate the cross sections for ^{8}B and ^{7}Be neutrinos.

The rate of the transitions to higher-energy excited states of ^{205}Pb has been estimated using preliminary (p,n) data generously made available by Krofcheck (1987) and his collaborators. This data cannot be used in the traditional way [see Goodman (1985)] to obtain the cross sections for the first forbidden transition to the low-lying 0.0023 MeV excited state of ^{205}Pb, the only transition estimated in previous calculations. It is possible to perform, however, an illustrative calculation of the cross section for absorption of ^{8}B neutrinos because allowed transitions to higher-energy states of ^{205}Pb dominate for this neutrino source and the usual calibration procedure is appropriate. The result is of interest even though it is preliminary since it shows that the capture rate may not be dominated by pp neutrinos, as has been assumed in all previous discussions.

The total absorption cross section for ^{8}B neutrinos incident on ^{205}Tl estimated in this way is $\sigma \approx 8 \times 10^{-42}$ cm^2 [Bahcall and Ulrich (1988)]. Given the preliminary nature of the data, this cross section must be considered to be uncertain by at least a factor of two and probably more. The cross sections for all the other solar neutrino sources were reestimated by Bahcall and Ulrich (1988) using the value of $\log ft_{1/2}$ of 5.7 that was suggested by Braun and Talmi (1986) as possibly appropriate for the transitions to the lowest-lying $1/2^-$ state in ^{205}Pb (0.0023 MeV excitation). The $\log ft_{1/2}$-value of Braun and Talmi implies that each of the "nominal" cross sections of Bahcall (1978) should be reduced by a factor of about 2.5.

With the above cross sections, the total rate predicted by the standard model is [Bahcall and Ulrich (1988)]:

$$\sum_i \phi_i \sigma_i = 263 \text{ SNU}. \tag{12.14}$$

Bahcall and Ulrich (1988) refrained from quoting a formal uncertainty on the above rate because the neutrino absorption cross sections are very uncertain. The nominal contributions of pp, ^{7}Be, and ^{8}B neutrinos to the total rate are, respectively, 173 SNU, 34 SNU, and 46 SNU. Given the large cross section uncertainties, one cannot rule out the possibility that the ^{7}Be and ^{8}B neutrinos contribute an amount that is comparable to the pp neutrinos.

A large collaboration of West German and Yugoslavian scientists has been formed [see Ernst (1984), Nolte and Pavićević (1988)] to investigate the possibility of performing a ^{205}Tl experiment in the Alchar mine in Macedonia. A determination of the extent of the shielding of the ore from cosmic rays over geological times and the development of techniques for separating tiny amounts of ^{205}Pb from ordinary lead are major problems that must be solved in order for the experiment to be feasible.

12.3 ^{7}Li

A lithium detector could provide crucial information about the pep and CNO neutrinos. Moreover, the interaction cross sections are well known because of a unique situation: The strength of both the ground state and the excited state neutrino transitions can be inferred accurately [Bahcall (1978)] from laboratory measurements on the electron capture decay of ^{7}Be (to both the ground and first excited states of the isotopic analogue states in ^{7}Li).

For many years, ^{7}Li has appeared to be an important and necessary part of a program of neutrino spectroscopy of the solar interior [see Bahcall (1969)]. It is an especially clean and interesting experiment from a theoretical point of view.[†] Also, Rowley (1978) has shown that the chemical extraction can be performed with confidence and that the backgrounds are manageable. The required amount of material is relatively cheap. All that is needed is a practical way of counting the radioactive ^{7}Be that is produced. A number of suggestions have been made in the literature [see the review by Rowley (1978) and the important recent Soviet work, Veretyonkin *et al.* (1985) and Gavrin and Yanovich (1987)], but so far no one has demonstrated an efficient scheme. The problem is that about 90% of the decays of ^{7}Be lead to the ground state of ^{7}Li, which emits only a 50 eV Auger electron. There is not much to work with unless one can detect the ^{7}Be atoms before they decay.

The reaction that registers the neutrinos is

$$\nu_e + {}^7\text{Li} \rightarrow e^- + {}^7\text{Be}, \quad E_{\text{th}} = 0.862 \text{ MeV}. \tag{12.15}$$

[†]The theoretical calculations for a ^{7}Li detector require a unique treatment. Because of the near equality between the energy of the emitted neutrinos (ground state to ground state transition) and the absorption threshold, the thermal motion in the Sun of both ^{7}Be ions and of electrons must be taken into account in calculating the absorption cross section for the ^{7}Be neutrinos [see Bahcall (1978)].

Table 12.2.   Capture rates
predicted by the standard solar
model for a ^{7}Li detector.

Neutrino source	Capture rate (SNU)
pp	0.0
pep	9.2
hep	0.06
^{7}Be	4.5
^{8}B	22.5
^{13}N	2.6
^{15}O	12.8
^{17}F	0.1
Total	51.8 SNU

The inverse reaction, the electron capture decay of ^{7}Be, has an experimentally convenient half-life of 53.4 days.

The capture rate predicted for a ^{7}Li detector by the standard model is [Bahcall and Ulrich (1988)]:

$$\sum_i \phi_i \sigma_i = 51.8 \left(1 \pm 0.31\right) \text{ SNU.} \qquad (12.16)$$

Table 12.2 shows the individual contributions to the capture rate predicted by the standard model. For this conventional picture, the ^{7}Li experiment is mostly sensitive to neutrinos from ^{8}B decay, although much less so than the ^{37}Cl or ^{98}Mo experiments. The threshold energy for neutrino absorption is 0.862 MeV, which means that only the pp neutrinos do not contribute. Approximately 43% (22.5 SNU) of the predicted event rate is contributed by ^{8}B neutrinos; the next largest contributions are the 25% (13 SNU) from ^{15}O neutrinos and the 18% (9 SNU) from pep neutrinos. About half of the total calculated uncertainty is related to the ^{8}B neutrino flux.

In fact, we know that the ^{8}B solar neutrino flux incident on Earth is at least a factor of two less than the value predicted by the standard model (see Section 10.4A), provided the shape of the spectrum is unchanged by neutrino oscillations or decay. For some nonstandard solar models and many MSW solutions, the CNO neutrinos are expected to be the largest contributors to the measured event rate.

For the illustrative case in which no ^{8}B neutrinos reach the Earth (see the discussion of the "No ^{8}B" model in Section 5.16), the predicted capture rate is 29.3 ± 8.5 SNU, of which 12.8 SNU is from ^{15}O and 2.6 SNU is from ^{13}N. Thus for this extreme (but conceptually possible) case, the CNO neutrinos contribute more than half (53%) of the counting rate. The pep neutrinos, which are simply related to the basic pp neutrinos (cf., Chapter 6), contribute about 31% of the total.

12.4 ^{81}Br

A ^{81}Br detector is as appealing to experimentalists as is the ^{7}Li detector to theorists. The detection reaction is

$$\nu_e + {}^{81}\text{Br} \rightarrow e^- + {}^{81}\text{Kr}, \quad E_{\text{th}} = 470 \text{ keV}. \tag{12.17}$$

The effective absorption threshold of 470 keV permits detection of all of the most important solar neutrino sources except the pp neutrinos. The radioactive half-life for the ^{81}Kr that is produced is 2×10^5 yr.

The chemical extraction of ^{81}Kr is simple and is similar to the process used in the ^{37}Cl experiment to extract ^{37}Ar. Also, the detector material is cheap and abundant. In fact, the entire experiment could be performed in the existing ^{37}Cl tank in the Homestake Mine. A detailed proposal and a feasibility demonstration for a radiochemical experiment have been presented by Hurst et al. (1984, 1985). This radiochemical experiment would be sensitive to the contemporary flux of solar neutrinos. The possibilities for carrying out a geochemical experiment, sensitive to the average neutrino flux over the past $10^{5.5}$ years, using deep underground water with a high bromine content has been reviewed by Kuzminov, Pomansky, and Chihladze (1988).

The capture rate predicted by the standard model for a ^{81}Br detector is [Bahcall and Ulrich (1988)]:

$$\sum_i \phi_i \sigma_i = 27.8^{+17}_{-11} \text{ SNU}. \tag{12.18}$$

The contributions of different neutrino sources to the total capture rate are shown in Table 12.3. In all the early discussions of a ^{81}Br detector [Scott (1976), Hampel and Kirsten (1978), Bahcall (1978), Bahcall (1981), Haxton (1981), Itoh and Kohyama (1981), Hurst et al. (1984, 1985), and Davis (1986)], it was assumed that only the

Table 12.3. Capture rates
predicted by the standard solar
model for a ^{81}Br detector.

Neutrino source	Capture rate (SNU)
pp	0.0
pep	1.1
hep	0.07
^{7}Be	8.6
^{8}B	15.3
^{13}N	0.9
^{15}O	1.9
^{17}F	0.02
Total	27.8 SNU

low-lying states in the daughter nucleus ^{81}Kr would be populated
strongly and therefore that ^{7}Be neutrinos would dominate the capture rate. Unfortunately, when (p,n) experiments were performed on
^{81}Br, we found that – at least for the standard solar model – ^{8}B neutrinos contribute about 55% of the expected capture rate [Krofcheck
et al. (1987)]. The reason for this important change in the predicted
nature of the detector sensitivity is the large measured Gamow–
Teller strength to highly excited levels of ^{81}Kr that was found in the
(p,n) measurements.

The next most important source is ^{7}Be, whose neutrinos contribute
31% of the total capture rate predicted by the standard model. The
^{8}B neutrino flux may be reduced by a much greater factor than
the ^{7}Be neutrino flux, as in the adiabatic MSW solution (see Section 9.2F). In the "No ^{8}B" model discussed in Section 5.16, the ^{7}Be
neutrinos contribute 69% of the predicted capture rate in the ^{81}Br
experiment. A ^{81}Br detector would give valuable information about
the flux of ^{7}Be neutrinos if the ^{8}B neutrino flux is depleted much
below the standard model value.

The dominant uncertainty in the predictions is caused by the transitions to excited states whose matrix elements must be inferred
from (p,n) measurements [cf., discussion in Section V.D of Bahcall
et al. (1982) and in Section IV of Bahcall and Ulrich (1988), as well

as Section 7.3 of this book]. The total uncertainty due to neutrino absorption cross sections is [Bahcall and Ulrich (1988)] +16 SNU or −8 SNU. If we ignore all of the uncertainties associated with absorption cross sections, then the remaining total theoretical uncertainty is only 7 SNU (or 25% of the total capture rate).

If ^{7}Be neutrinos dominate the capture rate, the total uncertainty from the absorption cross sections is only about 30% and could be significantly reduced by more precise laboratory experiments involving the decay of the 190 keV metastable state of ^{81}Kr [see Davids *et al.* (1987) and Lowry *et al.* (1987)]. Even more importantly, the ^{81}Br detector could be calibrated using a ^{51}Cr source [see Section IV.C of Bahcall and Ulrich (1988)]. A ^{51}Cr source experiment would be especially useful in determining the matrix element for the transition to the possibly important 457 keV excited state of ^{81}Kr [see Bahcall (1981)], which cannot be measured by β-decay experiments.

The expected backgrounds in a ^{81}Br experiment have been discussed in detail in Hurst *et al.* (1984, 1985) and do not present any major difficulties. The counting requires a special technique, but is feasible. The lifetime of the small number of ^{81}Kr atoms that are produced by solar neutrinos is too long to be counted efficiently with standard techniques of nuclear radioactivity. However, Hurst and his colleagues have developed a method of counting accurately the few hundred atoms of ^{81}Kr that would be produced per year by neutrino capture. They have shown [Hurst (1984, 1985)] that they can count with confidence the produced atoms using the techniques of resonance ionization spectroscopy.

A bromine detector would be a natural successor to the ^{37}Cl experiment since it could be performed inexpensively using the same tank and would involve similar chemical procedures [Hurst *et al.* (1984, 1985) and Davis (1986)]. The experiment is feasible: The background events and the counting are understood and satisfactory with existing techniques. A ^{81}Br detector could provide valuable independent constraints on both the ^{8}B and the ^{7}Be neutrino fluxes. It should be done.

12.5 ^{127}I

An iodine-bearing liquid could be used as a detector of solar (or supernova) neutrinos via the reaction

$$\nu_e + \,^{127}\text{I} \to e^- + \,^{127}\text{Xe}, \quad E_{\text{th}} = 0.789 \text{ MeV}. \tag{12.19}$$

An iodine detector is sensitive primarily to ^{7}Be and ^{8}B neutrinos, the same sources that produce nearly all of the events in the ^{37}Cl experiment. Haxton (1988) has discussed the advantages of using an iodine detector, stressing the large estimated absorption cross sections and the similarity of the chemical extraction and processing to what is used in the ^{37}Cl experiment. He argues that the larger estimated cross sections make an iodine detector useful for testing whether solar neutrino fluxes are variable. The counting and decay lifetime (36 days for ^{127}Xe) are similar to the ^{37}Cl experiment.

The main problem with an ^{127}I experiment is that the absorption cross sections cannot be calculated accurately. A ^{51}Cr source is not sufficiently energetic to calibrate the transitions caused by ^{7}Be or ^{8}B neutrinos. A calibration of the low-lying states populated by ^{7}Be might be possible with a ^{65}Zn source although not enough is known at present about the excited states of ^{127}Xe to determine whether a ^{65}Zn source would populate states not accessible (because of energetic reasons) to the ^{7}Be neutrinos. The ^{8}B neutrinos cause transitions to many excited states; the cross sections for these transitions can only be estimated approximately [ultimately with an uncertainty of order a factor of two from (p,n) measurements, see Section 7.3]. The uncertainty in the absorption cross section prevents one from using an iodine detector for an accurate absolute measurement of solar neutrino fluxes.

If the neutrino absorption cross sections for ^{127}I are [as suggested by Haxton (1988)] an order of magnitude larger than for ^{37}Cl, then the iodine detector could be used to test for anomalous runs or time variability of the fluxes, independent of the absolute value of the neutrino cross sections. However, direct counting experiments discussed in Chapters 13 and 14 have much better time resolution (microseconds instead of months) and more than adequate sensitivity to observe the time dependence suggested by Davis (1986). In addition, the interaction cross sections for the direct counting detectors are known better than for ^{127}I.

Bibliographical Notes

Cowan, G.A. and W.C. Haxton (1982) *Science*, **216**, 51. Classical paper proposing the ^{98}Mo experiment and discussing with great insight both the experimental and theoretical aspects.

Freedman, M.S. (1978) in *Proceedings of Informal Conference on Status and Future of Solar Neutrino Research,* edited by G. Friedlander (Brookhaven National Laboratory) Report No. 50879, Vol. 1, p. 313. Fascinating "real time" reaction to disappointing news about neutrino absorption cross sections by the proposer of the ^{205}Tl experiment.

Hurst, G.S., *et al.* (1984) *Phys. Rev. Lett.,* **53**, 116. A convincing proposal for a ^{81}Br experiment.

13

Neutrino–electron
scattering experiments

Summary

Neutrino–electron scattering experiments, which are based upon the reaction

$$\nu + e \rightarrow \nu' + e', \qquad (13.1)$$

provide additional information that is not available from radiochemical experiments.

Scattering experiments can show directly that the detected signal comes from the Sun. The recoil electrons in Eq. (13.1) are primarily scattered in the forward *direction* in which the neutrinos are arriving. If solar neutrinos produce scattering events in a detector, reconstruction of electron tracks must determine a vector that points back to the Sun.

Scattering experiments provide the exact *arrival times* for individual events. This accurate time keeping makes practical precise tests of possible correlations of neutrino events with time-dependent sources.

Neutrino–electron scattering is sensitive to neutrinos of any flavor, although the cross sections for nonelectron type neutrinos are much less than for electron type neutrinos. The absolute number of the events in scattering experiments, when combined with the event rates for absorption experiments that are sensitive only to electron type neutrinos, can determine the *flavor* of the incoming neutrinos.

The energy distribution of the recoil electrons reflects to some extent the *spectrum* of the incident beam. Therefore scattering experiments can distinguish between some explanations of the solar neutrino problem that predict different energy spectra, but the same total event rate in radiochemical experiments.

The Kamiokande II experiment, which is being performed in a mine in the Japanese Alps, is the second solar neutrino experiment to yield a definitive result. Measurements with the Kamiokande II water Cerenkov detector confirm the result, found with the ^{37}Cl detector, that the flux of ^{8}B solar neutrinos is less than predicted by the standard model.

Two new major neutrino–electron scattering experiments, SNO and ICARUS, are being developed as integral parts of programs that will study solar neutrinos by both absorption and scattering techniques. The SNO experiment will use a D_2O (and normal water) detector in a Canadian mine. The ICARUS experiment will use a liquid argon detector in the Gran Sasso Underground Laboratory in Italy. A water detector with somewhat different capabilities, SUN-LAB, is being developed in Australia. A large volume detector, LVD, that measures scintillation light is being installed in the Gran Sasso Laboratory.

Low-temperature detectors are being developed in order to study the fundamental, but difficult to observe, pp neutrinos. The basic principle is that a little energy has a big effect at low temperatures, perhaps making possible in the next decade accurate measurements of the electron recoil spectrum from scattering by pp neutrinos.

The first section of this chapter, 13.1, describes some characteristics of neutrino–electron scattering experiments that use water as a detector. The next two sections discuss experiments that are in progress: the Japanese Kamiokande II experiment (Section 13.2) and the Australian SUNLAB experiment (Section 13.3). Both of these experiments register neutrinos by measuring the Cerenkov light from recoil electrons. Section 13.4 describes the large volume liquid scintillator detector, LVD, that is being placed in the Gran Sasso Laboratory. The last sections, 13.5 to 13.8, discuss next-generation experiments including the large water detector Super Kamiokande (Section 13.5), the heavy water detector, SNO (Section 13.6), the liquid argon time projection chamber, ICARUS, for the Gran Sasso Laboratory (Section 13.7), and the low-temperature bolometric detectors for observing pp neutrinos (Section 13.8).

The expected rate of events in the different detectors are presented without multiplying by the stated detection efficiencies in order to avoid having to revise the rates with each new experimental improvement. The word **ton** is used to represent 10^6 gm.

The theoretical expectations for neutrino–electron scattering are described in Section 8.2, Bahcall (1987), and Bahcall, Gelb, and Rosen (1987). The earliest discussions of neutrino–electron scattering experiments for solar neutrinos were given by Reines and Kropp (1964) and Bahcall (1964).

Some aspects of the theoretical discussion of neutrino–electron scattering are especially important in the following discussion of specific experiments. Since this chapter concentrates on the experiments themselves, the reader may find it useful to know where the relevant theoretical description is located in Chapter 8. The absolute values of the scattering cross sections for ^{8}B neutrinos is about two orders of magnitude less than for absorption cross sections in the best detectors (see Tables 8.2 and 8.6). Sections 8.2B and 8.2C contain tables of the neutrino–electron scattering cross sections for different neutrino sources, flavors, and threshold energies for recoil electrons. The angular distribution of the recoil electrons is discussed in Section 8.2E. The strong forward peaking can be used to enhance the signal to noise ratio in a detector of a given mass and to show that the neutrinos originate from the Sun. The scattering reaction, Eq. (13.1), is sensitive to neutrinos of different flavors, although the cross sections for electron neutrinos to interact with electrons in the detector are typically a factor of seven higher than for neutrinos of other flavors (see Section 8.2B). The shape of the recoil spectrum is relatively insensitive to flavor, but can distinguish neutrino energy spectra that have very different shapes (see Sections 8.2D and 8.2G). Figure 8.8 compares the recoil electron energy spectra for a ^{8}B incident neutrino spectrum and for a hypothetical spectrum that increases linearly with energy; the differences between the recoil spectra are sufficiently large that they can be discerned experimentally.

13.1   Water detectors

The reasons for using water as a detector are simple. Water is cheap and can serve both as the target and as the detector for neutrino–electron scattering. The purest samples of water have low concentrations of radioactive contaminants. Moreover, the detection of recoil electrons by their Čerenkov radiation (see below) is insensitive to low-energy α-particles, making the water detectors relatively immune to low-energy α-decays.

On the other hand, there are also disadvantages in using water as the detector. There is no characteristic signature in neutrino scattering that defines a neutrino event; any incident neutral particle capable of scattering an electron (e.g., a gamma ray) can produce candidate neutrino events. Rare β-decays from radioactive elements can also give rise to false candidates. In order to keep the background at a manageable level, the energy threshold above which recoil electrons are counted must be set relatively high. For the experiments under consideration here, the threshold is expected to be in the range 5 MeV to 8 MeV, depending upon the final background levels that are achieved.[†]

The relatively high-threshold energies for water Cerenkov experiments means that these detectors are sensitive only to ^{8}B neutrinos and, for the most massive targets, hep neutrinos (see Section 6.1) The rates are relatively low because of the small cross sections (see Section 8.2). The rate of scattering events in which electrons with more than E MeV kinetic energy of recoil are counted with 100% efficiency is

$$(\phi\sigma)_{^8\mathrm{B}}\,(> E\ \mathrm{MeV}) = 610\left(\frac{\sigma\,(> E\ \mathrm{MeV})}{10^{-44}\ \mathrm{cm}^2}\right)$$

$$\times\left(\frac{\phi(^8\mathrm{B})}{5.8\times10^6\ \mathrm{cm}^{-2}\ \mathrm{s}^{-1}}\right)\ \mathrm{events}\ \mathrm{kton}^{-1}\,\mathrm{yr}^{-1}. \qquad (13.2)$$

The number of electrons is $N_\mathrm{e} \simeq 3.335 \times 10^{32}$ for a kiloton of water. Per target atom, this rate is low (0.04 SNU). The reason for the low rate is that the cross section for neutrino–electron scattering is small compared to the cross sections for absorption of neutrinos by (favorable) targets.

For a minimum electron recoil energy of 7 MeV, the number of ^{8}B events predicted by the standard model per kiloton year of water

---

[†] All of the above advantages and disadvantages, with the exception of the cheapness of the target material, also apply to the Sudbury deuterium experiment that is described below (Sections 13.5A and 15.1). The deuterium experiment has significant additional advantages, including sensitivity to pure neutral currents and good spectrum discrimination from the absorption process, that compensate for the much greater cost.

observations is 483 if the neutrinos are of the electron type. The rate is about a factor of seven times smaller (see Section 8.2B) if the neutrinos that reach the Earth are of a different flavor (muon or tau neutrinos). The rates are much smaller for hep neutrinos. For example,

$$(\phi\sigma)_{\phi(\text{hep})} (> 7 \text{ MeV})$$

$$= 2.3 \left(\frac{\phi(\text{hep})}{8 \times 10^3 \text{ cm}^{-2} \text{ s}^{-1}} \right) \text{ events kton}^{-1} \text{ yr}^{-1}, \quad (13.3)$$

for electron type neutrinos and a factor of seven smaller for muon or tau neutrinos with the same energy distribution.

The measured rates in scattering experiments depend upon the efficiency of the detector as a function of the energy of the recoil electrons that are produced. Therefore, one must determine accurately for each experiment the energy threshold, E_{th}, the sensitivity of the detector as a function of energy, $\epsilon(E')$, and the resolution function, $\rho(E, E')$. Here $\rho(E, E') \, dE$ is the probability that an electron of true energy E' is measured to have an energy E within an interval dE.

The interpretation of the measured rate is usually made with the aid of an elaborate Monte Carlo program that describes accurately the experimental constraints and features. This method has the advantage of yielding a precise number, but conceals somewhat the dependence of the final answer upon the input quantities and assumptions. In order to understand what is being measured, it is often easier to consider an analytic approximation, described below, to the Monte Carlo studies.

The observed rate can be represented by a double integral which has the form

$$\text{Rate} = \int_{E_{\text{th}}}^{E_{\text{max}}} dE \int_0^\infty dE' \epsilon(E') \rho(E, E') \left(\frac{d\sigma}{dE'} \right). \quad (13.4)$$

Here the quantity $d\sigma/dE'$ is the differential cross section, averaged over the incoming spectrum, for the production of an electron with true relativistic electron energy E'. For many purposes, the resolution function can be represented satisfactorily by a Gaussian,

$$\rho(E, E') = c(E') \times \exp\left[-(E - E')^2/2\sigma^2(E')\right]. \quad (13.5)$$

The standard deviation of the energy measurement is given approximately by

$$\frac{\sigma\left(E\right)}{E} \approx a \left(\frac{10 \text{ MeV}}{E}\right)^{1/2}. \tag{13.6}$$

For relatively high thresholds, one has the following simplifications: $[\sigma(E')/E' \ll 1]$, $c(E') = 1/\sqrt{2\pi}\sigma(E')$. The coefficient a in Eq. (13.6) is ≈ 0.2 for the Kamiokande II detector, giving a 20% energy resolution at 10 MeV.

Charged particles are detected by measuring Cerenkov photons. A charged particle traversing water with velocity larger than c/n, where n is the index of refraction of water (1.344), emits Cerenkov light. The radiation is emitted in a cone of half angle θ, measured from the direction of the particle track, where

$$\cos\theta = (n \times \beta)^{-1}. \tag{13.7}$$

Here $\beta \times c$ is the particle speed. For $\beta = 1.0$, $\theta = 42°$. For relativistic particles, the Cerenkov light amounts to about 340 photons per cm of path length in the water in the wavelength region in which the photomultipliers are sensitive.

13.2   Kamiokande II

The Kamiokande II experiment is the second solar neutrino experiment to yield definitive results, confirming the deficit in ^{8}B neutrinos found with the ^{37}Cl experiment. This experiment is also the first to provide the advantages of ν–e scattering measurements, including accurate times for all events, directionality of the recoil electrons, and some spectral sensitivity. A larger experiment, Super Kamiokande (which is discussed in Section 13.5), will be necessary to make accurate spectral measurements.

The Kamiokande II experiment has been described in a number of informative articles. The technical and scientific details are available in Beier (1987), Nakahata et al. (1986), Suzuki (1986), Hirata et al. (1987a,b), Totsuka (1987), and Nakahata (1988). The earlier phase of the experiment has been described by Arisaka et al. (1985) and Arisaka (1985).

Table 13.1 describes some characteristics of the detector.

Table 13.1.   Characteristics of the Kamiokande II detector.

Location	Kamioka (Japanese Alps)
Depth	1 km (2700 m.w.e.)
Detector fluid	H_2O
Fiducial mass	680 tons
Electron threshold	9 MeV
Number of electrons	2.27×10^{32}
Neutrino sensitivity	^8B
Detectors	20 in. diameter PMT Cerenkov light

Equations (13.3) to (13.5) give an approximate analytic representation of the observed event rate and the theoretical quantities (flux and cross section) if $\epsilon \approx 0.7$.

The laboratory is located 1000 m underground (2700 m of water equivalent) in the Kamioka metal mine of Mitsui Mining and Smelting Co. This mine is situated about 300 km west of Tokyo in the Japanese Alps at 36.4° N, 137.3° E, and 25.8° N geomagnetic latitude. The water detector is contained in a cylindrical tank, 15.6 m in diameter by 16 m in height. The steel walls of the tank are 12 mm thick at the bottom and 4.5 mm thick at the top. The inner surfaces of the tank were painted black with a specially prepared epoxy.

The entire volume of water weighs 3000 metric tons. Only the inner 680 tons is used for solar neutrino experiments because of the stringent background requirements. A larger volume, 2.1 ktons, was used to observe neutrinos from the bright supernova in the Large Magellanic Cloud, SN1987A (see Section 15.5). Large (20 in. diameter) photomultipliers constructed especially for Kamiokande constitute a key element of the experiment; the photomultipliers cover approximately 20% of the entire inner surface of the tank.

The Kamioka mine has several advantages. It is the deepest of all active mines in Japan, sufficiently deep to reduce the cosmic ray background to an acceptable level. The rock is also hard and stable enough to permit excavation of the large cavity needed to house an underground laboratory. Clean water is available from a natural, underground water source near the cavity. Horizontal access is provided by a train that is operated regularly by the mining company.

In addition, members of the Kamiokande collaboration have maintained a good working arrangement with the mine management and staff, a vital nonscientific component of any successful underground experiment.

The etymology of "Kamiokande" reflects important history. The world-famous name "Kamiokande" is derived from the two roots: Kamioka (the mine) and NDE (an acronym for nucleon decay experiment). The detector was originally designed and operated as a facility for studying nucleon decay.

In the Kamiokande facility, charged particles are detected by measuring Cerenkov light [see Eq. (13.7)] with specially constructed photomultipliers. The photomultipliers have several remarkable characteristics including a large area, high photoelectron yield (about three photoelectrons per MeV deposited), and good energy resolution (about 22% at 10 MeV). The detection efficiency is 50% at 7.6 MeV and increases rapidly at higher energies, becoming 90% at 10 MeV. The experimental characteristics are therefore similar to smaller photomultipliers but the detection area is much larger. The photomultipliers were developed for the Kamiokande experiment by the Hamamatsu Photonics Co. in cooperation with members of the collaboration [see Kume et al. (1983)]. Figure 13.1 shows the phototubes being installed in the Kamiokande detector.

The excavation of the laboratory area started early in 1982 and continued until the fall of that year. Subsequently, concrete was sprayed on the entire surface of the main cavity to prevent rock falling. From July 1983 to July 1985, 474 days' exposure time (1.11 kiloton yr) was accumulated. This phase is called Kamiokande I. A number of important scientific investigations were carried out using Kamiokande I data, including a search for nucleon decay into charged leptons plus mesons [Arisaka et al. (1985)], a search for nucleon decays into antineutrinos plus mesons [Kajita et al. (1986)], a search for nucleon decays catalyzed by magnetic monopoles [Kajita et al. (1985)], and a search for high-energy muons from Cygnus X-3 [Oyama et al. (1986)].

In order to detect solar neutrinos, the Kamiokande detector has been steadily improved since late 1984. In 1985, a 4π anticounter layer and a new electronics system were installed. After the installation, the upgraded experiment was called Kamiokande II. The anticounter is also a water Cerenkov detector that surrounds the inner detector with about 1.5 m thickness of water. The upgraded elec-

Figure 13.1 Installation of Kamiokande II phototubes   The figure
shows workers installing the phototubes on the walls of the Kamiokande
water Cerenkov detector in 1983. (Photograph courtesy of Y. Totsuka.)

tronic system records the timing information of Cerenkov photons
for each photomultiplier as well as pulse height information. Figure 13.2 shows workers installing the anticounter in the Kamiokande
II detector.

In the Kamiokande II detector, there are 948 20-in. diameter photomultiplier tubes that are uniformly placed on the inner surface
facing inward. The anticounter, which is a water Cerenkov counter
with a mean thickness of 1.5 m, surrounds the inner detector and
is viewed by 123 photomultipliers. The anticounter layer is useful
for shielding gammas and neutrons that would otherwise enter the
detector, as well as for vetoing cosmic ray muons. The electronics
system was improved for Kamiokande II, with separate time and
charge measurement devices being constructed for each photomul-

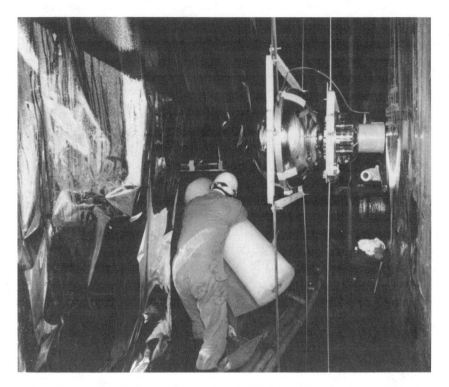

Figure 13.2   Installation of Kamiokande II side anticounters   This figure shows workers installing in 1985 anticounters on the side of the Kamiokande detector in order to make possible low background solar neutrino measurements. The photomultiplier in the foreground of the picture is supported by a steel belt. The workers are standing in an area between the tank and the outer wall that is now filled with water and is used to provide an anticoincidence signal for events that are not caused by solar neutrinos. (Photograph courtesy of Y. Totsuka.)

tiplier. The system records signals larger than 0.35 photoelectron. Currently, the trigger is formed by at least 20 photomultipliers being activated within 100 ns, which corresponds to 7.5 MeV for electrons. The timing information is useful for reconstructing the vertex position, especially for low-energy events. The 1σ error in reconstructing the vertex position of a 10 MeV electron is about 1.7 m. The angular resolution is about $28°$ for a 10 MeV electron. Special ion exchangers have been added to the water purification system. They are useful for removing radioactive impurities, the largest potential backgrounds for recoil electrons below 9 MeV.

In 1985, the principal effort began to reduce the background rate at the relatively low energies required to perform a solar neutrino experiment. With an 8 MeV threshold, the trigger rate was 10^3 counts per second (1 kHz)! The primary source of background was ^{222}Rn contained in the water. The detector was mainly counting β-decay electrons from ^{214}Bi, which is a daughter element of ^{222}Rn. Since the half-life of ^{222}Rn is 3.8 days, sealing of the tank and recirculation of the water eliminated almost all of the ^{222}Rn, and the trigger rate went down to several Hz after a few months, with ^{214}Bi still dominating the rate. Other sources of radioactive background are ^{226}Ra and ^{238}U. The ion exchanger system eliminated the ^{238}U. The trigger rate was reduced to 0.6 Hz in the spring of 1987, of which 0.37 Hz is due to cosmic ray muons. Only about 0.23 Hz is caused by radioactivity, an amount which will decrease with time and further experimental effort.

Gamma rays from the surrounding rock constitute another serious potential background source. Most of these gammas are absorbed in the anticounter layers. The remaining gamma rays have to be reduced by defining a fiducial volume that is farther inside the tank and therefore shielded from the outside source of radioactivity.

The third background source is radioactive fragments produced by cosmic ray muons interacting with ^{16}O nuclei in the water, especially ^{12}N and ^{12}B. These spallation products give off energetic electrons when they decay. The background from radioactive fragments is removed easily off line using their geometrical correlation with parent muons, the relatively short lifetimes of the β-decays, and the large energy deposition of parent muons.

In approximately one year, the low-energy background in the Kamiokande II experiment was reduced by more than a factor of a thousand, an extraordinary achievement. This background reduction made possible a direct experimental confirmation of the solar neutrino problem and the historic, first ever detection of supernova neutrinos.

The energy calibration of the detector is performed using three calibration sources: gamma rays from nickel–neutron reactions; decay electrons from stopping cosmic ray muons; and β-decay electrons from spallation products induced by cosmic ray muons. The absolute energy scale is believed to be known to an accuracy of about

Figure 13.3   Three generations of Kamiokande neutrino experimentalists
From left to right, Y. Totsuka (principal investigator), N. Nakahata (number one solar neutrino student), and Y. Totsuka (founding father). (Photograph in 1987 courtesy of Y. Totsuka.)

3% in the relevant energy region. Details are given in the papers by Hirata *et al.* (1987b), Totsuka (1987), and Nakahata (1988).

Kamiokande II started taking data in December 1985. The first analysis was based upon 128 days of low background data that were accumulated from December 1986 to May 1987. The average trigger rate was 0.7 Hz with 0.37 Hz due to cosmic ray muons. The expected event rate from solar neutrino interactions is 0.3 events/day/680 tons (> 10 MeV). The trigger rate is more than 10^5 larger than the expected signal!

Accurate subtraction of the background is essential; precise timing and track recognition are required. The identification of real ν–e scattering events involves four main steps: selection of low-energy events, making a fiducial volume cut, removing events induced by interactions of cosmic ray muons with ^{16}O, and determining the correlation with the direction of the Sun. The fiducial volume cut eliminates low-energy events that originate near the edge of the detector and that are due to gamma rays or neutrons that enter the target. The fiducial volume is a cylinder of radius 4.0 m and height 5.2 m, which corresponds to a mass of 680 tons. The muon-induced events are removed by making use of empirical relations [for the details of the criteria, see Nakahata (1988)]. The final reduction in the background is obtained by considering the angular distribution of the recoil electrons.

The initial report from the Kamiokande II experiment set a 90% confidence upper limit, $\phi(^8\text{B}) \leq 3.1 \times 10^6$ cm^{-2} s^{-1}, on the flux of ^{8}B solar neutrinos that is incident upon the surface of the Earth is [Totsuka (1987) and Nakahata (1988)]. In terms of the flux, $\phi(^8\text{B})_{\text{SSM}}$, that is computed for the standard solar model (see Table 6.6), the upper limit is

$$\phi(^8\text{B}) \leq 0.55 \, \phi(^8\text{B})_{\text{SSM}}. \tag{13.8}$$

This upper limit is inconsistent, at the 90% confidence level, with the range of values allowed by the recognized uncertainties in the calculation of the flux (see Table 6.5). The Kamiokande II result is of great importance since for two decades all of the observational results on solar neutrinos had come from a single experiment, the ^{37}Cl detector of Davis and his collaborators.

The regeneration of ν_e by passage through the Earth (see Section 9.2E) can cause the day and the night rates to differ. The observed upper limits are [Nakahata (1988)]:

$$\text{Day flux} < 0.80 \, \phi(^8\text{B})_{\text{SSM}}, \tag{13.9a}$$

and

$$\text{Night flux} < 0.64 \, \phi(^8\text{B})_{\text{SSM}}. \tag{13.9b}$$

Here "night" is defined as the period when the cosine of the zenith angle of the Sun is positive. The limit on the night flux alone is also below the calculated total theoretical range of the standard model.

An improved measurement with a lower average background rate and increased statistics yields [Totsuka (1988)] a preliminary detection of the ^{8}B neutrino flux

$$\phi(^8\text{B}) = (0.45 \pm 0.15) \, \phi(^8\text{B})_{\text{SSM}}, \tag{13.10}$$

1σ uncertainty. At the 3σ level of confidence, the measured ^{8}B neutrino flux is different from zero and is inconsistent with the best estimate from the standard solar model. The result given in Eq. (13.10) applies for a minimum total electron energy of 9.3 MeV.

If the uncertainties in the Kamiokande II measurement can be reduced by another factor of two or three, the detector will either confirm or rule out an unambiguous prediction of the most attractive MSW solutions. A value of $0.45\,\phi(^8B)_{SSM}$ for neutrinos above 9 MeV would, if confirmed by more accurate measurements, rule out the adiabatic (i.e., horizontal) solution of the MSW effect (see discussion of Figure 9.6 for the definition of the various MSW solutions). The scattering of neutrinos of nonelectron flavor is about a factor of seven less probable than for electron type neutrinos. A total reduction of about 10 (including theoretical errors) relative to the best estimate standard solar model is the maximum possible in a straightforward MSW explanation.

13.3   SUNLAB

Sydney University is developing a different ν_e–e scattering experiment, appropriately called SUNLAB, in Australia [see Bakich and Peak (1985)]. The detector is located in an underground chamber of the North Mine, which is near the city of Broken Hill, about 1100 km west of Sydney. The North Mine is an active mine producing mainly silver, lead, and zinc. The experimental area is at a depth of 1230 m, about 3300 m of water equivalent, and is well isolated from the main mine workings. The cosmic ray background in the chamber is negligible compared to the radioactive background from the surrounding rock [Bakich et al. (1984)]. Indeed, the residual cosmic ray muons will be useful for calibrating the detector.

The prototype detector module of the SUNLAB experiment was installed in late 1987 in the experimental chamber. The module has a fiducial target volume of approximately 10 tons of pure water and is designed to detect the Cerenkov radiation from electrons that are elastically scattered by solar neutrinos.

Just as for the Kamiokande experiment, the background events originate from a number of sources, but they all finally manifest themselves as gamma rays which will Compton scatter or pair produce electrons in the target, an identical result to the signal being sought. The angular distribution of the recoil electrons is therefore a crucial diagnostic of signal versus background: Recoil electrons produced by solar neutrinos move predominantly in the forward direction of the Earth–Sun axis. Unfortunately, the directionality is severely degraded by the multiple scattering of the low-energy electrons in the water target, as well as by the characteristic angle of emission of the Cerenkov radiation. The resultant photon pattern is spread rather broadly with about 70% of the photons lying within a 50° forward cone. However, directionality is in the end the special characteristic that enables the small neutrino signal to be detected amongst the larger background of similar events.

The basic building blocks of the detection system are the photon collection units. The Cerenkov yield of photons is concentrated towards the ultraviolet. The SUNLAB photomultipliers have an enhanced ultraviolet response to take account of the increased Cerenkov light at short wavelengths. This is an important difference from the photomultipliers used by the Kamiokande II collaboration.

The photomultipliers have been designed to optimize the angular resolution and photon detection efficiency [see Bakich *et al.* (1986)]. Each photon collection unit consists of an EMI 9623B photomultiplier (17.7 cm diameter S-1 photocathode) optically coupled to a 1.25 cm thick Bicron BC-480 acrylic wavelength shifting panel which measures 0.7 m². The edges of the panel are covered by reflecting aluminized mylar tape, which helps retain the converted photons in the panel, while the back of the panel is similarly covered to give an effective double thickness of the panel to ultraviolet Cerenkov photons entering the panel. Each unit is provided with a ballast of about 5 kg of lead to give it approximately neutral buoyancy when submerged in the water target. As well as enlarging the effective size of the photomultiplier, the wavelength shifting panel partially avoids the mismatch between the frequency of the emitted Cerenkov radiation, which peaks in the ultraviolet, and the response of the photocathode, which peaks in the visible. Detailed simulations of the angular response of the photon collection units indicate that a threshold as low as 6 MeV for the scattered electron energy could be attainable.

The module contains 54 photon collection units arranged in an inward facing cube of side 2.12 m (9 units per side). The units are supported by an anodized aluminum frame, and fully enclose a target of approximately 9.53 tons of water. The small length scale was chosen to minimize attenuation of the ultraviolet photons in the water, the most important factor in determining the minimum threshold for electron energies (and hence neutrino energies) that is accessible. Enclosing the detector is a 27,000 liter upright cylindrical fiberglass water tank, 3 m high and 3.4 m in diameter. Preliminary measure-

ments of the radiation background using a small crystal scintillator
have indicated that an optimum shield is 30 cm of Pb [Peak and
Bakich (1985)].

Water is circulated through the tank at the rate of about 6 liters
per minute and will be cooled by a refrigerator to about 5° C. This
has the dual benefits of reducing the photomultiplier noise and in-
hibiting the growth of undesirable organisms in the water. The ambient temperature in the mine is about 30° C.

It is planned to complete the pilot experiment in the next few years
after adding 30 cm of lead shielding that will completely surround the
module, requiring a total of about 300 tons of lead. After a series of
tests of the shielded module, the system will be expanded to a target
mass of at least 250 tons by the addition of identical modules. It
is hoped that the full scale detector will be in operation sometime
after 1990. The event rate predicted by the standard model is about
50 events per year per 100 tons of detector for recoil electrons with
energies in excess of 7 MeV. The event rate would be larger by
more than a factor of two if the threshold were as low as 5 MeV
(see Table 8.6).

13.4 LVD

A different type of ν–e scattering detector is being installed in the
Gran Sasso Laboratory, namely, a large volume detector (LVD) in
which the recoil electrons are observed via the scintillation light they
produce. This detector has the potential advantage of providing a
lower threshold energy than Kamiokande II and a larger number
of target electrons, but has the disadvantage that the direction of
the recoil electrons (and hence the incident flux) is not measurable.
The characteristics of this detector have been described by Bari et
$al.$ (1988).

The basic concept of the LVD equipment is a modular design consisting of 190 essentially identical modules. Each module contains
9.6 tons of liquid scintillator $C_n H_{2n+2} (\bar{n} = 10)$ and 6.7 tons of steel
and is covered on the top and one side by limited streamer chambers
which make up the tracking system. The liquid scintillator will contain of order 6×10^{32} electrons. The total weight of the detector,
including the support structure and the tracking system, is about
3600 tons. Each module is 6.2 m long by 2.2 m wide by 1.5 m high.

The 190 modules are placed in an array that is 40 m long by 12 m wide and reaches a height of 13.2 m at its maximum.

Each module contains eight scintillation counters. The liquid scintillator has a density of 0.8 g cm^{-3} and a light output equivalent to a standard plastic scintillator. The Soviet members of the LVD experiment manufactured the scintillator and the modular counters. Each counter consists of a stainless steel tank, 1 m by 1 m in cross section by 1.5 cm long. This tank is viewed by three 15 cm phototubes and one 5 cm phototube.

Seventy-two of these counters have been constructed by the Russian scientists [see Volvodsky, Dadykin, and Ryazhskaya (1970)] and installed inside the Mont Blanc Laboratory as part of the LVD experiment [see Badino et al. (1984) and Aglietta et al. (1987a,b)]. The laboratory is located in a cavity inside the road tunnel linking Italy and France. Similar detectors have been operating for some years in the Soviet Union: a 100 ton detector at Artemovsk [Beresnev et al. (1979)] and a 330 ton detector in the Baksan Neutrino Observatory in the North Caucasus Mountains [Alekseev et al. (1979)].

The LVD collaboration expects to reach a relatively low-threshold energy which will include most of the ^{8}B and hep neutrinos that the standard model predicts are produced in the Sun, although the threshold will still be above the maximum energy of all of the other neutrino sources discussed in Chapter 6. The trigger threshold at the Mont Blanc installation is 7 MeV, limited by the natural radioactivity in the Mont Blanc Laboratory. A somewhat lower threshold may be achievable in the Gran Sasso Laboratory.

The large volume and relatively low-energy threshold of LVD should lead to a high counting rate, $\simeq 10^2$ to 10^3 detected events per year, if the ^{8}B ν_e flux is not greatly overestimated by the standard model. The energy resolution of individual events is expected to be of order of 20%. Since events produced by solar neutrinos will not have a unique character in the scintillator detector, Monte Carlo simulations must be performed in order to determine at what energies the solar signal will stand out clearly above the background. If the simulations are as encouraging as some members of the collaboration expect, LVD should be able to determine well the spectrum of electron recoils, which can be compared with the distribution predicted by the standard model.

13.5   Super Kamiokande: water

A new underground facility, Super Kamiokande, is under development [Totsuka (1987)] in the same mine, Kamioka, as the operating Kamiokande II experiment. The primary purpose of this much larger detector is to perform a more sensitive nucleon decay experiment, although it will also be a superb astronomical laboratory. Present plans call for the installation of a water Cerenkov experiment with a total mass of 45 kilotons, about half of which can be used for solar neutrino detection.

According to the existing design, the target and detectors will be contained in a large steel tank (41 m high by 38 m diameter) filled with pure water. Approximately 11000 20-in. diameter photomultipliers will be placed uniformly on the whole inner surface, corresponding to approximately 40% photosensitive coverage, a factor of two greater than for the Kamiokande II detector. The increased coverage is expected to improve the energy resolution to 15% at 10 MeV. The total mass of water inside the photomultiplier surface (sensitive volume) will be 22 ktons, more than a full order of magnitude larger than Kamiokande's 0.68 ktons fiducial volume for solar neutrinos. The fiducial volume that will be available for solar neutrino detection with Super Kamiokande cannot be determined with certainty until the backgrounds are measured with the experiment operating, but may be of order 15 to 20 ktons. Anticoincidence layers of at least 2 m thickness of water will cover the sensitive volume. About 500 photomultipliers will view the anticoincidence layers. The angular resolution, including multiple scattering and finite detector resolution, is expected to be of order 40° for electrons of 7 MeV.

The standard solar model predicts, from ^{8}B neutrinos, 7×10^3 events per year with a recoil electron energy about 7 MeV, in a 15 kton detector. The number of predicted hep neutrino events is ~ 35 hep events above 7 MeV per year in 15 kilotons.

13.6   Sudbury neutrino observatory: deuterium

The Sudbury Neutrino Collaboration has proposed building a neutrino detector using 1000 tons of heavy water, D_2O (see references given in Section 14.1). The facility would be located more than 2000 m below ground in the Creighton Mine, near Sudbury, Ontario.

The detector would provide a three-pronged attack on the solar neutrino problem using: neutrino absorption (via the charged current); neutrino disintegration of deuterium (via neutral currents); and neutrino scattering (via charged and neutral currents). Detection of individual scattering (and absorption) events would be made by photomultipliers observing Cerenkov light, just as for the ordinary water detectors (Kamiokande and SUNLAB that were discussed previously). Only the neutrino scattering events will be discussed here; the Sudbury Neutrino Observatory will be described more fully in Section 14.1. The background rate is expected to be exceptionally low. If this low rate is achieved, the SNO detector may be the most sensitive neutrino–electron scattering experiment to be built in the near future.

The standard model predicts [see Eq. (13.2)] about 435 scattering events above 7 MeV recoil energy in a kiloton year of D_2O observations (3.01×10^{32} electrons per kiloton of D_2O) if the ^8B neutrinos that reach the Earth are of the electron flavor. About two events per kiloton year would be expected from hep neutrinos. These numbers should be reduced by a factor of seven for neutrinos of a different flavor.

13.7   ICARUS I: liquid argon

ICARUS I is a new type of detector, optimized for solar neutrino studies, that will be placed in the Gran Sasso Laboratory in Italy by the ICARUS Collaboration (1988), which consists of scientists from CERN, Italy, and the United States. The ICARUS I detector is a liquid argon time-projection chamber, the characteristics of which are described in Section 14.2 and in the proposal by the ICARUS I collaboration [see also Rubbia (1977) and Bahcall et al. (1986)]. The solar neutrino detector will contain a fiducial volume of 200 tons of liquid argon and is the forerunner of a much larger detector that will be used for a variety of low background experiments. The relatively small solar neutrino detector is designed to have very high spatial resolution (±1 mm) and exceptionally good energy resolution (estimated by the collaboration to be about 3% for 1 MeV electrons and gamma rays, improving first with energy and then becoming somewhat worse when intense cascading starts to take place). The expected threshold energy for solar neutrinos is about 5 MeV.

The lower threshold will enable the ICARUS I detector to study the important region near the peak of the ^{8}B energy spectrum (see Section 6.2) that is not accessible to the Kamiokande II experiment. This lower threshold compensates for the fact that the number of electrons in the ICARUS I fiducial volume is about one-fourth the number in the Kamiokande II fiducial volume (there are 2.71×10^{32} electrons per kiloton of liquid argon).

The number of events occurring in the detector is

$(\phi\sigma)_{^8\mathrm{B}}\,(> 5\;\mathrm{MeV})$

$$= 179 \left(\frac{\phi(^8\mathrm{B})}{5.8 \times 10^6\;\mathrm{cm}^{-2}\;\mathrm{s}^{-1}} \right) \text{ events kton}^{-1}\;\mathrm{yr}^{-1}, \quad (13.11)$$

for electron type neutrinos and a factor of seven smaller for muon or tau neutrinos with the same energy distribution. For a ^{8}B neutrino flux one-third of the standard model value of 5.8×10^6 cm^{-2} s^{-1}, the detector should experience about 60 forward peaked electron recoil events per year from ν_e's alone, sufficient to provide valuable information about the solar neutrino spectrum.

13.8   Low-temperature detectors

Several imaginative proposals have been made to detect low-energy solar neutrinos, especially the fundamental pp neutrinos, using low-temperature detectors [see especially Drukier and Stodolsky (1984), Cabrera, Krauss, and Wilczek (1985), Lanou, Maris, and Seidel (1987), and the collection of articles in Pretzl, Schmitz, and Stodolsky (1987)].

The basic principle is that a little energy has a big effect at low temperatures. This work is in an early stage of development, but has great promise and may extend solar neutrino research to real-time spectrum measurements of the pp and ^{7}Be neutrino fluxes. The background problems are difficult because the maximum electron recoil energy is only 260 keV for the electrons produced by scattering of pp neutrinos; many nonneutrino sources can produce recoil electrons of this energy. Much effort is currently devoted to developing the technology to allow the detection of elastic scattering in neutrino–nucleus interactions, which are also low-energy events ($\sim 10^2$ eV) [for summary of this work see Pretzl, Schmitz, and Stodolsky (1987)].

A group at Stanford University is concentrating on silicon detectors that register energy deposited in the form of ballistic phonons

[Cabrera, Krauss, and Wilczek (1985), Cabrera (1986), and Martoff (1987)]. Multikilogram silicon single crystals at a temperature of a few hundred millikelvins would be instrumented to detect the phonons created by electrons recoiling from single elastic scatterings of neutrinos. This scheme can potentially provide good temporal, spatial, and energy resolution.

Direct counting of solar pp neutrinos, interacting in silicon primarily by ν–e scattering, may be possible if several conditions are met [Martoff (1987)]. The silicon must be prepared from inorganic silica originating deep underground. Exposure to cosmic rays during refining must be held to a minimum. In addition, the high-energy background in such detectors must be made much lower than is obtained by scaling existing laboratory results. The required background level may be achieved either by improving the radio purity or by developing new electronic techniques for background rejection.

A group at Brown University [Lanou, Maris, and Seidel (1987)] has proposed using liquid helium as a calorimeter to observe ν–e scattering by incident pp neutrinos, detecting the energy deposited in the helium by the recoil electrons. As far as radioactive background is concerned, liquid helium is the ideal detector material. At low temperatures ($T < 1$ K), all impurities freeze out on the walls of the container. However, the specific heat per unit mass is very large for a helium detector at any reasonable temperature. For example, below 0.5 K the specific heat of helium is $\sim 10^5$ times larger than that of crystalline silicon at the same temperature. Therefore, a conventional calorimetric detection of the recoil energy of the electron is not possible.

Lanou et al. (1987) have suggested an approach which avoids the problem of the large specific heat. The large mass of superfluid helium would be held at a temperature of $\simeq 20$ mK. In neutrino scattering off an atomic electron in the helium, a fraction f of the electron recoil energy, E_{rec}, is converted within a short time into low-energy elementary excitations of the helium, primarily rotons. The detection of these rotons is difficult since the volume of the helium is large. However, rotons in superfluid helium have two remarkable properties which make their detection possible. At temperatures below $\simeq 0.1$ K where the density of thermal excitations is negligible, rotons are stable excitations. Thus, the rotons that are produced will propagate ballistically through the liquid without decay. Second, rotons induce evaporation of helium atoms when they reach the free surface of the liquid. Measurements show [Hope (1984) and Wyatt (1984)] that a roton incident on the liquid surface has a probability of $\simeq \frac{1}{3}$ of inducing the evaporation of a helium atom. The evaporated atoms can be detected by silicon wafers suspended a few millimeters above the helium surface. Lanou *et al.* (1987) suggest that it should be straightforward to measure Δt at 20 mK with a noise signal of less than 0.02 mK with the aid of superconducting bolometers. This would give a determination of the electron energy with an uncertainty of $\simeq 1.5$ keV, assuming a well-defined relation between energy received by the bolometer and the recoil energy of the electron.

A full scale detector of pp neutrinos based upon these physical principles could consist of an underground volume of 10 tons of liquid ^{4}He ($\simeq 70$ m^3), approximately two-thirds being used as a fiducial volume, and the remaining outer section used to establish background (expected to be low-energy gamma rays and neutrons) entering the helium from the outside. The helium would be divided into 10^3 to 10^4 cells, each with its own silicon wafer (or wafers). To prevent rotons from passing from one cell to the next, thin plastic sheets can be used as dividers; thus, apart from the silicon wafers no large mass other than helium need be in the fiducial volume. The rate predicted by the standard model from pp neutrinos is about 8 per day, a healthy signal by the standards of solar neutrino experiments. Moreover, the pp rate is immune to many of the uncertainties that afflict the ^{8}B signal studied in water (and heavy water) Cerenkov detectors.

Bibliographical Notes

Cabrera, B., L.M. Krauss, and F. Wilczek (1985) *Phys. Rev. Lett.*, **55**, 25; Lanou, R.E., H.J. Maris, and G.M. Seidel (1987) *Phys. Rev. Lett.*, **58**, 2498. Two exciting proposals to measure low-energy solar neutrinos.

Drukier, A.L. and L. Stodolsky (1984) *Phys. Rev.*, **30**, 2295. Proposed the use of bolometric detectors for observing low-energy solar neutrinos.

Nakahata, M. (1988) PhD Thesis, *Search for ^{8}B Solar Neutrinos at Kamiokande II*, Dept. of Physics, Univ. of Tokyo (preprint UT-ICEPP-88-01). A beautifully written thesis containing a wealth of information and wisdom about solar neutrino detection and the Kamiokande II experiment. Should have received an A$^+$.

14

Direct counting experiments

Summary

The next generation of solar neutrino experiments will utilize, in most cases, electronic techniques of modern accelerator physics in order to make real-time measurements of the energy spectrum of the recoil electrons, the direction of origin of the incoming beam, and, if current plans are successful, neutrino flavor. Large detectors are being designed to measure neutrino absorption and scattering as well as excitations that are unique to the neutral-current interactions. Absorption reactions permit the measurement of individual neutrino energies, E_ν, using the relation $E_\nu = E_e + \text{constant}$, where E_e is the energy of the electron that is produced and the constant is equal to the difference of initial and final nuclear masses. This simple relation is valid because nuclei absorb momentum but very little energy in capturing solar neutrinos, since nuclei are much heavier than electrons and neutrinos.

A one kiloton detector of heavy water, the SNO observatory, is planned for installation in a mine in Ontario, Canada. The heavy water detector will be capable of measuring the spectrum of ^{8}B neutrinos via charged-current absorption by deuterons and the total number of incident neutrinos irrespective of flavor using the neutral-current disintegration of deuterium. A 0.2 kiloton detector of liquid argon, ICARUS, will be installed in the Gran Sasso Underground Laboratory as the first stage in a program of low background fundamental experiments in physics. ICARUS I will be optimized for solar neutrino studies providing high spatial resolution and low backgrounds. The liquid argon experiment will be a time projection chamber with the ability to measure the spectrum of high-energy ^{8}B neutrinos using absorption to an analogue state in ^{40}K. The subsequent gamma decay of the excited analogue state will provide a unique signature of the absorption. A neutrino "beam-off" test may ultimately be possible by replacing argon with methane. Both the deuterium and the liquid argon experiments can measure the direction of origin of the neutrinos using neutrino–electron scattering (see Sections 13.6 and 13.7).

This chapter reviews the plans for a multipurpose heavy water detector, SNO (Section 14.1), a liquid argon time-projection chamber, ICARUS (Section 14.2), an ^{115}In experiment that can detect pp or ^{7}Be neutrinos (Section 14.3), a many-faceted ^{11}B detector (Section 14.4), ^{19}F, an efficient detector of ^{8}B neutrinos (Section 14.5), and detectors of coherent scattering by nuclei (Section 14.6).

14.1 SNO: D₂O

An international collaboration is developing a heavy water observatory whose primary physics objective is to study ^{8}B and hep neutrinos from the core of the Sun. The Sudbury Neutrino Observatory (SNO) will be four experiments in one setting: neutrino absorption by deuterons (purely charged current), neutrino disintegration of ^{2}H (purely neutral current), neutrino–electron scattering (combination of charged and neutral currents), and, for comparison, neutrino–electron scattering in ordinary water (charged and neutral currents). The detector will contain 1 kiloton of D₂O and will be placed deep underground in the Creighton Mine (owned by INCO Limited), which is located near Sudbury, Ontario. Canada currently has a surplus of heavy water (due to the depressed market in heavy water reactors), making the experiment economically feasible. The event rates, detection efficiencies, and backgrounds in the proposed one kiloton detector have all been extensively studied and show that the proposed experiment is feasible and will provide crucial information about the energy spectrum, the distribution of flavors, and the time dependence of incident solar neutrinos.

The experimental and theoretical aspects of SNO have been described in a number of important and detailed publications including Davidson et al. (1984), Chen (1985), Sinclair et al. (1986), Boivin et al. (1986), Earle et al. (1986), and Aardsma et al. (1987). A com-

Table 14.1. Characteristics of the SNO detector.

Location	Creighton Mine, Sudbury, Ontario
Depth	6800 ft (5900 m.w.e.)
Tank	acrylic vessel
Detector fluid	D_2O
Total weight of fluid	1 kiloton
Threshold	~ 5 MeV
D atoms	6.02×10^{31}
Neutrino sensitivity	^8B, hep

prehensive discussion of all aspects of the proposed observatory is contained in the Sudbury Neutrino Proposal of Ewan *et al.* (1987). Some of the characteristics of the detector are summarized in Table 14.1.

The detector is to be located in a cavity in the Creighton Mine at a depth underground of 2070 m.[†] At this depth, the only surviving cosmic ray components are muons and neutrinos. The muon intensity at the SNO detector site is a factor of 200 lower than that at Kamiokande II. Cosmic ray backgrounds should therefore be unimportant. This site offers the advantages of excellent shielding from cosmic rays, a stable homogeneous rock formation, a manageable level of background from terrestrial radioactivity and easy access due to proximity to an active, fully serviced mining area. The underground laboratory will consist of a cavity of about 20 m in diameter and 32 m high for the D_2O neutrino detector plus rooms for experimental activities.

The detector will consist of 1000 tons of 99.92% pure D_2O in a transparent acrylic vessel surrounded by 5000 tons of H_2O (of which about 2000 tons will be viewed by photomultiplier tubes). This vessel, constructed from 5 cm thick panels, will transmit ultraviolet light and will be surrounded by 4 m of high purity H_2O and 0.9

[†]At 7137 ft the Creighton No. 9 shaft is the second deepest continuous mine shaft in the western world. The INCO management expects to continue mining at and below the 7000 ft level for the next few decades and a high level of support services will be available in the mine. The neutrino detector will be located at the 6800 ft level in the host rock (norite), far enough away from the mining activities that they will neither interfere with the stability of the cavity nor the detector operation.

m of low activity concrete. Mounted uniformly around the acrylic tank, at a distance of 2.5 m from the vessel surface, will be about 2000 or more photomultiplier tubes of 50 cm diameter, providing at least 40% photocathode coverage. The H$_2$O shields the D$_2$O from radioactivity in the surrounding rock and in the photomultipliers. The photomultiplier array is sensitive to Cerenkov light produced by relativistic electrons and muons in the central regions of the detector. The D$_2$O will be supplied on loan by AECL (Atomic Energy of Canada Limited) who must exchange their surplus "low tritium" D$_2$O for virgin Ontario Hydro D$_2$O since the tritium content in the AECL D$_2$O will give an undesirable background in the experiment.

The operation of SNO will parallel the procedures at Kamiokande II (see Section 13.2), although there will be important scientific and technical differences between the two experiments. The depth of the SNO detector will be more than twice that of the Kamiokande II detector, resulting in a much reduced cosmic ray background for the Sudbury experiment. Both experiments use large photomultipliers to detect recoil electrons from neutrino–electron scattering, but the SNO detector will be sensitive in addition to neutrino absorption [see Eq. (14.1) below]. Finally, the SNO detector has a flavor-blind mode that is sensitive to pure neutral-current interactions [see Eq. (14.3) below]. Neutrino–electron scattering, which can be studied in both SNO and Kamiokande II, is sensitive to neutrinos of all flavors, but is typically a factor of seven less sensitive to nonelectron flavors compared to the electron flavor.

The bread and butter reaction in the SNO detector is neutrino absorption by deuterium (d) via the charged current,

$$\nu_e + d \rightarrow p + p + e^-. \tag{14.1}$$

In the original discussions of an ^{2}H solar neutrino experiment, this was the only reaction considered [Jenkins (1966), Bahcall (1966), Kelly and Uberall (1966), Ellis and Bahcall (1968) and Fainberg (1978)]. Given the high expected counting rates (in the thousands per year, see below), this reaction should provide a direct determination of the energy spectrum of the detected ν_e. The incident neutrino energy, E_ν, is related to the energy of the electron, E_e, that is produced by $E_\nu = E_e + 1.442$ MeV. The angular distribution of the recoil electrons with respect to the incident neutrino direction is given by Eq. (8.6a) with $\alpha = -1/3$, which implies a two to one backward to forward asymmetry.

The SNO detector will also be sensitive to neutrino–electron elastic scattering:

$$\nu + e \to \nu' + e'. \tag{14.2}$$

The use of SNO to study neutrino–electron scattering has already been discussed in Section 13.6. The yield from neutrino scattering [Eq. (14.2)] is an order of magnitude smaller than from absorption [Eq. (14.1)]. However, the scattered electrons are strongly peaked in the forward direction and can be separated from absorption events by their angular dependence.

The SNO collaboration will measure the background for the absorption reaction by an initial run with light water in which only neutrino–electron scattering occurs. The signal from neutrino–electron scattering will be strongly peaked in the forward direction; the number of events at large angles (with respect to the solar direction) is a direct measurement of the background that is expected for the absorption process.

The experiment with light water can be viewed as a "beam-off" experiment since the absorption reaction, Eq. (14.1), will not occur with ν_e when the deuterons are replaced by free protons. Free protons absorb $\bar{\nu}_e$, not ν_e.

The total neutrino flux, irrespective of neutrino type, can be measured by the neutral-current reaction,

$$\nu + d \to \nu + p + n, \tag{14.3}$$

which has a threshold of 2.225 MeV. The cross section for reaction Eq. (14.3) is independent of the flavor of the incident neutrino. The signal that deuterium has been disintegrated, Eq. (14.3), is the release of a free neutron, which will produce a gamma ray when captured in the target material. One of the most promising methods for detecting the neutrons is by observing the gamma rays produced by their enhanced capture on ^{35}Cl, which can be introduced into the D_2O in the form of NaCl. Unfortunately, the production of a free neutron is not a very specific signal and the SNO collaboration has had to devise clever methods for ensuring that only a small number of background events can produce neutrons.

Chen (1985) and Weinberg (1987) have stressed that the neutral-current reaction is particularly important for resolving the solar neutrino problem because it gives a flavor-blind measure of the total neutrino flux, providing a test of solar models even if neutrino oscillations occur. The ratio of the number of absorption events to neutral-current disintegrations provides a constraint on neutrino propagation characteristics, independent of the value of the total solar neutrino flux. Fortunately, the ratio of the neutral current to the charged-current cross sections has been calculated to an accuracy of $\pm 1/2\%$ [see Bahcall, Kubodera, and Nozawa (1988)].

The ^{2}H detector is sensitive only to ^{8}B and hep neutrinos. The predicted capture rate depends upon the adopted minimum recoil energy of the electrons that are produced, although not very strongly (see Table 8.3). For the plausible threshold of 5 MeV kinetic energy, the standard model predicts [Bahcall and Ulrich (1988)]:

$$\sum_i \phi_i \sigma_i = 6.01 \,(1 \pm 0.38) \text{ SNU}, \qquad (14.4)$$

of which all but 0.02 SNU comes from ^{8}B neutrinos. The calculated rate would increase only by 11% if the threshold were lowered to 3 MeV.

The cross section for neutrino absorption on ^{2}H can be calculated accurately (see Section 8.1D). Therefore, almost all of the uncertainty in the predicted capture rate is from the 37% uncertainty in the estimated ^{8}B neutrino flux (see Chapter 6, especially Table 6.5).

How many events are expected to occur in the one kiloton detector of heavy water? Assuming a threshold kinetic energy of 5 MeV for detecting recoil electrons and that nothing happens to the neutrinos from the time they are created until they reach the Earth,

$$(\text{Rate})_{^8\text{B}} \,(> 5 \text{ MeV})$$
$$= 1.2 \times 10^4 \,[\phi(^8\text{B})/6 \times 10^6 \text{ cm}^{-2} \text{ s}^{-1}] \text{ events kiloton}^{-1} \text{ yr}^{-1}. (14.5)$$

For the same threshold, the number of neutrino–electron scattering events is about 9% as large as the absorption rate, that is, with the same assumptions approximately 1.0×10^3 scattering events should occur in the one kiloton detector per year for the standard model flux.

If a significant fraction of the higher-energy ^{8}B neutrinos reach the Earth unaffected, one can measure well the shape of the ^{8}B neutrino spectrum from neutrino absorption, Eq. (14.1). If the MSW effect converts most of the higher-energy electron neutrinos into neutrinos of a different flavor, then the largest event rate will be from neutral-current disintegration of the deuteron.

The SNO collaboration expects to reach a sensitivity to ^{8}B neutrinos of 4×10^4 cm^{-2} s^{-1} in approximately one kiloton year of operations, a factor of 10^2 less than the flux predicted by the standard model.

In the initial stages of the experiment, before the NaCl is added to the D_2O to facilitate neutron detection, what is the *minimum* event rate that might be expected? If all the ^8B neutrinos are converted to a different flavor, then the absorption process will not occur and only neutrino–electron scattering will be observed. In this pessimistic case, the expected rate is reduced to 1.5×10^2 scattering events per kiloton per year (see Table 8.6 for the appropriate scattering cross section). This rate can be considered as a plausible lower limit to be expected in the proposed deuterium experiment provided only that the standard solar model is not disastrously incorrect.

Will the SNO collaboration be able to detect unambiguously the hep neutrinos? For the standard solar model predictions, the answer is "yes" provided that a significant fraction of the higher-energy solar neutrinos do not change their flavor on the way to the Earth and that the re-measurement of the thermal neutron cross section on ^3He does not lead (see Chapter 3) to a much reduced cross section for the production of these highest-energy neutrinos. Consider a threshold electron recoil energy that is sufficiently large that the expected number of events from hep neutrinos exceeds by a big factor the expected number of events from ^8B neutrinos. One can find such a threshold because the absorption cross sections depend sensitively, for a given neutrino energy spectrum, on the difference between the minimum accepted recoil energy and the maximum energy of the neutrino spectrum (see Table 8.3); the hep neutrinos have a larger end-point energy than do the ^8B neutrinos. For a minimum electron recoil kinetic energy of 13.5 MeV, hep neutrinos are expected to cause many more events than do ^8B neutrinos, assuming the correctness of the cross section factor, S_0, given in row 10 of Table 3.2,[†] the neutrino fluxes from the standard model given in Table 6.5, and the neutrino absorption cross sections given in Table 8.3. In this relatively clean region of the recoil spectrum, the standard solar model predicts 10 events per kiloton per year from absorption of hep neutrinos and

---

[†]The cross section value calculated by Werntz and Brennan (1973) has been used here. The cross section for the hep reaction is approximately proportional to the radiative capture cross section for thermal neutrons by ^3He [see Tegnér and Bargholtz (1983)]. The thermal neutron capture experiment should be repeated with greater accuracy.

2 events from the absorption of ^{8}B neutrinos.[†] Given the above assumptions, neutrino–electron scattering is expected to be too rare to be detectable in this energy range.

Suppose the standard solar model is disastrously wrong and that the flux of neutrinos from ^{8}B decay that is produced in the Sun is less than the calculated value by orders of magnitude. Will one still be able to observe the hep neutrinos? Yes, if the diminution is not caused by something happening to the ^{8}B neutrinos after they are produced. About 43 absorption events and 9 scattering events with electron recoil energies above 5 MeV are calculated to occur per kiloton per year in a heavy water detector exposed to the flux of hep neutrinos given by the standard solar model. Since the hep flux is relatively insensitive to changes in the solar model, one can be confident that a total of about 50 hep events should occur per kiloton year with electron recoil energies above 5 MeV unless MSW effects or some other weak-interaction process removes the high-energy neutrinos of electron flavor.

The rate of neutral-current disintegrations of deuterium on the basis of the standard model is [Bahcall, Kubodera, and Nozawa (1988)]

$$\sum_i (\phi_i \sigma_i) = 2.4(1 \pm 0.38) \text{ SNU.} \qquad (14.6)$$

Almost all of the neutral-current events are caused by ^{8}B neutrinos. The total number of neutral-current interactions is $4.5(1 \pm 0.38) \times 10^3$ events per kiloton year of observation.

14.2 ICARUS: ^{40}Ar

A 0.2 kiloton liquid argon detector will be placed in the Gran Sasso Underground Laboratory in northern Italy. This detector, known as ICARUS I, is optimized to observe solar neutrinos and will be installed sometime in 1990 if current plans are realized [see ICARUS Collaboration (1988)]. ICARUS I will be the initial stage of an underground laboratory, ICARUS (from Imaging Cosmic And Rare Underground Signals), which is designed to perform a variety of low background experiments in fundamental physics. The ultimate

---

[†]The ^{37}Cl and Kamiokande experiments show that the actual contamination by ^{8}B neutrinos will be at least a factor of two less than the above estimate, which is based on fluxes from the standard solar model.

ICARUS detector is envisioned to have a mass of several kilotons, an analyzing magnetic field, and the possibility of replacing argon partially or totally with methane.

The ICARUS detectors are radically different from the water detectors (heavy and light) that use photomultipliers to identify neutrinos via the Cerenkov light of the recoil electrons. Instead, the recoil electrons will be detected with ICARUS using a liquid argon time-projection chamber, as first proposed for neutrino observations by Rubbia (1977).

The ICARUS detector will form a three-dimensional electronic image by drifting electrons in a homogeneous electric field onto a read-out plane, where the charge is recorded. The arrival positions of the drifting electrons on the read-out plane give two spatial coordinates of points on the original track. The drift time determines the third coordinate. The detector will have good energy and angular resolution, and real-time analysis capability. The fundamental requirement for the liquid argon imaging detector is the ability to drift ionization electrons over large distance. Tests carried out at CERN have established the feasibility of the ICARUS I detector [see ICARUS Collaboration (1988) and Rubbia et al. (1985)]. Experimental feasibility was demonstrated previously on smaller samples of liquid argon [see, e.g., Mahler, Doe, and Chen (1983) and Giboni (1984)]. Background effects may require further study.

The larger-mass ICARUS detector can be used to measure, via neutrino absorption, the incident spectrum of the higher-energy ^{8}B and hep neutrinos. The dominant absorption transition is from the ground state of the ^{40}Ar nucleus to the analogue excited state in ^{40}K [see Bahcall et al. (1986)],

$$\nu_e + {}^{40}\text{Ar} \rightarrow e^- + {}^{40}\text{K}^*, \quad E_{\text{th}} = 5.885 \text{ MeV}. \quad (14.7)$$

Since the analogue state lies at an excitation energy of 4.38 MeV[†] and the difference in nuclear masses is 1.505 MeV, the effective threshold for absorption is rather high, 5.9 MeV$(E_\nu = E_e + 5.9 \text{ MeV})$.

The angular distribution of the recoil electrons from the reaction in Eq. (14.7) is slowly varying [see Eq. (8.6a)]:

$$P(\theta) = \left[1 + \left(\frac{v_e}{c}\right)\cos\theta\right], \quad (14.8)$$

[†]The importance of this transition was emphasized by Raghavan (1979).

where v_e is the recoil velocity of the electron and θ is the angle between the incident neutrino direction and the momentum of the recoil electron that is created. This broad angular distribution is different from the sharply peaked forward distribution that results from neutrino–electron scattering.

The analogue state of ^{40}K decays by emitting a 2.09 MeV gamma ray (branching ratio 76%) or a 1.65 MeV gamma, followed by characteristic lower-energy gammas. The total gamma ray energy is equal to the excitation energy of the analogue state, 4.38 MeV. In about 65% of the events, the recoil electron is emitted in coincidence with a photon cascade of total energy equal to 2.74 MeV that is followed in delayed coincidence by a single 1.644 MeV photon. In most cases, the photons are expected to produce a cluster of Compton tracks well removed from that of the recoil electron, yielding the electron energy without ambiguity. The maximum energy of a Compton track from any one of the gamma rays is less than 2.1 MeV. Raghavan (1986) concludes that an unambiguous identification of the prompt electron track and extraction of the value of the neutrino energy is assured for neutrino energies greater than 8 MeV. A modest spatial and time resolution is thus sufficient [see Raghavan (1986)] to define the coincidence (or delayed) events, separate them from background, characterize the ν_e capture, and measure the energy of the incident neutrino.

The gamma decays represent the "smoking gun" that indicates that a neutrino absorption to the analogue state has occurred.

The large mass ICARUS detector will permit a decisive test of the experiment. One can perform an effective "beam-off" test of the absorption process [Bahcall *et al.* (1986)] by replacing the liquid argon in the detector by methane. The threshold for neutrino absorption by ^{12}C is 17.3 MeV, well beyond the nominal (14 MeV) endpoint energy of ^{8}B neutrinos. An explicit calculation using the detailed ^{8}B neutrino spectrum of Bahcall and Holstein (1986) shows that the absorption cross section is negligibly small ($\ll 10^{-47}$ cm^2) even when the broad character of the ^{8}Be final state is included. Also, the cross section for absorption of ^{8}B neutrinos by ^{13}C is 7×10^{-43} cm^2 for $W_{cutoff} = 5$ MeV. Practically no solar neutrino absorption events would be produced in the time-projection chamber if the liquid argon were replaced by methane.

For the absorption process, the replacement of argon by methane would be equivalent to the Sun ceasing to shine in neutrinos.

For typical values of the minimum counted recoil electron energy, $T_{\min}$, the sum of $E_{\mathrm{th}} + T_{\min}$ is well beyond the peak, 6.4 MeV of the energy spectrum of the ^{8}B neutrinos. The predicted capture rate therefore depends sensitively (see Table 8.3) upon the adopted threshold, $T_{\min}$. For the plausible threshold of 5 MeV kinetic energy, the standard model predicts [Bahcall and Ulrich (1988)]:

$$\sum_i \phi_i \sigma_i = 1.70 \, (1 \pm 0.38) \ \mathrm{SNU}, \qquad (14.9)$$

of which all but 0.02 SNU comes from ^{8}B neutrinos. The cross section for neutrino absorption on ^{40}Ar can be calculated accurately (see Section 8.1D). Therefore, almost all of the uncertainty in the predicted capture rate is from the 37% uncertainty in the estimated neutrino flux.

The rate of absorption events predicted by the standard model from ^{8}B neutrinos is

$$(\phi\sigma)_{^8\mathrm{B}} \, (> 5 \, \mathrm{MeV})$$
$$= 831 \, [\phi(^8\mathrm{B})/6 \times 10^6 \ \mathrm{cm}^{-2} \ \mathrm{s}^{-1}] \ \mathrm{events \ kiloton}^{-1} \ \mathrm{yr}^{-1}. \quad (14.10)$$

If the observed flux is one-third of the standard model value, about 55 events per year should occur in the ICARUS I detector. After one or two years of operation, the ICARUS I results should be sufficient for making a significant statement about the shape of the incident neutrino spectrum.

The relatively high threshold for the dominant mode of neutrino absorption causes the number of neutrino–electron scattering events to exceed the number of absorption events for most reasonable values of the minimum accepted electron recoil energy [Bahcall *et al.* (1986)]. For $T_{\min} = 5$ MeV, the rate of neutrino–electron scattering events is about 11% larger than the absorption rate, that is, approximately 3×10^3 scattering events should occur in a 3 kiloton detector per year. The ratio of scattering to absorption events increases rapidly with increasing $T_{\min}$; the ratio is about 13.5 for the relatively moderate value of $T_{\min} = 8$ MeV.

Section 13.7 discusses ICARUS as a detector of solar neutrinos using neutrino–electron scattering.

What happens if some interaction process (e.g., the MSW effect, see Section 9.2) transforms the flavor of solar neutrinos? Assuming complete flavor conversion of all of the ^{8}B neutrinos, the expected scattering rate is reduced, for $T_{\min} = 5$ MeV, by a factor of 6.64

(see Table 8.6) to about 140 scattering events per kiloton per year. This value may be regarded as a plausible lower limit for the expected event rate provided that the standard solar model is not disastrously in error.

Hep neutrinos will produce a unique signature in the ICARUS detector. For a minimum electron recoil energy of 9 MeV, the expected number of absorption events per kiloton year is 3.4 from hep neutrinos and only 0.8 from ^{8}B neutrinos. Thus one expects an observable, although not large, signal from hep neutrinos in the ultimate ICARUS detector provided that the highest-energy solar neutrinos do not change their flavor on the way to the Earth and that the thermal neutron cross section on ^{3}He has approximately the value given in Table 3.1. Even if the standard solar model is off by an order of magnitude or more in the flux of the rare ^{8}B neutrinos, one may still expect to observe the much less model-sensitive flux of hep neutrinos. The standard model predicts about seven hep absorption events per kiloton year with electron recoil kinetic energies above 5 MeV.

14.3 ^{115}In

Two major projects are under way to develop an ^{115}In solar neutrino experiment [see Booth, Salmon, and Hunkin (1985), Bellefon, Espigat, and Waysand (1985), Booth (1987), Booth et al. (1987), and Evetts et al. (1988)]. This experiment presents major technological challenges with a big payoff, the measurement of the energy spectrum of the incident pp and/or ^{7}Be neutrinos. The current work is inspired by the pioneering papers of Raghavan and his collaborators [Raghavan (1976, 1978a, 1981) and Pfeiffer, Mills, and Chandross (1978)].

The reaction that registers the neutrinos is

$$\nu_e + {}^{115}\text{In} \rightarrow {}^{115}\text{Sn}^{**} + e^-, \quad E_{\text{th}} = 0.119 \text{ MeV}. \quad (14.11)$$

Neutrino absorption causes in most cases (see below) a Gamow–Teller transition from the $9/2^+$ ground state of ^{115}In to the $7/2^+$ second excited state of ^{115}Sn at an excitation energy of 0.61 MeV. The effective threshold for this reaction is 119 keV. The angular distribution of the recoil electrons is given by Eq. (8.6) with $\alpha = -1/3$.

The isotope ^{115}In is naturally radioactive, decaying with the emission of an electron of maximum energy 485 keV to ^{115}Sn with a half-life of 4×10^{14} yr. This radioactive decay is responsible for the largest background in the experiment.

The 0.61 MeV excited state of ^{115}Sn that is populated by neutrino absorption lives for about 3 μs and then emits two gamma rays in coincidence with energies of 115 keV and 498 keV. In about one-half of the captures a 115 keV gamma ray is internally converted, emitting a 90 keV electron plus characteristic X-rays. Thus, the unique signature for neutrino detection is a pulse from the electron, followed by, on average 3 μs later, two coincident pulses, one spatially very close to the electron with energy 115 keV and the second of energy 498 keV.

The basic problem to be overcome in an ^{115}In experiment is to reject the high background from the natural β-decay to ^{115}Sn. The ratio of the number of natural β-decays to the number of solar neutrino absorptions is enormous,

$$\frac{\text{Natural radioactivity}}{\text{Solar neutrinos}} = 6 \times 10^6. \tag{14.12}$$

Eighty percent of the β's from the radioactive decay overlap the energy range of electrons from neutrinos from pp fusion. Therefore one must make full use of the characteristic delayed time-coincidence signature of ^{115}In reaction. The signature can be recognized above background by dividing the detector into 10^4 or more segments, by having moderately good time resolution between the two gamma rays and by having good energy resolution [see Evetts *et al.* (1988) and references therein as well as Raghavan (1978a)].

Different experimental configurations have been explored including an indium-loaded liquid scintillator [Pfeiffer *et al.* (1978)] and superconducting samples of indium [see Booth *et al.* (1985), Bellefon *et al.* (1985), and Evetts *et al.* (1988)].

Drukier and Nest (1985) proposed an experiment in which the ^{7}Be and higher-energy neutrinos are detected and the pp neutrinos are not. Adopting this reduced goal would eliminate much of the background since the recoil electrons from ^{7}Be have energies in excess of the most intense natural background. This idea should be considered seriously because the ^{7}Be neutrinos cannot be observed separately in any of the other experiments that are being developed. The detection of the ^{7}Be neutrinos, which produce a healthy capture rate (116 SNU) in ^{115}In (see Table 14.2 below), would provide important diagnostic information about nuclear reactions in the Sun and about

Table 14.2. Capture rates
predicted by the standard solar
model for an ^{115}In detector.

Neutrino source	Capture rate (SNU)
pp	468
pep	8.1
hep	0.05
^{7}Be	116
^{8}B	14.4
^{13}N	13.6
^{15}O	18.5
^{17}F	0.2
Total	639 SNU

the propagation characteristics of neutrinos of a much lower energy
than the ^{8}B neutrinos.

The capture rate predicted by the standard model [Bahcall and
Ulrich (1988)] for an ^{115}In detector is

$$\sum_i \phi_i \sigma_i = 639^{+640}_{-321} \text{ SNU.} \tag{14.13}$$

The contributions of different neutrino sources that are predicted
by the standard model are shown in Table 14.2. The basic pp reaction supplies about 73% of the capture rate predicted by the standard
solar model, the largest percentage for any of the targets we have
discussed in this paper.

Nearly all of the theoretical uncertainties are small for an ^{115}In
detector. The uncertainties from all sources other than neutrino
absorption cross sections amount to only 28 SNU. The total uncertainty in the predictions is dominated therefore by the uncertainty
in the neutrino absorption cross sections, which at present must be
estimated with the aid of (p,n) reactions.

All of the allowed absorption transitions are to excited states [see
Bahcall (1978) for a discussion of the difficulties of calculating the
capture rates for ^{115}In]. The best estimate furnished by the (p,n)
experiments [Rapaport et al. (1985)] for the matrix element to the

lowest excited state at 0.61 MeV in ^{115}Sn is in remarkably good agreement with an insightful estimate made by Raghavan (1976) in his original paper on this subject. However, the available (p,n) measurements for all of the relevant transitions have large measurement errors, 1σ uncertainties of order 30%. In addition, the calibration is uncertain (see Section 7.3B) Thus the total theoretical range in the predicted capture rate for this detector due to the neutrino cross sections is about a factor of two.

Calibration of the detection sensitivity using a radioactive source is essential if the full potential of the experiment is to be realized, since the uncertainties in the absorption cross sections are much larger than any of the astrophysical uncertainties. Fortunately, a calibration experiment with either a ^{51}Cr or an ^{37}Ar source would provide exactly the required information. Either source of laboratory neutrinos would excite only the 0.61 MeV level of ^{115}Sn in the allowed approximation. The pp and ^{7}Be neutrinos, which according to the standard solar neutrino spectrum provide about 91% of the expected capture rate in a ^{115}In detector, also excite only the 0.61 MeV level. The neutrino sources that provide the remaining 9% of the expected capture rate only occasionally excite levels above the 0.61 MeV state. The total contribution of states above 0.61 MeV to the capture rate is primarily from ^{8}B neutrinos and is calculated to be of order 2% for the standard solar neutrino spectrum. A measurement of the ^{51}Cr or ^{37}Ar neutrino absorption cross section on ^{115}In would also provide a valuable test of the validity of the (p,n) method for determining GT matrix elements. Bahcall and Ulrich (1988) estimate a cross section of

$$\sigma(^{51}\text{Cr on }^{115}\text{In}) = 222^{+222}_{-111} \times 10^{-46} \text{ cm}^2, \qquad (14.14)$$

using the (p,n) data of Rapaport *et al.* (1985). A similar calculation for an ^{37}Ar source yields a cross section of

$$\sigma(^{37}\text{Ar on }^{115}\text{In}) = 257^{+257}_{-129} \times 10^{-46} \text{ cm}^2. \qquad (14.15)$$

An ^{115}In experiment can provide unique and important diagnostic information provided that the neutrino absorption cross sections can be accurately determined. A calibration experiment with ^{51}Cr or ^{37}Ar is the most direct and potentially accurate method of determining the cross section.

14.4   BOREX: ^{11}B

A ^{11}B target could provide the opportunity of performing several simultaneous experiments with ^{8}B solar neutrinos [Raghavan, Pakvasa, and Brown (1986) and Raghavan and Pakvasa (1988)]. In the same detector, one can study: neutral-current excitations to three excited nuclear states in ^{11}B and charged-current absorptions to four nuclear levels in the mirror nucleus ^{11}C.

The neutral-current transitions were estimated [Raghavan et al. 1987] from shell-model calculations and electromagnetic excitation experiments. The all-flavor excitations will be signified by monoenergetic gamma rays of 4.5 MeV and 5 MeV.

The charged-current transitions can determine the energy spectrum of the electron-flavor ^{8}B neutrinos. The thresholds for charged-current reactions range from 2 MeV to 7 MeV, thereby providing different samplings of the electron-flavor ^{8}B neutrino spectrum. The ratios of corresponding neutral-current and charged-current transitions should indicate the relation between the spectrum of electron-flavor neutrinos to the spectrum of all neutrinos (which contribute to the neutral-current excitations).

Raghavan et al. (1988) have suggested a conceptual design for a ^{11}B solar neutrino experiment. They propose using a boron loaded liquid scintillator which would be viewed by a large number ($\sim 10^3$) of phototubes. A steel tank would contain some 2 kilotons of borated liquid scintillator.

The recoil electrons will be detected by their scintillator light, as in LVD (see Section 13.3D), instead of their Cerenkov light, as in Kamiokande II and SNO. The radio purity of the scintillator will determine the extent to which the physics goals of the experiment can be realized in practice. If very low levels of radio purity can be achieved, it may be possible to observe the lower-energy solar neutrinos from ^{7}Be electron capture (see Chapter 6). An international collaboration of American and Italian scientists, BOREX, has been formed to investigate the extent to which the scientific goals outlined above can be achieved.

14.5 ^{19}F

The possibility of using ^{19}F to make an accurate measurement of of the incident energy spectrum of ν_e from ^{8}B solar neutrinos has

Table 14.3   Absorption cross sections for ^{8}B and hep neutrinos incident on ^{19}F. The cross sections are given for different values of the minimum accepted kinetic energy, T_{min}, of the recoil electrons. The unit for T_{min} is MeV and the unit for the cross sections, σ, is 10^{-42} cm^2.

T_{min}	$\sigma(^8\text{B})$	$\sigma(\text{hep})$	T_{min}	$\sigma(^8\text{B})$	$\sigma(\text{hep})$
0.0	1.825	4.77	8.0	0.36	3.29
1.0	1.82	4.77	9.0	0.15	2.73
2.0	1.78	4.75	10.0	0.035	2.11
3.0	1.69	4.70	11.0	0.0032	1.48
4.0	1.52	4.60	11.5	0.0007	1.19
5.0	1.27	4.41	12.0	0.0001	0.92
6.0	0.97	4.14	13.0	0.0000	0.46
7.0	0.65	3.77	13.5	0.0000	0.29

been considered by groups in the Soviet Union and in the United States [Barabanov, Domagatsky, and Zatsepin (1989) and Robertson (1986, 1988)]. Fluorine has several advantages as a neutrino target which justify further study of the of the feasibility of using ^{19}F in a next-generation detector. One attractive feature is that ^{19}F has a large and accurately known neutrino capture cross section ($\sigma_0 = 69.9 \times 10^{-46}$ cm^2, see Section 8.1B) because the ground states of ^{19}Ne and ^{19}F are isotopic analogues.[†] Other advantages are that fluorine has only one stable isotope, is inexpensive compared to some of the proposed neutrino targets, and could be incorporated into scintillators as the major component. A detailed study of the practicality of constructing a fluorine neutrino detector would be highly desirable.

The dominant absorption transition is superallowed and occurs between the ground states of ^{19}F and ^{19}Ne. The reaction is

$$\nu_e + {}^{19}\text{F} \rightarrow e^- + {}^{19}\text{Ne}, \quad E_{th} = 3.238 \text{ MeV}. \tag{14.16}$$

[†]An allowed Gamow–Teller transition can occur to the $3/2^+$ state of ^{19}Ne at an excitation energy of 1.536 MeV. This transition has a small cross section factor of $\sigma_0 = 0.230 \times 10^{-46}$cm^2 and contributes only 0.1% of the capture rate for ^{8}B and hep neutrinos.

The product of neutrino absorption, ^{19}Ne, is radioactive with a lifetime of 19 s. The decay of ^{19}Ne by positron emission is correlated in time and in position (within a scintillating target) with the absorption event that produces a prompt energetic electron; the combination of prompt electron production and delayed decay positron emission could be used to reject many troublesome backgrounds. The correlation of delayed decay with prompt electrons constitutes a characteristic signal that could be used to identify events induced by neutrinos. It seems likely that a ^{19}F detector could be developed which would determine the energy spectrum of ν_e above 4 MeV or 5 MeV incident neutrino energy, providing a lower neutrino threshold than most of the proposed neutrino experiments.

The angular distribution of the recoil electrons from the reaction in Eq. (14.16) is essentially isotropic [see Eq. (8.6a)]:

$$P(\theta) = \left[1 + 0.04\left(\frac{v_e}{c}\right)\cos\theta\right], \qquad (14.17)$$

where v_e is the recoil velocity of the electron and θ is the angle between the incident neutrino direction and the momentum of the recoil electron that is created.

For convenience in evaluating the feasibility of a fluorine detector, Table 14.3 gives the absorption cross sections for ^{8}B and hep neutrinos. The cross sections given in Table 14.3 correspond to a high event rate, 10^4 events per kiloton of fluorine per year for the standard model flux of ^{8}B electron neutrinos. The detector could also be used as an efficient telescope to observe neutrinos from stellar collapse, as indicated by the cross sections for individual neutrino energies which are given in Table 14.4.

14.6 Coherent scattering detectors

Several European and American groups are involved in ambitious programs to use the coherent scattering of neutrinos by nuclei to develop sensitive detectors of solar neutrinos and dark matter [see, e.g., Drukier and Stodolsky (1984), Cabrera, Krauss, and Wilczek (1985), Martoff (1987), Pretzl, Schmitz, and Stodolsky (1987), and Sadoulet et al. (1988)]. This process can be represented by the reaction

$$\nu + A \rightarrow \nu + A. \qquad (14.18)$$

Table 14.4   Absorption cross sections for neutrinos of specific energies incident on ^{19}F. The unit for neutrino energy, q, is MeV and for neutrino cross sections is 10^{-42} cm^2. The cross sections refer to transitions between the ground states of ^{19}F and ^{19}Ne, which are isotopic analogue states.

q	σ	q	σ	q	σ	q	σ	q	σ
4.0	0.12	8.0	2.21	12.0	6.77	16.0	13.7	25.0	37.2
5.0	0.41	9.0	3.12	13.0	8.29	18.0	18.0	30.0	54.2
6.0	0.85	10.0	4.18	14.0	9.95	20.0	22.9		
7.0	1.45	11.0	5.40	15.0	11.8	22.0	28.2		

The cross section for coherent scattering is approximately proportional to N^2, the square of the total number of neutrons in the target nucleus (see Section 8.3A). In coherent scattering all of the neutrons act together. For practical cases, the cross section can be a factor of 10^3 larger than for the corresponding neutrino absorption and neutrino–electron scattering processes. Because the energy deposition by Eq. (14.18) is small, most of the experimental proposals are intended to be carried out at low temperatures.

Coherent scattering is equally sensitive to all neutrinos described by the standard electroweak model, providing another method of achieving a flavor-blind neutrino detector (in addition to the disintegration of deuterium and the excitation of ^{11}B).

The rate of coherent scattering per target nucleus is much larger than the rates for the incoherent processes discussed previously (neutrino absorption, neutrino–electron scattering). The large cross section implies that relatively small quantities of material ($\gtrsim 0.1$ kiloton) can be used for a solar neutrino experiment; the possibilities of using different medium or heavy nuclei are being explored. The detectors could be calibrated conveniently using relatively small targets that would detect antineutrinos from a reactor. The major difficulty in constructing such a neutral-current detector is that the observable signal is nuclear recoil, which has a low-energy (typically keV or lower, hence the low temperature of the suggested detectors) and is relatively unspecific (which means that the background must be understood and kept low). In the original proposal to observe coherent scattering [Drukier and Stodolsky (1984)] the change of state of a superconducting grain would signal the occurrence of a neutrino-induced nuclear recoil. In this case, the neutrino event could be distinguished from many of the background events because one, and only one, grain would change its state.

In principle, coherent scattering could be used to detect all of the solar neutrino sources. In practice, the higher-energy ^{8}B neutrinos may be the easiest to detect since both the coherent scattering cross section and the magnitude of the nuclear recoil are proportional to the square of the neutrino energy.

Bibliographical Notes

Booth, N.E. (1987) *Sci. Prog. Oxf.*, **71**, 563. An excellent summary of progress being made toward an ^{115}In experiment.

Ewan, G.T., *et al.* (1987) in *Sudbury Neutrino Observatory Proposal* (Sudbury Neutrino Observatory Collaboration: Queen's University at Kinston) Pub. No. SNO-87-12. Detailed and convincing discussion of solar neutrino experiments and theoretical issues.

Fainberg, A.M. (1978) in *Proceedings of Informal Conference on Status and Future of Solar Neutrino Research,* edited by G. Friedlander (Brookhaven National Laboratory) Report No. 50879, Vol. 2, p. 93. Discussed solar neutrino absorption experiment with D_2O using Cerenkov detectors.

Jenkins, T.L. (1966) *A Proposed Experiment for the Detection of Solar Neutrinos,* Case University proposal (unpublished). This experiment was attempted using 2 kiloliters of D_2O. The recoil electrons were to be detected by their Cerenkov light, just as in the modern experiments. The Case experiment was abandoned because of the high background that was encountered. The detector was located only 610 m deep in a salt mine. [For some details of this early experiment, see Reines (1967).]

Raghavan, R.S. (1976) *Phys. Rev. Lett.*, **37**, 259. Original proposal of an ^{115}In experiment. A masterpiece of experimental creativity.

Raghavan, R.S. (1979) Bell Laboratory Technical Report TM #79-1131-31. Insightful proposal to use a liquid argon detector for observation of neutrinos from meson factories and collapsing stars.

Rubbia, C. (1977) EP Internal Report 77-8, CERN. Suggested the basic features of a liquid argon time projection chamber to be used as a novel and massive detector of neutrinos.

15

Stellar collapse

Summary

Old massive stars do not fade away; they collapse with a bang. Stars evolve peacefully and quasistatically for millions of years, passing through recognizable stages of development, until they finally run out of nuclear fuel and can no longer support themselves. When this happens, the core of the star collapses under its own weight in less than a second. In the final stages of the collapse, the stars are expected to radiate nearly all of their binding energy in the form of neutrinos (and antineutrinos) of different flavors. This impulsive release of neutrinos can give rise to a detectable pulse in neutrino detectors on Earth, as was first observed in the astronomical Event of the Decade, Supernova 1987A.

Numerical calculations of great precision and complexity are required in order to simulate what happens during a stellar collapse. The physical conditions that are relevant are extreme; the stellar density in the core may exceed nuclear densities and the temperatures may rise above a billion degrees. Unlike main-sequence evolution, which is characterized by moderate changes on time scales of order a billion years, the dramatic changes that occur in a stellar collapse occur in less time than it takes to blink an eye. Nonequilibrium processes are crucial, with shock waves playing an important role. Of necessity, the existing dynamical simulations of stellar collapse and explosions use approximate descriptions of matter at nuclear densities and usually neglect the role of convection, magnetic fields, rotation, and departures from spherical symmetry.

The detailed results of the stellar collapse calculations for neutrinos can be summarized by a few observable quantities. Simple physical arguments are sufficient to derive approximate numerical values of the total energy radiated, the neutrino temperatures, and the duration of the radiation pulse.

By extraordinarily good fortune, two proton decay detectors (IMB and Kamiokande II) were upgraded just in time to permit observations of SN1987A, which exploded in the nearby galaxy, the Large Magellanic Cloud. One of the detectors, Kamiokande II, was made sensitive to solar neutrinos, which have approximately the same energies as supernova neutrinos. The energies of solar and supernova neutrinos are much lower than the energies that are relevant for studying nucleon decay.

Preexisting calculations are in agreement with the observations of neutrinos from SN1987A, representing a great triumph for 50 years of astrophysical theory and speculation. However, important aspects of the theory could not be tested because of the small number (~ 20) of neutrinos that were detected. Most of the energy is believed to have been emitted in higher temperature muon and tau neutrinos and antineutrinos, but only $\bar{\nu}_e$ were detected with certainty. Solar neutrino experiments that are in progress could make more diagnostic measurements if a stellar collapse were to occur in the Galaxy while the detectors are operating.

Stellar collapses may occur without the optical fireworks that announce a supernova. The collapse rate exceeds, by an unknown but possibly large factor, the collapse rate of observable optical supernova. Only long-term neutrino monitoring can determine the rate at which neutrino-producing stellar collapses occur.

Observations of neutrinos from stellar collapses in other galaxies are more difficult and, for nearby galaxies, less frequent than observations of collapses in the Galaxy.

The detection of neutrinos from SN1987A led to important inferences about neutrino properties. New limits were obtained on the mass, charge, magnetic moment, decay rate, limiting velocity, and geodesics of electron neutrinos, as well as on the total number of neutrino flavors.

The first section of this chapter, 15.1, outlines the general ideas of stellar collapse, relying heavily on the qualitative description of

Bethe and Brown (1985). The next section, 15.2, summarizes the results of numerical studies by different theoretical groups. Table 15.1 gives the parameters for neutrino emission from a "standard stellar collapse." Section 15.3 estimates the rate of occurrence of "neutrino bombs," stellar collapses without the emission of a large amount of light (photons), in the Galaxy. The following section, 15.4, summarizes the expected rate and fluences (number per cm^2) for neutrinos emitted from stellar collapses in other galaxies in the Local Group. Section 15.5 describes the Event of the Decade, the detection of neutrinos from supernova SN1987A. The first parts of Section 15.5 describe the observations by the Kamiokande II, IMB, and Baksan neutrino detectors and summarize the phenomenological analyses that yield, among other quantities, an estimated neutrino temperature, total energy, and cooling time. The last part of Section 15.5 discusses limits that can be placed on neutrino properties from the observations of SN1987A. The concluding section, 15.6, presents the estimated number of events that would be observed, by the planned solar neutrino detectors, from a standard stellar collapse in the Galaxy.

The physics of a stellar collapse is described in a clear and nontechnical fashion by Bethe and Brown (1985) and Burrows (1987a). There are also excellent technical summaries by a number of authors including Arnett (1977), Bowers and Wilson (1982), Nomoto (1984), Bruenn (1985), Burrows and Lattimer (1986), Woosley and Weaver (1986), Bethe (1986), Mayle, Wilson, and Schramm (1987), and Cooperstein (1988b) (see also the proceedings of the NATO Advanced Study Institute, *Supernova: A Survey of Current Research*, on current supernova research, listed in the Bibliographical Notes at the end of this chapter).

15.1   How a star gets into trouble

During nearly all of the lifetime of a star, the heat given off by the sequence of nuclear fusion reactions creates a thermal pressure, which counteracts the gravitational attraction that would otherwise make the star collapse. Previous chapters in this book have discussed only the first series of fusion reactions in which four protons are converted to an α-particle, accompanied by two positrons and two electron-type neutrinos [see Eq. (1.3)]. This proton burning is energetically favorable: The mass of the helium atom is slightly less than the combined masses of the four hydrogen atoms, and the energy equivalent of the excess mass is released as heat.

Proton burning continues until the hydrogen in the stellar core is used up. The core then contracts, since gravitation is no longer opposed by energy production, and as a result both the core and the surrounding material are heated. Hydrogen fusion then begins in the surrounding layers. Meanwhile the core becomes hot enough to ignite other fusion reactions, burning helium to form carbon, then burning the carbon to form neon, oxygen, and finally silicon. Each of these reactions releases energy. The last cycle of fusion combines silicon nuclei to form iron, specifically the common iron isotope ^{56}Fe. Iron is the final stage for spontaneous fusion. The ^{56}Fe nucleus is the most strongly bound of all nuclei, and further fusion would absorb, rather than release, energy.[†]

The star has an onion-like structure at this stage. A core of iron and related elements is surrounded by a shell of silicon and sulfur, and beyond this are shells of oxygen, neon, carbon, and helium. The outer envelope is mostly hydrogen.

Only the largest stars proceed all the way to the final iron-core stage of the evolutionary sequence. A star the size of the Sun gets no further than helium burning, and the smallest stars stop with hydrogen fusion. A larger star also consumes its stock of fuel much sooner, even though there is initially more material to burn. Because the internal pressure and temperature are higher in a larger star, the fuel burns faster. The calculated main-sequence lifetime of the Sun is about 10^{10} yr, whereas a star 10 times as massive can complete its evolution in only a few times 10^7 yr. Regardless of how long it takes, all the usable fuel in the core will eventually be exhausted. At that point heat production in the core ends and the star must contract.

When fusion ends in a small star, the star slowly shrinks, becoming a **white dwarf**: a burned-out star, supported against gravity by the degeneracy pressure of electrons, that emits only a faint glow of radiation.

Stellar collapses that emit observable amounts of neutrinos, and especially **Type II supernovae**, are believed to arise from the collapse

---

[†]The importance of ^{56}Fe as an end product of nuclear fusion has been dramatically illustrated by the light curve of SN1987A after the first few months, which shows a beautiful fit to the decay lifetime for ^{56}Ni $\rightarrow$ ^{56}Co $\rightarrow$ ^{56}Fe.

of more massive stars. The lower limit is now thought to be of order eight solar masses.

In the final day before a Type II supernova explodes, the fusion of silicon to form iron first becomes possible at the center of the star. At this point the star has already passed through successive stages of burning hydrogen, helium, carbon, oxygen, and neon and has taken on an onion-like structure with the innermost regions containing the heaviest nuclei and the outer surface containing virgin hydrogen. Fusion continues at the boundary between the iron core and the silicon shell, steadily adding mass to the core. Within the core, however, there is no longer any production of energy by nuclear reactions; the core is an inert sphere under great pressure. A typical massive star takes several million years from its birth on the main sequence until it has developed an iron core.

The mass of the inner core containing mostly iron-like elements is near the **Chandrasekhar mass limit**, the maximum mass a star can have and support itself against gravity by the degeneracy pressure of electrons. Numerically, $M_{Ch} = 5.8\, Y_e^2 M_\odot \approx 1.4 M_\odot$ with $Y_e \approx Z/A$ the electron ratio per baryon. The temperature at the center of the core is ≤ 0.7 MeV. The core is supported by the Fermi pressure of the highly degenerate electron gas, $E_F \approx 8$ MeV, and its highest density is near 10^{10} g cm^{-3}, some four orders of magnitude below that of normal nuclear matter.

The core is now compressed by its own gravity. This compression raises the central temperature, which might be expected to raise the pressure and slow the collapse. However, at these extreme temperatures, the heating has the opposite effect.

Pressure is determined by two factors: the number of particles in a system and their average energy. In the stellar core, electrons, nuclei, and photons contribute to the pressure. When the core is heated, a fraction of the iron nuclei are broken up into smaller nuclei, increasing the number of nuclear particles and raising the nuclear component of the pressure. At the same time, however, the dissociation of the nuclei absorbs energy from the photons; since energy is released when an iron nucleus is formed, the same quantity of energy must be supplied in order to break the nucleus apart. The loss in pressure that is required to disassociate the iron nuclei is greater than the gain in nuclear pressure. The net result is that the collapse accelerates.

The core begins to collapse because of the net pressure loss due to the photodisintegration of iron [see Burbidge *et al.* (1957)]. The collapse is accelerated by the pressure drop due to rapid capture of free electrons on nuclei and on free protons [as suggested by Gamow and Schönberg (1941)]. Recently calculated presupernova models are destabilized primarily by electron capture, rather than photodissociation (which dominated in the models of several years ago). The speed of collapse increases to an appreciable fraction of free-fall, densities rise, nuclei merge into nuclear matter, which, because of the electron capture, contains approximately two neutrons for every proton. The collapse is halted by the resistance of nuclear matter to further compression; the core would collapse directly into a black hole if nuclear matter were not sufficiently stiff.

The clash of incoming matter with outgoing matter at the moment of bounce back generates an outward moving shock which ultimately produces the visible fireworks that we call a "supernova."

The roughly equal partition between different light neutrino and antineutrino species is easy to understand. A **neutron star**, formed as a result of stellar collapse, consists primarily of neutrons at close to nuclear density and is supported by the degeneracy pressure of the neutrons. The degenerate star has about 2×10^{57} baryons. Since the matter begins as one-half neutrons and one-half protons and winds up as neutron matter, somewhat less than 10^{57} ν_e's must be emitted to shed the lepton number. Even for a generous assumed average energy of 15 MeV, this amounts to only about 25×10^{51} ergs, or about 10% of the total energy emitted. Most of these ν_e's diffuse out from the center of the star, where they have degeneracy energies ~ 150 MeV. Thus, their degradation in energy as they escape from the star produces a cascade of about 10 pair neutrinos for each ν_e that carries away one unit of lepton number.

The earliest neutrinos produced in the collapse are believed to be entirely electron capture neutrinos. There are too few positrons available to produce any significant amount of electron antineutrinos and the temperatures are much too low to produce thermally any neutrinos of other flavors. These early ν_e arise both from electron captures on free protons and captures on protons bound in nuclei.

Neutrinos are produced continuously once the star reaches a density of about 10^{10} g cm^{-3}, but the capture rates increase rapidly once the density has risen close to $\rho_{\text{trap}} \sim 10^{12}$ g cm^{-3}, where the neutrino mean free path becomes much smaller than the size of the inner core of the star.

The total energy released in neutrinos during the infall stage depends upon details of the presupernova model and upon the physics employed in the simulation, but is generally estimated to be of order of 1% of the binding energy of the neutron star, with a typical energy of 10 MeV [see, e.g., Bowers and Wilson (1982) and Bruenn (1985, 1986)].

Since the infall energy fraction is small, it will not be easy to identify unambiguously events associated with this stage. However, these infall neutrinos have a unique characteristic: They are all ν_e (no other flavors are produced in abundance at this stage). Thus solar neutrino detectors that register neutrino absorption in real time, such as the SNO and ICARUS detectors described in Sections 14.1 and 14.2, may highlight the initial burst of infall neutrinos.

Many of the details of what occurs subsequent to the infall are not well understood because of the intertwining of weak interaction, nuclear, hydrodynamical, and radiation physics that occurs during the collapse and explosion. There are numerous and fundamental complications including the effects of the nuclear equation of state, the great variety of nuclear and weak interaction processes that occur, the incompletely understood roles of convection, magnetic fields, rotation and departures from spherical symmetry, as well as the technical difficulties of following all of the exchanges of energy and momentum that occur as the neutrinos interact with matter on their way out of the stellar core.

Fortunately, many of these complications are not crucial, as long as we do not demand high accuracy in our theoretical expectations.[†] The general expectations for detectable neutrino emission are independent of the most difficult physics. Different theoretical groups predict similar observable neutrino fluxes.

The next section concentrates on a simplified quantitative description of the neutrino emission from a stellar collapse that leads to a neutron star. These results can be used as a basis for evaluating what one might hope to see with solar neutrino and other underground experiments.

[†] The complications are decisive for the question of how the stellar envelope is blown off in a supernova explosion.

15.2   Neutrino emission from a standard stellar collapse

Most of the binding energy that is released when a neutron star is formed is believed to be emitted in the form of neutrinos. The calculated binding energy for a neutron star is, in order of magnitude, about 10% of the gravitational mass of the star. The precise value depends upon this mass and the nuclear equation of state that is assumed. Observations and theory both suggest that the mass of the residue star is between about $1M_\odot$ and $2M_\odot$ [see, e.g., Bahcall (1978), Taylor and Weisberg (1982), Shapiro and Teukolsky (1983), and Joss and Rappaport (1984)]. For this range of assumed gravitational masses, the total binding energy of the neutron star lies in the conservative range [see Arnett and Bowers (1977) and Cooperstein (1988a)]

$$E_b = (2.5 \pm 1.5) \times 10^{53} \text{ ergs}, \qquad (15.1)$$

which is several hundred times the energy the Sun will emit in its entire main-sequence lifetime (10^{10} yr).

What should we expect to see when this huge amount of energy is released in neutrinos? What are the characteristics of the emission that will determine the event rates in neutrino detectors?

Many different parameters are required to describe the neutrino emission. For each of the six types of neutrinos (three flavors, neutrinos and antineutrinos), there is a function which describes the shape of the energy spectrum as well as an absolute value that fixes the total intensity of that type: altogether six unknown functions and six absolute values. The shapes of the spectra are established by the complicated physics of the explosion and are affected by nonequilibrium as well as equilibrium processes, including nuclear, weak interaction, hydrodynamical, and radiative aspects of the production and transport of the neutrinos. Different authors calculate somewhat different shapes depending upon the numerical scheme that is used and the physics that is included. By contrast, the shape of solar neutrino spectra in the standard solar model are well determined by laboratory nuclear physics (see Section 6.3).

The detailed results of stellar collapse calculations may be summarized approximately in a few physically meaningful quantities. [For state-of-the-art calculations of neutrino emission from stellar collapse, published before the observations of SN1987A see, e.g., Burrows and Lattimer (1986), Woosley, Wilson, and Mayle (1986), and Mayle, Wilson, and Schramm (1987).] The total number of events that will be observed in an operating detector is determined by the **fluences** of each type of neutrino, the number per unit area received during the entire supernova pulse, as well as by their typical energies. The neutrinos that result from a stellar collapse may be divided crudely into two classes: the initial neutrinos produced dynamically in the collapse and early rebound phases (duration $\sim$ 10 ms) and the much more numerous neutrinos produced in longer thermal cooling (of order seconds). The neutrinos from the dynamical phase contain only a small fraction, ϵ, of the total binding energy of the star. Most published models yield $\epsilon \simeq 0.01$. Nearly all of the remaining fraction of the binding energy $(1 - \epsilon)$ is emitted in neutrinos that thermally cool the star over a period of seconds. The shape of the energy spectra may be represented approximately by a Boltzmann distribution with a fixed temperature T,

$$\frac{\mathrm{d}F}{\mathrm{d}q} = \frac{F_{\text{total}}}{2T^3} \, q^2 \exp\left(-q/T\right) \mathrm{d}q, \tag{15.2}$$

where $F(q)$ is the fluence (number per cm^2) as a function of the neutrino energy q. The distribution is an adequate approximation to the overall spectrum shape computed from detailed numerical models of supernova collapse. Some authors prefer to represent the energy spectrum by a Fermi–Dirac distribution. The detailed shape of the spectrum cannot be calculated with confidence since the result depends upon physical processes, such as neutrino convection [see Burrows (1987a)], that are not well understood. For simplicity, the spectra are described in this chapter by Boltzmann distributions.

While the star is cooling, the neutrino temperatures decrease somewhat, typically by of order 20% over a period of 5 to 10 seconds. To the accuracy that is required for approximate estimates of counting rates in future experiments, the temperatures can be assumed to be constant.

For fluences of neutrinos that are produced in the dominant thermal phase, there is considerable agreement among different groups that calculate numerically stellar collapses. The results can be parameterized as

$$F\left(\nu_j, T_j\right) = \frac{f_j \left(1 - \epsilon\right) E_{\text{b}}}{12\pi D^2 \, kT_j} \, , \quad j = 1, \ldots, 6, \tag{15.3}$$

where f_j is the fraction of the (noninfall) binding energy that is carried off by the six types of neutrinos (three flavors of neutrinos and antineutrinos) and D is the distance of the collapsing star.

For the dynamical phase, $1 - \epsilon$ in Eq. (15.3) is replaced by ϵ, with essentially all of the energy being emitted in the form of ν_e's. As a rough approximation, the expected event rates in different detectors can be estimated by assuming that the temperature of ν_e in the dynamical phase is about the same as in the cooling phase.

The simplest approximation to the detailed numerical calculations suggests that each type of neutrino carries away from the thermal cooling phase approximately the same amount of energy. Within the likely accuracy of future experiments, one expects that the temperature of ν_e and $\bar{\nu}_e$ are about the same, $T(\nu_e)$, and that the temperatures of all the nonelectron-flavor neutrinos ($j = 3, ..., 6$) are about the same, $T(\nu_\mu)$. Thus one can make plausible estimates of expected event rates in neutrino detectors by assuming

$$f_{\text{cool},j} \equiv \frac{1}{6} \; , \; T_{\text{cool}}(\nu_e) \equiv T_{\text{cool}}(\bar{\nu}_e) \; ,$$

and

$$T_{\text{cool},j} \equiv T_{\text{cool},\mu} \; , \; j = \nu_\mu, ..., \bar{\nu}_\tau. \tag{15.4}$$

Expressing the fluences in terms of characteristic values of the parameters for a galactic supernova (see Sections 15.3 and 15.4), one may write

$$F_{\text{cool}}(\nu_j)$$

$$= 1.5 \times 10^{11} \; \text{cm}^{-2} \left(\frac{10 \; \text{kpc}}{D}\right)^2 \left(\frac{5 \; \text{MeV}}{T_{\text{cool},j}}\right) \left(6 f_{\text{cool},j}\right). \tag{15.5}$$

The much smaller flux of neutrinos emitted during the dynamical phase is

$$F(\nu_e)_{\text{dyn}} = 9 \times 10^9 \; \text{cm}^{-2} \left(\frac{10 \; \text{kpc}}{D}\right)^2 \left(\frac{5 \; \text{MeV}}{T_{\text{dynamic},j}}\right). \tag{15.6}$$

Table 15.1 summarizes the main features of the neutrino emission from the simplified model outlined above for stellar collapse.

There is a simple physical argument [see Schramm (1987) and Dar (1988a)] that gives a useful numerical estimate for the temperature of antineutrinos of the electron type, which are most easily observed in the existing water detectors. The mean free path for neutrino

Table 15.1.   Parameters of a stellar collapse.

Binding energy (E_{b})	$(2.5 \pm 1.5) \times 10^{53}$ ergs
Infall energy released in ν_{e}	$0.01 E_{\mathrm{b}}$
Average energy of infall ν_{e}	10 MeV
Energy released in cooling	$0.99 E_{\mathrm{b}}/6$
by each of six neutrino types	
$T_{\mathrm{cool}}\,(\nu_{\mathrm{e}}) = T_{\mathrm{cool}}\,(\bar{\nu}_{\mathrm{e}})$	5 MeV
$T_{\mathrm{cool}}\,(\mu) = T_{\mathrm{cool}}(\tau)$	10 MeV
Δ_{cool}	1 to 10 s$'$

absorption at the temperature and density at which the $\bar{\nu}_{\mathrm{e}}$'s finally escape is defined to be some fraction α of the neutron star radius R, that is,

$$\frac{1}{n_{\mathrm{p}}\sigma_{\mathrm{abs}}} \equiv \alpha R, \qquad (15.7)$$

where n_{p} is the proton number density. The main source of opacity for these $\bar{\nu}_{\mathrm{e}}$'s is absorption by protons,

$$\bar{\nu}_{\mathrm{e}} + \mathrm{p} \to \mathrm{n} + \mathrm{e}^{+}. \qquad (15.8)$$

The cross section for $\bar{\nu}_{\mathrm{e}}$ absorption is[†]

$$\sigma\,(\bar{\nu}_{\mathrm{e}} + \mathrm{p}) \;=\; 2.23 \times 10^{-44}\ \mathrm{cm}^{2}\,(pW/m_{\mathrm{e}}c^{3})\,, \qquad (15.9a)$$

where p, W are the momentum and energy of the positron that is created [cf., Eq. (15.8)]. In more conventional units, this result is

$$\sigma\,(\bar{\nu}_{\mathrm{e}} + \mathrm{p}) \;=\; 8.5 \times 10^{-42}\ \mathrm{cm}^{2}\,(W/10\ \mathrm{MeV})^{2}\,, \qquad (15.9b)$$

for positron energies much greater than an MeV. Here the incident neutrino energy, q, is related to the observed positron energy via

$$q \;=\; 1.293\ \mathrm{MeV} + W. \qquad (15.9c)$$

Averaging the absorption cross section over a Boltzmann thermal spectrum, one obtains

$$\sigma = 12\sigma_{0}\,(T/10\ \mathrm{MeV})^{2}\,, \qquad (15.9d)$$

[†]The numerical coefficients in Eqs. (15.9a) and (15.9b) assumes that $G_{\mathrm{A}}/G_{\mathrm{V}} = 1.245$ [see Bopp et al. (1986)].

where $\sigma_0 = 8.5 \times 10^{-42}$ cm^2. In the region of the star from which the $\bar{\nu}_e$'s escape to the outer world, the relation between the matter density and temperature is $\rho \simeq 10^{12}$ g cm^{-3} $(T/4.46 \text{ MeV})^3$. By definition, the proton number density is $Y_p \rho N_A$, where N_A is Avogadro's number. Thus $n_p \sigma \propto T^5$. Inserting typical values for the different variables, one obtains

$$T \cong 4.8 \text{ MeV} \times \left[\left(\frac{0.1}{\alpha}\right)\left(\frac{20 \text{ km}}{R}\right)\left(\frac{0.3}{Y_p}\right)^{1/5}\right]. \qquad (15.10)$$

The main lesson to be learned from Eq. (15.10) is that a simple physical argument gives approximately the same temperature for the $\bar{\nu}_e$'s as is obtained by detailed calculations. One also sees from Eq. (15.10) that the final inferred temperature is insensitive to the input parameters because the temperature depends on the 1/5th power of the indicated numerical parameters.[†] A factor of two change in one of the parameters only changes the inferred temperature by about 15%. A similar argument yields a temperature for the ν_μ and ν_τ neutrinos of $\sim$ 7 MeV if one uses in the previous equation the applicable cross sections for neutral-current scattering of neutrinos by nucleons [see Dar (1988b)].

The time scale, Δt, over which the neutrinos are emitted is determined by the diffusion time out of the interior of the collapsed star. The neutrino opacity is sufficiently large [see Bahcall (1964)] that the neutrinos are trapped inside the neutron star. The observed duration of the antineutrino pulse extends over many collapse times because the neutrinos trapped inside the star and only gradually diffuse outward [see, e.g., Mazurek (1974), Sato (1975), Arnett (1977, 1987), Burrows, Mazurek, and Lattimer (1981), and Burrows and Lattimer (1986)]. For typical numerical models,

$$\Delta t \cong 1 \text{ to } 10 \text{ s}. \qquad (15.11)$$

Burrows (1987b) has performed some calculations of collapses of massive stars that lead directly to black holes and has shown that the total emission can be somewhat less ($\sim$ 50% to 90% of the binding energy) than when neutron stars are formed ($\sim$ 99% of the binding

---

[†]The exponent found in detailed models may differ somewhat from the 1/5th power, depending upon the exact dependence of the density upon temperature near ρ_{esc}.

energy). In the simulations of black hole formation which Burrows has carried out, the neutrino emission is terminated rather abruptly after a few seconds.

15.3 Neutrino bombs: how often and where?

There are two crucial questions that need to be answered in order to plan experiments to detect neutrinos from stellar collapses. These questions are: (1) At what distance from the Sun is the next collapse likely to be? (2) How often do stellar collapses occur? A definite answer, with acceptable reliability, can be given to the first question. The second question is more difficult and more important, since different experimental approaches are preferred if it is known that the next stellar collapse will occur within a year instead of within one hundred years. Unfortunately, the best guess may be somewhere between these two extremes, with an uncomfortably large uncertainty.

Most of the published answers to the questions of how often and where stellar collapses occur are based upon the assumption that stellar collapse and supernova production are synonymous. This may not be the case: Many stellar collapses may lead to only modest optical emission while the neutrinos carry away essentially all of the binding energy. Over the years, theoreticians have had great difficulty in making stellar models explode, even when they found collapses with enormous amounts of neutrino emission. Perhaps, the problem is not in their computer codes. Maybe only a small fraction of the stellar deaths that occur lead to visible explosions with enough optical fireworks to be called supernovae.

Two recent and respected estimates for the rate of galactic Type II supernova place the value in the range of 1 every 30 years [see Tammann (1982)] or 1 every 80 years [see van den Bergh, McClure, and Evans (1987)]. In an earlier estimate, Tammann (1982) suggests that the expected time between supernovae events in the Galaxy could be about 25 years and may be as short as 10 years if the Galaxy is a spiral of type Sbc. These estimates are based upon observations of the rate of occurrence of supernovae in other galaxies; the extragalactic rates embody large corrections to account for the uncertain effects of obscuration (dust and gas could hide many supernovae) and of viewing angle (supernovae are easier to discover in face on spiral galaxies in which the light does not have to traverse much interstellar medium in order to escape). Also, the inferred rate for the Galaxy depends upon the square of the uncertain value of the Hubble constant, H_0^2, since the galactic rate is computed by normalizing observations of other galaxies per unit of luminosity (which depends upon their distance and therefore H_0). Finally, the probability that a supernova has a specified luminosity is not known. Even systematic statistical surveys, of which there are few, cannot properly correct their inferred rates to take account of optically faint (but neutrino bright) stellar collapses, since the relative numbers of bright and faint supernovae are unknown. Altogether, the uncertainties in estimating the galactic supernova rate from observations made on other galaxies amount to at least a factor of a few.

The rate of stellar collapses can also be estimated from the rate of formation of galactic pulsars, which are rotating, magnetized neutron stars. In the most recent systematic study, Narayan (1987) found a mean time, P, between the formation of neutron stars which, depending upon the adopted model was:

$$20\,\text{yr} \lesssim P \lesssim 60\,\text{yr}, \tag{15.12}$$

with a preferred value of 56 yr. The highest formation rate for pulsars that has been inferred from a systematic analysis of complete surveys is one every six years [see Taylor and Manchester (1977)]. The rate of stellar collapses that is inferred from the observed characteristics of pulsars is subject to many important selection effects, including the fraction of neutron stars that are are observable pulsars (which depends in an unknown way upon pulsar period) and the periods with which pulsars are born.

A different method of calculating the rate of stellar collapses makes use of the known distribution of stars in the Galaxy and the relatively well understood rate at which they are evolving. This calculation avoids some of the uncertainties involved in extragalactic supernova searches, for example, the value of H_0 is irrelevant, and there are no corrections to be made for viewing angle or for optically faint supernova. Unfortunately, there are different uncertainties.

The stellar death rate (the number of stellar collapses per unit of time in the Galaxy), $D(M_{\text{min}})$, can be written [Bahcall and Piran (1983)]:

$$D(M_{\text{min}}) = N \int_{M_{\text{min}}}^{\infty} \frac{\text{d}M \; q[R(M)]f(M)\Phi(M)}{T(M)}. \tag{15.13}$$

Here M_{min} is the minimum stellar mass at which stellar collapse occurs with a majority of the binding energy of the remnant being emitted in neutrinos. This mass must be determined from theoretical calculations and remains one of the major uncertainties in the estimate, although there is surprising agreement currently on the numerical value of M_{min}. The function $q(R)$ is the fraction of stars in the galactic disk that lie within a distance R of the Sun, N is the total number of disk stars ($\sim 7 \times 10^{10}$), $f(M)$ is the fraction of stars with mass M that collapse, $\Phi(M)\,dM$ is the probability that a disk star has a mass between M and $M + dM$, and $T(M)$ is the evolutionary lifetime of stars of mass M. In using Eq. (15.13), it is assumed that the distribution of galactic stars has remained approximately constant for the lifetimes of the stars (typically 10^7 to 10^8 yr) that give the greatest contribution to the integral.

The fraction $q(R)$ gives, in this model, the probability distribution for the distances of stellar collapses. Table 15.2 gives the numerical values for $q(R)$ that are predicted by the Bahcall and Soneira (1980) Galaxy model. This phenomenological Galaxy model summarizes what is known about the population of stars in the Galaxy on the basis of star counts and optical spectra. Calculations with the model are in agreement with observations of stars made in many different directions, colors, and brightness ranges [see Bahcall (1986)].

The mean distance expected for a galactic supernova, which corresponds to $q = 0.5$, is about 10 kpc, slightly larger than the distance to the galactic center. This result answers the first question posed at the beginning of this section.

Evolutionary lifetimes for stars, $T(M)$, have been computed using many different numerical codes. Bahcall and Piran (1983) give convenient formulae that summarize the published values.

The shape of the mass function, $\Phi(M)$, is not accurately known for the masses of interest here ($\sim 5M_\odot$ to $15M_\odot$). The function used by Bahcall, Soneira, and Piran is taken from the observational summaries of McCuskey (1966), Luyten (1968), and Wielen (1974).

The total collapse rate calculated from the Bahcall–Soneira Galaxy Model is [Bahcall and Piran (1983)], $M_{min} \simeq 10M_\odot$,

$$D_{\text{Galaxy}} \cong 0.09 \text{ yr}^{-1}, \tag{15.14}$$

assuming that a collapse anywhere in the Galaxy could be detected ($f = q = 1$). For the large direct counting experiments discussed in Chapters 13 and 14, any standard stellar collapse (see Table 15.1)

Table 15.2.   Fraction of stars in the galactic disk within a specified distance of the Sun. The fraction, q, of stars in the galactic disk with a distance R from the Sun is given for galactic model of Bahcall and Soneira (1980). For small $R \leq 3$ kpc, $q = 0.0028 \times (R/\text{kpc})^2$.

R (kpc)	q ($\leq R$)	R (kpc)	q ($\leq R$)
1.0	0.003	10.0	0.53
2.0	0.012	12.0	0.68
3.0	0.028	14.0	0.79
4.0	0.056	16.0	0.87
5.0	0.10	18.0	0.92
6.0	0.16	20.0	0.95
7.0	0.24	25.0	0.98
8.0	0.34	30.0	1.0

within the Galaxy could be detected easily; therefore Eq. (15.14) applies to these experiments. Equation (15.14) provides an answer obtained from optical observations of stars and the theory of stellar evolution to the second question posed at the beginning of this section. The total death rate of stars in the Galaxy is estimated to be 1 every 11 years. Since this rate is considerably larger than is estimated for visual supernovae, the majority of stellar collapses may be neutrino bombs that do not have any remarkable optical counterparts. Ratnatunga and van den Bergh (1989) have repeated the calculation of Bahcall and Piran (1983), using a different assumed mass distribution $\Phi(M)$, and have obtained a collapse rate of between 1 and 2 per century, in agreement with the estimated rate of galactic supernovae by van den Bergh *et al.* (1987). The difference between the two calculations is a measure of the uncertainty in the input data. The data on bright massive stars could be improved by modern optical observations, although many of the massive stars could be in molecular clouds and therefore hidden from optical detectors.

The best available upper limit on the rate of galactic stellar collapses is from the Baksan Underground Neutrino Observatory (sometimes called UNO), which is described in Section 15.5B [see also Alekseev *et al.* (1987)]. This scintillation detector has operated for a total of 5.5 years without seeing a neutrino pulse in excess of seven events (in the active mass of about 200 tons). A standard neutrino collapse at 10 kpc would produce in this detector about 73 absorption events by protons for an antineutrino temperature of 5 MeV. Therefore, the detector is sensitive to stellar collapses anywhere in the Galaxy. The 90% confidence upper limit implied by the Baksan observations is

$$D_{\text{Galaxy}} \lesssim 0.42 \text{ yr}^{-1}, \qquad (15.15)$$

which is within a factor of four of the collapse rate given in Eq. (15.14).

The stellar collapse rate in the Galaxy is not determined accurately by traditional astronomical observations of photons. Plausible estimates give a collapse rate somewhere between 1 every 10 years and 1 every 100 years. A prudent experimentalist should design a detector that will give interesting observational results (on nonexplosive sources) if there is no stellar collapse within the operational lifetime of the detector but also will be capable of making diagnostic measurements if a supernova occurs shortly after the apparatus is turned on.

15.4   Neutrinos from other galaxies

How observable are neutrinos from stellar collapses in other galaxies?

Table 15.3 shows the ratio of the neutrino fluxes from the luminous galaxies in the Local Group to the fluxes from a standard collapse in the Large Magellanic Cloud (see Section 15.5 for a discussion of the neutrinos from SN1987A). The fluxes given in Table 15.3 are proportional to the inverse squares of the distances [taken from Hodge (1987)]. The number of stellar collapses was computed assuming that the collapse rate is proportional to the total blue luminosity corrected for obscurations and inclination [see de Vaucouleurs, de Vaucouleurs, and Corwin (1976)] with the total blue magnitude of the Galaxy taken to be −20.1 [Bahcall and Soneira (1980)]. This list does not include the approximately two dozen other Local Group members that are too faint to contribute a significant chance for observation with detectors of the kind discussed in this book. The less luminous galaxies of the local group would experience a stellar collapse only once every few hundred years.

**Table 15.3. Neutrino
fluxes and interval
between stellar collapses
for galaxies in the Local
Group.** The neutrino
fluxes are denoted by ϕ
and the average interval
between stellar collapses
by $\langle t \rangle$.

Name	ϕ/ϕ_{LMC}	$\langle t \rangle$ (yr)
Milky Way	25	10
LMC	1	40
SMC	0.5	91
M31	8×10^{-3}	6
M33	6×10^{-3}	33

The observation of neutrinos from the Andromeda Nebula (M31)
would require neutrino detectors that are $\sim 10^2$ times more sensitive than existing detectors. The observation of stellar collapses in
galaxies other than those listed in Table 15.3 does not appear possible with existing techniques.

15.5 SN1987A

The optically bright supernova in the Large Magellanic Cloud,
SN1987A, resulted from the only stellar collapse from which neutrinos have been observed [see the historic papers of Hirata *et al.* (1987)
and Bionta *et al.* (1987), cf., also Alekseev *et al.* (1987)]. As we shall
see below, the results are in satisfactory agreement with the conventional notions of a "standard stellar collapse" that are summarized
in Section 15.2 and Table 15.1. The approximate distance to the
Large Magellanic Cloud is 50 kpc, five times further than the average distance expected for a galactic supernova (see Section 15.3).

This section reviews the main inferences from SN1987A.

A Detectors

The first detection of extrasolar system neutrinos was registered by the Kamiokande II [Hirata *et al.* (1987)] and IMB [Bionta *et al.* (1987)] water Cerenkov detectors with a possibly supporting observation by the Baksan liquid scintillation telescope [Alekseev *et al.* (1987)].

The Kamiokande II detector was discussed in Section 13.2 in connection with the observation of solar neutrinos via neutrino–electron scattering. The characteristics of this detector that are relevant for the study of solar neutrinos are summarized in Table 13.1. Fortunately, supernovae emit neutrinos of similar energies (~ 10 MeV) to the energies of the ^{8}B solar neutrinos that Kamiokande II was studying when SN1987A exploded. The only change in the experimental sensitivity that occurs when switching from solar to supernova neutrinos is that the instantaneous flux of supernova neutrinos is enormously higher (by about a factor of 10^3 for SN1987A) than the steady flux of higher-energy solar neutrinos.

Background problems are not as important in studying an intense burst of collapse neutrinos as for a low steady flux of solar neutrinos. A larger volume, 2.14 ktons (1.4×10^{32} free protons), could be used for the SN1987A observations; only the inner 0.7 ktons is used for solar neutrinos.

The IMB detector [Bionta *et al.* (1983), Haines (1986), and Matthews (1988)] was designed, like the original Kamioka detector, to search for proton decay. The IMB detector is located in the Morton-Thiokol salt mine near Fairport, Ohio ($41.7°$ N, $81.3°$ W) at a depth of 1570 m of water equivalent. A rectangular tank is filled with purified water, the six sides of which are instrumented with 2048 8-in. photomultiplier tubes arranged on an approximate 1 m grid. Just as for the Kamiokande II experiment, the water serves both as a target for incoming neutrinos and as a Cerenkov radiator for the charged products of such reactions. The timing, pulse height, and geometry of photomultiplier hits are used to reconstruct the vertex, direction, and energy of charged particle tracks.

The main differences between the Kamiokande II and IMB detectors are in the usable masses and in the thresholds for detecting neutrino events. The IMB experiment has the larger mass that can be used for detection, which is 6.8 kilotons when studying a pulsed source like SN1987A. However, the IMB detector is less sensitive to low-energy events because of higher background rates and lower efficiency in the collection of Cerenkov light. The detection threshold with IMB for pulsed neutrino events like those from SN1987A is about 20 MeV (which also indicates why the IMB detector has not been applied to the study of solar neutrinos).

The Institute for Nuclear Research (INR) of the USSR Academy of Sciences has had a neutrino telescope operating since June 1980 in search of stellar collapses in the Galaxy [see Alekseev *et al.* (1987, 1988)]. This telescope, which was mentioned in Section 13.4 in connection with ν–e solar neutrino scattering experiments, is located in the North Caucasus Mountains, under Mount Andyrchi, at a depth of 850 m water equivalent. The detector consists of approximately 330 tons of an oil-based liquid scintillator that is contained in 3156 standard units, each viewed by a single photomultiplier sensitive to Cerenkov light. The recoil electrons or positrons are contained within a single unit. Therefore, events in which only one unit is activated are relatively free from background. A total target mass of about 200 tons, 1.9×10^{31} protons, could be used to detect SN1987A.

An Italian–Soviet collaboration has been operating since late 1984 a liquid scintillator detector, known as LSD, in the Mont Blanc Laboratory. The LSD neutrino telescope was designed to observe stellar collapses anywhere in the Galaxy and is located deep underground (shielded by 5200 hg/cm^2 of standard rock). The detector has a total active mass of 90 tons of liquid scintillator containing 8.4×10^{30} free protons. The scintillator is contained in 72 separate counters, each of which is observed by 3 photomultipliers. The facility has been described in detail by Badino *et al.* (1984) and Aglietta *et al.* (1986).

The dominant reaction by which neutrinos are detected in all three detectors is antineutrino absorption by protons, as described in Eq. (15.8). The cross section for this absorption reaction, which is given in Eq. (15.9) is about two orders of magnitude larger than the cross section for neutrino–electron scattering [see Eq. (8.39) and Table 8.8],

$$\nu + e \rightarrow \nu' + e'. \tag{15.16}$$

At higher temperatures ($T > 5$ MeV), ν_e and $\bar{\nu}_e$ absorption on ^{16}O is comparable to neutrino–electron scattering [Haxton (1987b)].

B Data

Kamiokande II and IMB. Table 15.4 gives the measured properties of the neutrino events from SN1987A that were obtained with the water Cerenkov detectors, Kamiokande II and IMB, including the event time, electron recoil energy, and direction of the electron's momentum with respect to the vector connecting the Large Magellanic Cloud and the Earth. The indicated errors are the 1σ errors given in the original papers [Hirata *et al.* (1987) and Bionta *et al.* (1987)] and included corrections made in later studies [Hirata *et al.* (1988a) and Bratton *et al.* (1988)]. The first events detected in Kamiokande II and IMB were simultaneous to within the accuracy of the known zero point of time (about 1 minute). The detection efficiencies, energy thresholds, and other experimental characteristics are described in the observational papers.

A detailed follow-up study by the Kamiokande II collaboration [Hirata *et al.* (1988a)] showed that there were no other statistically significant neutrino bursts in a 10 hour period from 2:27 UT to 12:27 UT on February 23. Moreover, there was no evidence for a much larger signal at energies below the originally determined threshold. When the threshold was decreased to 5.6 MeV, the observed number of candidate events was 138 ± 12 compared to an expected value of 127 obtained from the average background trigger rate. One significant revision resulted from the reanalysis, namely, the angle that the second event makes with the LMC was increased from $15°$ to $40°$. The suggestion that ν_e scattering caused some of the observed events is less compelling as a result of this revision of the Kamiokande II data.

Kamiokande II observations for the six months after the discovery of SN1987A have set stringent limits on the flux of very high-energy neutrinos ($q > 10^{19}$ eV) that were produced by the supernova [see Oyama (1987)]. By searching for upward going muons produced by high-energy neutrinos, an experimental upper limit of between 10^{41} erg s^{-1} and $10^{42.5}$ erg s^{-1} was set on the luminosity of high-energy neutrinos. The precise value of the upper limit depends upon the spectral index and the cutoff energy of the assumed neutrino energy spectrum; the larger upper limit applies for an assumed index $\gamma = 2.7$ and a cutoff energy of 10^{12} eV while the smaller upper limit applies for $\gamma = 2.1$ and a cutoff energy of 10^{15} eV. Some models for

**Table 15.4.   Measured properties
of neutrino events from SN1987A
observed in water Cerenkov detectors.**
The first events were detected on
February 23, 1987 at about 7 hr 36
m UT. The angle in the last column
is relative to the direction of the
LMC. The errors are estimated 1σ
uncertainties.

Event	Event time (s)	Electron energy (MeV)	Electron angle (degrees)
Kamiokande II:			
1	0	20.0 ± 2.9	18 ± 18
2	0.107	13.5 ± 3.2	40 ± 27
3	0.303	7.5 ± 2.0	108 ± 32
4	0.324	9.2 ± 2.7	70 ± 30
5	0.507	12.8 ± 2.9	135 ± 23
6	0.686	6.3 ± 1.7	68 ± 77
7	1.541	35.4 ± 8.0	32 ± 16
8	1.728	21.0 ± 4.2	30 ± 18
9	1.915	19.8 ± 3.2	38 ± 22
10	9.219	8.6 ± 2.7	122 ± 30
11	10.433	13.0 ± 2.6	49 ± 26
12	12.439	8.9 ± 1.9	91 ± 39
IMB:			
1	0	38 ± 7	80 ± 10
2	0.412	37 ± 7	44 ± 15
3	0.650	28 ± 6	56 ± 20
4	1.141	39 ± 7	65 ± 20
5	1.562	36 ± 9	33 ± 15
6	2.684	36 ± 6	52 ± 10
7	5.010	19 ± 5	42 ± 20
8	5.582	22 ± 5	104 ± 20

the production of cosmic rays in supernova remnants predict a high-energy neutrino luminosity in excess of 10^{43} erg s^{-1} [see Gaisser and Stanev (1987)]. Antineutrinos from relic supernovae in other galaxies are potentially observable since on some models the largest flux is expected from galaxies with moderate redshifts ($Z \lesssim 1$). Using Kamiokande II observations taken between January 1986 and January 1988, Zhang *et al.* (1988) set an upper limit on the incident flux of antineutrinos of

$$\Phi \leq 212 \text{ cm}^{-2} \text{ s}^{-1} \tag{15.17}$$

for antineutrinos in the energy interval between 19 MeV and 35 MeV. This result can be translated into an interesting upper limit on the thermal flux of antineutrinos, depending upon the assumed temperature of the antineutrinos. For an effective temperature of 2 MeV, the flux limit corresponds to 5×10^4 cm^{-2} s^{-1}. Values between 10^0 cm^{-2} s^{-1} and 10^5 cm^{-2} s^{-1} have been estimated by different authors using a variety of assumptions [see Zeldovich and Guseinov (1965), Domogatskii (1984), Bisnovatyi-Kogan and Seidov (1982), Krauss, Glashow, and Schramm (1984), and Woosley, Wilson, and Mayle (1986)].

Other detectors. The Baksan neutrino telescope reported a burst of six events that was originally believed [see Alekseev *et al.* (1987)] to have occurred 25 seconds later than the first IMB event, with an absolute uncertainty of 2 seconds in the Baksan time and much less than a second of uncertainty in the IMB time. An additional uncertainty of 54 seconds in the direction of reconciling the discrepancy was subsequently found in the time measurements for the Baksan detector [see Alekseev *et al.* (1988)]. This detector had an active mass (200 tons with 1.9 $\times 10^{31}$ free protons) that is an order of magnitude smaller than the Kamiokande II and IMB detectors; the background rate in the Soviet detector was much higher than in the Kamiokande or IMB experiments. The Baksan team estimates that approximately one event in the burst is caused by background and suggests, without convincing or detailed justification, that the first event they recorded (which had an energy of 17.5 MeV) was caused by background in the detector. The five events that the team regarded as real had arrival times (measured in seconds from the first pulse) and positron energies (in MeV) of, respectively, 0.0, 12 ± 2.4; 0.45, 18 ± 3.6; 1.73, 23.3 ± 4.7; 7.75, 17 ± 3.4; and 9.12, 20.1 ± 4.0.

For the ^{37}Cl detector, there were no events above the solar neutrino background that were attributable to SN1987A. The lack of observation translates into a 90% confidence upper limit of about two ^{37}Ar atoms produced by the explosion [Davis (1988)].

The LSD scintillation detector with 90 tons of active mass, located in a tunnel underneath Mont Blanc, reported five events which they suggested might be associated with the SN1987A [see Aglietta *et al.* (1987a,b)]. The events were detected at 2h 53m (UT), about 4.7 hours before the simultaneous observations by the Kamiokande II and IMB experiments.

The burst of events observed in the Mont Blanc detector is unusual and occurred relatively close in time to the stellar collapse detected by Kamiokande II and IMB, but I believe that the Mont Blanc events are not associated with SN1987A. My reasons for this belief are

(1) No neutrino events were observed in the much larger Kamiokande II and IMB detectors at the earlier time reported by the Cerenkov detectors [Hirata *et al.* (1987), Bionta *et al.* (1987), and especially Hirata *et al.* (1988a)]. The number of free protons in the Mont Blanc telescope, 0.08×10^{32}, is more than an order of magnitude less than in the Kamiokande II detector (1.4×10^{32} protons) and the IMB detector (4.5×10^{32} protons).

(2) The expected number of events in the Mont Blanc detector for a standard stellar collapse (see Table 15.1) is only $\sim$ one event, assuming a 100% detection efficiency [Bahcall, Dar, and Piran (1987)]. The satisfactory agreement between the *a priori* model predictions and the observations made with the Kamiokande II and IMB detectors strengthens this argument.

(3) The reported events have energies that are close to the threshold energy for detection, which is between 5 MeV and 7 MeV [depending upon which counters were excited, see Aglietta *et al.* (1987a)]. The measured energies are, in MeV: 7, 8, 11, 7, and 9. Theoretically, one expects a greater spread in energy since the absorption cross section increases with the square of the neutrino energy for charged-current absorption and the numerical models predict an average antineutrino energy of more than 10 MeV.

(4) No plausible astrophysical scenario has been suggested for two distinct neutrino bursts [cf., de Rújula (1987), Berezinsky *et al.* (1988)].

(5) It is difficult to obtain a satisfactory light curve for the visual supernova if the earlier time indicated by the scintillation experiments is adopted as the time at which the star collapsed [see Arnett (1988) and Woosley (1988)].

C *Phenomenological analyses*

Many different papers have appeared analyzing the observed neutrino events from SN1987A. A representative sample of this large set of papers is listed here: Arafune and Fukugita (1987), Arafune *et al.* (1987), Bahcall, Dar, and Piran (1987), Burrows (1987a), Burrows and Lattimer (1987), Gaisser and Stanev (1987), Krauss (1987), Lamb, Melia, and Loredo (1987), Sato and Suzuki (1987a,b), and Schaeffer, Declais, and Jullian (1987).

The reader can find similar calculations in other papers that are referred to in the cited references.

A significant fraction of the papers cited above draw far reaching conclusions based upon two aspects of the Kamiokande II data, the angular distribution and the time dependence. Since the first two recoil electrons moved in the forward direction, many authors assumed that these events were caused by ν_e–e scattering, not absorption of $\bar{\nu}_e$. There is no strong statistical basis for this assumption, (see below) especially since the direction of the second event was revised to be 40° instead of 15° [see Hirata *et al.* (1988a)]. Both the IMB and Kamiokande II detections show events in the forward direction than would be expected on the average. Monte Carlo simulations yield a probability of the order of percent for the joint angular distribution, tantalizingly small but insufficient to establish that something beyond the conventional physical description is required.

A number of papers discuss the implications of assuming that there is a temporal "gap" between the last three events observed by Kamiokande II and all the other events (cf., Table 15.4). The number of observed events is too small to give any high significance to specific features of the time sequence of the registered pulses [see, e.g., Bahcall, Spergel, and Press (1987) and Piran and Spergel (1988)]. Most of the phenomenological inferences described in the remainder of this chapter are based upon the results of Bahcall *et al.* (1987), Spergel *et al.* (1987), and Bahcall, Spergel, and Press (1988). These studies use rigorous statistical techniques to infer conservative limits on the model parameters.

For simplicity, consider first a constant temperature analysis of all the observed eight events detected by the IMB collaboration and the first eight events seen by the Kamiokande detector. The subsequent three Kamiokande events had a somewhat lower average energy, which may reflect a cooling of the neutrino photosphere with time.[†]

Unfortunately, a "minimum" model[‡] has proved adequate to describe the sparse amount of data. In this over-simplified model, the (muon and tau) "scattering" neutrinos and antineutrinos have the same temperature (with no high-energy cutoff). The six types of neutrinos are separated into only two incident model fluxes. The electron antineutrinos, $\bar{\nu}_e$'s, have the largest interaction cross sections (see Section 15.2) in the water Cerenkov detectors. At the temperatures considered here, the ν_e's as well as the ν_μ's, ν_τ's, and their antiparticles can only scatter off the electrons in the water targets. The scattering cross sections given in Table 8.8 and Eq. (8.39) for ν_e and $\bar{\nu}_e$ are about a factor of 10^2 less than the absorption cross section for $\bar{\nu}_e$ on protons (which is given in Section 15.2); for muon and tau neutrinos and antineutrinos the scattering cross sections are almost a factor of 10^3 less than the $\bar{\nu}_e$ absorption cross section. Since scattering is much less probable than $\bar{\nu}_e$ absorption, everything except $\bar{\nu}_e$ is lumped into one flux of "average" scatterers. In most detailed model calculations, $F_{\nu_e} \sim 0.5 \times F_{\text{scatt}}$.

The best constant value for the temperature that is deduced with this simplified two-component model is [Bahcall *et al.* (1987)]:

$$T = 4.1 \text{ MeV}. \tag{15.18}$$

The corresponding value of $F_{\bar{\nu}_e} = 0.5 \times 10^{10} \text{ cm}^{-2}$.

The single temperature model represents a satisfactory fit to the observations. For $T = 4.1$ MeV and $F_{\text{scatt}}/F_{\bar{\nu}_e} = 10$, the probability

---

[†]All 19 of the events are included in the "cooling solution" that will be discussed in the next section; the two analyses give essentially the same results for the total energy emitted and the peak temperature.

[‡]Following Bahcall *et al.* (1987).

of obtaining a worse Kolmogorov–Smirnov (hereafter KS) measure than was found for the observed energy distribution is 33% for the Kamiokande data and 65% for the IMB data. Monte Carlo simulations show that between three and seven of the first eight events in Kamiokande were due to absorption of electron antineutrinos (95% confidence level). The KS test can only reject the possibility of zero scattering events at 2% significance, which is small but not completely negligible.[†] There were between six and eight absorption events in IMB (95% confidence level).

The statistical analysis shows [Bahcall *et al.* (1987)] that there is a well-defined range of antineutrino fluences,

$$F_{\bar{\nu}_e} = (0.15 \text{ to } 0.7) \times 10^{10} \text{ cm}^{-2} \qquad (15.19)$$

(95% confidence level) that is consistent with both the observed event rates. If we require that the count rates in IMB and Kamiokande are compatible and assume that the detector efficiencies are known accurately, then the neutrino temperature must exceed 3.7 MeV.

For the temperature and flux ranges inferred from the Kamiokande II and IMB results, one would not expect a signal well above background in either the Mont Blanc or the Baksan scintillator detectors. Assuming a detection efficiency of 100%, Bahcall *et al.* (1987) estimated a total number of antineutrino absorption events in 100 tons of liquid scintillator equal to

$$\text{Total neutrino events} = 0.8 \times (T/4.1 \text{ MeV})^2$$
$$\times \left(F_{\bar{\nu}_e}/0.5 \times 10^{10} \text{ cm}^{-2}\right) \text{ per } 0.1 \text{ kton.} \qquad (15.20)$$

According to the above equation, the most probable signal in the Mont Blanc detector was less than of order one event and less than of order two events in the Baksan detector.

The flux of ν_{scatter} is poorly determined since it is constrained experimentally only by the small number of events in the forward peak of the Kamiokande II and IMB detectors. Of course, one expects that the ν_{scatter} flux is harder to determine than the $\bar{\nu}_e$ flux

---

[†]In my opinion, 2% is a sufficiently large probability that one should not base any conclusion about the nature of neutrinos on the inference that at least one of the observed events was from neutrino–electron scattering. This attitude makes me skeptical of the many fascinating papers that have been written assuming that one or more of the observed neutrino events was from neutrino–electron scattering.

because the scattering cross section is much smaller than the absorption cross section. The strongest experimental limit on F_{scatt} comes from the detection of one to five (forward peaked) scattering events in Kamiokande. This corresponds to

$$F_{scatt} = (0.1 \text{ to } 5) \times 10^{10} \text{ cm}^{-2}. \tag{15.21}$$

For the ^{37}Cl detector of Davis, the corresponding number of supernova-induced events is small. The calculated number of ^{37}Cl events varies from 0.02 to 2, depending upon how one manipulates the limits in the above equation. The *a priori* estimate [Bahcall, Dar, and Piran (1987)] for the ^{37}Cl detector was one ^{37}Ar atom produced for a standard stellar collapse. According to these estimates, the background from solar neutrinos should have swamped Davis's supernova signal![†]

There is one unexpected feature of the LMC supernova neutrino data: the 7.3 seconds gap between the first eight and the last three events in the Kamiokande II data. The Kamiokande II detector observed eight events in the first 1.9 seconds, followed by a quiet period of 7.3 seconds, and then three events were detected within 3.2 seconds.

However, similar gaps are found with appreciable frequency in Monte Carlo simulations of the sparse data with "events" that were produced by random sampling of a distribution that has a smooth time dependence [see Bahcall, Spergel, and Press (1988) and Lattimer and Yahil (1988)]. The simulated data contains remarkable and amusing patterns, including a temperature that increases with time, huge deleptonization pulses, evidence for periodicity in the arrival times of the neutrino events, and gaps that correspond to a mass of the tau neutrino that just closes the universe [see Figures 1 to 3 of Bahcall *et al.* (1988)].

The IMB detector observed six events in the first 2.7 seconds, followed by a quiet period of 2.4 seconds, and then two events were detected within 0.6 seconds. There are two IMB events in the Kamiokande "time gap."

[†]For years, solar neutrino astronomers have suffered from low counting rates, while optical and radio astronomers observe countless photons per year. Finally, the photon-counting scientists do their best with a spectacular visual supernova and the fireworks are overwhelmed, for the only operating detector of low-energy ν_e absorption events, by solar neutrinos.

Figure 1 of Spergel *et al.* (1987) shows that the average energy of an event declines with time and demonstrates the agreement of the data with a cooling black body model. A cooling hot neutron star model fits well all the observed data and provides an estimate of the radius of the hot neutron star [see also Lamb *et al.* (1988) and Burrows (1988b)].

Spergel *et al.* (1987) adopted a simplified model in which the neutrino source is a black body with an exponentially decaying temperature: $T = T_0 \exp(-t/4\tau)$. (The energy density at the surface is proportional to T^4, thus τ is the cooling time scale for the hot neutron star.) The joint (Kamiokande II and IMB) likelihood function is maximized at $T_0 = 4.2$ MeV, $\tau = 4.6$ s, and a total fluence $F = 1.3 \times 10^{10}\bar{\nu}_e$ cm^{-2}. Using the multidimensional KS test and Monte Carlo simulations to determine the significance of this solution, Spergel *et al.* (1987) find that the observed KS measure is better than 55% of the cases obtained from the synthetic data. At 95% confidence level, they obtain:

$$T_0 = 4.2^{+1.2}_{-0.8} \text{ MeV}; \quad \tau = 4.5^{+1.7}_{-2.0} \text{ s} \ ;$$

$$F_0 T_0^2 = 4.0^{+2.4}_{-2.0} \times 10^{10} \text{ cm}^{-2} \text{ MeV}^2 \text{ s}^{-1}. \tag{15.22}$$

There is a strong correlation between the inferred flux and temperature, which is why the fluence is given only in the combination $F_0 T_0^2$. The Kamiokande data alone yield a peak temperature of $2.9^{+1.3}_{-0.5}$ MeV and the IMB data alone yield a peak temperature of $4.9^{+4.2}_{-1.9}$ MeV.

The rate at which absorption events occur in Kamiokande II and IMB detectors is proportional to a high power of the temperature when the effects of the cross section and of the detector efficiencies are included. Since the event rate drops much faster than the temperature, the simplest constant temperature model [cf., preceding discussion] gives a reasonable fit to the data.

The luminosity of a black body of radius R can be equated to the product of the detected flux, F, and the average energy of a neutrino, $\langle E \rangle$,

$$L_{\bar{\nu}_e} = 4\pi D^2 F \langle E \rangle, \tag{15.23}$$

where D is the distance to the LMC. This yields a crude estimate of the radius of the cooling neutrino sphere,

$$R = 27^{+17}_{-15} D_{50} \text{ km}, \tag{15.24}$$

where $D_{50} \equiv D/50$ kpc. The radius will decrease as the neutron star cools. The range for the radius R given above is in agreement with expectations from published numerical models of stellar collapse [see, e.g., Cooperstein (1988ab)].

The total thermal energy emitted by the hot neutron star is

$$\int L_\nu \mathrm{d}t = N_\nu F_0 (3.15 T_0)(4\pi D^2)\tau$$
$$= 6.1^{+3.5}_{-3.6} \times 10^{52} N_{\mathrm{all}} D_{50}^2 \text{ erg}, \qquad (15.25)$$

where N_{all} is the ratio of energy emitted in all neutrino species to the energy emitted in electron antineutrinos. The quantity N_{all} is expected to be of order six.

The results for the temperature, decay time scale, and the $\bar{\nu}_e$ flux are consistent with the standard picture of stellar collapse that is based upon detailed numerical models and on analytic arguments (cf., Table 15.1). The success of this simplified "standard" model suggests that it will be difficult to use the neutrino events observed from SN1987A to establish more detailed models. The observations of SN1987A have confirmed the general picture of core collapse; however, the data are not sufficient to discriminate between equations of state or to validate specific detailed models. There is no need to evoke new particle physics or complicated astrophysical scenarios. When a supernova is observed in the Galaxy, neutrino detectors should record many hundreds of events and neutrino astronomy may then reveal surprises about stellar collapses and weak interaction physics.

D   Neutrino properties

Mass of ν_e. The observations of neutrinos from the LMC supernova place an interesting upper limit on the mass of the electron's neutrino, m_{ν_e}. The basic idea was discussed first by Zatsepin (1968), who pointed out that if neutrinos have a finite mass, the higher-energy neutrinos from a supernova explosion arrive before the more slowly moving, lower energy neutrinos. The extra time, Δt, that a finite mass neutrino requires to reach the Earth compared to a zero mass particle is

$$\Delta t_i = 2.57 \text{ s} \left(\frac{D}{50 \text{ kpc}} \right) \left(\frac{10 \text{ MeV}}{E_i} \right)^2 \left(\frac{m_{\nu_e}}{10 \text{ eV}} \right)^2, \qquad (15.26)$$

where E_i is the energy of the i-th neutrino. A nonzero mass will cause particles of different energies to arrive at different times, even if they are emitted simultaneously. This dispersion with energy will typically stretch out a burst in time, with the lowest-energy particles arriving last, unless there are unusual cancellations or special initial conditions.

One cannot deduce a mass limit without some assumption. For any assumed value of m_{ν_e}, the dispersion relation for Δt_i determines only the emission time for each of the observed neutrinos. Unless we make a supporting statistical or physical assumption, we cannot infer a mass limit. The situation is similar to many laboratory experiments in which the final answer must be determined by Monte Carlo simulations.

Immediately after the observations were made public, Arnett and Rosner (1987) and Bahcall and Glashow (1987) derived an upper limit of about 11 eV for m_{ν_e} by assuming that nature was not satanic and therefore the observed 2 second half width of the observed neutrino pulse was not narrowed in transit by more than a factor of two. The argument is simple, but does not provide a statistical confidence level for the inferred mass limit. If the delay time due to finite neutrino mass becomes comparable to the characteristic width of the pulse, the expected distribution of events should show a correlation of energy with time. This correlation is not present, which suggests in a model-independent way that the observed data do not imply a measurable mass for ν_e. Bahcall and Glashow pointed out that more stringent limits could be inferred if one were willing to interpret the observed substructure of the neutrino burst in terms of specific physical processes and indeed many authors have obtained more stringent limits or specific mass values by identifying a substructure in the neutrino pulse (or by neglecting some possible way that nature could fool us). The interested reader can find different treatments of this problem in, for example, Sato and Suzuki (1987a), Burrows and Lattimer (1987), Kolb, Stebbins, and Turner (1987), Arafune and Fukugita (1987), Takahara and Sato (1987), Arafune *et al.* (1987), and Burrows (1988a).

Spergel and Bahcall (1988) carried out a comprehensive statistical treatment of the mass limits that is based upon an extensive set of Monte Carlo simulations. They calculated confidence limits by performing Monte Carlo simulations that took account of the complexity of the actual experimental measurements, including the detector efficiencies and the uncertainties in the measured energies. This modeling took account of the possible satanic tendency of nature by allowing the supernova temperature to first rise and then fall as a function of time, permitting low-energy neutrinos to be emitted first. The model parameters were determined separately for each realization of a simulated observation and for each model by maximizing the joint likelihood function, which is a product over all events observed in the Kamiokande and the IMB detectors.

The basic assumption used in these simulations is that the neutrino emission temperature is a smoothly varying function of time. The functional forms that were used represent well the published values for the neutrino emissivity calculated from detailed models of supernova explosions, but were more general and therefore allow a wider range of acceptable neutrino masses. Six classes of models were investigated, most of which had six parameters that were constrained by the available observations. The range of models and parameters included in these simulations is larger than in the other discussions cited above, which causes the inferred mass limit to be somewhat conservative.

The fraction of the finite mass simulations that fit better than the observed data drops well below 5% at 16 eV in all of the models that were investigated. Spergel and Bahcall conclude that

$$m_{\nu_e} \leq 16 \text{ eV}, \tag{15.27}$$

at the 5% significance level. The result given above is inconsistent with some quoted laboratory measurements of the neutrino mass using tritium decay [Boris *et al.* (1987)].

The mass limit from SN1987A, Eq. (15.27), shows that ν_e's cannot supply enough missing matter to close the universe.

Can the limit stated in Eq. (15.27) be avoided, following Cowsik (1988) and Huzita (1987), by fitting the observed time dependence of the events using two different neutrino masses? One might think that with additional parameters the data might be easier to fit. However, the total energy emitted in neutrinos is a severe constraint. If, as suggested by Huzita (1987) and Cowsik (1988), the earlier events were caused by $\bar{\nu}_e$ absorption [with a mass $m(\bar{\nu}_e) \sim 3$ eV], then the later arriving neutrinos, $\bar{\nu}_x$, will contain only a small component of $\bar{\nu}_e$. Since the absorption cross section, Eq. (15.10), is more than two orders of magnitude larger than the relevant scattering cross sections, the later events would occur via $\bar{\nu}_e$ absorption from a small component of $\bar{\nu}_x$. Using considerations of neutrino mixing similar to those described in Section 9.2, one finds that the ratio of energy emitted in $\bar{\nu}_x$ to energy in $\bar{\nu}_e$ is

$$\frac{E(\bar{\nu}_x)}{E(\bar{\nu}_e)} = \frac{T(\bar{\nu}_x)}{T(\bar{\nu}_e)} \times \frac{\phi(\bar{\nu}_x)}{\phi(\bar{\nu}_e)}, \qquad (15.28)$$

where $T(\bar{\nu}_x)$ represents the temperature of $\bar{\nu}_x$ and $\phi(\bar{\nu}_x)$ represents the flux of $\bar{\nu}_x$. Let r be the ratio of the observed number of events caused by $\bar{\nu}_x$ to those caused by $\bar{\nu}_e$ and recall that the average cross section is proportional to T^2 while the energy carried by the neutrinos is proportional to T, then the energy ratio can be written

$$\frac{E(\bar{\nu}_x)}{E(\bar{\nu}_e)} = r\,\mathrm{ctn}^2\,\theta_V\frac{T(\bar{\nu}_e)}{T(\bar{\nu}_x)}. \qquad (15.29)$$

In the mass range of interest here, the mixing angles are known to be very small. For $\nu_\mu \to \nu_e$ mixing, $\sin^2 2\theta_V$ is known to be less than 0.01 [Blümer and Kleinknecht (1985) and Boehm and Vogel (1987)]. The weakest applicable limit is for $\nu_\tau \to \nu_e$, for which the limit is $\sin^2 2\theta_V \lesssim 0.13$. The ratio of $\bar{\nu}_x$ to $\bar{\nu}_e$ temperatures is approximately a factor of two (see Section 15.2 and Table 15.1) and the "best-fitting" ratio r of events is about unity [see Cowsik (1988) and Huzita (1987)]. The weakest limit is obtained by considering ν_e–ν_τ mixing. Substituting the numbers given above in Eq. (15.29), one finds

$$\frac{E(\bar{\nu}_x)}{E(\bar{\nu}_e)} \gtrsim 14. \qquad (15.30)$$

The ratio would have been at least 140 had we considered the mixing of ν_μ and ν_e. The minimum ratio given in Eq. (15.30), obtained from the two mass solution, implies that the energy emitted in all types of neutrinos is an order of magnitude larger than the binding energy of a neutron star. Therefore one cannot avoid the mass limit for $\bar{\nu}_e$ given in Eq. (15.27) by hypothesizing a coupling to $\bar{\nu}_x$.

Charge of ν_e. The electric charge, Q, of an electron neutrino can be limited by a similar argument [Barbiellini and Cocconi (1987)]. If Q is different from zero, then neutrinos of different energies will have different paths in the galactic magnetic field. Higher-energy neutrinos would move along a straighter path and therefore arrive at Earth before lower-energy neutrinos, just as is the case for neutrinos that are assumed massive. In fact, the analogy can be made specific since both the fractional time delay for neutrinos of a finite charge,[†]

$$\frac{\Delta t}{t} = \frac{Q^2 B^2 x_G^2}{6 c^2 p_\perp^2},$$ (15.31)

and the fractional delay for neutrinos of a finite mass,

$$\frac{\Delta t}{t} = \frac{1}{2} \frac{m_\nu^2 c^4}{E^2},$$ (15.32)

depend upon the inverse square of the energy, E (or the inverse square of the perpendicular momentum, $p_\perp$). Here B is the galactic magnetic field and x_G is the path length in the field. Comparing Eqs. (15.31) and (15.32), we find that the upper limit (UL) on the charge Q (in units of the electron charge, e) can be written in terms of the upper limit on the neutrino mass: $(Q/e)_{UL} = \sqrt{3} (m_\nu c^2)_{UL} / eBx_G$. Using this isomorphism and the statistical analysis of Bahcall, Spergel, and Press (1988), one finds

$$\left(\frac{Q}{e}\right)_{UL} \leq 3 \times 10^{-17} \left[\left(\frac{1\ \mu G}{B}\right) \left(\frac{1\ kpc}{x_G}\right) \right].$$ (15.33)

The above limit is expressed in terms of conservative estimates: A typical value for the ordered galactic magnetic field in the solar vicinity may be of order 2 or 3 μG over an effective path length of order a kpc or more [see Heiles (1987)].

Lifetime. Many authors noted immediately upon the detection of neutrinos from SN1987A that their observation implies that the lifetime of a 10 MeV $\bar{\nu}_e$ must exceed 10^5 yr. More formally, one can write

$$t_0 \gtrsim 5 \times 10^5 \text{ s } \left(\frac{m_\nu}{1\ eV}\right),$$ (15.34)

[†]The formula given by Barbiellini and Cocconi (1987) may not apply to the case of interest since it assumes that the neutrinos move in a constant magnetic field along their entire path from the Large Magellanic Cloud to the Earth. A more conservative assumption is that the magnetic field is effective for a limited path within the Galaxy. The formulae given above apply to this more conservative case. The quantity $x_G B$ that appears in the equation for the fractional time delay caused by a finite charge is: $x_G B = \int_0^{z_{max}} dz' B(z') / \sin|b|$, where z is the height above the plane (z_{max}, the maximum extent of the ordered galactic field $\sim$ a kpc) and b is the galactic latitude of the LMC ($b \approx -33°$).

where t_0 is the $\bar{\nu}_e$ lifetime at rest. There is a loophole in this argument. If, for example, $\bar{\nu}_e$ is a linear combination of two mass eigenstates, then the heavier state might decay and the lighter state could be stable and reach the Earth from SN1987A. Frieman, Haber, and Freese (1988) have discussed models in which fast decays of ν_e are allowed when all the effects of MSW mixing are included.

Limiting velocity. The speed of light plays a dual role in the theory of special relativity. On the one hand, it is the limiting velocity for all objects, regardless of their nature. On the other hand, it is the velocity of a particular particle, the photon. Stodolsky (1988) has pointed out that the approximate equality of the arrival times of the neutrinos and the photons from SN1987A allows an accurate check on the hypothesis of special relativity that the limiting velocity for all forms of radiation is the same. With a generous estimate of 10 hours for the uncertainty of the coincidence in arrival time between the first photons and the intense burst of neutrinos, Stodolsky noted that the speed of photons and of neutrinos cannot differ by more than 1 part in 10^8, which is close to the accuracy with which the speed of light is known.

Geodesics. The observation of a neutrino burst within a few hours of the associated optical burst from SN1987A provides a new test of the weak equivalence principle, demonstrating that neutrinos and photons follow the same trajectories in the gravitational field of the Galaxy [Longo (1988) and Krauss and Tremaine (1988)]. A maximum possible violation can be calculated by assuming (implausibly) that the difference in arrival time for neutrinos and photons was caused solely by the passage through the galactic gravitational field. The neutrinos were detected at February 23.32 UT [Hirata *et al.* (1987) and Bionta *et al.* (1987)] and the first optical brightening was observed at February 23.443 UT [McNaught (1987)], less than 3 hours later. Longo (1988) and Krauss and Tremaine (1988) considered a range of gravitational potentials for the Galaxy that bracket the possibilities allowed by measurements of the galactic rotation curve. Their results show that photons and neutrinos move along the same trajectories to an accuracy of about 0.5% or better.

Number of neutrino flavors. The supernova observations give, in principle, a limit on the number of different flavors of neutrinos that cooled the neutron star since the total energy carried off by all the flavors cannot exceed the binding energy of a neutron star. Combining Eqs. (15.1) and (15.22), one obtains

$$N_{\text{flavors}} \cong \frac{E_{\text{binding}}}{2E_{\text{observed}}(\bar{\nu}_e)} \lesssim \frac{(2.5 \pm 1.5) \times 10^1}{2 \left(6.1^{+3.5}_{-3.6}\right) D_{50}^2}, \qquad (15.35)$$

where N_{flavors} is the number of neutrino flavors (counting ν and $\bar{\nu}$ as one flavor). Using the extreme limits in order to be conservative, one finds that there are solutions for all values of N_{flavor} between 1 and 8. This conclusion is more pessimistic than many discussions given in the literature.

Dynamical constraints. Several authors have derived stringent limits on the magnetic moment of an electron neutrino by arguing that larger moments would have led to interactions within the star that can be excluded on the basis of the existing observations. These arguments assume that the conditions inside SN1987A, at the time of collapse and neutrino production, are sufficiently well understood to use the parameters in the stellar core to make approximate calculations of the rates of spin flip interactions. If this assumption is valid, then [see, e.g., Goldman *et al.* (1988), Lattimer and Cooperstein (1988), and Barbieri and Mohapatra (1988a)]

$$\mu_{\text{astrop.}} \lesssim 10^{-12} \mu_{\text{B}}. \qquad (15.36)$$

Raffelt and Seckel (1988) [see also Barbieri and Mohapatra (1988b)] have argued that the observed duration of the neutrino cooling time scale for SN1987A can be used to set strong limits on the ratio of the coupling constant for right-handed weak interactions, G_{RH}, to the usual Fermi coupling constant for left-handed currents, G_{F}. If the cooling of neutron stars is well understood, then these arguments show that

$$\epsilon = \frac{G_{\text{RH}}}{G_{\text{F}}} \lesssim 10^{-4}, \qquad (15.37)$$

for right-handed interactions involving either neutral or charged currents.

Until the solar neutrino problem is solved, conclusions like Eqs. (15.36) and (15.37) that are based upon an assumed understanding of the dynamics of stellar collapse will be treated with caution by some workers. Solar evolution (on time scales of 10^{10} yr and densities

Table 15.5. Some thermally averaged neutrino cross sections. The temperature, T, is given in MeV and the cross sections are given in units of 10^{-46} cm^2.

T	$\sigma(^2\text{H})_{\text{abs}}$	$\sigma(^2\text{H})_{\text{nc}}$	$\sigma(^{37}\text{Cl})_{\text{abs}}$	$\sigma(^{40}\text{Ar})_{\text{abs}}$	$\sigma(^{71}\text{Ga})_{\text{abs}}$
3.0	3.3E+04	1.3E+04	4.5E+04	3.8E+04	8.5E+04
4.0	6.5E+04	2.7E+04	9.9E+04	8.4E+04	1.7E+05
5.0	1.0E+05	4.3E+04	1.7E+05	1.4E+05	2.6E+05
6.0	1.4E+05	5.7E+04	2.5E+05	2.1E+05	3.5E+05
4.1	6.8E+04	2.9E+04	1.05E+05	8.9E+04	1.75E+05

of 10^2 g cm^{-3}) is mild compared to stellar collapse (on time scales of milliseconds and nuclear densities). If we do not understand main-sequence solar evolution, we may not understand nonequilibrium stellar collapses.

15.6   Expected number of events in detectors

How many neutrino events would be observed from a standard stellar collapse within the Galaxy? The information required to answer this question is available in the previous chapters and sections, which describe how to calculate the neutrino cross sections (Chapter 8), the detectors that are likely to be operating in the future (Chapters 10 to 14), and the parameters of a standard stellar collapse (Section 15.1).

Table 15.5 contains neutrino absorption cross sections that are calculated for several different neutrino temperatures (including the best-estimate value for SN1987A of 4.1 MeV). These cross sections were obtained by averaging over an assumed Boltzmann spectrum. For the higher temperatures, energies well above 30 MeV must be considered in order to obtain an accurate value for the total cross sections; therefore, the final results for ^{37}Cl, ^{40}Ar, and ^{71}Ga are uncertain because of forbidden weak interaction corrections [see Section IV.D of Bahcall and Ulrich (1988)]. The uncertainties for ^{37}Cl and ^{40}Ar are probably less than or of order 25% while the uncertainties in the ^{71}Ga cross sections are at least a factor of two. The absorption cross section on deuterium is relatively accurately known (see Section 8.1D), as is also the antineutrino absorption cross section in normal water.

Table 15.6. Neutrino and antineutrino events from a standard stellar collapse (see Table 15.1). The collapse is assumed to occur at 10 kpc. The LVD and H_2O (Super Kamiokande) detectors are primarily sensitive to $\bar{\nu}_e$'s; the other detectors are mainly sensitive to ν_e. Neutral-current events are not included.

Detector	Mass (kton)	Events
^{71}Ga	0.03	0.4
^{37}Cl	0.62	6
D_2O	1.0	1E+02
LVD	1.8	6E+02[†]
^{40}Ar	3.0	10
H_2O	22	6E+03[†]

[†] Mainly antineutrinos.

Table 15.6 shows the total number of events that a galactic neutrino bomb would produce in some of the solar neutrino detectors[†] discussed in the previous chapters. The parameters for the collapse were taken from Table 15.1, assuming that the collapse occurred at a distance of 10 kpc from the Sun (see Table 15.2) and had a temperature of 5 MeV for the ν_e's.

Super Kamiokande and LVD would provide the most events; these two detectors are primarily sensitive to antineutrinos which are observed via absorption by protons. For the near future, the ^{37}Cl detector remains the only neutrino experiment that can give a clear signal from ν_e's produced by a stellar collapse in the Galaxy.

[†] Absorption cross sections are much larger than the scattering cross sections, for the energies of interest here. Therefore, one must calculate the number of proton absorbers for LVD and Super Kamiokande from the stated number of electrons given in Chapter 14.

What about the ν_μ's and ν_τ's, as well as their antiparticles? These particles will contribute a small additional signal from ν_e scattering in the H_2O, D_2O, and liquid argon experiments, but even the high degree of directionality associated with neutrino–electron scattering may not be sufficient to provide a cleanly separated signal.

The best hope for observing a clean signal from the ν_μ's and ν_τ's may be the neutral-current disintegration of the deuteron that is described in Eq. (14.3). A distinctive high-energy gamma ray is produced when the recoil neutron is captured by, for example, ^{35}Cl that would be introduced into the tank (see the discussion of neutral-current detection in Section 14.1). Bahcall, Kubodera, and Nozawa (1988) estimate that a standard stellar collapse in the Galaxy would give rise to approximately 10^3 neutral-current events in a kiloton deuterium detector, nearly all of which would be from muon and tau neutrinos and antineutrinos. Masses of the mu and tau neutrinos could be revealed, if either exceeds 100 eV, by a long, distinctive tail in the occurrence times of events.

Bibliographical Notes

Alekseev, E.N., *et al.* (1987) *JETP Lett.*, **45**, 589. Report of an observation of a possible neutrino burst with the Baksan Neutrino Telescope. The signal was originally believed by many astronomers and physicists not to be associated with SN1987A because the time of the burst disagreed with the times observed in the larger IMB experiment. Subsequent investigations revealed that the Baksan clock may have been incorrect.

Baade, W. and F. Zwicky (1934) *Proc. Nat'l. Acad. Sci. USA*, **20**, 254. An astounding proposal that supernovae represent the collapse of a normal star to a star composed of neutrons, published two years after the discovery of the neutron. Correctly estimated the energy released by the formation of a neutron star..

Bethe, H.A. and G. Brown (1985) *Sci. Am.*, **252**, 60. A clear physical description of how a supernova explodes, presented in beautiful and entertaining prose. What every physicist wants to know and what every astronomer should know.

Burbidge, E.M., G.R. Burbidge, W.A. Fowler, and F. Hoyle (1957) *Rev. Mod. Phys.*, **40**, 547. The bible of nuclear astrophysics.

Hirata, K., *et al.* (1987) *Phys. Rev. Lett.*, **58**, 1490; Bionta, R.M., *et al.* (1987) *Phys. Rev. Lett.*, **58**, 1494. Historic detection of supernova neutrinos by two water Cerenkov detectors.

Shapiro, S.L. and S.A. Teukolsky (1983) *Black Holes, White Dwarfs, and Neutron Stars* (New York: Wiley). An excellent physical discussion of much of the relevant high-energy astrophysics.

Supernovae: A Survey of Current Research, edited by M.J. Rees and R.J. Stonehan (Dordrecht: Reidel). A treasure. A collection of insightful papers by most of the leaders in the field.

Zatsepin, G.I. (1968) *Sov. JETP*, **8**, 205. The original paper pointing out that antineutrino detectors could set limits on masses of neutrinos produced in supernovae. Simple and beautiful.

16

Synopsis and future directions

Summary

Neutrino astrophysics is in a transition stage. The number of operating solar neutrino experiments will increase to at least five within the next few years, which is to be compared to two experiments over the past quarter century. The new data may establish whether the discrepancy between observation and theory is caused by faulty astrophysics or new physics. The energy spectrum of the incident neutrinos is the key experimental feature that will discriminate between alternative classes of explanations. Information about the spectrum will be obtained by comparing the results obtained with radiochemical experiments that have different energy sensitivities and by direct observations of electrons produced by absorption of energetic neutrinos.

Much progress has been made over the past 25 years. What has been achieved, however, is small compared to what remains to be accomplished.

This chapter gives a synopsis of some of the areas in which there have been major developments and highlights problems that need to be solved in the future.

16.1   What is needed

Neutrino astrophysics needs more experiments.

Precision absorption experiments are required in order to measure the energy spectra of individual solar neutrino sources. Accurate

461

neutral-current measurements are necessary to determine the total flux, independent of neutrino flavor. Neutrino–electron scattering experiments can demonstrate, by the angular distribution of the recoil electrons, that the neutrinos come from the Sun and can help establish the fluxes of different neutrino flavors. Radiochemical experiments with different energy thresholds must be used to obtain information about separate neutrino reactions that provide diagnostics of the solar interior. Geochemical experiments offer unique opportunities to constrain the time dependence of the neutrino fluxes on astronomical time scales.

There are many quantities to measure. For each neutrino source (e.g., pp, ^{7}Be, or ^{8}B), one would like to measure the total number of ν_e's that are incident on the Earth, the spectrum of neutrino energies, the flavor content of the incident beam, and the time dependence. In addition, there should be redundancy in the solar neutrino measurements in order to prevent misinterpretations resulting from unknown systematic effects that may afflict individual detectors or techniques. So far, only two solar neutrino experiments have been performed in the quarter century history of the subject. During this time, a single supernova was observed serendipitously. These pioneering experiments represent great progress (sufficient to justify writing a whole book on the subject), but are only a drop in the bucket compared to what should be done.

Contrast this with the favorable situation with respect to laboratory measurements of the mass of ν_e. There are approximately 20 high-precision experiments in progress or planned that will be sensitive to a mass for ν_e in the range of 5 eV to 50 eV. Many other experiments have been performed in the past. A large fraction of these experiments are comparable in difficulty to carrying out a solar neutrino experiment. In the laboratory studies, one physical parameter will be measured or constrained: m_{ν_e}. Nevertheless, the experimental situation remains controversial and important new laboratory techniques are being developed.

Numerous other experiments that require great ingenuity and skill have been performed to search for proton decay or neutrino-less double β-decay. Experiments in both these fields share some physics goals and techniques with solar neutrino observations.

For solar neutrinos, there are many physical parameters that may be important besides the mass of ν_e. The masses and mixing angles of at least three neutrino flavors may affect the observations. The relevant mass range spans at least four orders of magnitude, from 10^{-2} eV to 10^{-6} eV, and the relevant mixing angles span almost two orders of magnitude. Even neutrino electromagnetic moments and decay rates are conceivably relevant. Additionally, measurements of the fluxes and spectra from individual neutrino sources are required for astronomical diagnostics.

In order for solar neutrino astrophysics to fulfill its promise, in order to measure well some of the important physical and astrophysical parameters, a variety of experiments must be performed. To answer the already recognized questions about solar neutrinos and to measure some of the most important physical and astrophysical parameters, the experiments must include the current generation of radiochemical experiments, electron scattering measurements with good angular resolution, spectral measurements for ^{8}B and pp neutrino sources, neutral-current measurements of the flux of high-energy neutrinos and of low-energy neutrinos, a precise measurement of the ^{7}Be flux and line shape, tests for temporal variations, and an accurate measurement of the flux and energy spectrum of pp neutrinos. If different techniques are to be used to measure in more than one way some of the same quantities, then the above list indicates that a minimum of 10 experiments must be completed.

To take advantage of new experimental techniques that are developed in the next decade and of the unique opportunities that solar neutrinos provide for learning about both weak interaction physics and stellar interiors, a vigorous observational program should include 20 or more experiments by the end of the century.

16.2   Solar neutrino detectors

Measurements of the ν_e energy spectrum (via charged-current absorption experiments) and of the total neutrino flux (via neutral currents) are a key part of the future program in solar neutrino research. On a longer time scale, detectors can be developed to observe individual lower-energy sources such as the pp, pep, and ^{7}Be neutrinos. Observations of these individual neutrino sources will provide incisive diagnostics of the solar interior and establish specific constraints on possible physics beyond the standard electroweak model.

A   Spectral measurements

Measurements of the energies of electrons produced by neutrino absorption can determine the spectrum of incident neutrino energies. The electron takes away essentially all of the available energy (above threshold) since the baryonic absorbers are much heavier than the electron. The standard shape of the neutrino energy spectrum from each nuclear source can be calculated accurately. Nonstandard solar physics can alter the flux from a given neutrino source, but not the shape of the energy spectrum.

If the measured shape of the energy spectrum from ^{8}B neutrinos differs from the standard shape, then physics beyond the standard electroweak model must be occurring.

The SNO (heavy water), ICARUS (liquid argon), BOREX (^{11}B), and ^{115}In experiments could measure the spectrum of the ^{8}B neutrinos at electron recoil energies above several MeV. It seems likely that two or more of these experiments will be performed over the next decade. At least one high signal to noise measurement of the neutrino energy spectrum, in addition to the four proposed experiments mentioned above, should be made using a different detector and new observational techniques. The possibility of using ^{19}F as a target should be investigated, since a fluorine detector has a number of advantages, experimental and theoretical.

All of the experiments with demonstrated feasibility to measure an energy spectrum concentrate on the high-energy tail (above 5 MeV) of the solar neutrino energy spectrum. There are no experiments with demonstrated feasibility that can measure the energy spectrum of the lower-energy solar neutrino sources. Measurement of the energy spectrum of lower-energy neutrino sources represents the most difficult, but perhaps the most potentially rewarding, experimental challenge.

B   Neutral-current measurements

Neutral-current measurements are flavor blind (the ideal Equal Opportunity Experiments), sensitive to the total number of neutrinos of a given energy that are incident upon the detector. Even though the flavor of a neutrino beam is changed by oscillations in vacuum or in matter, the spectrum of energies incident on the Earth will be the same as the spectrum when the neutrinos are created. The event rate induced by neutral currents is the same for a beam that has undergone complete mixing as for a beam that has not suffered any changes. On the other hand, the rates in absorption or scattering experiments can change dramatically as a result of changes in the flavor content of the beam.

Neutral-current experiments, when combined with absorption or scattering measurements to determine the normalization of the ν_e beam, test whether physics beyond the standard electroweak model is occurring.

Two neutral-current experiments are being developed, SNO and BOREX. The SNO heavy water detector can reveal the total number of ^{8}B neutrinos that disintegrate deuterium. The neutral-current disintegration can be calibrated using neutrino absorption by deuterium; the ratio of disintegration to absorption cross sections for deuterium can be calculated to an accuracy of about 1/2%. The absorption and disintegration cross sections can be calculated separately to an accuracy of 10%. The BOREX experiment can study both the neutral-current excitations of ^{11}B as well as the charged-current absorptions. The cross sections must be estimated in some cases using shell model calculations. No angular or energy distributions can be measured in either neutral-current experiment.

Coherent nuclear recoil is a promising method, which is being actively developed, for studying neutral currents. For typical targets under consideration, the cross sections for coherent scattering can be 10^3 times larger than for incoherent absorption or excitation cross sections. Several promising schemes are being explored which would work at low temperatures and could be calibrated using reactor antineutrinos.

Additional neutral-current experiments would be highly desirable.

C pp and pep neutrino detectors

The ^{71}Ga detectors that will be used by the GALLEX and SAGE collaborations are the only experiments in progress that can detect the fundamental pp neutrinos. The standard model predicts that about half of the counting rate in the gallium experiments is from pp and pep neutrinos. Most nonstandard solar models predict capture rates for the ^{71}Ga experiments that are similar to the value predicted by the standard model. The minimum predicted rate consistent with nuclear fusion currently supplying the solar luminosity is 61% of the standard solar model prediction, provided that no physics beyond the standard electroweak model affects the neutrinos. On the other hand, some parameter choices for the MSW effect lead to a rate for the ^{71}Ga experiment of less than 10% of the rate of the standard model.

The results of the two gallium experiments will eliminate a number of suggested solutions of the solar neutrino problem that are discussed in this book. We have only to wait a few years to know which sections should be torn out.

The flux of low-energy pp and pep neutrinos can be calculated with a precision of a few percent using the standard model. This precise prediction deserves a precision test.

The gallium experiments are sensitive to all of the solar neutrino sources and are expected to have significant contributions from ^{7}Be and ^{8}B neutrinos, in addition to the pp neutrinos. The response of the ^{71}Ga detectors to ^{8}B neutrinos is uncertain due to the difficulty of calculating the absorption cross sections for transitions to excited states of ^{71}Ge. The total theoretical uncertainty is, for a variety of assumed neutrino spectra, $\sim 10\%$ for a ^{71}Ga detector. In order to carry out a precise measurement of the pp flux, a low-energy detector must be developed that has the ability to measure energies of individual neutrino-initiated events or which is sensitive only to pp neutrinos.

Low-temperature detectors are being studied that may be able to measure the total flux of pp neutrinos that reach the Earth and perhaps measure their energy spectrum. The basic principle is that a little energy has a big effect at low temperatures. Great progress has been made in studying detectors that would be sensitive to coherent neutrino–nucleus scattering. In a parallel series of studies, prototype silicon detectors are being investigated that could register energy deposited by neutrino–electron scattering in the form of ballistic phonons. Superfluid helium is a promising detector of the scattering by electrons of pp neutrinos.

Low-temperature detectors offer the promise of a major breakthrough, although it may be a decade before their use leads to a completed solar neutrino experiment.

An ^{115}In experiment could detect both the pp and pep neutrinos, although the background from the natural radioactivity of indium is difficult to overcome. The neutrino cross sections for ^{115}In should be calibrated with an intense laboratory source of ^{51}Cr.

D ^7Be neutrino detectors

The 0.86 MeV neutrino line carries a lot of information. A measurement of the intensity of the ^{7}Be neutrino flux would determine the relative frequency of the two major modes of terminating the pp chain in the solar interior. The temperature and density distributions in the solar core determine the width of the neutrino line. In turn, the width of the line determines the shape of the spectrum of energies of recoil electrons near the maximum recoil energy. The normalization of the ^{7}Be flux would be a key diagnostic in sorting out which, if any, modifications of the standard electroweak model are favored. The spread in energies is small compared to the mean energy in the neutrino line. Sharp constraints could be imposed on modifications of electroweak theory if a measurement of the ^{7}Be ν_e flux were available, since one does not have to average the predicted effect over a broad spectrum of energies (as is the case for the ^{8}B neutrinos).

An ^{115}In absorption detector could determine the total ^{7}Be neutrino flux. The recoil electrons from ^{7}Be neutrino absorption are more energetic than the intense background from the natural radioactivity of ^{115}In. The absorption cross section for ^{7}Be neutrinos could be calibrated using a laboratory source of ^{51}Cr.

A radiochemical measurement using ^{81}Br is feasible and not expensive (compared to other solar neutrino experiments). The relative contribution of ^{7}Be and ^{8}B neutrinos to the experimental rate depends upon the unknown ratio of their fluxes at Earth. If the ^{8}B flux is suppressed by a factor of four below the standard value and the ^{7}Be flux is not appreciably decreased, then the rate of ^{7}Be events in a ^{81}Br detector is twice the rate of the ^{8}B events. In order to interpret the results with precision, the nuclear matrix elements would have to be determined in laboratory experiments using a ^{51}Cr calibration source and accurate studies of the 190 keV metastable state of ^{81}Kr.

Neutrino–electron scattering experiments could reveal the characteristic knee shape of the electron recoil distribution near the maximum recoil energy of 0.665 MeV. Instead of the discontinuous drop to zero intensity beyond 0.665 MeV that is shown in the illustrative figure (Figure 8.5) in this book, the distribution of recoil energies has a small thermal width of order a keV that conceivably could be revealed by a future low-temperature experiment.

E   CNO neutrino detectors

The CNO fusion reactions are responsible for only 1.6% of the energy generation in the standard solar model, but they are believed to be the principal source of main sequence fusion energy in stars somewhat heavier than the Sun. A measurement of one or more of the CNO fluxes would be a useful test of the standard solar model. Even an experimental upper limit on the fluxes more stringent than the one implied by the ^{37}Cl observations would be a valuable constraint on stellar models. The flux of ^{17}F neutrinos is of great interest since it is proportional to the initial oxygen abundance, but there is no known observational technique that is capable of measuring this flux.

A ^{7}Li detector could provide a valuable measurement of the CNO neutrino fluxes, especially the flux of ^{15}O neutrinos. For a lithium target, the neutrino absorption cross sections can be calculated accurately, the chemistry is simple, and the detector is inexpensive. The difficult experimental challenge is to develop a way of counting the ^{7}Be atoms that are produced. The relative contribution of CNO and ^{8}B neutrinos to the absorption rate by ^{7}Li depends upon which flux is more strongly depressed. If, for example, the ^{8}B flux of ν_e is decreased by a factor of four with respect to the standard model and the (lower-energy) CNO neutrinos are not depleted, then CNO neutrinos will contribute about a factor of three more events than will ^{8}B neutrinos.

F   Temporal variations

Observations with the 0.6 kiloton chlorine detector suggest the possibility that the capture rate may be varying with time, perhaps with an 11 year cycle. Unfortunately, the existing detector is small and does not permit the accumulation of a large number of events in a year (at best, of order 10^2 events per year). Careful monitoring of the capture rate in the existing detector should be carried out for at least another half a solar cycle in order to determine if the intriguing suggestions of variability are indicative of true variations. If variations are established, nearly all of the interpretations of the average 2 SNU event rate will have to be revised. The only proposed theoretical explanation of the solar neutrino problem that predicts a variation on time scales of years is based upon the assumption that the neutrino has a large magnetic moment, many orders of magnitude larger than is expected on the basis of standard particle physics ideas and also larger than suggested by astrophysical limits.

Detectors that have a high counting rate are required in order to perform stringent tests of variability. The predicted rate for the proposed SNO detector is of order 10^4 events per year for absorption of ^{8}B neutrinos by deuterium, a rate that is large enough to make accurate tests for time dependences. Some of the next-generation neutrino–electron scattering experiments (e.g., LVD and Super Kamiokande) and some of the direct counting absorption experiments (e.g., the bigger ICARUS detector and BOREX) may also have sufficiently high rates to study variability accurately. Unfortunately, Kamiokande II and the gallium detectors (GALLEX and SAGE) are rather small for variability studies, although Kamiokande II can provide limits that are comparable to or superior to the existing ^{37}Cl detector. A much larger ^{37}Cl detector has been under development for some time by the Soviet solar neutrino group; a larger detector would serve as a useful check on the existing experiment and a valuable monitor of possible time dependences.

Regeneration of ν_e in the Earth via the MSW effect might cause the Sun to appear brightest at night when observed with electron neutrinos. This effect can be searched for effectively using real-time detectors that record individual events, neutrino–electron scattering experiments and absorption experiments, as well as with the ^{37}Cl experiment (by purging the tank twice a day). The *a priori* probability is not large for observing temporal effects due to regeneration in the Earth since observable variations occur only for a small fraction of the possible solution space for the MSW parameters. However, the product of probability times importance is large. If day–night or seasonal effects due to regeneration were observed, then the neutrino parameters (mixing angles and mass differences) for the MSW effect would be well determined.

There is always the possibility of surprises. Real-time detectors can perform interesting searches for possible correlations of neutrino events with known astronomical phenomena, including flaring on the Sun and in galactic stellar systems.

16.3 Stellar collapses

The rate of stellar collapses in the Galaxy is uncertain and can only be determined observationally from long-term neutrino studies. The average number of collapses may be anywhere from one every 10 years to one every 100 years, depending upon what fraction of the collapses produce optically bright phenomena and upon the uncertain number of massive stars in the Galaxy.

The observation of stellar collapses out to the distance of Messier 31 (the Andromeda nebula) may be possible, but only by increasing the sensitivity by two orders of magnitude over currently operating detectors. Even if the next-generation detectors are able to observe stellar collapses in the Andromeda nebula, the total expected frequency of detectable collapses will still remain uncomfortably long, perhaps of the order of one collapse every 5 to 50 years.

Because of the uncertain rate at which events may be observed, a detector of stellar collapses should be part of a more general underground observational facility. Ideally, an underground observatory can make possible low background experiments in different fields, including particle physics studies (such as proton decay or monopole detection), cosmic ray investigations, searches for high-energy gamma ray or neutrino sources, tests for oscillations using neutrinos from cosmic ray secondaries, and of course solar neutrinos.

The historic observations of SN1987A confirmed three key estimates based upon standard models of stellar collapse. The observations show that the total energy emitted in $\bar{\nu}_e$'s ($\sim 10^{53}$ ergs), the average energy per particle (15 MeV), and the time scale for thermal emission (seconds) are all approximately what was expected from the standard model of stellar collapses. Altogether, about 20 $\bar{\nu}_e$'s were observed. This meager data set also yielded valuable limits on the mass and charge of ν_e, as well as informative tests of the postulates of special relativity and of general relativity. However, the only neutrinos observed with certainty were $\bar{\nu}_e$'s. Most of the energy is expected to be emitted in the form of ν_μ's and ν_τ's and their antiparticles. The only detector operating in the near future that will be sensitive to ν_e (instead of $\bar{\nu}_e$) is the ^{37}Cl detector.

Future detectors should make possible observations of ν_μ's and ν_τ's (and their antiparticles) via neutral-current detectors or by electron–neutrino scattering (in a detector in which absorption of $\bar{\nu}_e$ does not overwhelm electron–neutrino scattering). The neutral-current mode of the SNO deuterium detector would be sensitive to all neutrino flavors and could reveal masses of the ν_μ's and ν_τ's as low as 100 eV.

Massive detectors may be able to resolve some of the dynamical details of stellar collapse by observing enough events to reveal the time evolution of the explosion. The collapse and early rebound phases are expected to give rise to a weak but distinctive signal. The initial phase may contain of order 1% ($10^{51.5}$ ergs) of the total energy release, all of which would be released in a brief burst of approximately 10 ms and which would contain only ν_e's. The remaining 99% of the total energy released may be emitted on a much longer time scale of order 10 s and is believed to be shared approximately equally between all flavors of neutrinos and antineutrinos.

Relic antineutrinos from past supernovae in the Galaxy and in more distant galaxies may be observable with future detectors that measure the energies of individual recoil positrons (e.g., SNO, ICARUS, BOREX, Super Kamiokande, and LVD). The expected isotropy of the relic antineutrinos may provide a valuable experimental discriminant against backgrounds of terrestrial or atmospheric origin. An upper limit of order 2×10^2 cm^{-2} s^{-1} $\bar{\nu}_e$'s with energies between 19 MeV and 35 MeV has been established by Kamiokande II using approximately one year of data taken under excellent operating conditions. The theoretical calculations of the expected flux are uncertain and depend upon the unknown rate of stellar collapses in the past in other galaxies and the high-energy tail of the antineutrino spectrum emitted from different collapses (e.g., from the contemporary formation of neutron stars and black holes, from stellar collapses at earlier epochs and during the period of galaxy formation). Fluxes of order 10^0 cm^{-2} s^{-1} to 10^5 cm^{-2} s^{-1} have been calculated by different authors, with typical energies of order 10 MeV. The upper limit provided by Kamiokande II is sufficient to stifle some theoretical speculations.

16.4   Stellar evolution

A   Radiative opacity

The opacity of matter to radiation determines in part the temperature gradient in the solar center. If the opacity is smaller, the thermal energy created by fusion can escape more readily and a lower temperature gradient is established. If the opacity is raised, the radiation requires a larger temperature gradient in order to escape and to supply the observed surface luminosity. A higher-temperature gradient is associated with a higher central temperature and a more intense ^{8}B neutrino flux. The ^{8}B neutrino flux is sensitive to the temperature gradient because the fusion reaction that produces ^{8}B must overcome a large Coulomb barrier; ions in the tail of the distribution of thermal energies initiate the p + ^{7}Be reaction.

About 55% of the opacity in the central regions of the Sun is produced by inverse bremsstrahlung on ionized hydrogen and helium and by scattering by free electrons. The contributions of these processes to the opacity can be calculated to an accuracy of about 10%.

The remaining opacity ($\sim$ 45% of the total) is caused by bound-free scattering by heavier elements (elements heavier than helium). The opacity of heavy elements causes a significant but not dominant uncertainty in the predictions of solar neutrino fluxes made with the standard solar model. The uncertainties in the heavy element opacity result from the approximate nature of the atomic physics models of the solar plasma and from the observational uncertainties, discussed in the following subsection, in determining the heavy element abundances on the solar surface.

The lack of certainty regarding numerical values of the radiative opacity is a significant obstacle to progress in a number of problems in the theory of stellar evolution. For decades, astronomers have depended upon one primary supplier of radiative opacities; detailed and accurate codes have been developed in stages by a talented group at Los Alamos National Laboratory. A new and sophisticated opacity code has been under development for several years at Livermore National Laboratory. Astronomers have felt unable to compete with the national research and weapons laboratories which have large resources to devote to the development of accurate opacity codes. After decades of development, the atomic and plasma physics codes that support the opacity calculations contain detailed physical descriptions of the plasma, the participating ions, and the atomic physics data.

Different groups should use independent approaches in order to develop their own computer codes to calculate opacities. To date, essentially all opacity calculations have concentrated on deriving the best possible numerical values, not the range of permitted values. The new calculations could produce tables of opacities that are calculated with different simplifying approximations in the physical description or input data. Astrophysicists would then be able to use the opacities calculated with different assumptions to estimate the range of uncertainties in their astronomical studies.

In this book, the uncertainties in opacities were assumed equal to the differences between the best Los Alamos opacities computed in 1982 and a corresponding preliminary set of opacities computed independently and at the same time by a group at Livermore National Laboratory. This 1982 evaluation is the only available detailed comparison of the radiative opacity computed by two approximately state-of-the-art but independent codes for the conditions and element composition of the solar interior.

Based upon the comparison of the Los Alamos and Livermore calculations, the uncertainties in calculating the radiative opacity correspond to an uncertainty of about 0.5 SNU out of a total predicted rate of 7.9 SNU for the ^{37}Cl detector. For comparison, the total uncertainty in the prediction of the standard model is 2.6 SNU. The estimated uncertainty in opacity amounts to 6% of the rate predicted by the standard model for the ^{37}Cl detector. The calculation of opacity causes an uncertainty in the predictions for other solar neutrino experiments that are discussed in this book, ranging from less than 1% for detectors that are sensitive only to pp neutrinos to up to 8% for detectors, like Kamiokande II, that are sensitive just to ^{8}B neutrinos.

B Heavy element abundances

The present chemical composition of the solar surface is assumed, in constructing a standard model, to reflect the initial composition of the entire Sun for all elements that are heavier than helium. The relative abundances of hydrogen and helium are changed by nuclear burning in the solar interior, but the abundances of all of the heavier elements are assumed to be unaffected by solar evolution.

The assumed abundances of the heavy elements influence strongly the calculated values of the radiative opacity. The most significant elements in determining the calculated neutrino fluxes are, in order of their importance: iron, oxygen, and magnesium. For example, a factor of two change in the assumed abundance of iron relative to hydrogen would imply about a 33% change in the predicted ^{8}B neutrino flux.

The major uncertainties in the determination of heavy element abundances are systematic, not statistical. In this book, the fractional uncertainty in the ratio of total heavy elements to hydrogen,

Z/X, was assumed to be 19%. This estimate is equal to the change over the past decade in the best value of Z/X, which is obtained by experts from systematic reviews of the careful measurements of individual abundances that have been made by many different groups. The error estimate used here is slightly larger than the uncertainty that is obtained if the quoted uncertainties given in the literature (which are often not defined) for individual abundances are multiplied by three and assumed to add incoherently.

With the error estimate adopted here, the adopted value of the initial heavy element to hydrogen ratio causes an uncertainty of 25% in the calculated ^{8}B neutrino flux. The corresponding uncertainty is 11% for the ^{7}Be neutrino flux and is 1% for the pp neutrino flux. Great progress has been made over the past two decades in improving our knowledge of the abundances of heavy elements on the solar surface. But, the estimated errors still cause large uncertainties in the calculation of the ^{8}B neutrino flux.

C   Nonstandard solar models

Any quantitative test of stellar models can help clarify the solar neutrino problem. If the physical description of the solar interior is incorrect, then there may be observable consequences for other stars. The relative number of stars that are observed in different evolutionary phases, the spread in luminosity or color at a given phase, the ages and chemical compositions of stars, all of these observable quantities are the result of stellar evolution and can indicate to what extent modifications in standard evolutionary calculations are required or permitted.

Nonstandard solar models should be calculated with the same precision as standard solar models so that both types of models can be tested by traditional astronomical observations, as well as by future solar neutrino experiments. The best available input parameters and physical descriptions should be included in all the model calculations. The range of predicted neutrino fluxes and other measurable quantities should be determined. This has not been done for most of the nonstandard models, because the suggested modifications are not sufficiently specific or detailed to define a unique solar model. However, observable consequences can be calculated with nonstandard models by allowing the altered parameters to vary over a broad range.

Nonstandard solar models, if calculated accurately, can be used to explore what aspects of stellar evolution are critical for understanding existing observations and what aspects may require modifications. Constructing precise nonstandard solar models is not a glamorous occupation, but it is important and should be pursued more actively.

D Helioseismology

The development of helioseismology has made solar structure a precise laboratory science. Thousands of accurate frequencies of solar acoustic (p-mode) oscillations are measured, providing information about the interior density and chemical composition. The observations are in agreement with the standard solar model to a typical accuracy of better than 1% of the frequency of a given normal mode. The pattern of normal modes predicted by the standard solar model, with eigenfunctions that depend upon the radial and angular quantum numbers n and l, represents well the observed pattern. Rotation and magnetic fields break the degeneracy of the spherically symmetric angular eigenfunctions (specified by n and l) and introduce characteristic splittings that are measurable.

The p-mode oscillations are primarily sensitive to the outer region of the Sun, with the greatest amplitudes lying within the solar convective zone. Both turbulence and convection are important in the outer region of the Sun and could be responsible for small discrepancies, at the level of several tenths of a percent of the eigenfrequencies, between the calculated and predicted oscillation periods.

Sophisticated theoretical analyses have been developed to deconvolve the observed frequencies and to produce, for example, an inferred rotational speed as a function of depth in the Sun. The results are consistent with the standard solar model described in this book, although some small splittings suggest that the model may need fine tuning in order to obtain agreement with all of the observations. There is no observation available yet that establishes an unambiguous discrepancy with the standard model description of the solar interior.

The relation between nonstandard solar models and helioseismology could be deepened by calculating accurately the p-mode oscillation frequencies for different nonstandard models. The results of the theoretical calculations could be compared directly with the rich body of precise observational data. For example, one could assume a rotation speed in the solar interior that is a thousand times faster than the surface rotation speed and compute the resulting p-mode splittings. Similar calculations could be performed for nonstandard solar models with strong interior magnetic fields. The observational data may already exist that could eliminate some nonstandard models discussed in this book, but the accurate helioseismological theoretical calculations have not yet been performed.

The frequencies of g-mode oscillations are sensitive to conditions in the stellar interior and can test nonstandard and standard models of the Sun's nuclear burning region. Many attempts have been made to observe the solar g-modes and several claimed detections have been made. However, there is no consensus among observers that g-modes have been detected.

E   Diffusion

The standard solar models described in this book do not include the effect of diffusive processes that change slightly the distribution of element compositions with radius. A numerical formalism has been developed that permits the inclusion of diffusive processes in the most precise standard solar models, but complete calculations have not yet been carried out. Two preliminary estimates show the effect of diffusion on the neutrino fluxes will in general be small, but may increase the calculated flux of ^{8}B neutrinos by of order 10%. Standard solar models that are calculated in the future should include diffusive processes, since they are known to occur and can be described with satisfactory accuracy using equations that are not difficult to incorporate in an evolutionary code.

16.5   Nuclear physics

A   Low-energy reaction cross sections

The Sun shines by virtue of nuclear fusion reactions in the solar core. The cross sections for these nuclear reactions determine to a large extent the calculated solar neutrino fluxes. For example, the flux of ^{8}B neutrinos obtained from the standard solar model is proportional to the rate of the ^{7}Be$(p, \gamma)^8$B reaction. The flux of pp neutrinos depends upon the ratio of the cross section for the ^{3}He$(^3$He, 2p)^{4}He reaction to the cross section for the $^3\mathrm{He}(^4\mathrm{He}, \gamma)^7\mathrm{Be}$ reaction. If the fusion of four protons to form an α-particle were always completed by the $^3\mathrm{He}(^4\mathrm{He}, \gamma)^7\mathrm{Be}$ reaction, then the calculated pp neutrino flux would be about one-half the value that is estimated using the standard solar model and the measured values for the fusion cross sections.

Over the past two decades, many beautiful nuclear physics experiments have been performed to measure accurately the low-energy cross sections for the fusion reactions that are most important in the Sun. These experiments have shown that uncertainties in the nuclear reaction rates are not responsible for the solar neutrino problem. However, significant uncertainties in the reaction rates still limit the amount of information that can be obtained from solar neutrino experiments about the Sun and about weak interactions.

The $^7Be(p, \gamma)^8B$ reaction. The most important nuclear reaction for solar neutrino astronomy is the most difficult to measure and the most susceptible to systematic uncertainties. The rates are small for the $^7\mathrm{Be}(\mathrm{p}, \gamma)^8\mathrm{B}$ reaction at low energies and the target is radioactive. Six difficult experiments have yielded a result with a 1σ uncertainty of 7.4%. The 1σ uncertainty in this reaction must be reduced to 3% in order for the measurements of the nuclear cross section not to limit the interpretation of solar neutrino experiments. This is a Herculean task of great importance.

Thermal neutron absorption by ^3He. The flux of hep neutrinos can be observed in massive next-generation solar neutrino detectors. The rate of hep production is approximately proportional to the cross section for thermal neutron absorption by $^3\mathrm{He}$. A precise measurement of this cross section is needed and is feasible.

Other reactions in the pp chain. The cross sections for the important reactions $^3\mathrm{He}(^3\mathrm{He}, 2\mathrm{p})^4\mathrm{He}$ and $^3\mathrm{He}(^4\mathrm{He}, \gamma)^7\mathrm{Be}$ are relatively well determined after two decades of heroic experiments. The recognized uncertainties in these reactions correspond to 3σ uncertainties of 18% and 13%, respectively, in the interpretation of solar neutrino experiments that are sensitive primarily to $^8\mathrm{B}$ neutrinos. Additional reductions in the errors in the cross sections for these reactions are difficult to achieve, but important.

B   Validating (p,n) determinations of GT matrix elements

The demonstration of an approximate proportionality between the
cross sections for (p,n) reactions at moderate energies ($\sim$ 100 MeV
to 200 MeV) and the allowed Gamow–Teller weak interaction matrix elements is one of the most important contributions to neutrino
astrophysics over the past decade. The (p,n) measurements permit
one, in principle, to determine experimentally the rates of neutrino-
induced transitions to excited nuclear levels. The experimental determinations are important for Gamow–Teller transitions in which
neutrinos populate excited nuclear states whose transition matrix
elements can not be calculated theoretically using isotopic spin in-
variance (for transitions between analogue states). For a number
of the solar neutrino detectors discussed in this book (^{71}Ga, ^{81}Br,
^{98}Mo, ^{127}I, and ^{205}Tl), Gamow–Teller transitions to excited states
represent a significant contribution to the total event rate.

The statistical uncertainties in the (p,n) measurements performed
so far are uncomfortably large, 1σ uncertainties of order 10% to
30%. The experiments are difficult and require large amounts of
time at major accelerator facilities, but the statistical errors could
be reduced by additional measurements of the same type.

Systematic errors may be more important than the purely statistical uncertainties. Because of the paucity of calibration experiments
at large mass numbers, there is at present no rigorous way of establishing 3σ uncertainties on the GT matrix elements determined from
(p,n) measurements for solar neutrino targets. Plausible arguments
suggest that the systematic uncertainties might be approximately
comparable to the current statistical errors, but this is not possible
to demonstrate with existing data.

No fundamental theoretical derivation has been published that
yields the empirical relation, with the observed departures from the
average correlation, between (p,n) cross sections and GT matrix
elements. The absence of an established theory implies that the
uncertainties must be derived by performing many experiments in
which the same matrix elements can be obtained accurately both
from β-decay and from (p,n) measurements.

No precise calibrations of transitions of the approximate strength
that are relevant for solar neutrino experiments have been published
for nuclei with mass numbers in the range of interest. The neutrino

targets for which the (p,n) experiments are most important have mass numbers between 71 and 205.

The most urgently needed (p,n) experiments are measurements in the mass range from $A \sim 70$ to $A \sim 130$ for cases in which the weak interaction matrix elements are known from β-decay. An empirical relation has been established at smaller mass numbers between Gamow–Teller and Fermi matrix elements; it is important to extend this relation to the larger mass numbers of special interest for solar neutrino experiments. A calibration sample that included of order 10 targets with mass numbers larger than 70 would provide a much needed empirical basis for estimating the errors.

In the absence of a rigorous method of establishing the errors, a factor of two uncertainty has been adopted in this book for all neutrino cross sections that are estimated from (p,n) reactions.

16.6 Particle physics

Experiments that use astrophysical neutrino beams provide information about particle physics that cannot be obtained with laboratory neutrino beams. The long path lengths and low energies associated with astronomical neutrino sources provide sensitive tests for neutrino mixing, masses, and electromagnetic moments. This section discusses neutrino oscillations in vacuum and in matter, neutrino magnetic moments, and WIMPs. In addition, two purely theoretical problems are described: the calculation of electroweak radiative corrections to neutrino–electron scattering and the estimation of mesonic corrections to the rate of the pp reaction.

A Vacuum oscillations

Oscillations can occur between different neutrino states if the flavor eigenstates are linear combinations of at least two mass eigenstates and at least one of the mass eigenstates has a nonzero mass. For solar neutrino experiments, the relevant range of neutrino masses is from 10^{-1} eV to 10^{-6} eV.

The probability that a neutrino's flavor is changed in transit is independent of energy for vacuum oscillations, unless the oscillation parameters (neutrino masses, energies, and the Earth–Sun distance) are adjusted precisely, that is, fine tuned.

Measurements of the energy spectrum of solar neutrinos will provide the key test of the hypothesis that vacuum oscillations between neutrinos of different flavors cause the discrepancy between the standard model prediction and the observations. The shape of the spectrum will have the standard calculated form if vacuum oscillations are occurring (without fine tuning); the normalization of the solar neutrino flux will be decreased by a factor of at least 2.5. All experiments should show the same numerical discrepancy, within experimental errors, between the measured and the predicted standard model rate.

This prediction of a large energy-independent deficit will be tested first by the neutrino–electron scattering experiment, Kamiokande II. If vacuum oscillations are occurring, the flux of ^{8}B neutrinos seen by this detector should have the standard energy spectrum but be decreased in amplitude by a factor of between 2.5 and 5. If a sufficiently high signal to noise ratio is achieved, the Kamiokande II experiment may be able to place constraints on the shape of the incident neutrino energy spectrum as well as on the total number of high-energy electron neutrinos.

In the next several years, the two ^{71}Ga experiments will provide a decisive test of the vacuum oscillation hypothesis. The gallium detectors are most sensitive to low-energy neutrinos, unlike the ^{37}Cl and Kamiokande II experiments. If vacuum oscillations are occurring, then the observed capture rate in the gallium detectors will be decreased by the same factor as is observed in the ^{37}Cl and Kamiokande II detectors. The capture rate in the gallium experiment that is predicted by the hypothesis of vacuum neutrino oscillations is lower than is predicted by any stellar model, standard or nonstandard, that is discussed in this book.

Fine tuning of the oscillation parameters has been suggested as a possible explanation for a large reduction of the observed ^{37}Cl rate relative to the predicted standard rate. If the required fine tuning exists, then there will be a semiannual periodicity, caused by the changing Earth–Sun distance, in both the event rate and the shape of the energy spectrum. For high signal to noise experiments, this periodicity may be measurable.

In the next decade, experiments that measure neutral-current interactions may provide unambiguous tests of the hypothesis of vacuum neutrino oscillations. The neutral-current experiments are sensitive to the total number of neutrinos in a given energy range, independent of their flavor. Vacuum oscillations, whether fine tuned or not, will not change the event rates predicted by the standard model for neutral-current detectors.

B MSW effect

The MSW effect can convert neutrinos of one flavor to nearly all neutrinos of a different flavor. The mass eigenstate that the electron neutrino most closely resembles must have a mass that is less than some other neutrino mass eigenstate to which the electron neutrino is also coupled. The mixing angles can be small and no fine tuning need occur. The neutrino mixing angles can vary by more than an order of magnitude (from $\sim 1°$ to $45°$) and the neutrino masses can vary by two orders of magnitude (from 10^{-2} eV to almost 10^{-4} eV) and still solve the solar neutrino problem as defined by existing observational data and solar model calculations.

In the simplest case, in which the electron neutrino is coupled to only one other neutrino flavor, there are three types of solutions. The high-energy (^{8}B) neutrinos may be depleted and the low-energy (pp) neutrinos may be practically unaffected. Or, the low-energy neutrinos may be strongly depleted and the higher-energy neutrinos present with essentially the standard model flux. There is a third solution, similar to vacuum oscillations, in which the depletion factor is practically independent of energy.

The key diagnostic in the near future will be the spectrum of neutrino energies that reach the Earth. Will the Kamiokande II collaboration be able to reduce the errors in the measurement of the high-energy (^{8}B) neutrinos and show that the flux of ν_e's is (or is not) within a factor of a few of that predicted by the standard solar model? Will the low-energy pp neutrinos be detected, at approximately the standard rate, in the ^{71}Ga experiments? The answers to these questions will determine which, if any, versions of the MSW effect survive.

The space of possible MSW solutions becomes large if we consider the mixing angles and mass differences between three different neutrino flavors. The orders of magnitude required to explain a given set of observations are not changed by generalizing the discussion to include three neutrino flavors, but the parameters associated with each neutrino flavor can be affected greatly.

Multiple solutions are possible for the experiments in progress. A given set of hypothetical results with the ^{37}Cl, Kamiokande II, and ^{71}Ga experiments can, in many examples, be explained by invoking one set of neutrino parameters for two coupled neutrinos or by supposing that a coupling between ν_e and a particular mass eigenstate explains the ^{37}Cl and Kamiokande II results (referring mostly to ^{8}B neutrinos) and a separate coupling between ν_e and a different mass eigenstate explains the ^{71}Ga results (referring preferentially to pp neutrinos).

Experiments in progress or planned for the next decade have the potential to demonstrate that the MSW effect operates powerfully upon solar neutrinos and that the electron neutrino interacts with at least one heavier neutrino via a mass difference $\Delta m \sim 10^{-3\pm1}$ eV and a mixing angle $\theta_V \gtrsim 1°$. A particularly dramatic demonstration could occur if solar neutrinos passing through the Earth are reconverted to electron neutrinos after having their flavor changed in the Sun. This terrestrial regeneration could give rise to an observable day–night or seasonal modulation of events produced by solar neutrinos.

Given the limited number of experiments in progress and the unavoidable measuring errors, we probably will not be able to establish in the near future the identity of the ν_e's interaction partner, ν_x, nor even whether the interaction occurs with only one other neutrino flavor. Theoretical prejudice may produce a consensus but the number of experimental facts will be too small to constrain rigorously the MSW solutions to a small parameter domain.

C   Magnetic moment of ν_e

A neutrino with a large magnetic moment ($\sim 10^{-10}\mu_B$) can interact with strong magnetic fields that may be present in the Sun, causing the spin of the neutrino to flip from left-handed to right-handed helicity. The right-handed neutrino would not be observed in solar neutrino experiments since the weak interaction coupling constant for right-handed currents is very small.

The spin flip could occur in the presence of moderately large fields ($\sim 10^3$ gauss) in the solar convective zone. If spin flip does occur in the convective zone, then there will be an observable semiannual periodicity in the ^{8}B neutrino flux. The semiannual periodicity results from the combined effects of the inclination of the plane of the ecliptic to the solar equator, the weakening of the solar magnetic field near the equator, and the small size of the production region of ^{8}B neutrinos. There would also be a an 11 year periodicity that is caused by the waxing and waning of the magnetic field with the phase of the solar cycle. The 11 year periodicity would be in the same sense as the empirical correlation that has been suggested between SNUs and SPOTs; the left-handed ^{8}B neutrino flux should be highest when the field is weakest, that is, when the sunspot number is smallest.

On the other hand, spin flip could occur resonantly at high electron densities in the deep solar interior, requiring somewhat larger magnetic fields ($> 10^4$ gauss). The process is analogous to the MSW effect with additional interactions being possible because of the hypothesized large electromagnetic moments. If spin conversion occurs in the deep solar interior, then no correlation is predicted between the flux of left-handed ^{8}B neutrinos and the sunspot number.

Some astrophysical limits appear to be inconsistent with a magnetic moment as large as $10^{-10} \mu_B$. The strongest limits have been inferred by comparing observations of neutrinos from SN1987A with models of the collapse and physical evolution of a neutron star. The conditions for main sequence solar evolution are milder than the conditions for stellar collapse: evolutionary time scales of order of billions of years (instead of milliseconds), densities of order 10^2 g cm^{-3} (instead of 10^{14} g cm^{-3}) and temperatures of order keV (instead of MeV). Some physicists have questioned whether the astrophysical understanding of the Sun is sufficient to justify worrying about a deficit of solar neutrinos. These same physicists may wonder if the understanding of stellar collapse is sufficient to place a strong limit on the magnetic moment of the neutrino.

D   *Laboratory searches for WIMPs*

Weakly interacting massive particles (WIMPs) that are captured gravitationally by the Sun from the halo of the Galaxy can in principle solve both the solar neutrino problem and the dark matter problem. However, stringent requirements must be met by WIMPs in order to solve these puzzles. The mass of the particle must be approximately in the range 4 GeV to 10 GeV. The scattering cross section must be of order 10^3 times the cross section due to Z^0 exchange and the annihilation rate must be suppressed with respect to the scattering rate.

Laboratory experiments can be performed which search with low background detectors for WIMPs from the galactic halo. Future generations of these detectors have the potential to detect some candidate WIMPs that might be present in the hypothesized abundance. The motion of the Earth around the Sun causes a periodic modulation of the detection rate of WIMPs because of the Earth's changing velocity with respect to the halo particles. The modulation provides a unique signature that can be searched for with high sensitivity.

E   Radiative corrections to neutrino–electron scattering

Several solar neutrino experiments use neutrino–electron scattering as the detection reaction. The cross sections for this process can be calculated accurately in standard electroweak theory. So far, however, only the lowest-order diagrams have been included. The radiative corrections to neutrino–electron scattering should be calculated for energies ranging from 0.1 MeV to 15 MeV. The corrections may be of order several percent.

F   Mesonic corrections to pp reaction rate

Corrections due to the exchange of mesons represent one of the main uncertainties ($\approx 4\%$) in the calculation of the pp reaction rate, corresponding to about a 10% uncertainty in the predicted ^{8}B neutrino flux. The existing calculations could be improved by evaluating the corrections with different theoretical models in order to estimate the uncertainties.

16.7   The bottom line

The discrepancy between the predicted and the observed event rates in the ^{37}Cl and the Kamiokande II solar neutrino experiments can not be explained by a "likely" fluctuation in input parameters using the best estimates of different quantities and their associated uncertainties given in this book. Whatever is the correct solution to the solar neutrino problem, it is unlikely to be a "trivial" error. If the cause of the discrepancy is that the neutrino fluxes from the standard solar model have been calculated incorrectly, then the identification of this astrophysical error will probably have important consequences for the theory of stellar evolution and therefore for many branches of astronomy. If new electroweak physics is the correct explanation, then the many scientists who have helped to establish the solar neutrino problem will have been fortunate in discovering something fundamental about microscopic physics while attempting to test a basic macroscopic theory, stellar evolution.

Appendix

An Account of the Development of the Solar Neutrino Problem

John N. Bahcall and Raymond Davis, Jr. [reprinted from: *Essays in Nuclear Astrophysics* (1982) edited by C.A. Barnes, D.D. Clayton, and D. Schramm (Cambridge: Cambridge University Press) p. 243.]

Some readers may be interested in the history of the solar neutrino problem, a subject not discussed in detail in the main text. The following reprint is an informal account of some aspects of the early development of solar neutrino research. Readers may enjoy learning about the origin of the ^{37}Cl experiment, the steps in the improvement of the observational sensitivity, the struggles to refine the theory and the input data to the models, and the initial reactions of the community. The photograph reproduced below is a nostalgic reminder of a time when a few individuals could, with wise and timely support, conceive of and execute an experiment in neutrino astrophysics.

Long ago. Shortly after the proposal in 1964 that a ^{37}Cl solar neutrino experiment was feasible, three of the people most involved were photographed in front of a small version of the chlorine tank. From right to left, they are Raymond Davis, Jr., John Bahcall, and Don Harmer. [photograph courtesy of R. Davis, Jr.]

12

An account of the development of the solar neutrino problem

JOHN N.BAHCALL AND RAYMOND DAVIS, JR.

Introduction

This chapter is a summary of some of the ideas and events that have led to what has come to be known as the solar neutrino problem. The account given here is based upon recollections of many years past and therefore probably contains many inadvertent errors. We hope, nevertheless, that our memories of pleasant and exciting times may be of some interest to students of nuclear astrophysics and especially to friends of Willy Fowler. At every stage of the story described below, Willy provided encouragement, wise advice, and above all, unequaled enthusiasm and a sense of fun. He has stressed by example that the human aspects of science are at least as important as the strictly technical aspects.

Theory and observation depend upon each other for their significance in solar neutrino research. Without a well-defined predicted counting rate the observed number of captures per day loses most of its meaning. Similarly, the theoretical work derives its motivation from the possibility of observational tests. The calculations required for this problem are detailed, precise, and specific; they are not necessary in making the general comparisons with observations that are appropriate for most other work in stellar evolution research. This synergism between theory and observation in solar neutrino work can be contrasted with the situation in a number of other astronomical fields whose initial development occurred during the period described here. The discoveries of quasars, infrared sources, radio pulsars, x-ray sources, and interstellar molecules all had immediate and obvious significance independent of previous theoretical work. The interdependence of solar neutrino theory and observation has been clearly recognized by the funding agencies. Because of this interdependence, we have found it natural to describe the combined history of the subject as we remember it.

We adopt an unconventional format for this narrative. We list in chronological order the highlights of each year as we remember them. We make no attempt to be complete in reciting references or developments; this would be a complicated task for us to undertake now and inappropriate for the present volume. It would also deprive the story of whatever interest it may possess. Naturally enough, we concentrate on events in which we participated since we know these best. For further discussions and many additional references, the reader may wish to consult: Tombrello (1967), Kavanagh (1972), Rolfs and Trautvetter (1978), Barnes (1980), and Chapters 8 and 9 in this volume for a detailed account of the low-energy nuclear physics experiments; Reines (1967) for a description of some of the solar neutrino experiments that were not continued; and Bahcall and Sears (1972), Kuchowicz (1976), and Rood (1978) for summaries of a variety of nonstandard solar models.

Prior to 1962

It is interesting to note that the early literature on nuclear fusion as the basis of stellar energy production did not mention the possibility of testing the ideas by observing neutrinos. In the great papers by Bethe, neutrinos were not included explicitly in the nuclear reactions (see, e.g., Bethe 1939; Bethe and Critchfield 1938). When these works were written, the Fermi–Pauli theory of β decay was more than five years old. However, the principle of lepton conservation was not clearly articulated and one was not required to balance leptons as well as baryons. One of the earliest discussions of the Sun as a source of neutrinos was the stimulating review article by H. R. Crane (1948), a graduate student colleague of Willy's at Caltech in earlier days. Crane used geophysical evidence concerning the rate of heat production in the Earth to exclude neutrino absorption cross sections in the range 10^{-32} to 10^{-35} cm^2.

In the early 1950s, a radiochemical neutrino detector was developed at Brookhaven by R. Davis, based on the reaction $^{37}Cl(\nu,e^-)^{37}Ar$. This method was suggested by B. Pontecorvo (1946) when he was working at Chalk River, and was later studied independently by L. Alvarez (1949). Pontecorvo gave persuasive reasons why chlorine (or bromine) would be a good detector of neutrinos and why a reactor experiment with either of these detectors might be feasible (Pontecorvo presciently dismissed solar neutrinos as not sufficiently energetic).

The report by Alvarez is a remarkable combination of theoretical and experimental insights. It could easily serve as a model of how to write proposals to do experiments in basic physics – in this case, testing the theory of beta radioactivity with a chlorine detector near a pile. Alvarez made specific suggestions on the chemical procedures, expected neutrino capture cross sections, and estimated background effects. Alvarez stressed that ". . . the most important experimental problems lie in the elimination of the various types of background" (Alvarez 1949), a statement that applies equally well today. It is interesting to note that at the Irvine Solar Neutrino Conference (see Reines and Trimble 1972) Alvarez mentioned that even in 1949 he had considered using a chlorine detector for observing solar neutrinos.

Note that the Pontecorvo and Alvarez proposals to use chlorine as a detector for reactor neutrinos explicitly assumed that neutrinos and antineutrinos were equivalent. In 1948, the double-beta-decay experiment of Ed Fireman on ^{115}Sn indicated that neutrinos and antineutrinos could be Majorana particles. This experiment was discussed in Alvarez's proposal. The original experiment by Fireman was later shown to be invalid by a more specific search for neutrinoless double-beta-decay by Fireman, among others.

In the course of developing the detector, a 3800 l tank of CCl_4 was buried 19 ft below the sandy soil at Brookhaven in order to reduce the cosmic ray background. This experiment gave a crude upper limit to the solar neutrino flux if the Sun operated on the CNO cycle [the limit being 10^{14} neutrinos-cm^{-2}-s^{-1} (Davis 1955)]. In more modern terms, this amounts to an upper limit of about 40,000 SNU. A reviewer of Davis's paper made the following critical but amusing comment:

> Any experiment such as this, which does not have the requisite sensitivity, really has no bearing on the question of the existence of neutrinos. To illustrate my point, one would not write a scientific paper describing an experiment in which an experimenter stood on a mountain and reached for the moon, and concluded that the moon was more than eight feet from the top of the mountain.

It was generally believed by astrophysicists in the 1950s that the Sun operates predominantly on the p–p chain and that the only significant production of neutrinos was from the proton–proton reaction that initiates this chain. These neutrinos have a maximum energy of only 0.4 MeV. Since the chlorine detector has a threshold of 0.86 MeV, it is incapable of detecting these p–p neutrinos. The only neutrinos expected to come from the Sun with sufficient energy to be absorbed by chlorine were those from ^{13}N and ^{15}O in the CNO cycle. Although observing neutrinos appeared to be hopeless, the topic was discussed by Davis with Alastair Cameron, Clyde Cowan, Willy Fowler, and Fred Reines, among others, at Gordon Conferences in the 1950s.

A dramatic event occurred in early 1958 that altered the picture completely. At the New York meeting of the American Physical Society, Holmgren and Johnston (1958) reported that the ^{3}He(α,γ)^{7}Be cross section had been measured at the Naval Research Laboratory and was a thousand times higher than expected! The consequences of this result were pointed out immediately in two letters received by Davis from Willy Fowler and Al Cameron. They suggested, following Bethe's original discussion, that if ^{7}Be was produced, it could capture a proton yielding ^{8}B, which Fowler and Cameron stressed would promptly decay emitting energetic neutrinos. (Both Willy and Al remember that Willy first pointed out the importance of the Holmgren–Johnston experiment in an informal conversation.) The important question was the lifetime of ^{7}Be in the Sun, which would be determined by the total rate of electron capture and proton capture. Both Willy and Al were hopeful that the cross section for proton capture would be large and that the energetic ^{8}B neutrinos might be observed with a chlorine neutrino detector. A handwritten postscript at the bottom of Willy's letter expressed his usual optimism: "It may be possible to use your results to calculate how many neutrinos are emitted by the Sun and thus determine a lower limit on the cross section for ^{7}Be(p,γ)^{8}B astrophysically!" Cameron reported his views in a contributed paper to the American Physical Society at the New York meeting (Cameron 1958a), in a summary in the *Annual Review of Nuclear Science* (Cameron 1958b), and in more detail in a supplement to his Chalk River report on stellar evolution, nuclear astrophysics, and nucleogenesis (Cameron 1958c). Willy discussed the implications of a large ^{3}He(α,γ)^{7}Be cross section in a paper that contained a detailed quantitative treatment of the reaction networks (see Fowler 1958). Davis sent a reply to both letters that contained a calculation of expected rates. The expected capture rate for ^{8}B neutrinos was calculated using an expression obtained from Ed Kelley, a postdoctoral physicist at Brookhaven National Laboratory a few years earlier. If the cycle gave the full yield of ^{8}B (4.3×10^{10}/cm^2/s), there would be 7.7 captures per day per 1000 gal of C$_2$Cl$_4$, that is, 3900 SNU! The crest of this wave of optimism was soon to pass, as will be described later, but for a time the wave rolled on and set the immediate course for the future.

At the time these letters were received, a 1000 gal C$_2$Cl$_4$ experiment was essentially completed at the reactor site of the Savannah River plant. These experiments were performed under 25 m water equivalent of cosmic ray shielding, and the background of 26 ^{37}Ar atoms per day could be explained by cosmic ray interactions. Clearly, it was necessary to move the chlorine detector to a mine if one wished to observe the solar neutrino signal. However, this could not be done immediately because the work at Savannah River consumed the entire experimental effort. Don Harmer and Davis were building a new 3000 gal experiment that was designed to distinguish between the four component neutrino theory (of Mayer, Telegdi and Preston) and the two component theory. They had been urged to build the larger 3000 gal experiment by W. Pauli in a letter to Maria Mayer.

The 1000 gal tank used in the initial experiment at Savannah River was taken to Brookhaven at the end of 1959. After some minor improvements were made, it was moved to the Barberton Limestone Mine (in Ohio) of the Columbia-Southern Chemical Co. This mine was 2300 ft deep and had an enormous excavated volume, nearly a square mile with 32 ft high ceilings. John Galvin and Davis completed the installation in July, 1960, and completed the first experiments in October. Immediately, they found that the solar neutrino capture rate was low, 3 ± 5 ^{37}Ar atoms/day (<4000 SNU), but by late 1960 we expected a low flux of ^{8}B neutrinos!

The critical reaction ^{7}Be(p,γ)^{8}B was studied by Ralph Kavanagh (1960) and the low value (0.027 keV-b) that he found was very disappointing. Kavanagh measured the cross section at two energies, 800 and 1400 keV, by observing the energetic positrons from ^{8}B decay. The whole attitude of Davis (and others) on the possibility of observing solar neutrinos was greatly influenced by this measurement. It was abundantly clear that the detection of solar neutrinos was indeed a difficult problem. The last sentence in the review by

Reines (1960) reflected the general view: "The probability of a negative result even with detectors of thousands or possibly hundreds
of thousands of gallons of C_2Cl_4 tends to dissuade experimentalists
from making the attempt."

1962

Our collaboration began in 1962. Characteristically, it was initiated
by Willy. He was the referee for the paper on beta decay in stellar
interiors by Bahcall (1962a) where it was pointed out that electron
capture rates in stars could be very different from the terrestrial values that had been used previously in most nuclear astrophysics calculations. Calculations of capture rates from continuum orbits were
made including Coulomb effects and the exclusion principle. Willy
described these calculations to Davis who then wrote to Bahcall (in
February, 1962) asking about the rate of electron capture by ^{7}Be in
the Sun. The results for the capture rate of ^{7}Be appeared in Bahcall
(1962b). We have been asking each other questions ever since.

1963

The first calculation of the neutrino fluxes obtained from a detailed
model of the Sun was presented by Bahcall, Fowler, Iben, and
Sears (1963), who evaluated the expected ^{7}Be and ^{8}B neutrino
fluxes with the aid of a quantitative model for an evolved Sun.
These fluxes corresponded to a capture rate of only 0.01 captures
per day in the 1000 gal tank experiment in the Barberton Mine (i.e.,
5 SNU with the presently known values for only the *ground-state*
transitions). This calculation did not provide any encouragement to
build a larger experiment, because even 100,000 gal would only
capture about one neutrino per day according to this estimate.

The collaboration on the calculation of the neutrino fluxes was
typical of the strong interactions that characterized science (and
partying) at the Kellogg Radiation Laboratory. The model was computed by Dick Sears using an energy-generation subroutine and
opacity code originally developed by Icko Iben. The energy-
generation routine was improved by Bahcall and Fowler. The neutrino fluxes were computed by hand by Bahcall from the detailed
model results. Bahcall had come to Kellogg in the summer of 1962
with the idea of stimulating a collaboration that would make use of

the new and more accurate values for the ^{7}Be production and destruction rates (Kavanagh 1960; Bahcall 1962b; and Parker and Kavanagh 1963). Willy was especially important in seeing that the work was actually done. Bahcall, Iben, and Sears were all research fellows in the Kellogg Laboratory, run by Charlie Lauritsen with the active assistance of Willy and Tommy Lauritsen. Iben and Sears were, along with most other astronomers and astrophysicists, more interested in studying evolved stars than in making models of the Sun.

High quality measurements by Parker and Kavanagh (1963) of the reaction ^{3}He$(\alpha,\gamma)^7$Be gave a cross section about a factor of two lower than was deduced from the initial results of Holmgren and Johnston. In an important companion paper, Tombrello and Parker (1963) developed a theoretical model for this reaction which underlies our current understanding of the process. These papers provided an important step in determining more accurately the nuclear physics parameters in the p–p chain.

Davis had been studying for some time the idea of carrying out a large scale solar neutrino experiment with ^{37}Cl. There was not much enthusiasm among astronomers for what was viewed as an expensive experiment and not too much reason to hope that an observation could be performed that would actually detect solar neutrinos.

Even though the prospects for observing solar neutrinos looked dim, Davis was eager to build a 100,000 gal experiment. He has often been asked why this particular size was chosen and the reasons may be of some interest. First, the size was picked to be a hundred times larger than the Barberton experiment because expansion by a factor of 100 appeared feasible. Davis felt that a tank this size could be processed in a reasonable time and trusted that it could be made sufficiently leakproof. The latter specification was necessary in order to prevent the inward leakage of atmospheric argon. The total volume of argon had to be kept small to permit using a proportional counter with a small internal volume (say, 0.5 cm^3). The cosmic ray background was an important consideration requiring a room large enough to contain a 100,000 gal tank at least 4000 ft below the surface. Davis did not know if a suitable mine existed and if so, whether it could be used for a scientific experiment. In early 1963, Blair Munhofen and Davis started looking for deep mines in the United States, even though the theoretical and funding prospects were dismal. James E. Hill of the Bureau of Mines recommended two: the Homestake Gold Mine and the Anaconda Copper Mine. Visits to these mines convinced the Brookhaven scientists that the rock at Homestake at the 4850 ft level would permit the opening up of a cavity large enough to hold the 100,000 gal tank, whereas the rock at the Anaconda would allow only a 14 ft diameter hole at their 4200 ft level. The Anaconda Copper Company was eager for their mine to be used and quoted a very reasonable cost for providing a concrete lined cylindrical hole. However, the Homestake Company estimated a very high cost for a suitable excavation, so it was decided to look for other mines. The Sunshine Silver Mine (Willy loved these names) in Kellogg, Idaho, was considered and their management expressed interest in the project. The depth at the Sunshine Mine was 5400 ft, the rock strength was satisfactory, and their cost estimate reasonable. It seemed that the Sunshine Mine was a suitable location. Thus, although there was no approval for a larger project, there was at least one place where a 10^5 gal experiment could be carried out.

The planning for a solar-neutrino experiment became a practical exercise after Bahcall showed that the expected capture rate for ^{8}B neutrinos was about twenty times larger than previously calculated due to transitions to excited states of ^{37}Ar (especially the superallowed transition from the ground state of ^{37}Cl to the $T = \frac{3}{2}$ state of ^{37}Ar at about 5 MeV excitation in ^{37}Ar). The idea of considering transitions to excited states was stimulated by a question by Ben Mottelson in a seminar Bahcall gave during the summer of 1963 at the Niels Bohr Institute in Copenhagen.

Our combined results suggesting the feasibility of a 10^5 gal ^{37}Cl experiment were first presented in November of 1963 at an international conference on stellar evolution organized by B. Stromgrem and A. G. W. Cameron at the Institute for Space Studies in New York [the proceedings appeared much later, see Bahcall and Davis (1966)]. It is indicative of the then (and perhaps still prevailing) preference by astronomers for studies of more exotic stages of stellar evolution that neither the possibility described by Bahcall and Davis of a solar neutrino experiment nor the solar models of Sears (1966) were mentioned in the conference summary.

Shortly after this conference, Bahcall visited Brookhaven National Laboratory to describe in a physics department seminar his new results on neutrino capture cross sections for ^{37}Cl, and to join

Davis in a crucial discussion with Maurice Goldhaber regarding the desirability of Brookhaven requesting funds to carry out a full scale solar neutrino experiment. Accounts of this meeting have already been published (see Goldhaber 1967; Bahcall 1967). Bahcall and Davis both remember being very worried about the meeting because Goldhaber was the director of Brookhaven and was known to be skeptical of the ability of astronomers to say anything correct about anything interesting. We planned to discuss two points. First, we hoped that the nuclear aspects of the new absorption cross sections might intrigue the director (both Goldhaber and his wife Trudy are distinguished nuclear physicists). Second, we tried to stress that a failure of the theory to predict the correct capture rate in a solar neutrino experiment would be the most scientifically interesting result possible because it would confirm his (Goldhaber's) conviction that astrophysicists did not really know what they were talking about. At the present time, we do not remember what, if anything, Maurice agreed to in this initial meeting. However, in a later published account, Goldhaber agreed publicly with the idea that we had not understood what we were talking about (see Reines and Trimble 1972, pp. d-1 and d-2). Nevertheless, Charlie Lauritsen, who was both one of America's most distinguished scientific statesmen and a good friend of Dick Dodson, then the chairman of the Chemistry Department at Brookhaven (and a research fellow in Kellogg in 1940), was successful in mobilizing important support.

The realization that neutrino capture to the analog state in ^{37}Ar greatly increased the total capture rate made an enormous difference in Davis's view of a 100,000 gal experiment. It seemed to him that the analog state was a beautiful new concept in the present context that *should* appeal to nuclear physicists. Moreover, the total expected capture rate was increased to about 4 to 9 per day, making the experiment seem more reasonable. Since only the neutrinos from ^{8}B were sufficiently energetic to feed the analog state, the experiment was very sensitive to the magnitude of the flux of ^{8}B neutrinos. The reactions producing ^{8}B were in turn very sensitive to the interior temperatures of the Sun and these temperatures depended (in the models) upon the solar composition and calculated opacities. Thus, the chlorine experiment could be considered as a way of measuring the central temperature of the Sun, which Davis felt should appeal to astrophysicists (Bahcall 1964a set an upper limit of 2×10^7 K for the central temperature of the Sun using the Barberton results and claimed that a measurement of the ^{8}B solar neutrino flux accurate to 50% would determine the central temperature of the Sun to 10%). The importance of the ^{8}B flux for the chlorine experiment also made clear that the ^{7}Be(p,γ)^{8}B and ^{3}He(^{3}He,2p)^{4}He cross sections had to be remeasured to obtain values of comparable accuracy to the ^{3}He(α,γ)^{7}Be cross section measured by Parker and Kavanagh.

1964

A large fraction of the most important theoretical ideas and suggestions for experiments were first described in this year.

We published companion papers in the March issue of *Physical Review Letters* (Bahcall 1964a, Davis 1964). These two articles were originally intended to be one paper but we could not squeeze all we wanted to say into one (letter-sized) article, so we divided the subject into theoretical and experimental aspects of the proposed solar neutrino experiment. Willy urged us to publish these papers to present our plans to the scientific community; he felt this was an essential step toward funding the project. Bahcall's paper described both the calibration of the suggested detector (i.e., the neutrino absorption cross sections) and the neutrino fluxes (from Bahcall, Fowler, Iben, and Sears 1963, and the very important discussion of uncertainties in the values of the calculated fluxes by Sears 1964). This permitted him to make the first quantitative prediction of the rate expected for the ^{37}Cl experiment: (40 ± 20)SNU, where we have used the acronym SNU for 10^{-36} capture per target atom per second. Davis reported the results of a pilot experiment that used 1000 gal of perchlorethylene. The measurements were carried out in the limestone mine of the Pittsburgh Plate Glass Company of Barberton, Ohio, at a depth of 1800 m water equivalent. The limit reported from the 1000 gal experiment was 300 SNU. Davis showed that a tank containing 100,000 gal of perchlorethylene would permit detection of the predicted capture rate and that the expected backgrounds would be small.

There was a flurry of activity and discussion in the United States following these analyses. In one of his letters to Bahcall from this period, Davis mentioned the sometimes unappreciated side benefits of publicity (an article in *Time* had just appeared) in finding a suitable mine for the experiment and procuring a satisfactory tank for the liquid detector: ". . . these tank people take us more seriously after the article in *Time.*"

Pontecorvo told us many years later (at a conference in Leningrad) that he reported on our two papers at a special seminar in the Soviet Union about this time. According to his account, he was the only person present who expressed the opinion that it would be a successful experiment.

Reines and Kropp (1964) reported in April an upper limit on the flux of ^{8}B neutrinos that was obtained from an experiment designed for other purposes. Their limit referred to elastic scattering of electrons by ^{8}B neutrinos. The upper limit was equivalent to 1000 SNU. Bahcall (1964b) then showed that the electron-scattering experiment proposed by Reines and Kropp could determine the direction of the incoming neutrinos to better than ten degrees.

Fred Reines and his associates were engaged in building a large scintillation counter system for observing cosmic-ray produced neutrinos. They were working in the deepest mine in the world (10,000 ft), the famous East Rand Proprietary Gold Mine near Johannesburg. Fred's team also built a 4000 gal scintillation counter in this location in order to observe solar neutrinos from ^{8}B decay (see Reines 1967), an impressive achievement.

A more detailed description of the nuclear physics calculations for the ^{37}Cl experiment was published in *Physical Review* (Bahcall 1964c). This paper contains also an extensive discussion of possible nuclear physics measurements that could reduce the uncertainties in the calculation of the neutrino absorption cross sections. Figure 1 of this paper is particularly interesting from a historical point of view because it shows that only a few nuclear states were then known in the mass-37 quartet (^{37}Cl, ^{37}Ar, ^{37}K, and ^{37}Ca); by now, hundreds of states are known. Bahcall predicted the existence of a particle-stable ^{37}Ca isotope that would decay by positron emission in the order of 130 ms.

Initially, it was suprisingly difficult to interest any experimental group, with the appropriate facilities, in searching for ^{37}Ca. Vigorous attempts by both of us and Willy were unsuccessful (and discouraging) until Charlie Barnes proposed during a discussion in the coffee room at Kellogg Laboratory, that ^{37}Ca could be best studied experimentally by searching for the delayed protons emitted by the highly excited states of ^{37}K that would be produced by ^{37}Ca positron decay. It was pointed out (Bahcall and Barnes 1964) that the matrix elements from the ground state of ^{37}Ca to the excited states of ^{37}K that are proton emitters are essentially the same matrix elements that are most important for calculating the capture cross section for ^{8}B neutrinos. Barnes's suggestion of searching for delayed protons provided the experimental twist that stimulated the experiments leading to the present secure estimate of the cross section.

July 1964 was a crucial month for the funding of the 100,000 gal experiment and Willy played an important role in obtaining approval for the project. On July 27, Dick Dodson wrote to Willy describing the budget considerations that were then underway in Washington and requested an authoritative statement from him regarding the importance of doing the experiment. Dick posed the problem in a way that says a lot about the climate of opinion at the time. He wrote, "I suppose one can reduce, somewhat crudely, the question we need to answer to: why spend a substantial sum trying to measure something which is calculated with great confidence by nuclear astrophysicists – and who cares about confirming the central temperature of the sun anyway?" Willy answered on July 31 with a department chairman's dream letter. He wrote, "The Brookhaven solar neutrino experiment has my enthusiastic support. . . . The observation of solar neutrinos and the detection of the flux at the earth is crucial to further progress in nuclear astrophysics and to related efforts in thermonuclear research and the space sciences." Willy went on to describe the relation of the solar neutrino experiment to ". . . a diverse set of terrestrial experiments and calculations which are of considerable practical importance." This formal letter was typed and addressed to "Dr. Dodson." In an accompanying handwritten note addressed to his old friend "Dick," Willy offered to supply further material if it were required.

Two systematic studies of uncertainties in the prediction of solar neutrino fluxes were also carried out during this year. Sears (1964) published a study of the effects of various uncertainties on the solar-model calculations. This was a very significant article that strongly influenced thinking regarding a new experiment. The first sentence of Sears's article contains an interesting apologetic disclaimer to his astronomical colleagues: "Theoretical models of the internal structure of the Sun are no longer at the frontier of the theory of stellar structure and evolution." He concluded that the flux of ^{8}B neutrinos could be estimated to within a factor of two, the primary uncertainty being the initial homogeneous solar composition. Sears calculated a ^{8}B flux of $(3 \pm 1) \times 10^7$ cm^{-2} s^{-1}. Pochoda and Reeves (1964) also published the results of a calculation of the neutrino fluxes from a solar model constructed by Martin Schwarzschild and Pochoda. In a note added in proof, they pointed out that when Bahcall's neutrino absorption cross sections were used, the capture rate corresponded to 38 SNU. In their excellent article, Pochoda and Reeves also noted that the calculated increase in solar luminosity with time (from the initial main sequence stage to the present) would have deep effects on the history of the solar system, a topic that was much discussed some ten or so years later. A detailed study of the termination of the proton–proton chain was performed by Parker, Bahcall, and Fowler (1964), who investigated a variety of deuterium and helium-burning reactions. Several skeletons in the nuclear closet were unearthed during the course of this later work, the most important being the systematic uncertainties in the then available data on the ^{3}He(^{3}He, 2p)^{4}He cross section.

Various aspects of the subject of solar neutrino astronomy were reviewed at the Second Texas Symposium on Relativistic Astrophysics in the middle of December (see the talks in the proceedings by Bahcall 1969a; Davis et al. 1969; and Reines 1969).

Plans to build the solar neutrino experiment in the Sunshine Silver Mine collapsed about a month before the 1964 Texas Conference. Funds were apparently available, but suddenly there was no suitable mine. During the conference, Blair Munhofen returned to the Homestake Mine and asked them to reconsider the project. They quickly reviewed the costs and presented Brookhaven with a very favorable estimate for excavation, $125,000. Homestake provided a detailed design of the rock cavity and was anxious to begin work in the spring of 1965. We were of course very pleased with their plans and their cooperative attitude. As a consequence of the larger facilities, it was necessary to have the tank fabricators, who were also pleased with the larger facilities, rebid for constructing the tank.

Following the Second Texas Conference, G. T. Zatsepin and A. E. Chudakov of the Lebedev Institute of Moscow visited Brookhaven and learned of the detailed plans for the Homestake experiments. These Soviet scientists were very interested in establishing a program of neutrino astronomy in the Soviet Union. They were developing a chlorine experiment and also large scintillation counters. During their visit, Zatsepin gave us a curve showing the calculated cosmic ray muon background for the chlorine experiment as a function of depth underground. It was both gratifying and useful to have an independent calculation of this important parameter (see O. Ryajskaya and G. Zatsepin 1965). The visit of Zatsepin and Chudakov was the first of a number of valuable discussions with these outstanding scientists about problems and developments in neutrino astronomy.

The discovery of ^{37}Ca was reported in companion *Physical Review Letters* in late December. Hardy and Verrall (1964) and Reeder, Poskanzer, and Esterlund (1964) reported independent experiments detecting the delayed protons following the positron decay of ^{37}Ca; the measured lifetime (170 ms) was in satisfactory agreement with the predicted (130 ms) value. Thus, in this one year the issue of the enhanced ^{8}B cross section was raised and settled. Bahcall remembers the phone call from Poskanzer (which appropriately enough came during another Kellogg coffee hour) notifying him of the detection of ^{37}Ca decay with approximately the predicted lifetime as the most exciting and satisfying single moment of his professional career.

In the enthusiasm of the moment, we discussed other possible experiments. Bahcall (1964d) proposed, near the end of the year, a program of neutrino spectroscopy of the solar interior that was to be carried out with a variety of targets. In a remarkable example of accidental prophecy, he suggested, "If no neutrinos are observed in the Davis–Harmer experiment, it will be even more desirable to try to observe the low-energy (p–p and ^{7}Be) neutrinos." The use of ^{7}Li was also advocated here for the first time.

1965–1967

This period was relatively quiet on the theoretical front. Most astrophysicists were concerned with quasars and other problems that have come to be called high energy astrophysics. If they took notice of solar neutrinos at all, theorists appeared to be waiting for the observations to confirm the predictions. A few independent solar models were published by Ezer and Cameron (1965, 1966) and by Weymann and Sears (1965). It is indicative of the mood of the time among astrophysicists that Weymann and Sears did not calculate neutrino fluxes from their improved solar model. The neutrino fluxes calculated by Ezer and Cameron (1965) for a particular solar model correspond to a capture rate of 15 SNU, using the then available estimates for neutrino absorption cross sections (Bahcall 1964c).

One new idea is worth noting, mainly because of its simplicity. Bahcall (1966) pointed out that the capture rate in SNU could be calculated accurately without the aid of solar models if the Sun were assumed to shine by the CNO cycle. Each conversion of four protons to an alpha particle results, in this case, in the production of one ^{13}N and one ^{15}O neutrino. At the time, Bahcall was concerned that the prediction (of order 30 SNU) from the CNO hypothesis agreed, within the errors, with the prediction obtained from detailed solar models that showed that the proton–proton chain was dominant. He recalculated neutrino fluxes from detailed printouts of solar models generously made available by Ezer and Cameron; Iben, Weymann, and Sears; and Sears, using a somewhat improved computer routine for calculating neutrino fluxes. His result was a capture rate that lay between 15 and 60 SNU, with a best estimate of 30 SNU. It appeared therefore, at the time, that even neutrino observations with a ^{37}Cl detector might not be able to distinguish between the CNO cycle and the p–p chain.

There were a number of important laboratory experiments that measured the low-energy behavior of nuclear reaction processes occurring in the proton–proton chain (see the discussion in Kavanagh 1972). Willy played a major role in encouraging and supporting all of these experiments.

Parker (1966) remeasured, in a classic experiment performed at Brookhaven, the crucial ^{7}Be(p,γ)^{8}B production cross section. Davis prepared the ^{7}Be source for this experiment by a new technique, and Parker employed in his experiment a method that was superior to the earlier procedure of Kavanagh; the new method involved flipping the target, after exposure to the beam, in front of a silicon detector that observed alphas following the decay of ^{8}B. Parker's method also permitted the measurement of the number of ^{7}Li atoms on the target by a d-p reaction on ^{7}Li, which also produced alphas. Both of us were in constant touch with Parker during the long series of tests involved in this experiment since the predicted capture rate for the ^{37}Cl experiment depends almost linearly on the rate of this reaction. Parker's value was about a factor of two greater than obtained previously in Kavanagh's (1960) pioneering study, an encouraging result for the would-be solar neutrino experimentalist. Much of the experimental and theoretical work during this period concentrated on the ^{3}He(^{3}He,2p)^{4}He reaction (following the realization by Parker et al. (1964) that it was highly uncertain). This work eventually led to improved measurements of the low-energy cross section factor, which led to a predicted neutrino flux from ^{8}B decays that was reduced by a factor of two from the original (1963 and 1964) results. Kavanagh (1972) summarized the experiments by Bacher and Tombrello (unpublished), Dwarakanath, Winkler, and others that ultimately yielded a cross section factor five times larger than the value recommended by Fowler (1954) on the basis of less accurate experiments; the 1954 value was used in the early calculations of solar-neutrino fluxes. Shaviv, Bahcall, and Fowler (1967) stressed that the value of the low-energy cross section factor for this reaction was the "major nuclear physics uncertainty" in the prediction of the neutrino capture rate. They computed solar models with values of this parameter that differed by as much as a factor of 50!

Iben, Kalata, and Schwartz (1967) computed the decay rate of ^{7}Be due to the capture of *bound* electrons, and showed that the bound electron capture rate increases the total electron capture rate in the Sun by about 20%, reducing the predicted proton capture rate by about the same percentage.

Kuzmin (1965) raised again the question of the possible role in the proton–proton chain of the reaction ^{3}He(p,e$^+$)^{4}He. This reaction is potentially very important because it emits energetic neutrinos (maximum energy = 18.6 MeV) that have a large cross section for absorption by ^{37}Cl (Bahcall 1964a). Moreover, the rate of this reaction is not as temperature sensitive as are the rates of the reactions that produce ^{8}B neutrinos. The possibility of an additional source of high-energy neutrinos was stimulating to the experimentalists, but the careful analysis of Werntz and Brennan (1967) showed that the cross section for proton capture by ^{3}He in the Sun was so low that the neutrinos produced would be too rare to be observed.

In a short but important paper, Kuzmin (1966) pointed out the advantages of a radiochemical solar neutrino experiment with ^{71}Ga as the detector (see also, Kuzmin and Zatsepin 1966). He stressed the importance of the low threshold which permits the detection of the fundamental proton–proton neutrinos, the large ground-state cross section, the experimentally convenient half life (11.4 days), and the relatively large K-capture energy (12 keV). He wisely did not discuss the availability or expense of the required large amount of gallium, the main problem today in carrying out a solar neutrino experiment with this detector.

The experimental solar neutrino effort was devoted during these two years to building the 100,000 gal chlorine detector. Already by the end of 1964, the Homestake Mining Company had agreed to let Brookhaven build the detector in their mine. Excavation of the rock cavity that they designed for the installation was started in May, 1965 and completed in about two months. It was an exciting day for Blair Munhofen and Davis when they were first shown the 30 × 60 ft room with a 32 ft ceiling. They were brought into the room and immediately started looking around with miner's lamps. Suddenly, the lights were turned on and they could see the enormous room with its walls covered with chain-link fencing, the concrete floor with pedestals for the tank supports, and the monorail for the lifting hoist 32 ft above. The Homestake people were very pleased with the room and delighted that it also pleased the Brookhaven experimenters.

Building of the tank by the Chicago Bridge and Iron Company (CBI) was started in the summer of 1965. Don Harmer, Blair Munhofen, and Ray Davis had visited their plant in Salt Lake City to check the steel plate for the surface alpha emission rate. The alpha yield of ^{37}Ar from perchlorethylene had been measured by dissolving ^{222}Rn in the liquid as an alpha source. Based on these measurements, limits were set on the total acceptable alpha emission rate from the tank walls and in the liquid itself. There was great concern about natural alphas producing ^{37}Ar in the detector by the sequential reactions ^{35}Cl$(\alpha,p)^{38}$Ar and ^{37}Cl$(p,n)^{37}$Ar. If sulfur were present in the liquid, ^{37}Ar could also be produced by the ^{34}S$(\alpha,n)^{37}$Ar reaction. The sulfur reaction was studied with ^{222}Rn dissolved in carbon disulfide (see Davis 1969). After selecting the steel, CBI fabricated the tank parts and shipped them to Homestake. All parts were designed to fit the shaft hoist and mine tunnels. CBI personnel said later that they ordinarily would not have been interested in building a small, rather conventional tank such as was required for the neutrino experiment but they were intrigued by the aims of the project and the unusual location. Another critical feature of the tank and pumping system is that it had to be absolutely leakproof to prevent the inward leakage of atmospheric argon. CBI engineers were experienced in making vacuum leak tests on large vessels with helium, using a mass spectrometer as a detector. They had built many space chambers and large Dewars for NASA. After the tank was complete, it was evacuated and leak tested. The two 500 gal/min liquid pumps were canned rotor pumps, designed and built especially for the experiment by the Chempump Company. Canned rotor pumps have the armature and impeller in a single, sealed can containing the liquid being pumped. This design avoids using a shaft seal, and can therefore be made permanently sealed and leakproof. The liquid is circulated by these pumps through a set of eductors that are used to force helium through the liquid. A system of 40 eductors provides vigorous agitation and a thorough mixing of the helium purge gas with the liquid. The eductor system was designed at Brookhaven and tested in the Brookhaven swimming pool.

During the period when the tank was being designed, Davis received a letter from B. Kuchowicz of the University of Warsaw suggesting that a ^{64}Cu neutrino source could be used to test the neutrino capture cross section calculations. This isotope has a half life of 12.5 h and is too short-lived to be suitable for a practical test. However, his letter prompted the Brookhaven group to modify the internal piping to allow a reentrant well to be installed in the center of the tank. This design feature may yet be very useful. In 1965, Bernard Manowitz of Brookhaven suggested that [152]Eu be used as a calibration source. He pointed out that its long half life and the fact that europium is used as a reactor control rod made it an attractive and available source. The possibility of using an intense neutrino source to test the calculated neutrino capture cross sections and the chemical extraction and counting procedures has been discussed many times since.

The vessel was completed in the summer of 1966. The final step was to thoroughly clean the inside of the tank by shot blasting and scrubbing with solvent. In order to insure that the alpha emission rate from the inside walls of the tank was below the acceptable levels, selected areas of the tank were checked with a windowless proportional alpha counter that could cover an area 60 × 180 cm. Next, the cover flange was installed and the filling began. Ten railroad tank cars of perchloroethylene were brought one by one to the head of the shaft. Dutch Stoenner of Brookhaven had previously checked the alpha content of the perchloroethylene in samples from each tank car before it left the Frontier Chemical Company's plant in Wichita, Kansas. The liquid was transported to the 100,000 gal tank nearly a mile underground by a set of three 650 gal tank cars designed to fit the hoist and mine rail system. The work was completed in five weeks with the aid of the Homestake hoistman and five Brookhavenites. Then, the processing system was installed and a long series of preliminary purges were carried out to remove dissolved air and reduce the amount of atmospheric argon present to less than a few tenths of a cm³. Once this was accomplished, it seemed clear that the detector would indeed work as planned and a sensitive measurement of the solar neutrino flux could be made.

In the course of removing atmospheric argon from the tank, a relatively small sample of argon of about 6 cm³ was finally obtained and brought back to Brookhaven. There was little interest in this sample because it was too large to put in the small counters. However, Dick Dodson suggested that the sample should be measured somehow, as it was the first sample from the tank. It was measured in a relatively large counter and surprisingly, the rate was 6 counts/min! The high level of activity was explained by the presence of ^{85}Kr in the sample from atmospheric krypton dissolved in the huge volume of perchloroethylene. The krypton was easily removed by gas chromatography. Dutch Stoenner showed Davis how to set up a simple gas chromatograph with a charcoal column, and this became an essential step in the gas purification procedure, not only to remove ^{85}Kr but also ^{222}Rn.

This part of the story would be grossly incomplete without at least some mention of some of the other people who helped to make it happen. During the design and building phase, Don Harmer spent a year at Brookhaven helping on the solar neutrino experiment while on leave from Georgia Tech. Kenneth Hoffman, a young but very experienced engineer, provided guidance through a number of problems. The Homestake Mining Company and the Chicago Bridge and Iron Company provided excellent cooperation throughout the building phase; the Homestake company has continued to be an active partner in the operation of the detector they helped build. The Homestake people that have been directly connected with the project are the mine supervisors Donald Delicate and Joel Waterland, the research and planning engineer Albert Gilles, and James Dunn of the public relations office. During the construction phase, Jim Dunn and Don Howe devoted four issues of the company magazine, *Sharp Bits*, to the project. All of these men contributed in an important way to the solar neutrino experiment by their enthusiastic support, valuable advice, and direct help.

Meanwhile, an active program was underway to develop direct counting detectors for observing the energetic neutrinos from ^{8}B decay in the Sun. One of these experiments, the 4000 l scintillation counter designed to detect solar neutrinos by elastic scattering, has already been mentioned. A second detector used ^{7}Li as a neutrino absorber in the form of a half ton of lithium metal (see Reines and Woods 1965). A third detector was built by Tom Jenkins and his associates at Case Western Reserve. This experiment contained 2000 l of D_2O and was designed to be operated as a Cerenkov detector for the electrons produced by neutrino capture by deuterium. All of these detectors were built (see Reines 1967) but were eventually abandoned after the chlorine experiment showed that the ^{8}B flux was low.

1968

The last systematic theoretical calculation of the solar neutrino fluxes to appear before the first experimental results were obtained was a detailed paper on the uncertainties in the predicted rate by Bahcall and Shaviv (1968). These authors varied all of the parameters within the limits that were then believed to be a plausible range of uncertainties and obtained a predicted capture rate (assuming the uncertainties combined as statistical errors) that lay between 8 SNU and 29 SNU. The lower values calculated in this paper were primarily the result of the much increased estimate for the cross-section factor for the ^{3}He–^{3}He reaction, as discussed above. In a short note, Bahcall, Bahcall, Fowler, and Shaviv (1968) pointed out that the cross-section factor for the ^{3}He$(\alpha,\gamma)^7$Be reaction was also highly uncertain and derived predicted capture rates that lay between 7 and 49 SNU. In a statement that could have been written today, these authors closed their paper with the following exhortation: "We urge that additional low-energy cross section measurements be made for the reactions ^{3}He$(\alpha,\gamma)^7$Be and ^{7}Be$(p,\gamma)^8$B in order to reduce the large uncertainty in the predictions for the neutrino experiments designed to test the theory of nuclear energy generation in stars."

The major results for both the observations and the revised theoretical estimates were presented again in a pair of papers in *Physical Review Letters* (Davis, Harmer, and Hoffman 1968; Bahcall, Bahcall, and Shaviv 1968). It is surprising to us, and perhaps more than a little disappointing, to realize that there has been very little quantitative change in either the observations or the standard theory since these papers appeared, despite a dozen years of reexamination and continuous effort to improve details (see Figs. 1–3).

The first results for the search for solar neutrinos by Davis, Harmer, and Hoffman (1968) yielded an upper limit of 3 SNU, based on the results of the initial two runs. The operating experimental system was described in this paper as well as various tests and limits on the backgrounds, the recovery efficiency, and the counting efficiency.

The accompanying theoretical paper by Bahcall, Bahcall, and Shaviv (1968) gave a most probable rate of (7.5 ± 3) SNU, with specified assumptions regarding the uncertainties in various parameters. Despite the subsequent careful examination of dozens of effects and parameters, documented in many highly detailed papers, the best estimate for the capture rate predicted by the current standard solar model has never fallen outside this range (although there have been many fluctuations up and down within the quoted range).

When these papers appeared, Ed Salpeter wrote an incisive review of the experimental and theoretical results in *Comments on Nuclear and Particle Physics* (Salpeter 1968). His review contained the wise and felicitously worded summary statement: "Thus, at the present time, we neither have a positive identification of solar neutrinos nor the morbid satisfaction of predicting a scandal in stellar evolution theory!"

It is instructive to compare the 1968 calculation with the earlier results from 1963 and 1964 (see Bahcall et al. 1963, Sears 1964; and Bahcall 1964a). The 1968 calculation took account of the larger cross-section factor for the ^{3}He–^{3}He reaction discussed above (a reduction factor of about 0.6); a more detailed calculation of the proton–proton reaction rate by Bahcall and May (1968) who also made use of a more accurate measurement of the lifetime of the neutron by Christensen et al. (1967) (all of which resulted in a re-duction factor of about 0.7); and an improved determination by Lambert and Warner (1968a,b) of the heavy element to hydrogen ratio on the surface of the Sun (a reduction factor of 0.5 using their new value of $Z/X = 0.02$). These three changes were all in the same direction and resulted in a net reduction factor of $0.6 \times 0.7 \times 0.5 = 0.2$, that is, a reduction from about 40 SNU to an estimated 7 or 8 SNU.

Another discussion of the first experimental results was given by Iben (1968), who constructed a large number of solar models in order to illustrate the parameter dependence of the neutrino fluxes. Iben used the primordial helium abundance by mass, Y, as a com-position parameter to be varied in obtaining consistency with the results of the solar neutrino experiment. He deduced an upper limit of $Y = 0.16$ for the primordial helium abundance by demanding consistency of the neutrino results with standard solar models and the accepted values of the nuclear parameters. His inferences dif-fered from the parallel investigation of Bahcall and Shaviv (1968) who chose the photospheric ratio of heavy element to hydrogen abundance to be their composition parameter (determined by ob-servation) and found values of the helium abundance consistent with other astronomical determinations (albeit in conflict with the solar neutrino observations for the best-guess model parameters).

Ezer and Cameron (1968) made the first serious proposal of a nonstandard solar model that would be consistent with the ob-served upper limit on the neutrino capture rate. They suggested that the Sun was thoroughly mixed, fresh hydrogen being contin-ually brought into the central regions. This process would permit the proton–proton chain to proceed at a lower central temperature than in standard models and could reduce the predicted capture rate in extreme cases to one-fourth the value calculated for the standard models. Arguments were immediately given against the likelihood of maintaining such extreme mixing (see Bahcall, Bah-call, and Ulrich 1968; Shaviv and Salpeter 1968), but the idea was an important one because it was the forerunner of many related suggestions.

G. Zatsepin, one of the earliest and most influential enthusiasts for solar neutrino experiments, organized an international con-ference on neutrino physics and astrophysics in Moscow in September, 1968. Bahcall (1966b), Davis (1969), and Kuzmin and

Zatsepin (1969) all wrote summaries of various aspects of the solar neutrino problem for this conference. The Moscow meeting was an occasion for discussing informally what to do next, given the recently discovered discrepancy between theory and observation. Davis gave a detailed account of the 100,000 gal experiment including the design of the detector, the tests of the recovery efficiency, the counting procedure, and the first observational results (capture rate less than 3 SNU). During the conference, a number of young Soviet physicists asked many questions about the details of the design. It was clear that Zatsepin's group was actively engaged in building a chlorine detector in the Soviet Union. Bahcall was not able to attend because his first child was born only nine days after the conference concluded (Bahcall's paper, including the jokes, was read to the conference somewhat uncomfortably by Davis). In lieu of being able to express an informal opinion in person, Bahcall wrote in his manuscript:

> It seems to me most likely that nature has been nasty to us and that some of the experimentally-measured parameters, S_{17}, S_{11}, Z, and perhaps others . . . are different than we originally believed. I feel especially uncertain about the extrapolated value for S_{17}. I think, however, Davis will ultimately measure (provided a lower sensitivity is possible) a capture rate between 1 SNU and 3 SNU; otherwise there will be a serious conflict with the theory of stellar interiors.

Kuzmin and Zatsepin (1969) expressed very similar attitudes, stressing the need to remove the uncertainties in the experimental values of S_{17} and S_{34} before conclusions could be drawn regarding the possible astrophysical importance of the discrepancy between the predicted capture rate and the observed limit. They drew special attention to the broad spread in values of S_{17} that resulted when this cross section factor was determined in different ways (cf. Kavanagh 1960; Parker 1968; and Tombrello 1965).

The conference proceedings also reflect some of the fun we have had with our subject and our colleagues. When asked by A. Wolfendale the cost of the experiment, Davis replied: "Ten minutes time on commercial television ($600,000)." Also, the text of a carefully drafted and detailed bet between the late Jon Mathews (Professor of Theoretical Physics, Caltech) and Bahcall was shown; Bahcall agreed to pay Mathews two dollars if an upper limit of less than 1 SNU was established in the ^{37}Cl experiment. It should be obvious from the above financial data that, at least in this subject, we have always valued experiments more highly than theory.

1969–1977

This period was devoted largely, both theoretically and experimentally, to the reexamination and validation of inferences whose basic outline had been established in the previous five years. For example, in response to the urgings of Bahcall, Bahcall, Fowler, and Shaviv (1968), Kavanagh and his collaborators at Kellogg remeasured the $^7Be(p,\gamma)^8B$ cross section in great detail and with improved precision down to a proton energy of 164 keV, extending and confirming Parker's earlier results and thereby greatly increasing our confidence in the low energy extrapolation of the rate of this crucial reaction. Shortly afterward, Dwarakanath (1974) returned to Kellogg and, in a real tour de force, managed to push the $^3He-^3He$ cross section measurements down to 33 keV (0.1 nb), showing that the earlier extrapolation continued smoothly to low energies and that there was no evidence for a low lying threshold resonance.

There were also many suggestions of possible solutions to the solar neutrino problem, none of which has been accepted generally and nearly all of which were either ad hoc or were discredited by further analysis, or both. It is possible that the correct solution is one of the suggestions that were made during this period, but if so it is still very difficult for us to guess which one this might be. Hence, we will content ourselves here with simply recalling some of the more interesting (or, in some cases, more amusing) proposals [for discussions of many of these nonstandard models, see Bahcall and Sears (1972) or Rood (1978)].

Proposals made during this period include: turbulent diffusion of ^3He (Schatzman 1969); neutrino oscillations (Gribov and Pontecorvo 1969; Wolfenstein 1978); an overabundance of ^3He in the present Sun (Kocharov and Starbunov 1970); the effect of a magnetic field (Abraham and Iben 1971; Bahcall and Ulrich 1971; Bartenwerfer 1973; and Parker 1974); a secular instability such that the presently observed solar luminosity does not equal the current energy-generation rate (Fowler 1968, 1972; Sheldon 1969); quark catalysis (Libby and Thomas 1969; Salpeter 1970); a very low heavy element abundance in the solar interior (Bahcall and Ulrich 1971); an appreciable magnetic moment for the neutrino (Cisneros 1971);

an instability of the Sun that makes now a special time (Fowler 1972; Dilke and Gough 1972); neutrino decay (Bahcall, Cabbibo and Yahil 1972); a low-energy resonance in the ^{3}He$-^3$He reaction (Fowler 1972; Fetisov and Kopysov 1972); rapid rotation of the solar interior (Demarque, Mengel, and Sweigert 1973; Roxburgh 1974; and Rood and Ulrich 1974); rotation plus magnetic fields (Snell, Wheeler, and Wilson 1976); a burned-out Sun with a helium core (Prentice 1973); a half-solar mass core of large heavy element abundance that survived the big bang and subsequently accreted another half solar mass at the time of the formation of the solar system (Hoyle 1975); a departure from the Maxwellian distribution (Clayton et al. 1975); a fractionation of the primordial hydrogen and helium (Wheeler and Cameron 1975); accretion onto a black hole in the center of the Sun (Clayton, Newman, and Talbot 1975); and multiplicative mass creation (Maeder 1977).

This list of published suggestions is certainly incomplete; we have not attempted a full literature search. In any event, it does show that during the period under consideration many astronomers and physicists thought seriously about the solar neutrino problem, a situation in marked contrast to what occurred in the first few active years of the subject (1963–1966).

A valuable conference on the solar neutrino problem was organized by Fred Reines at the Irvine campus of the University of California in February, 1972 (Reines and Trimble 1972; Trimble and Reines 1973). The emphasis of this conference was on the experimental aspects of the problem and there was a thorough examination of the details of the chlorine experiment. The prospects for future experiments were also discussed. The first day of the meeting was held in the conference room of President Nixon's Western White House; this was an unusual touch provided by Fred's always active imagination and skill with arrangements. Shortly after this conference, two interesting theoretical suggestions were published: Willy's speculation (Fowler 1972) that there might be a resonance in the ^{3}H$-^3$He reaction; and Al Cameron's analysis of the effect of a sudden mixing of the solar interior on the ^{8}B neutrino flux (Cameron 1973). Another important idea that may have been stimulated by this conference was the proposal by Luis Alvarez that the chlorine detection system could be tested by using an intense radioactive source of ^{65}Zn (Alvarez 1973). He pointed out that a strong monoenergetic source of neutrinos could be prepared by neutron activation of ^{64}Zn. (We are currently discussing a detailed plan to carry out this experiment.)

We next review briefly a few events that occurred during this period in which we were personally involved. We will depart from a strictly chronological order to more logically group some related developments.

The SNU was first introduced by Bahcall (1969c) in a paper that argued that another solar neutrino experiment (preferably ^{7}Li) was needed to decide if the discrepancy between theory and observation was due to a deficiency in our astrophysical understanding or to an unknown phenomenon affecting neurinos in transit from the Sun.

Gribov and Pontecorvo (1969) suggested that a factor of two discrepancy between theory and observation might be due to oscillations between electron– and muon–neutrino states. They presented the relevant equations for this two-component system, expanding upon the earlier, less formal discussion of Pontecorvo (1968). Gribov and Pontecorvo discussed two kinds of averages of the neutrino fluxes: an average over the emitting region of the Sun and an average (suggested by I. Pomeranchuk) over the time of reception.

Bahcall and Frautschi (1969) pointed out that another average, over the broad spectrum of neutrino energies produced by the Sun, was more important and that some of the variations discussed by Gribov and Pontecorvo would not occur. Bahcall and Frautschi recommended new solar neutrino experiments that were designed to detect p–p or p–e–p neutrinos for which the uncertainties in the predicted fluxes arising from astrophysical considerations are minimal. They argued that such experiments would be sensitive to neutrino masses of order 10^{-6} eV and hence, could discriminate between different elementary-particle explanations of neutrino mixing.

In a more contemporary context, Bilenky and Pontecorvo (1978) have argued that ". . . neutrino mixing is much more natural solution than any other that has been proposed either in terms of elementary particle physics or astrophysics." They also state: "From the elementary particle physics point of view, lepton mixing is a reasonably likely and quite attractive hypothesis." They go on to note that: ". . . neutrino oscillations were not invented ad hoc, for the sake of explaining the result of the experiment of Davis et al."

We would love to test the oscillation hypothesis with a gallium experiment.

The sensitivity of the solar-neutrino fluxes to small changes in opacity, the equation of state, and nuclear cross sections, solar age, and heavy-element abundance was the subject of a detailed study by Bahcall, Bahcall, and Ulrich (1969). The convenient formulae given in this paper for the dependences of the predicted capture rate on various quantities have been used often by us and others to make quick estimates of the possible importance of various uncertainties in parameters and/or of some proposed solutions to the solar neutrino problem. (This paper was the first in a long series of happy collaborations involving J. Bahcall and Roger Ulrich that is continuing even today.) Somewhat similar results were obtained by Torres-Peimbert, Simpson, and Ulrich (1969) using the Berkeley stellar-evolution program. These authors stressed that the primordial heavy element abundance of the Sun, Z, must be less than 0.02 in order for there to be any hope of obtaining with standard models a neutrino capture rate that was not in obvious disagreement with the observed upper limit.

Detailed studies of the effects of various changes in composition, as well as the importance of magnetic fields and turbulent diffusion, were considered also by Abraham and Iben (1971) and Bahcall and Ulrich (1971). The neutrino fluxes calculated by all the active groups using different stellar evolution computer codes were shown by Bahcall and Ulrich to give consistent answers when proper account was taken of the different choices of parameters. New radiative opacities calculated by the Los Alamos group were used by Bahcall, Huebner, Magee, Merts, and Ulrich (1973) to obtain a standard model that predicted a rather low neutrino capture rate, 5.5 SNU. This paper was the first one on solar neutrinos where the calculators of opacities (here, Huebner, Magee, and Merts), whose results had long been recognized as central to the whole subject, were coauthors of a paper specifically on the solar neutrino problem.

Various corrections to the stellar opacity were discussed by Watson (1969a, 1969b, 1969c), including an increased iron abundance. These corrections, among others, were included in the models of Bahcall and Ulrich (1970), who obtained a best estimate of 7.8 SNU.

In December, 1970, we both attended the Symposium on Relativistic Astrophysics in Austin, Texas. The main excitement at this symposium was related to the possibility that Joe Weber had detected pulses of gravitational radiation from distant sources. Searches for radio and microwave signals coincident with Weber pulses were described in detail in a number of interesting talks. During one of these talks, it occurred to us that the ^{37}Cl solar neutrino detector could provide a useful limit on the amount of neutrino energy reaching the Earth that might be associated with Weber pulses. Afterwards, we retreated to a table at a nearby coffee shop and were able to derive quickly a strong limit on the ratio of neutrino energy flux to gravitational energy flux (0.1% for 10 MeV neutrinos, see Bahcall and Davis 1971). The signal to noise ratio of the ^{37}Cl detector is so large (for energy fluxes comparable to those claimed in Weber's gravity wave experiments) that one does not need to perform coincidence experiments (which are necessary in the radio and microwave regions).

We also collaborated with John Evans in a similar study of a reported possible detection of antineutrinos from a stellar collapse by the University of Pennsylvania group (Lande et al. 1974). The absence of detectable neutrinos associated with the January 4 antineutrino event was shown (Evans, Davis, and Bahcall 1974) to be difficult to reconcile with the suggestion that a collapsed star had been detected. We continue to believe that solar neutrino experiments are good detectors for collapsing stars and that their use for this purpose complements the more specific experiments that are being carried out with gravitational wave detectors.

The close connection that has existed between theory and observation (or, perhaps more correctly, between cocktail-hour suggestions and observation) is illustrated by the genesis of the proposal (Bahcall, Cabbibo, and Yahil 1972) that neutrinos may be unstable (i.e., may decay to some other particles.) This idea was considered because Davis told Bahcall in a telephone conversation (and in a subsequent memorandum) in November, 1971 of the latest run that he was counting where *no* counts (either background or signal) had been observed in the counter for two months. This result suggested that the production rate in the tank might well be shown to be zero when counters with sufficiently low background rates were generally available. This possibility made it natural to consider various Lagrangians for which neutrino decay was allowed. The idea of neutrino decay as an explanation for a low neutrino rate has somewhat fallen out of favor in recent years because it appears that a finite production rate in the tank may have been observed (the November, 1971 result was a statistical fluctuation).

There were a number of experimental developments in this period that ultimately increased dramatically the sensitivity of the experiment. Of course, the detector size was already fixed and the chemical recovery was nearly quantitative. The only possibility of increasing the sensitivity of the Homestake detector was to somehow reduce backgrounds. The background counting rate was 10 counts per month in the ^{37}Ar region (full width at half maximum) and a nearly zero background rate was needed for a really substantial improvement in sensitivity.

The crucial suggestion that led to a dramatic improvement was made by Gordon Garmire. After a seminar at Caltech, we went for a swim at the campus pool. Lounging around the side of the pool, we started talking to Gordon. He pointed out that x-ray astronomers had developed pulse rise-time techniques for proportional counters that enabled them to observe and characterize x-ray events in the presence of a high flux of cosmic rays. He suggested that this same technique could be used for characterizing ^{37}Ar decay events in the small Brookhaven counters. When Davis first asked the Brookhaven electronics engineers if this might be possible, they pointed out that their amplifiers were not fast enough to be used for this purpose with the small counters. However, about a year later, they developed amplifiers and pulse stretchers with sufficient speed to be applicable. The first working system was developed at Brookhaven by Robert L. Chase, Veljko Radeka, and Lee C. Rogers, and was used in run number 18 (cf. Fig. 3) in late 1970. This improvement reduced the background counting rate for events simulating ^{37}Ar to one event per month.

With this reduction in counter background, the background ^{37}Ar produced in the tank became an important consideration. The production of ^{37}Ar by fast neutrons from the surrounding rock wall was very small, approximately 0.04 ^{37}Ar atoms per day. This background effect was easily eliminated by flooding the rock cavity with water. The water was added for run number 21 (cf. Fig. 3) in the summer of 1971. The water shield has remained in place since that time, except for a six month period in 1975 when it was removed in order to inspect the tank for corrosion and to give it a new coat of paint. (The water shield has been converted to an active water Cerenkov par-

ticle detector by Ken Lande's group from the University of Pennsylvania. Their 250 ton detector is being used currently to search
for baryon decay, to observe cosmic rays, and to try to detect neutrinos from collapsing stars.)

The cosmic ray background was known to be small, but by now
we had also learned that the solar neutrino signal was very low.
Evaluating the cosmic ray background was a difficult problem. It
required measurements of the ^{37}Ar production rate by cosmic ray
muons as a function of depth and a valid extrapolation of these measured rates to the full depth of the chlorine detector. Measurements
were performed at depths from 30 to 108 kg/cm². Arnold Wolfendale and E. C. M. Young from Durham analyzed the results to give
the extrapolated ^{37}Ar production rate in the detector at 440 kg/cm²
(Wolfendale et al. 1972). Later, an independent analysis was made
by George Cassidy of the University of Utah; he obtained similar
results (Cassidy 1973).

In recent years, Ed Fireman has developed an independent technique for determining the muon background by using ^{39}K as a target
and detecting events in which ^{37}Ar and a neutron and a proton are
produced. These important measurements are being carried out in
the Homestake Mine.

Still another method of scaling the background effect with depth
makes use of a radiochemical neutron detector based upon the
^{40}Ca(α,n)^{37}Ar reaction. The Brookhaven group is currently using a
2000 l tank of calcium nitrate solution exposed at various levels in
the mine to measure the neutron production rate by fast muons. In
reporting the results of the chlorine solar neutrino detector, a
cosmic-ray background of 0.08 ^{37}Ar atoms/day is used in all of the
recent analyses; this is the value resulting from the original studies
by Wolfendale, Young, and Cassidy, based on the measurements
made with perchloroethylene.

The ^{37}Ar counting system was moved to the Homestake Mine in
1977. Moving the system underground did not reduce counter
background as much as was hoped. However, the underground
muon flux is negligibly small in the counters, and this permitted the
measurement of the environmental gamma-ray background effect
on the counters. These measurements of the gamma-ray background may eventually lead to further reduction of the counter
backgrounds.

There have been worries expressed by physicists and astron-

omers that there could be something wrong with the radiochemical procedures used for extracting a few tens of atoms of ^{37}Ar from
a large volume of perchloroethylene, a typical concentration of one
atom in 10,000 l. Some individuals speculated that the ^{37}Ar produced by neutrino captures ends up in a chemical state that is non-
volatile and thus, is not removed by a helium purge. Some specific
suggestions were advanced by Kenneth Jacobs; he proposed
molecule-ions and radiation induced polymerization traps (Jacobs
1973).

Although these suggestions were not based upon sound chemistry, we felt that an experiment should be performed to test these
unlikely possibilities. To this end, an experiment was performed
with ^{36}Cl-labeled perchloroethylene. This isotope decays by β^-
emission to produce ^{36}Ar. The dynamics of the decay process is essentially identical to a neutrino capture and electron emission. In
an experiment performed by Herman Vera-Ruiz, John Evans, and
Ray Davis, it was found that the yield of ^{36}Ar recovered from perchloroethylene by a helium purge was quantitative. This experiment and the other argon efficiency tests made with the 100,000 gal
tank show that ^{37}Ar is recovered with high efficiency.

The Soviet solar neutrino project has developed into a major program under the leadership of G. T. Zatsepin. We first learned of the
magnitude of their effort at a lunch table discussion during the
Neutrino 1974 conference at Balatonfured, Hungary. A group of
interested Americans, Fred Reines, Ken Lande, John Bahcall, and
Ray Davis, asked to hear about the Soviet plans. Answers were provided by Ya Chudakov, A. Pomanski, V. A. Kuzmin, and B. Pontecorvo. They outlined their plans to dig a 4 km long tunnel under a
mountain in the Caucasus to contain a number of neutrino detectors, including a chlorine detector about five times larger than
the Homestake experiment and a 1000 ton scintillation detector for
observing collapsing stars. (This ambitious program is well advanced at the present time. The Neutrino 1977 conference was held
on site and at that time the tunnel was about 1.7 km deep. At the
Neutrino 1980 conference, the Soviet group indicated that a 50 ton
gallium experiment would be operating in 1983!)

A summary of the solar neutrino problem, as we saw it after 15
years of work in collaboration with many colleagues, was published
in *Science* (Bahcall and Davis 1976). The origin of this paper is
somewhat unusual; it was originally solicited by the editor of the

British journal, *The New Scientist,* but the manuscript we produced was rejected by the editor as unsuitable for his readership. We then submitted the manuscript to the editor of *Science* who graciously accepted it.

1978 to present

Our efforts, along with those of many colleagues, have concentrated in recent years on bringing about a new solar neutrino experiment. The most promising targets at present appear to be ^{7}Li, ^{37}Cl (enlarged experiment), ^{71}Ga, ^{81}Br, ^{115}In, and electron–neutrino scattering. The current status of the subject and discussions of each of these targets are summarized in the remarks by the various speakers whose talks are recorded in the *Proceedings of the Informal Conference on the Status and Future of Solar Neutrino Research* (Friedlander 1978), which took place at Brookhaven National Laboratory in January, 1978. This meeting was an occasion for examining the present status of the subject and for informally discussing what ought to be done next. Davis (1978) opened the conference by describing the technical details of the ^{37}Cl experiment, including tests which showed that any argon produced in the tank was extracted, as well as an explanation of the counting and data analysis techniques. The experimental rate reported was 2.2 ± 0.3 SNU. There was much lively discussion following this and the other talks. The theoretical calibration of each of the possible new targets was discussed by Bahcall at the conference (Bahcall 1978a), and more completely in a detailed paper (Bahcall 1978b) that gives the best estimates of the neutrino absorption cross sections for the various targets, the estimated uncertainties in the cross sections, and an analysis of what can be learned about astrophysics (or physics) by using each target. In some cases, experiments that had been the subject of much previous work were dropped as a result of Bahcall's analysis, which showed that for several otherwise useful neutrino targets, the absorption cross sections were inherently uncertain. Willy skilfully guided a panel discussion on the final day among several of the participants; the discussion showed remarkable unanimity in supporting a new experiment that would be sensitive primarily to neutrinos from the proton–proton reaction.

The important story of the proposed ^{205}Tl experiment has been told in a complete and dramatic manner by Mel Freedman (1978), who suggested and worked out the observational details of this clever detection scheme in collaboration with his colleagues at Argonne (see Freedman et al. 1976). The basic idea is to use ^{205}Tl as a measure of the *average* neutrino flux from p–p reactions (or almost equivalently, the solar luminosity) for the past ten million years. The ^{205}Tl must be obtained from geological deposits. A recent reinvestigation of this suggested experiment by Rowley, Cleveland, Davis, Hampel, and Kirsten (1980) has confirmed Freedman's analysis that background effects are expected to be small and that this detector could provide in principle otherwise inaccessible information regarding the average luminosity of the Sun over timescales that some theorists (including Willy) have speculated may be relevant to the internal history of the Sun and to the solar neutrino problem. The principal difficulty with the proposed experiment is that Bahcall (1978a,b) has shown that the neutrino absorption cross sections cannot be calculated accurately for this target and are essentially unknowable to the desired precision (a factor of two or better).

The basic reason that a new experiment is required is to establish whether the origin of the present discrepancy between theory and observation (with the ^{37}Cl detector) is due to errors in our understanding of astrophysics (stellar models) or physics (e.g., properties of neutrinos). Detectors such as ^{71}Ga and ^{115}In (Raghavan 1976) that are primarily sensitive to p–p neutrinos are preferable for this purpose. If the Sun is currently producing its average luminosity by virtue of nuclear fusion reactions in its interior, then the flux of p–p neutrinos can be calculated essentially from the observed optical luminosity of the Sun. One need only assume for this calculation that the branches involving ^{7}Be are relatively rare (less than, or of order of 10%), an assumption that can be justified on the basis of either stellar models or the ^{37}Cl experiment.

A modular experiment that uses ^{71}Ga as a detector is currently underway. It involves a collaboration between individuals at Brookhaven National Laboratory, the Institute for Advanced Study, the University of Pennsylvania, the Max Planck Institute for Nuclear Physics at Heidelberg, and the Weizman Institute in Rehovot. The rationale and procedures for this project were summarized in an article in *Physical Review Letters* (Bahcall et al. 1978) and in the proposal submitted to the *Max Planck Gesellschaft zur Forderung der Wissenschaften* (September, 1978). The extraction procedure has been tested successfully in the spring of 1980 on a 1.3 ton sample of gallium by Bruce Cleveland, Israel Dostrovsky, Gerhart Friedlander, and Davis; the procedures for counting ^{71}Ge efficiently have been developed by the group at the Max Planck Institute for Nuclear Physics in Heidelberg under the leadership of Til Kirsten and Wolfgang Hampel. This experiment could be completed in three or four years if support were forthcoming from the U.S. Department of Energy to supply, together with the Max Planck Institutes, the required amount of gallium (about 50 tons total).

The next stage in the gallium experiment is to use a ^{65}Zn source to calibrate the detector throughput and the neutrino absorption cross sections, in a manner first proposed by Luis Alvarez (1973) for the ^{37}Cl detector. About 10 tons of gallium are required for this intermediate step. The recent suggestions that neutrinos may have been observed to oscillate on scales observable in the laboratory (see, e.g., Barger et al. 1980; Reines et al. 1980) make this experiment of special interest. Only the 1.343 MeV neutrino line of ^{65}Zn (from electron capture) contributes significantly to the observable neutrino capture rate (Bahcall 1978b). If oscillations do occur at the suggested level (neutrino masses of order 1 eV), then the oscillation parameters could be determined by using the effectively monoenergetic ^{65}Zn neutrinos and varying the source-absorber distance.

Other experiments are also being actively investigated at present. These include the following targets: ^{7}Li (K. Rowley, S. Hurst, S. D. Kramer, R. Davis, A. M. Bakich, L. S. Peak); ^{115}In (R. S. Raghavan and M. Deutsch); and ^{81}Br (S. Hurst, Bahcall and Davis, following the recent important experiment on the beta-decay rate of the 190 keV excited state of ^{81}Kr by Benett et al. 1981).

Retrospective

It is instructive to look back over the history of this subject to see how the observational and theoretical values have changed with time. This is shown in Figs. 1 and 2.

Figure 1 is an overall pictoral history of the subject as it looked in 1970 (when this drawing was originally used by Davis in a public lecture). A few of the major events are indicated on the figure at the period corresponding to the time they occurred. It is interesting to note that the only change that would have to be made to bring it up to date ten years later is to lower the experimental upper limit by about a factor of two.

Figure 2 shows all the published values in which we have participated, with the exception of the observational limits obtained in 1955 and 1964 (These earliest upper limits of 4000 and 160 SNU would not fit conveniently in Fig. 2, which unlike Fig. 1, has linear scales).

A few remarks need to be made about the theoretical error bars in Fig. 2. These uncertainties are more "experimental" than "theoretical" since the basic theory has not changed since 1964. What have changed are the best estimates for many different input parameters (see the earlier discussion under 1968). The error bars shown in Fig. 2 for the theoretical points were taken in all cases from the original papers (see caption to Fig. 2), and represent the range of capture rates that were obtained from standard solar models when the various nuclear and atomic parameters were allowed to vary over the range conventionally regarded as acceptable at the time the calculations were made. A number of detailed theoretical studies and improvements have been introduced into the stellar model calculations over the past fifteen years, at great expense in personal effort and computing time, but these theoretical refinements have had only relatively minor effects on the calculated cap-

Figure 1. Some of the principal events in the development of the solar neutrino problem. The experimental upper limit is indicated by the thin black curve and the range of theoretical values (after 1964) by the cross-hatched region. The units are captures per target atom per second (10^{-36} capture/target atom/s = 1 SNU).

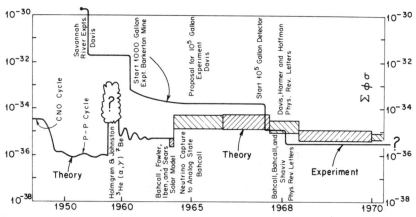

Figure 2. Published values of the predicted and observed neutrino capture rates from 1964 to 1980. The observational results are from Davis, Harmer, and Hoffman (1968); Davis (1970); Davis, Rogers, and Radeka (1971); Davis (1971); Davis, Evans, Radeka, and Rogers (1972); Davis and Evans (1973); Davis and Evans (1974); Davis and Evans (1976); Rowley, Cleveland, Davis, and Evans (1977); Davis (1978); and Rowley, Cleveland, Davis, Hampel, and Kirsten (1980). The theoretical values are from Bahcall (1964a); Bahcall (1966); Bahcall and Shaviv (1968); Bahcall, Bahcall, Fowler, and Shaviv (1968); Bahcall, Bahcall, and Shaviv (1968); Bahcall (1969b); Bahcall and Ulrich (1970); Bahcall and Ulrich (1971); Bahcall, Huebner, Magee, Merts, and Ulrich (1973); Bahcall (1977); and Bahcall, Huebner, Lubow, Magee, Merts, Parker, Rozsnyai, Ulrich, and Argo (1980). Similar results by other authors are mentioned in the text.

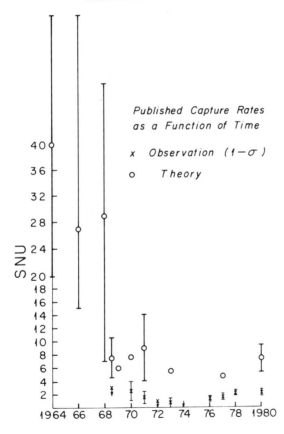

ture rates compared to the rather large changes produced by new measurements of experimental parameters. The various ups and downs in the best estimate theoretical values since 1968 represent the largely statistical variations in the uncertainties in the many input parameters. The current theoretical estimate is (7.5 ± 1.5) SNU, where the quoted uncertainty takes account of known uncertainties in opacities, primordial chemical composition, and nuclear reaction parameters (Bahcall et al. 1980).

The procedures for analyzing the data have evolved with time; the techniques are discussed fully in the report by Davis (1978). All of the published capture rates prior to 1977 were described in the original papers (see caption to Fig. 2) as one-standard-deviation upper limits. The sensitivity of the experiment has greatly improved with time as experience has been gained with the operating system and the extremely low count rates. The measurement of both the rise time (as first suggested by Gordon Garmire) and the pulse height of the proportional-counter events allows one to discriminate strongly against noise pulses. Measuring the rise time was introduced in run 18 (1970); it greatly reduced the number of background events. Bruce Cleveland has developed a maximum likelihood method of analyzing the data that utilizes the time of occurrence of all the events detected in the counters; this procedure is unbiased and gives a best estimate and uncertainty for both the background and the ^{37}Ar production rate. Using this method, it has been possible to establish that the ^{37}Ar production rate in the tank, although small, is actually not zero. Another way of demonstrating this fact is to use all of the events from the different runs and show that they collectively decay with the lifetime of ^{37}Ar; the resulting cumulative decay curve is a dramatic and convincing way of seeing that the experiment is actually detecting ^{37}Ar (see Davis 1978). The present best estimate for the production rate is 2.2 ± 0.4 SNU (Rowley et al. 1980).

It appears from Fig. 2 that the published estimates for the capture rate were at a minimum during 1972–4. This effect is due almost entirely to the change in the method of analyzing the data (see Davis 1978); all of the later points include the earlier data as well. In order to check this interpretation, Bruce Cleveland has reanalyzed the data using his maximum likelihood method. For the data available in 1972, Cleveland finds 1.3 ± 1 SNU (compared to the earlier published value of 0.2 SNU) and for the 1974 data Cleve-

land finds 2.0 ± 0.4 SNU (compared to the earlier published value
of 1.3 SNU). The main difference between the present analyses
and the earlier calculations is due to the fact that the statistical uncertainty for a very small number of events is now properly taken into
account.

All of the high quality data presently available are shown in
Fig. 3, which has been assembled by Cleveland and Davis. The average production rate of ^{37}Ar in the tank from all these data is
2.2 ± 0.4 SNU.

The current difference between theory and observation using the
best available estimates for the parameters is about a factor of three.
Experiments to remeasure at low energies and with the most modern techniques [see Rolfs and Trautvetter (1978) and Barnes (1981)]
the cross section factors for the ^{3}He–^{3}He, ^{3}He–^{4}He, and ^{7}Be–p
reactions are urgently needed. Experiments are underway to remeasure the second of these reactions, which is being studied
by Claus Rolfs and his associates in Germany and also by an enthusiastic crew at the Kellogg Laboratories. Of the total 7.8
SNU predicted by the current best estimate model, 6.3 SNU is
from the ^{7}Be(p,γ)^{8}B reaction, last studied in detail in 1969 by
Ralph Kavanagh and his associates in an unpublished investiga-

Figure 3. Summary of ^{37}Ar production rates for individual experimental runs, 1970–9.

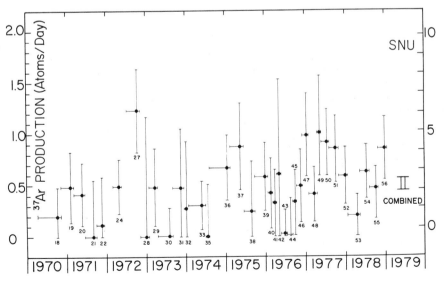

tion (see Kavanagh 1972). It is worth stressing again that the entire difference between the theoretical and observational values in Fig. 2 is due to neutrinos from ^{8}B produced in the ^{7}Be(p,γ) reaction. The total capture rate also depends sensitively upon the ^{3}He(α,γ)^{7}Be reaction, approximately as the (cross section factor)$^{0.8}$.

It would not be surprising if Willy once again used his exceptional powers of persuasion to see that the above experiments were repeated expeditiously. After all, he has been telling all of us for many years what we ought to be doing; we have profited scientifically by his advice and have had fun in the process.

In conclusion, we believe that, whatever the solution of the solar neutrino problem turns out to be, the combined efforts of many people (chemists, nuclear physicists, astrophysicists, geophysicists, and elementary particle physicists) over the past two decades will ultimately result in a greater understanding of both the solar interior and the limitations of our present knowledge. In the interim, many parameters have been determined more accurately and many theoretical possibilities have been rendered implausible. Future solar neutrino experiments must delineate more clearly what is the missing element in our present understanding and even whether it is primarily in the realm of physics or astrophysics.

This research was supported in part by NSF Grant No. PHY79-19884 and in part by the Energy Research and Development Administration.

References

Abraham, Z., and Iben, I., Jr. 1971, *Astrophys. J.*, **170**, 157.
Alvarez, L. W. 1949, University of California Radiation Laboratory Report UCRL-328.
– 1973, Physics Notes, Mem. No. 767, Lawrence Radiation Laboratory (March 23).
Bahcall, J. N. 1962a, *Phys. Rev.*, **126**, 1143.
– 1962b, *Phys. Rev.*, **128**, 1297.
– 1964a, *Phys. Rev. Lett.*, **12**, 300.
– 1964b, *Phys. Rev. B*, **136**, 1164.
– 1964c, *Phys. Rev. B*, **135**, 137.
– 1964d, *Phys. Lett.*, **13**, 332.
– 1966, *Phys. Rev. Lett.*, **17**, 398.
– 1967, *Proceedings of the Second International Conference on High-Energy Physics and Nuclear Structure*, ed. G. Alexander, 232.
– 1969a, *Quasars and High-Energy Astronomy*, eds. K. N. Douglas, I. Robinson, A. Schild, E. L. Schucking, J. A. Wheeler, and N. J. Woolf (New York: Gordon and Breach), p. 321.

- 1969b, *Proceedings of the International Conference on Neutrino Physics and Astrophysics* (Moscow), **2**, 133 (USSR: F. I. Academy Science).
- 1969c, *Phys. Rev. Lett.*, **23**, 251.
- 1977, *Astrophys. J. Lett.*, **216**, L115.
- 1978a, *Proceedings of the Informal Conference on Status and Future of Solar Neutrino Research*, Brookhaven National Laboratory 50879, **1**, 223.
- 1978b, *Rev. Mod. Phys.*, **50**, 881.
Bahcall, J. N., Bahcall, N. A., Fowler, W. A., and Shaviv, G. 1968, *Phys. Lett.*, **26B**, 359.
Bahcall, J. N., Bahcall, N. A., and Shaviv, G. 1968, *Phys. Rev. Lett.*, **20**, 1209.
Bahcall, J. N., Bahcall, N. A., and Ulrich, R. K. 1968, *Astrophys. Lett.*, **2**, 91.
- 1969, *Astrophys. J.*, **156**, 559.
Bahcall, J. N., and Barnes, C. A. 1964, *Phys. Lett.*, **12**, 48.
Bahcall, J. N., Cabibbo, N., and Yahil, A. 1972, *Phys. Rev. Lett.*, **28**, 316.
Bahcall, J. N., Cleveland, B. T., Davis, R., Jr., Dostrovsky, I., Evans, J. C., Jr., Frati, W., Friedlander, G., Lande, K., Rowley, J. K., Stoenner, R. W., and Weneser, J. 1978, *Phys. Rev. Lett.* **40**, 1351.
Bahcall, J. N., and Davis, R., Jr. 1966, *Stellar Evolution*, ed. R. F. Stein and A. G. W. Cameron (New York: Plenum Press), p. 241.
- 1971, *Phys. Rev. Lett.*, **26**, 662.
- 1976, *Science*, **191**, 264.
Bahcall, J. N., Fowler, W. A., Iben, I., and Sears, R. L. 1963, *Astrophys. J.*, **137**, 344.
Bahcall, J. N., and Frautschi, S. C. 1969, *Phys. Lett.*, **29B**, 623.
Bahcall, J. N., Huebner, W. F., Lubow, S. H., Magee, N. H., Jr., Merts, A. L., Parker, P. D., Rozsnyai, B., Ulrich, R. K., and Argo, M. F. 1980, *Phys. Rev. Lett.*, **45**, 945.
Bahcall, J. N., Huebner, W. F., McGee, N. H., Merts, A. L., and Ulrich, R. K. 1973, *Astrophys. J.*, **184**, 1.
Bahcall, J. N., and May, R. M. 1968, *Astrophys. J. Lett.*, **152**, L17.
Bahcall, J. N., and Sears, R. L. 1972, *Ann. Rev. Astron. Astrophys.*, **10**, 25.
Bahcall, J. N., and Shaviv, G. 1968, *Astrophys. J.*, **153**, 113.
Bahcall, J. N., and Ulrich, R. K. 1970, *Astrophys. J. Lett.*, **160**, L57.
- 1971, *Astrophys. J.*, **170**, 593.
Barger, V., Whisnat, K., and Phillips, R. J. N. 1980, *Phys. Rev. Lett.*, **45**, 2084.
Barnes, C. A. 1981, "Laboratory Approaches to Nuclear Astrophysics," three lectures in *Progress in Particle and Nuclear Physics*, ed. Sir Denys Wilkinson (Oxford: Pergamon Press), to be published.
Bartenwerfer, D. 1973, *Astron. Astrophys.*, **25**, 455.
Benett, C. 1981, to be published.
Bethe, H. A. 1939, *Phys. Rev.*, **55**, 434.
Bethe, H. A., and Critchfield, L. 1938, *Phys. Rev.*, **54**, 248.
Bilenky, S. M., and Pontecorvo, B. 1978, *Phys. Reports*, **41**, 225.
Burbidge, E. M., Burbidge, G. R., Fowler, W. A., and Hoyle, F. 1957, *Rev. Mod. Phys.*, **29**, 547.
Cameron, A. G. W. 1958a, *Bull. Am. Phys. Soc. II*, **3**, 227.
- 1958b, *Ann. Rev. Nucl. Sci.*, **8**, 299.
- 1958c, *Chalk River Report* CRL-41, 2nd ed., unpublished.
- 1973, *Explosive Nucleosynthesis* ed. D. N. Schramm and W. D. Arnett (Austin: University of Texas Press).
- 1973, *Rev. Geophys. Space Phys.*, **11**, 505.

Cassidy, G. A. 1973, *Proceedings of the Thirteenth International Conference on Cosmic Rays*, Denver, vol. 13, p. 1958.

Christensen, C. J., Nielsen, A., Bahnsen, A., Brown, W. K., and Rustad, B. M. 1967, *Phys. Lett.*, **26B**, 11.

Cisneros, A. 1971, *Space Sci.*, **10**, 87.

Clayton, D. D., Dwek, E., Newman, M. J., and Talbot, R. J. 1975, *Astrophys. J.*, **201**, 489.

Crane, H. R. 1948, *Rev. Mod. Phys.*, **20**, 278.

Davis, R., Jr. 1955, *Phys. Rev.*, **97**, 766.

– 1964, *Phys. Rev. Lett.*, **12**, 303.

– 1969, *Proceedings of the International Conference on Neutrino Physics and Astrophysics* (Moscow), **2**, 99 (USSR: F. I. Academy Science).

– 1970, *Hungaricae*, Suppl. 4, **29**, 371.

– 1971, *Acad. Naz. Lincei*, **157**, 59.

– 1978, *Proceedings of the Informal Conference on the Status and Future of Solar Neutrino Research*, ed. G. Friedlander, Brookhaven National Laboratory 50879, **1**, 1.

Davis, R., Jr., and Evans, J. C. 1973, *Proceedings of the Thirteenth International Cosmic Ray Conference*, Denver, vol. 3, 2001.

– 1974, *Proceedings of International Conference on Particle Acceleration and Nuclear Reactions in Space*, Leningrad, August, 91.

– 1976, *Proceedings International Seminar on Active Processes on the Sun and the Solar Neutrino Problem*, Leningrad, September 22, Brookhaven National Laboratory 1984, p. 84.

Davis, R., Jr., Evans, J. C., Radeka, V., and Rogers, L. C. 1972, in *Neutrino '72*, **1**, 5, ed. A. Frenkel and G. Marx (OMDK – Technoinform, Budapest, June, 1972, Europhysics Conference).

Davis, R., Jr., Harmer, D. S., and Hoffman, K. C. 1968, *Phys. Rev. Lett.*, **20**, 1205.

Davis, R., Jr., Harmer, D. S., and Neely, F. 1969, in *Quasars and High-Energy Astronomy*, ed. K. N. Douglas, I. Robinson, A. Schild, E. L. Schucking, J. A. Wheeler, and N. J. Woolf (New York: Gordon and Breach), p. 287.

Davis, R., Jr., Rogers, C., and Rodeka, V. 1971, *Bull. Am. Phys. Soc. II*, **16**, 631.

Demarque, P., Mengel, J. G., and Sweigart, A. V. 1973, *Mon. Not. R. Astron. Soc.*, **165**, 19.

Dilke, F. W. W., and Gough, D. O. 1972, *Nature*, **240**, 262.

Dwarakanath, M. R. 1974, *Phys. Rev. C*, **9**, 805.

Evans, J. C., Davis, R., Jr., and Bahcall, J. N. 1974, *Nature*, **251**, 486.

Ezer, D., and Cameron, A. G. W. 1965, *Can. J. Phys.*, **43**, 1497.

– 1966, *Can. J. Phys.*, **44**, 593.

– 1968, *Astrophys. Lett.*, **1**, 177.

Fetisov, V. N., and Kopysov, Y. S. 1972, *Phys. Lett. B*, **40**, 602.

Fowler, W. A. 1954, *Mém. Soc. Roy. Sci. Liège*, **14**, 88.

– 1958, *Astrophys. J.*, **127**, 551.

– 1968, *Contemporary Physics: Trieste Symposium 1968*, vol. 1 (Vienna: International Atomic Energy Agency), p. 359.

– 1972, *Nature*, **238**, 24.

Freedman, M. S., Stevens, C. M., Horwitz, E. P., Fuchs, L. H., Lerner, J. L., Goodman, L. S., Childs, W. J., and Hessler, J. 1976, *Science*, **193**, 1117.

Freedman, M. S. 1978, *Proceedings of the Informal Conference on the Status and Future of Solar Neutrino Research*, Brookhaven National Laboratory 50879, p. 313.

Friedlander, G. 1978, ed. *Proceedings of the Informal Conference on the Status and Future of Solar Neutrino Research*, Brookhaven National Laboratory 50879, vols. 1 and 2.

Goldhaber, M. 1967, *Proceedings of the Second International Conference on High-Energy Physics and Nuclear Structure*, ed. G. Alexander, Amsterdam, p. 475.

Gribov, V., and Pontecorvo, B. 1969, *Phys. Lett.*, **28B**, 493.

Hardy, J. C., and Verrall, R. I. 1964, *Phys. Rev. Lett.*, **13**, 764.

Holmgren, H. P., and Johnston, R. 1958, *Bull. Am. Phys. Soc. II*, 3, 26.

Hoyle, F. 1975, *Astrophys. J. Lett.*, **197**, L127.

Iben, I., Jr. 1968, *Phys. Rev. Lett.*, **21**, 1208.

Iben, I., Jr., Kalata, K., and Schwartz, J. 1967, *Astrophys. J.*, **150**, 1001.

Jacobs, K. 1973, *Nature*, **256**, 560.

Kavanagh, R. W. 1960, *Nucl. Phys.*, **15**, 411.

– 1972, *Cosmology, Fusion, and Other Matters*, ed. F. Reines (Boulder: Colorado Associated University Press), p. 169.

Kocharov, G. E., and Starbunov, Yu. N. 1970, *Acta Phys. Acad. Sci. Hung.*, Suppl. to Vol. 29 (Proceedings of the Eleventh International Conference on Cosmic Rays, Budapest), **4**, 353.

Kuchowicz, B. 1976, *Rep. Prog. Phys.*, **39**, 291.

Kuzmin, V. A. 1966, *Sov. Phys. JETP*, **22**, 1051.

Kuzmin, V. A., and Zatsepin, G. T. 1966, *Proceedings of the International Conference on Cosmic Rays*, London, Vol. **2**, 1023.

– 1969, *Proceedings of the International Conference on Neutrino Physics and Astrophysics* (Moscow), vol. 2, (USSR: F. I. Academy Science), p. 135.

Lambert, D. L., and Warner, B. 1968a, *Mon. Not. R. Astron. Sci.*, **138**, 181.

– 1968b, *Mon. Not. R. Astron. Sci.*, **138**, 213.

Lande, K., Bozoki, G., Frati, W., Lee, C. K., Fenyes, E., and Saavedra, O. 1974, *Nature*, **251**, 485.

Libby, L. M., and Thomas, F. J. 1969, *Nature*, **222**, 1238.

Maeder, A. 1977, *Astr. Astrophys.*, **56**, 359.

Parker, E. N. 1974, *Astrophys. Space Sci.*, **31**, 261.

Parker, P. D. 1966, *Phys. Rev.*, **150**, 851.

– 1968, *Astrophys. J. Lett.*, **153**, L85.

Parker, P. D., Bahcall, J. N., and Fowler, W. A. 1964, *Astrophys. J.*, **139**, 602.

Parker, P. D., and Kavanagh, R. W. 1963, *Phys. Rev.*, **131**, 2578.

Pochoda, P., and Reeves, H. 1964, *Planet. Space Sci.*, **12**, 119.

Pontecorvo, B. 1946, Chalk River Laboratory Report PD-205.

– 1968, *Sov. Phys. JETP*, **26**, 984.

Prentice, A. J. R. 1973, *Mon. Not. R. Astron. Sci.*, **163**, 331.

Raghavan, R. S. 1976, *Phys. Rev. Lett.*, **37**, 259.

Reeder, P. L., Poskanzer, A. M., and Esterlund, R. A. 1964, *Phys. Rev. Lett.*, **13**, 767.

Reines, F. 1960, *Ann. Rev. Nucl. Sci.*, **10**, 25.

– 1967, *Proc. R. Soc. Ser. A*, **301**, 159.

– 1969, *Quasars and High-Energy Astronomy*, ed. K. N. Douglas, I. Robinson, A. Schild, E. L. Schucking, J. A. Wheeler, and N. J. Woolf (New York: Gordon and Breach), p. 295.

Reines, F., and Kropp, W. R. 1964, *Phys. Rev. Lett.*, **12**, 457.

Reines, F., Sobel, H., and Pasierb, E. 1980, *Phys. Rev. Lett.*, **45**, 1307.

Reines, F., and Trimble, V. 1972, *Proceedings of the Solar Neutrino Conference*, University of California, Irvine.

Reines, F., and Woods, R. M., Jr. 1965, *Phys. Rev. Lett.*, **14**, 201.

Rolfs, C., and Trautvetter, H. P. 1978, *Ann. Rev. Nucl. Part. Sci.*, **28**, 115.

Rood, R. T. 1978, *Proceedings of the Brookhaven Solar Neutrino Conference*, **1**, 175.

Rood, R. T., and Ulrich, R. K. 1974, *Nature*, **252**, 366.

Rowley, J. K., Cleveland, B. T., Davis, R., Jr., and Evans, J. C. 1977, *Proceedings of the International Conference on Neutrino Physics and Neutrino Astrophysics* (Neutrino '77, Elbrus, USSR), **1**, 15.

Rowley, J. K., Cleveland, B. T., Davis, R., Jr., Hampel, W., and Kirsten, T. 1980, Brookhaven National Laboratory Report 27190, to be published.

Roxburgh, I. W. 1974, *Nature*, **248**, 209.

Ryajskaya, O., and Zatsepin, G. 1965, *Proceedings of the International Conference on Cosmic Rays*, London, 987.

Salpeter, E. E. 1968, *Comments Nucl. Part. Phys.* **II**, 97.

– 1970, *Nature*, **225**, 165.

Schatzman, E. 1969, *Astrophys. Lett.*, **3**, 139.

Sears, R. L. 1964, *Astrophys. J.*, **140**, 477.

– 1966, *Stellar Evolution*, ed. R. F. Stein and A. G. W. Cameron (New York: Plenum Press), p. 245.

Shaviv, G., Bahcall, J. N., and Fowler, W. A. 1967, *Astrophys. J.*, **150**, 725.

Shaviv, G., and Salpeter, E. E. 1968, *Phys. Rev. Lett.*, **21**, 1602.

Sheldon, W. R. 1969, *Nature*, **221**, 650.

Snell, R. L., Wheeler, J. C., and Wilson, J. R. 1976, *Astrophys. Lett.*, **17**, 157.

Tombrello, T. A. 1965, *Nucl. Phys.*, **71**, 459.

– 1967, *Nuclear Research with Low Energy Accelerators*, ed. J. B. Marion and D. M. Palter (New York: Academic Press), p. 195.

Tombrello, T. A., and Parker, P. D. 1963, *Phys. Rev.*, **131**, 2582.

Torres-Peimbert, S., Ulrich, R. K., and Simpson, E. 1969, *Astrophys. J.*, **155**, 957.

Trimble, V., and Reines, F. 1973, *Rev. Mod. Phys.*, **45**, 1.

Watson, W. D. 1969a, *Astrophys. J.*, **157**, 375.

– 1969b, *Astrophys. J.*, **158**, 303.

– 1969c, *Astrophys. J.*, **158**, 303.

Werntz, C., and Brennan, J. G. 1967, *Phys. Rev.*, **157**, 759.

Weymann, R., and Sears, R. L. 1965, *Astrophys. J.*, **142**, 174.

Wheeler, J. C., and Cameron, A. G. W. 1975, *Astrophys. J.*, **196**, 601.

Wolfendale, A. F., Young, E. C. M., and Davis, R., Jr. 1972, *Nature Phys. Sci.*, **238**, 1301.

Wolfenstein, L. 1978, *Phys. Rev. D*, **17**, 2365.

References

Aardsma, G., *et al.* (1987) *Phys. Lett.*, **B194**, 321.

Abraham, Z. and I. Iben, Jr. (1971) *Ap. J.*, **170**, 157.

Adelberger, E.G. and W.C. Haxton (1987) *Phys. Rev. Lett.*, **C36**, 879.

Aglietta, M., *et al.* (1986) *Nuovo Cimento*, **9C**, 185.

Aglietta, M., *et al.* (1987a) *Europhys. Lett.*, **3**, 1315.

Aglietta, M., *et al.* (1987b) *Europhys. Lett.*, **3**, 1321.

Akhmedov, E.Kh. and M.Yu. Khlopov (1988) *Mod. Phys. Lett.*, **A3**, 451.

Alekseev, E.N. (1979) in *Proceedings of the Sixteenth International Cosmic Ray Conference*, Kyoto, Japan (Tokyo: University of Tokyo) p. 276.

Alekseev, E.N., L.N. Alekseev, V.I. Volchenko, and I.V. Krivosheina (1987) *JETP Lett.*, **45**, 589.

Alekseev, E.N., L.N. Alekseev, I.V. Krisvosheina, and V.I. Volchenko (1988) *Phys. Lett.*, **B205**, 209.

Alexander, T.K., G.C. Ball, W.N. Lennard, H. Geissel, and H.-B. Mak (1984) *Nucl. Phys.*, **A427**, 526.

Aller, L.H. (1986) in *Spectroscopy of Astrophysical Plasmas*, edited by A. Dalgarno and D. Layzer (Cambridge: Cambridge University Press) p. 89.

Alvarez, L.W. (1949) University of California Radiation Laboratory Report UCRL-328.

Alvarez, L.W. (1973) Physics Notes, Memo No. 767, Lawrence Radiation Lab. (March 23).

Ando, H. and Y. Osaki (1975) *Pub. Astro. Soc. Japan*, **27**, 581.

Arisaka, K. (1985) PhD Thesis, Faculty of Science, University of Tokyo UTICEPP-85-01 (unpublished).

Arisaka, K., *et al.* (1985) *J. Phys. Soc. Jpn.*, **54**, 3213.

Arafune, J. and M. Fukugita (1987) *Phys. Rev. Lett.*, **59**, 367.

Arafune, J., M. Fukugita, T. Yanagida, and M. Yoshimura (1987) *Phys. Lett.*, **B194**, 477.

Arnett, W.D. (1977) *Ap. J.*, **218**, 815.
Arnett, W.D. (1987) *Ap. J.*, **319**, 136.
Arnett, W.D. (1988) *Ap. J.*, **331**, 377.
Arnett, W.D. and L.B. Bowers (1977) *Ap. J. Supp.*, **33**, 415.
Arnett, W.D. and J.L. Rosner (1987) *Phys. Rev. Lett.*, **58**, 1906.
Assenbaum, H.J., K. Langanke, and C. Rolfs (1987) *Zeits für Physik A*, **327**, 461.
Avignone, F.T., *et al.* (1986) *Phys. Rev. D*, **34**, 97.
Baade, W. and F. Zwicky (1934) *Proc. Natl. Acad. Sci. USA*, **20**, 254.
Badino, G., *et al.* (1984) *Nuovo Cimento*, **7C**, 573.
Bahcall, J.N. (1961) *Phys. Rev.*, **124**, 495.
Bahcall, J.N. (1962a) *Phys. Rev.*, **126**, 1143.
Bahcall, J.N. (1962b) *Phys. Rev.*, **128**, 1297.
Bahcall, J.N. (1963) *Phys. Rev.*, **129**, 2683.
Bahcall, J.N. (1964a) *Phys. Rev. Lett.*, **12**, 300.
Bahcall, J.N. (1964b) *Phys. Rev. B*, **135**, 137.
Bahcall, J.N. (1964c) *Phys. Rev. B*, **136**, 1164.
Bahcall, J.N. (1964d) *Phys. Lett.*, **13**, 332.
Bahcall, J.N. (1965) *Nucl. Phys.*, **71**, 267.
Bahcall, J.N. (1966a) *Nucl. Phys.*, **75**, 10.
Bahcall, J.N. (1966b) *Phys. Rev. Lett.*, **17**, 398.
Bahcall, J.N. (1966c) *Ap. J.*, **143**, 259.
Bahcall, J.N. (1969) *Phys. Rev. Lett.*, **23**, 251.
Bahcall, J.N. (1977) *Ap. J. Lett.*, **216**, L115.
Bahcall, J.N. (1978a) *Rev. Mod. Phys.*, **50**, 881.
Bahcall, J.N. (1978b) *Ann. Rev. Astron. Astrophys.*, **16**, 241.
Bahcall, J.N. (1981a) *Phys. Rev. C*, **24**, 2216.
Bahcall, J.N. (1981b) in *Neutrino 81, Proceedings of the 1981 International Conference on Neutrino Physics and Astrophysics*, Honolulu, Hawaii, edited by R.J. Cence, E. Ma, and A. Roberts (Honolulu: University of Hawaii) Vol. 2, p. 253.
Bahcall, J.N. (1986) in *Weak and Electromagnetic Interactions in Nuclei, Proceedings of the International Symposium*, Heidelberg, Germany, edited by H.V. Klapdor (Berlin: Springer-Verlag) p. 705.
Bahcall, J.N. (1987) *Rev. Mod. Phys.*, **59**, 505.
Bahcall, J.N. (1988) *Phys. Rev. Lett.*, **61**, 2650.
Bahcall, J.N., N.A. Bahcall, W.A. Fowler, and G. Shaviv (1968) *Phys. Lett.*, **26**, B359.
Bahcall, J.N., N.A. Bahcall, and G. Shaviv (1968) *Phys. Rev. Lett.*, **20**, 1209.
Bahcall, J.N., N.A. Bahcall, and R.K. Ulrich (1968) *Astrophys. Lett.*, **2**, 91.
Bahcall, J.N., N.A. Bahcall, and R.K. Ulrich (1969) *Ap. J.*, **156**, 559.

Bahcall, J.N., M. Baldo-Ceolin, D.B. Cline, and C. Rubbia (1986) *Phys. Lett.*, **178**, B324.

Bahcall, J.N., N. Cabibbo, and A. Yahil (1972) *Phys. Rev. Lett.*, **28**, 316.

Bahcall, J.N., B.T. Cleveland, R. Davis, Jr., I. Dostrovsky, J.C. Evans, W. Frati, G. Friedlander, K. Lande, J.K. Rowley, R.W. Stoenner, and J. Weneser (1978) *Phys. Rev. Lett.*, **40**, 1351.

Bahcall, J.N., B.T. Cleveland, R. Davis, Jr., and J.K. Rowley (1985) *Ap. J. Lett.*, **292**, L79.

Bahcall, J.N. and R. Davis, Jr. (1976) *Science*, **191**, 264.

Bahcall, J.N. and R. Davis, Jr. (1982) in *Essays in Nuclear Astrophysics*, edited by C.A. Barnes, D.D. Clayton, and D. Schramm (Cambridge: Cambridge University Press) p. 243.

Bahcall, J.N., A. Dar, and T. Piran (1987) *Nature*, **326**, 135.

Bahcall, J.N., G.B. Field, and W.H. Press (1987) *Ap. J.*, **320**, L69.

Bahcall, J.N., W.A. Fowler, I. Iben, and R.L. Sears (1963) *Ap. J.*, **137**, 344.

Bahcall, J.N. and S.C. Frautschi (1964) *Phys. Rev. B*, **136**, 1547.

Bahcall, J.N. and S.C. Frautschi (1969) *Phys. Lett.*, **B29**, 263.

Bahcall, J.N., J. Gelb, and S.P. Rosen (1987) *Phys. Rev. D*, **35**, 2976.

Bahcall, J.N. and S. Glashow (1987) *Nature*, **326**, 476.

Bahcall, J.N. and W.C. Haxton (1989) *Phys. Rev. D*, , to be submitted.

Bahcall, J.N. and B.R. Holstein (1986) *Phys. Rev. C*, **33**, 2121.

Bahcall, J.N., W.F. Huebner, S.H. Lubow, P.D. Parker, and R.K. Ulrich, (1982) *Rev. Mod. Phys.*, **54**, 767.

Bahcall, J.N., K. Kubodera, and S. Nozawa (1988) *Phys. Rev. D*, **38**, 1030.

Bahcall, J.N., S.H. Lubow, W.F. Huebner, N.H. Magee, Jr., A.L. Merts, M.F. Argo, P.D. Parker, B. Rozsnyai, and R.K. Ulrich (1980) *Phys. Rev. Lett.*, **45**, 945.

Bahcall, J.N. and R.M. May (1969) *Ap. J.*, **155**, 501.

Bahcall, J.N. and C.P. Moeller (1969) *Ap. J.*, **155**, 511.

Bahcall, J.N., S.T. Petcov, S. Toshev, and J.W.F. Valle (1986) *Phys. Lett.*, **B181**, 369.

Bahcall, J.N. and T. Piran (1983) *Ap. J. Lett.*, **267**, L77.

Bahcall, J.N., T. Piran, W.H. Press, and D.N. Spergel (1987) *Nature*, **327**, 682.

Bahcall, J.N. and R.L. Sears (1972) *Ann. Rev. Astron. Astrophys.*, **10**, 25.

Bahcall, J.N. and G. Shaviv (1968) *Ap. J.*, **153**, 113.

Bahcall, J.N., D.N. Spergel, and W.H. Press, in *Supernova 1987A in the Large Magellanic Cloud, Proceedings of the Fourth George Mason Astronomy Workshop*, Fairfax, Virginia, edited by M. Kafatos and A. Michalitsianos (Cambridge: Cambridge University Press) p. 172

Bahcall, J.N. and R.M. Soneira (1980) *Ap. J. Supp.*, **44**, 73.

Bahcall, J.N. and R.K. Ulrich (1970) *Ap. J. Lett.*, **160**, L57.

Bahcall, J.N. and R.K. Ulrich (1971) *Ap. J.*, **170**, 593.

Bahcall, J.N. and R.K. Ulrich (1988) *Rev. Mod. Phys.*, **60**, 297.

Bahcall, J.N. and R.A. Wolf (1964) *Ap. J.*, **139**, 622.

Bakich, A.M., M. Omori, L.A. Peak, and N.T. Wearne (1984) *Nucl. Inst. Meth.*, **226**, 383.

Bakich, A.M. and L.S. Peak (1985) in *Solar Neutrinos and Neutrino Astronomy*, edited by M.L. Cherry, W.A. Fowler, and K. Lande (New York: American Institute of Physics) Conference Proceedings No. 126, p. 238

Bakich, A.M. and L.S. Peak (1986) *Nucl. Inst. Meth.*, **A247**, 334.

Bakich, A.M., L.S. Peak, and N.T. Wearne (1984) *Aust. J. Phys.*, **37**, 567.

Balin, D. (1983) *Weak Interactions* (Bristol: Hilger).

Baltz, A.J. and J. Weneser (1987) *Phys. Rev. D*, **35**, 528.

Baltz, A.J. and J. Weneser (1988) *Phys. Rev. D*, **37**, 3364.

Bambynek, W., H. Behrens, M.H. Chen, B. Craseman, M.L. Fitzpatrick, K.W.D. Ledingham, H. Genz, M. Mutterer, and R.L. Intermann (1977) *Rev. Mod. Phys.*, **49**, 78.

Barabanov, I.R., *et al.* (1980) *JETP Lett.*, **32**, 359.

Barabanov, I.R., *et al.* (1985) in *Solar Neutrinos and Neutrino Astronomy*, edited by M.L. Cherry, W.A. Fowler, and K. Lande (New York: American Institute of Physics) Conference Proceedings No. 126, p. 175.

Barabanov, I.R., G.V. Domagatsky, and G.T. Zatsepin (1989) *Proceedings of the Thirteenth International Conference on Neutrino Physics and Astrophysics* (to be published).

Barbiellini, G. and G. Cocconi (1987) *Nature*, **329**, 21.

Barbieri, R. and R.N. Mohapatra (1988) *Phys. Rev. Lett.*, **61**, 27.

Barger, V., K. Whisnant, S. Pakvasa, and R.J.N. Phillips (1980) *Phys. Rev. D*, **22**, 2718.

Barger, V., R.J.N. Phillips, and K. Whisnant (1986) *Phys. Rev. D*, **34**, 980.

Barger, V., K. Whisnant, and R.J.N. Phillips (1981) *Phys. Rev. D*, **24**, 538.

Bari, C., *et al.* (1988) *Meth. in Phys. Research*, **A264**, 5.

Barker, F.C. and R.H. Spear (1986) *Ap. J.*, **307**, 847.

Barnes, C.A. (1981) in *Progress in Particle and Nuclear Physics*, edited by D. Wilkinson (Oxford: Pergamon Press) Vol 6, p. 235.

Bartenwerfer, D. (1973) *Astron. Astrophys.*, **25**, 455.

Basu, D. (1982) *Solar Phys.*, **81**, 363.

Bazilevskaya, G.A., Yu.I Stozhkov and T.N. Charakhch'yan (1982) *JETP Lett.*, **35**, 341.

Bearden, J.A. and A.F. Burr (1967) *Rev. Mod. Phys.*, **39**, 125.

Beg, M.A.B, W.J. Marciano, and M. Ruderman (1978) *Phys. Rev. D*, **17**, 1935.

Behrens, H. and J. Janecke (1969) *Numerical Tables for Beta-Decay and Electron Capture* (Berlin: Springer-Verlag).

Beier, E.W. (1987) in *Proceedings of Seventh Workshop on Grand Unification, ICOBAN'86*, Toyama, Japan, edited by J. Arafune (Singapore: World Scientific) p. 79.

Bellefon, A., P. Espigat, and G. Waysand (1985) in *Solar Neutrinos and Neutrino Astronomy*, edited by M.L. Cherry, W.A. Fowler, and K. Lande (New York: American Institute of Physics) Conference Proceedings No. 126, p. 227.

Beresnev, V.I. *et al.* (1979) in *Proceedings of the Sixteenth International Cosmic Ray Conference*, Kyoto, Japan (Tokyo: University of Tokyo) p. 293.

Berezinsky, V.S., C. Castagnoli, V.I. Dokuchaev, and P. Galeotti (1988), *Nuovo Cimento*, **11C**, 287.

Bernstein, J., *et al.* (1963) *Phys. Rev.*, **132**, 1227.

Bethe, H.A. (1939) *Phys. Rev.*, **55**, 434.

Bethe, H.A. (1972) in *Nobel Lectures Physics 1963–1970* (Amsterdam: Elsevier) p. 209.

Bethe, H.A. (1986) *Phys. Rev. Lett.*, **56**, 1305.

Bethe, H.A. and G. Brown (1985) *Sci. Am.*, **252**, 60.

Bhalla, C.P. and M.E. Rose (1962) *Phys. Rev.*, **128**, 774.

Bilenky, S.M. and S.T. Petcov (1987) *Rev. Mod. Phys.*, **59**, 671.

Bilenky, S.M. and B.M. Pontecorvo (1978) *Phys. Reports*, **C41**, 225.

Bilenky, S.M. and B.M. Pontecorvo (1987) *Rev. Mod. Phys.*, **59**, 671.

Bionta, R.M., *et al.* (1983) *Phys. Rev. Lett.*, **51**, 27.

Bionta, R.M., *et al.* (1987) *Phys. Rev. Lett.*, **58**, 1494.

Bisnovatyi-Kogan, G.S. and Z.F. Seidov (1982) *Sov. Astron.*, **28**, 213.

Blümer H. and K. Kleinknecht (1985) *Phys. Lett.*, **161B**, 407.

Boehm F.A. and P. Vogel (1987) *Physics of Massive Neutrinos* (Cambridge: Cambridge University Press).

Boercker, D.B. (1987) *Ap. J.*, **316**, L95.

Boivin, L.P., W.F. Davidson, R.S. Storey, D. Sinclair, and E.D. Earle, (1986) *Appl. Optics*, **25**, 877.

Booth, N.E. (1987) *Sci. Prog. Oxf.*, **71**, 563.

Booth, N.E., G.L. Salmon, and D.A. Hunkin (1985) in *Solar Neutrinos and Neutrino Astronomy*, edited by M.L. Cherry, W.A. Fowler, and K. Lande (New York: American Institute of Physics) Conference Proceedings No. 126, p. 216.

Booth, N.E., *et al.* (1987) p. 74.

Bopp, P., D. Dubbers, L. Hornig, E. Klemt, J. Last, H. Schütze, S.J. Freedman, and O. Schärpf (1986) *Phys. Rev. Lett.*, **56**, 919. *Erratum*, **57**, 1192.

Boris, S.D., *et al.* (1987) *Phys. Rev. Lett.*, **58**, 2019.

Bowers, R. and J.R. Wilson (1982) *Ap. J.*, **263**, 366.

Bowers, R. and J.R. Wilson (1985) *Ap. J. Supp.*, **50**, 115.

Bouchez, J., M. Cribier, W. Hampel, J. Rich, M. Spiro, and D. Vignaud (1986) *Z. Phys.*, **C32**, 499.

Boyd, R.N., R.E. Turner, B. Sur, L. Rybarcyk, and C. Joseph (1985) in *Solar Neutrinos and Neutrino Astronomy*, edited by M.L. Cherry, W.A. Fowler, and K. Lande (New York: American Institute of Physics) Conference Proceedings No. 126, p. 145.

Boyd, R.N., R.E. Turner, M. Wiescher, and L. Rybarcyk (1983) *Phys. Rev. Lett.*, **51**, 609.

Boyd, R.N., R.E. Turner, L. Rybarcyk, and C. Joseph (1985) *Ap. J.*, **289**, 155.

Bratton, C.B., *et al.* (1988) *Phys. Rev. D*, **37**, 3361.

Braun, E. and I. Talmi (1986) in *Weak and Electromagnetic Interactions in Nuclei, Proceedings of the International Symposium*, Heidelberg, Germany, edited by H.V. Klapdor (Berlin: Springer-Verlag) p. 47

Brown, T.M., B.W. Mihalas, and E.J. Rhodes (1986) in *Physics of the Sun*, edited by P.A. Sturrock, T.E. Olzer, D.M. Mihalas, and R.K. Ulrich (Dordrecht: Reidel) Vol. I, p. 177.

Brown, T.M. and C.A. Morrow (1987) *Ap. J. Lett.*, **314**, L21.

Bruenn, S. (1985) *Ap. J. Supp.*, **58**, 771.

Bruenn, S. (1986) *Ap. J. Lett.*, **311**, L69.

Burbidge, E.M., G.R. Burbidge, W.A. Fowler, and F. Hoyle (1957) *Rev. Mod. Phys.*, **29**, 547.

Burrows, A. (1987a) *Ap. J. Lett.*, **318**, L57.

Burrows, A. (1987b) *Phys. Today*, **40**, 28.

Burrows, A. (1988a) in *Supernova 1987A in the Large Magellanic Cloud, Proceedings of the Fourth George Mason Astronomy Workshop*, Fairfax, Virginia, edited by M. Kafatos and A. Michalitsianos (Cambridge: Cambridge University Press) p. 161.

Burrows, A. (1988b) *Ap. J.*, **334**, 891.

Burrows, A. and J.M. Lattimer (1986) *Ap. J.*, **307**, 178.

Burrows, A. and J.M. Lattimer (1987) *Ap. J. Lett.*, **318**, L63.

Burrows, A., T.J. Mazurek, and J.M. Lattimer (1981) *Ap. J.*, **251**, 325.

Cabrera, B., L.M. Krauss, and F. Wilczek (1985) *Phys. Rev. Lett.*, **55**, 25.

Cabrera, B. (1986) in *'86 Massive Neutrinos in Astrophysics and Particle Physics, Proceedings of the Sixth Moriond Workshop*, edited by O. Fackler and J. Tran Thanh Van (Gif-sur-Yvette: Editions Frontiéres) p. 423.

Cahen, S., C. Doom, and M. Cassè (1986) in *'86 Massive Neutrinos in Astrophysics and Particle Physics, Proceedings of the Sixth Moriond Workshop*, edited by O. Fackler and J. Tran Thanh Van (Gif-sur-Yvette: Editions Frontiéres) p. 83.

Cahn, R.N. and S.L. Glashow (1981) *Science*, **213**, 607.

Cameron, A.G.W. (1957) *Chalk River Report CRL-41*.

Cameron, A.G.W. (1958) *Ann. Rev. Nucl. Sci.*, **8**, 299.

Carlson, E.D. (1986) *Phys. Rev. D*, **34**, 1454.

Carraro, C., A. Schäfer, and S.E. Koonin (1988) *Ap. J.*, **331**, 565.

Cassè, M., S. Cahen, and C. Doom (1986) in *Neutrinos and the Present Day Universe, Colloque de la Societé Française de Physique*, edited by T. Montmerle and M. Spiro (Gif-sur-Yvette: Commissariat 'a l'Energie Atomique) p. 49.

Caughlan, G.R. and W.A. Fowler (1962) *Ap. J.*, **136**, 453.

Champagne, A.E., G.E. Dodge, R.T. Kouzes, M.M. Lowry, A.B. MacDonald, and M.W. Roberson (1987) *Phys. Rev. C*, **38**, 900.

Chandrasekhar, S. (1939) *An Introduction to Stellar Structure* (Chicago: University of Chicago Press).

Chen, H.H. (1985) *Phys. Rev. Lett.*, **55**, 1534.

Chen, H.H. and J.F. Lathrop (1978) *Nucl. Inst. Meth.*, **150**, 585.

Chitre, S., D. Ezer, and R. Stothers (1973) *Astrophys. Lett.*, **14**, 37.

Christensen-Dalsgaard, J., *et al.* (1974) *Mon. Not. R. Astron. Soc.*, **169**, 429.

Christensen-Dalsgaard, J., T.L. Duvall, D.O. Gough, J.W. Harvey, and E.J. Rhodes, Jr. (1985) *Nature*, **300**, 242.

Christensen-Dalsgaard, J., D.O. Gough, and J.G. Morgan (1979) *Astron. Astrophys.*, **73**, 121.

Christensen-Dalsgaard, J., D.O. Gough, and J. Toomre (1985) *Science*, **229**, 923.

Cicernos, E.L. (1971) *Astro. Space Sci.*, **10**, 87.

Claverie, A., G.R. Isaak, C.P. McLeod, H.B. van de Raay, and T. Roca Cortez (1979) *Nature*, **282**, 591.

Clayton, D.D. (1974) *Nature*, **294**, 131.

Clayton, D.D. (1983) *Principles of Stellar Evolution and Nucleosynthesis* (Chicago: University of Chicago Press).

Clayton, D.D. (1986) *The Joshua Factor* (Austin: Texas Monthly Press).

Clayton, D.D., M.J. Newman, and R.J. Talbot, Jr. (1975a) *Ap. J.*, **201**, 489.

Clayton, D.D., E. Dwek, M.J. Newman, and R.J. Talbot, Jr. (1975b) *Ap. J.*, **199**, 494.

Cleveland, B.T. (1983) *Nucl. Inst. Meth.*, **214**, 451.

Cleveland, B.T., R. Davis, Jr., and J.K. Rowley (1984) in *Proceedings of the Second International Symposium on Resonance Ionization Spectroscopy and Its Applications*, Knoxville, Tennessee, edited by G.S. Hurst and M.G. Payne (Bristol: Hilger-IOP) p. 241.

Commins, E.D. and P. Bucksbaum (1983) *Weak Interactions of Leptons and Quarks* (Cambridge: Cambridge University Press).

Cooperstein, J. (1988a) *Phys. Rev. C*, **37**, 786.

Cooperstein, J. (1988b) *Physics Reports*, **163**, 95.

Cooperstein, J. and J. Wambach (1984) *Nucl. Phys.*, **A240**, 591.

Cowan, G.A. and W.C. Haxton (1982) *Science*, **216**, 51.

Cowling, T.G. (1941) *Mon. Not. R. Astron. Soc.*, **101**, 367.

Cowling, T.G. (1945) *Mon. Not. R. Astron. Soc.*, **105**, 166.

Cowling, T.G. (1965) in *Stellar Structure*, edited by L.H. Aller and D.B. McLaughlin (Chicago: University of Chicago Press), p. 425.

Cowsik, R. (1988) *Phys. Rev. D*, **37**, 1685.

Cox, A.N., J.A. Guzik, and R. B. Kidman, (1989) *Ap. J.* (to be published).

Cox, A.N., R.B. Kidman, and M. J. Newman (1985) in *Solar Neutrinos and Neutrino Astronomy*, edited by M.L. Cherry, W.A. Fowler, and K. Lande (New York: American Institute of Physics) Conference Proceedings No. 126, 93.

Cox, J.P. and R.T. Giuli (1968) *Principles of Stellar Structure* (New York: Gordon and Breach) Vols. I and II.

Cribier, M., W. Hampel, J. Rich, and D. Vignaud (1986) *Phys. Lett.*, **B182**, 89.

Cribier, M., J. Rich, M. Spiro, D. Vignaud, W. Hampel, and B.T. Cleveland (1987) *Phys. Lett.*, **B188**, 168.

Cribier, M., *et al.* (1988) *Nucl. Inst. Meth.*, **A265**, 574.

Däppen, W., R.L. Gilliland, and J. Christensen-Dalsgaard (1986) *Nature*, **321**, 229.

Dar, A. (1985) *Phys. Rev. Lett.*, **55**, 1422.

Dar, A. (1988a) in *Proceedings of International Conference on Extrasolar Neutrino Astronomy*, UCLA, USA (1987), edited by D. Cline (Los Angeles: UCLA).

Dar, A. (1988b) in *Proceedings of International Workshop on SN1987A – One Year Later* (La Thuile, Italy, February 1988) edited by G. Belletini, M. Greco, and P. Galeotti (Paris: Editions Frontiéres) p. 241.

Dar, A. and A. Mann (1987) *Nature*, **325**, 790.

Dar, A., A. Mann, Y. Melina, and D. Zajfman (1987) *Phys. Rev. D*, **35**, 3607.

Davids, C.N., T.F. Wang, I. Ahmad, R. Holzmann, and R.V.F. Janssens, (1987) *Phys. Rev. C*, **35**, 1114.

Davidson, W.F., P. Depommier, G.T. Ewan, and H.-B. Mak (1984) in *Proceedings of the International Conference on Baryon Non Conservation* (ICOBAN '84, Park City, Utah) p. 273.

Davis, R., Jr. (1964) *Phys. Rev. Lett.*, **12**, 303.

Davis, R., Jr. (1968) in *Proceedings of the International Seminar on Neutrino Physics and Neutrino Astrophysics*, Moscow, USSR, p. 99.

Davis, R., Jr. (1978) in *Proceedings of Informal Conference on Status and Future of Solar Neutrino Research*, edited by G. Friedlander, (Upton: Brookhaven National Labortory) Report No. 50879, Vol. 1, p. 1.

Davis, R., Jr. (1987) in *Proceedings of Seventh Workshop on Grand Unification, ICOBAN'86*, Toyama, Japan, edited by J. Arafune (Singapore: World Scientific) p. 237.

Davis, R., Jr. (1988) private communication.

Davis, R., Jr., J.S. Evans, and B.T. Cleveland (1978) in *Long-Distance Neutrino Detection-1978*, edited by A.W. Sáenz and H. Überall (New York: American Institute of Physics) Conference Proceedings No. 52, p. 17.

Davis, R., Jr., J.C., Evans, V. Radeka, and L.C. Rogers (1972) in *Neutrino '72: Proceedings of Europhysics Conference* (Baltonfüred Hungary) edited by A. Frenkel and G. Marx, (Budapest: OMKDK Technoinform) p. 5.

Davis, R., Jr., D.S. Harmer, and K.C. Hoffman (1968) *Phys. Rev. Lett.*, **20**, 1205.

Davis, R., Jr., D.S., Harmer, and F.H. Neely (1969) in *Quasars and High-Energy Astronomy*, edited by K.N. Douglas, I. Robinson, A. Schild, E.L. Schucking, J.A. Wheeler, and N.J. Woolf (New York: Gordon and Breach) p. 287.

Dearborn, D.S.P., G. Marx, and I. Ruff (1987) *Prog. Theor. Phys. Lett.*, **77**, 12.

de Braemaecker, J.C. (1985) in *Geophysics: The Earth's Interior* (New York: Wiley) p. 63.

Delache, P. and P. Scherrer (1983) *Nature*, **306**, 651.

de la Zerda-Lerner, A. and K. O'Brien (1987) *Nature*, **330**, 353.

Demarque, P., J.G. Mengel, and A.V. Sweigart (1973) *Ap. J.*, **183**, 997.

de Rújula, A. (1987) *Phys. Lett.*, **B193**, 514.

de Rújula, R.C. Giles, and R.L. Jaffe (1978) *Phys. Rev. D*, **17**, 285.

Deubner, F.-L. (1975) *Solar Phys.*, **44**, 371.

Deubner, F.-L. and D.O. Gough (1984) *Ann. Rev. Astron. Astrophys.*, **22**, 593.

de Vaucouleurs, G., A. de Vaucouleurs, and H.G. Corwin (1976) *Second Reference Catalogue of Bright Stars* (Austin: Univiversity of Texas Press).

Dicke, R.H., J.R. Kuhn, and K.G. Libbrecht (1985) *Nature*, **316**, 687.

Diesendorf, M.O. (1970) *Nature*, **227**, 266.

Dilke, F.W.W. and D.O. Gough (1972) *Nature*, **240**, 262.

Doe, P.J., H.-J. Mahler, and H.H. Chen (1982) *Nucl. Inst. Meth.*, **199**, 639.

Domogatskii, G.V. (1984) *Sov. Astron.***28**, 30.

Domogatskii, G.V. and D.K. Nadezhin (1971) *Sov. J. Nucl. Phys.*, **12**, 678.

Donnelly, T.W., D. Hitlin, M. Schwartz, J.D. Walecka, and S.J. Wiesner (1974) *Phys. Lett.*, **49B**, 8.

Dostrovsky, I. (1978) in *Proceedings of Informal Conference on Status and Future of Solar Neutrino Research*, edited by G. Friedlander, (Upton: Brookhaven National Labortory) Report No. 50879, Vol. 1, p. 231.

Drukier, A.K., K. Freese, and D.N. Spergel (1986) *Phys. Rev. D*, **33**, 3495.

Drukier, A.K. and R. Nest (1985) *Nucl. Inst. Meth.*, **A239**, 605.

Drukier, A. and L. Stodolsky (1984) *Phys. Rev. D*, **30**, 2295.

Durand, L. (1964) *Phys. Rev. B*, **135**, 310.

Duvall, T.L. and J.W. Harvey (1984) *Nature*, **310**, 19.

Duvall, T.L., W.A. Dziembowski, P.R. Goode, D.O. Gough, J.W. Harvey, and J.W. Leibacher (1984) *Nature*, **310**, 22.

Dwarakanath, M.R. and H. Winkler (1971) *Phys. Rev. C*, **4**, 1532.

Earle, G.T., *et al.* (1986) in *Proceedings of the Second Conference on the Intersections between Particle Physics*, edited by D.F. Geesaman (New York: American Institute of Physics) Conference Proceedings No. 150, p. 1094.

Eddington, A.S. (1920) *Observatory*, **43**, 341.

Eddington, A.S. (1926) *The Internal Constitution of the Stars* (Cambridge: Cambridge University Press).

Eder, G. (1966) *Nucl. Phys.*, **78**, 657.

Ehrlich, R. (1982) *Phys. Rev. D*, **25**, 2282.

Ellis, S.D. and J.N. Bahcall (1968) *Nucl. Phys.*, **A114**, 636.

Elton, L.R.B. (1961) *Nuclear Sizes* (Oxford: Oxford University).

Engstler, S. *et al.* (1988) *Phys. Lett.*, **B202**, 179.

Ernst, H. *et al.* (1984) *Nucl. Instr. Meth.*, **B5**, 426.

Evetts, J.E., *et al.* (1988) *Nucl. Inst. Meth.*, **A264**, 41.

Ewan, G.T., *et al.* (1987) in *Sudbury Neutrino Observatory Proposal* (Sudbury Neutrino Observatory Collaboration: Queen's University at Kingston) Pub. No. SNO-87-12. .

Ezer, D. and A.G.W. Cameron (1965) *Can. J. Phys.*, **43**, 1497.

Ezer, D. and A.G.W. Cameron (1968) *Astrophys. Lett.*, **1**, 177.

Fainberg, A.M. (1978) in *Proceedings of Informal Conference on Status and Future of Solar Neutrino Research*, edited by G. Friedlander, (Upton: Brookhaven National Labortory) Report No. 50879, Vol. 2, p. 93.

Faulkner, J. and R.L. Gilliland (1985) *Ap. J.*, **299**, 994.

Faulkner, J., D.O. Gough, and M.N. Vahia (1986) *Nature*, **321**, 226.

Feinberg, G. and M. Goldhaber (1959) *Proc. Natl. Acad. Sci. USA*, **45**, 1301.

Filippone, B.W. and D.N. Schramm (1982) *Ap. J.*, **253**, 393.

Filippone, B.W., A.J. Elwyn, C.N. Davids, and D.D. Koetke (1983) *Phys. Rev. C*, **28**, 2222.

Fireman, E.L., B.T. Cleveland, R. Davis, Jr., and J.K. Rowley (1985) in *Solar Neutrinos and Neutrino Astronomy*, edited by M.L. Cherry, W.A. Fowler, and K. Lande (New York: American Institute of Physics) Conference Proceedings No. 126, p. 22.

Fowler, W.A. (1958) *Ap. J.*, **127**, 551.

Fowler, W.A. (1972) *Nature*, **238**, 24.

Fowler, W.A. (1984) *Rev. Mod. Phys.*, **56**, 149.

Fowler, W.A. (1987) private communication.

Fowler, W.A., G.R. Caughlan, and W.A. Zimmerman (1967) *Ann. Rev. Astron. Astrophys.*, **5**, 525.

Fowler, W.A., G.R. Caughlan, and W.A. Zimmerman (1975) *Ann. Rev. Astron. Astrophys.*, **13**, 69.

Freedman, D.Z. (1974) *Phys. Rev. D*, **9**, 1389.

Freedman, D.Z., D.N. Schramm, and D.L. Tubbs (1977) *Ann. Rev. Nucl. Sci.*, **27**, 167.

Freedman, M.S. (1978) in *Proceedings of Informal Conference on Status and Future of Solar Neutrino Research*, edited by G. Friedlander, (Upton: Brookhaven National Labortory) Report No. 50879, p. 313.

Freedman, M.S. (1988) *Nucl. Inst. Meth.*, **A271**, 267.

Freedman, M.S., *et al.* (1986) *Science*, **193**, 1117.

Frieman, J., H.E. Haber, and K. Freese (1988) *Phys. Lett.*, **B200**, 115.

Fujikawa, K. and R.E. Shrock (1980) *Phys. Rev. Lett.*, **45**, 963.

Fukugita, M. (1988) private communication.

Fukugita, M. and T. Yanagida (1987) *Phys. Rev. Lett.*, **58**, 1807.

Gaisser, T.K. and T. Stanev, (1985) in *Solar Neutrinos and Neutrino Astronomy,* edited by M.L. Cherry, W.A. Fowler, and K. Lande (New York: American Institute of Physics) Conference Proceedings No. 126, p. 277.

Gaisser, T.K. and T. Stanev (1987) *Phys. Rev. Lett.*, **58**, 1695. *Erratum*, **59**, 844.

Gamow, G. (1938) *Phys. Rev.*, **53**, 595.

Gamow, G. and M. Schönberg (1941) *Phys. Rev.*, **59**, 539.

Gari, M. (1978) in *Proceedings of Informal Conference on Status and Future of Solar Neutrino Research*, edited by G. Friedlander, (Upton: Brookhaven National Labortory) Report No. 50879, Vol. 1, p. 137.

Garrett, C. and Munk (1979) *Ann. Rev. Fluid Mech.* **11**, 339.

Gavrin, V.N., A.V. Kopylov, and N.T. Makeev (1982) *JETP Lett.*, **35**, 608.

Gavrin, V.N., A.V. Kopylov, and A.V. Streltsov (1985) in *Solar Neutrinos and Neutrino Astronomy,* edited by M.L. Cherry, W.A. Fowler, and K. Lande (New York: American Institute of Physics) Conference Proceedings No. 126, p. 185.

Gavrin, V.W. and E.A. Yanovich (1987) *Sov. Phys. - JETP*, **51**, 191.

Gelmini, G.B., L.J. Hall, and M.J. Lin (1987) *Nucl. Phys.*, **B281**, 726..

Gell-Mann, M., P. Ramond, and R. Slansky (1979) in *Supergravity: Proceedings of Supergravity Workshop* Stony Brook, New York, edited by P. Van Nieuwenhuizen and D.Z. Freedman (Amsterdam: North-Holland) p. 315.

Giboni, K.L. (1984) *Nucl. Inst. Meth.*, **225**, 579.

Gilliland, R.L. (1982) *Ap. J.*, **253**, 399.

Gilliland, R.L., J. Faulkner, W.H. Press, and D.N. Spergel (1986) *Ap. J.*, **306**, 703.

Gliese, W. (1983) in *Proceedings of IAU Colloquium 76* edited by A. G. Davis Philip and A.R. Upgren (Schenectady: L. Davis Press) p. 5.

Glashow, S. (1961) *Nucl. Phys.*, **22**, 579.

Glashow, S.L. and L.M. Krauss (1987) *Phys. Lett.*, **B190**, 199.

Goldhaber, M. (1975) *Proc. Am. Philos. Soc.* **119**, 24.

Goldman, I., Y. Aharonov, G. Alexander, and S. Nussinov (1988) *Phys. Rev. Lett.*, **60**, 1789.

Goodman, C.D. (1985) in *Solar Neutrinos and Neutrino Astronomy*, edited by M.L. Cherry, W.A. Fowler, and K. Lande (New York: American Institute of Physics) Conference Proceedings No. 126, p. 109.

Goodman, M.W. and E. Witten (1985) *Phys. Rev. D*, **31**, 3059.

Gough, D.O. (1977) in *The Energy Balance and Hydrodynamics of the Solar Chromosphere and Corona*, edited by R.M. Bonnet and P. Delanch (de Bussex: Claremont-Ferrand) p. 3.

Gough, D.O. (1985) *Solar Phys.*, **100**, 65.

Gould, A. (1987a) *Ap. J.*, **321**, 560.

Gould, A. (1987b) *Ap. J.*, **321**, 571.

Gramlich, J.W. and L.A. Machlan (1985) *Analytical Chemistry*, **57**, 1788.

Grec, G., E. Fossat, and M.A. Pomerantz (1980) *Nature*, **288**, 541.

Grevesse, N. (1984) *Physica Scripta*, **T8**, 49.

Gribov, V. and B. Pontecorvo (1969) *Phys. Lett.*, **B28**, 493.

Griest, K. and D. Seckel (1987) *Nucl. Phys.*, **B283**, 681.

Griffiths, G.M., M. Lal, and C.D. Scarfe (1963) *Can. J. Phys.*, **41**, 495.

Guzik, J.A., L.A. Willson, and W.M. Brunish (1987) *Ap. J.*, **319**, 957.

Hailes, C. (1987) in *Proceedings of Symposium on Interstellar Processes*, Grand Teton National Park, Wyomming, edited by D.J. Hollenbach and H.A. Thronson, Jr. (Dordecht: Reidel) p. 171 (see especially Section 3.1).

Haines, T.J., *et al.* (1986) *Phys. Rev. Lett.*, **57**, 1986.

Halprin, A. (1986) *Phys. Rev. D*, **34**, 3462.

Halzen, F. and A.D. Martin (1984) *Quarks and Leptons: An Introductory Course in Modern Particle Physics* (New York: Wiley).

Hampel, W. (1981) in *Neutrino 81, Proceedings of the 1981 International Conference on Neutrino Physics and Astrophysics*, Honolulu, Hawaii, edited by R.J. Cence, E. Ma, and A. Roberts (Honolulu: University of Hawaii) p. 6.

Hampel, W. (1985) in *Solar Neutrinos and Neutrino Astronomy*, edited by M.L. Cherry, W.A. Fowler, and K. Lande (New York: American Institute of Physics) Conference Proceedings No. 126, p. 162.

Hampel, W. (1986) in *Weak and Electromagnetic Interactions in Nuclei, Proceedings of the International Symposium*, Heidelberg, Germany, edited by H.V. Klapdor (Berlin: Springer-Verlag) p. 718.

Harari, H. and Y. Nir (1987) *Nucl. Phys.*, **B292**, 251.

Hardie, R.E. *et al.* (1984) *Phys. Rev. C*, **29**, 1199.

Haubold, H.J. and E. Gerth (1983) *Astron. Nacht*, **304**, 299.

Hawking, S. (1971) *Mon. Not. R. Astron. Soc.*, **152**, 75.

Haxton, W.C. (1981) *Nucl. Phys.*, **A367**, 517.

Haxton, W.C. (1986) *Phys. Rev. Lett.*, **57**, 1271.

Haxton, W.C. (1987a) *Phys. Rev. D*, **35**, 2352.

Haxton, W.C. (1987b) *Phys. Rev. D*, **36**, 2283.

Haxton, W.C. (1988a) *Phys. Rev. Lett.*, **60**, 768.

Haxton, W.C. (1988b) *Phys. Rev. C*, **38**, 2474.

Hayashi, C. (1961) *Pub. Astro. Soc. Japan*, **13**, 450.

Hayashi, C. (1966) *Ann. Rev. Astron. Astrophys.*, **4**, 171.

Heiles, C. (1987) in *Interstellar Processes,* edited by D.J. Hollenbach, and H.A. Thronson, Jr. (New York: Reidel) p. 171.

Henning, W., *et al.* (1985) in *Solar Neutrinos and Neutrino Astronomy,* edited by M.L. Cherry, W.A. Fowler, and K. Lande (New York: American Institute of Physics) Conference Proceedings No. 126, p. 203.

Hickey, J.R., L.L. Stowe, H. Jacobowitz, P. Pellegrino, R.H. Maschhoff, F. House, and T.H. Vonder Haar, *et al.* (1980) *Science*, **208**, 281.

Hilgemeir, M., *et al.* (1988) *Zeits für Physik A*, **329**, 243.

Hirata, K.S., T. Kajita, M. Koshiba, *et al.* (1987) *Phys. Rev. Lett.*, **58**, 1490.

Hirata, K.S., T. Kajita, M. Koshiba, *et al.* (1988a) *Phys. Rev. D*, **38**, 448.

Hirata, K.S., T. Kajita, T. Kifune, *et al.* (1988b) *Phys. Rev. Lett.*, **61**, 2653.

Hiroi, S., H. Sakuma, T. Yanagida, and M. Yoshimura (1987a) *Phys. Lett.*, **B198**, 403.

Hiroi, S., H. Sakuma, T. Yanagida, and M. Yoshimura (1987b) *Prog. Theor. Phys.*, **78**, 1428.

Hodge, P. (1987) *Galaxies* (Cambridge: Harvard University Press).

't Hooft, G. (1971) *Phys. Lett.*, **B37**, 195.

Holmgren, H.D. and R.I. Johnston (1958) *Bull. Am. Phys. Soc. II*, **3**, 26.

Holmgren, H.D. and R.I. Johnston (1959) *Phys. Rev.*, **113**, 1556.

Holjevic, S., B.A. Logan, and A. Ljubicic (1987) *Phys. Rev. C*, **35**, 341.

Hope, F. (1984) *Phys. Rev. Lett.*, **52**, 1528.

ICARUS Collaboration (1988) ICARUS I: An Optimized, Real Time Detector of Solar Neutrino's, CERN Preprint (March 21).

Huebner, W.F. (1986) in *Physics of the Sun,* edited by P.A. Sturrock, T.E. Holzer, D.M. Mihala, and R.K. Ulrich (Dordrecht: Reidel) Vol. I, p. 33.

Huebner, W.F., A.L. Merts, N.H. Magee, Jr., and M.F. Argo (1977) in *Astrophysical Opacity Library* (Los Alamos Scientific Laboratory Report LA-6760-M).

Huffman, W.A., J.M. LoSecco, and C. Rubbia (1979) *IEEE Trans. Nucl. Sci.*, **NS-26**, 64.

Hurst, G.S., *et al.* (1984) *Phys. Rev. Lett.*, **53**, 1116.

Hurst, G.S., C.H. Chen, S.D. Kramer, and S.L. Allman (1985) in *Solar Neutrinos and Neutrino Astronomy,* edited by M.L. Cherry, W.A. Fowler, and K. Lande (New York: American Institute of Physics) Conference Proceedings No. 126, p. 152.

Huzita, H. (1987) *Mod. Phys. Lett. A*, **2**, 905.

IAU Symposium 105 (1984) *Observational Tests of the Stellar Evolution Theory* (Geneva, Switzerland) edited by A. Maeder and A. Renzini (Dordrecht: Reidel).

Iben, I., Jr. (1967) *Ann. Rev. Astron. Astrophys.*, **5**, 571.

Iben, I., Jr. (1974) *Ann. Rev. Astron. Astrophys.*, **12**, 215.

Ichimaru, S., H. Iyetomi, and S. Tanaka (1987) *Phys. Reports*, **149**, 91.

Isaak, G.R., *et al.* (1984) *Mem. Soc. Astron. Ital.*, **55**, 99.

Itoh, N. and Y. Kohyama (1981) *Ap. J.*, **246**, 989.

Jenkins, T.L. (1966) *A Proposed Experiment for the Detectors of Solar Neutrinos*, Case University Proposal (unpublished).

Joseph, C.L. (1984) *Nature*, **311**, 254.

Joss, P.C. and S.A. Rappaport (1984) *Ann. Rev. Astron. Astrophys.*, **22**, 537.

Kajino, T., H. Toki, and S.M. Austin (1987) *Ap. J.*, **319**, 531.

Kajita, T., *et al.* (1985) *J. Phys. Soc. Jpn*, **54**, 4065.

Kajita, T., *et al.* (1986) *J. Phys. Soc. Jpn*, **55**, 711.

Kavanagh, R.W. (1972) in *Cosmology, Fusion, and Other Matters*, edited by F. Reines (Boulder: Colorado Associated University Press) p. 169.

Kayser, B. (1981) *Phys. Rev. D*, **24**, 110.

Kelly, F.J. and H. Uberall (1966) *Phys. Rev. Lett.*, **16**, 145.

Kienle, P. (1988) *Nucl. Inst. Meth.*, **A271**, 277.

Kim, C.W., J. Kim, and W.K. Sze (1988) *Phys. Rev. D*, **37**, 1072.

Kim, C.W., S. Nussinov, and W.K. Sze (1987) *Phys. Rev. Lett.*, **184B**, 403.

Kim, C.W., W.K. Sze, and S. Nussinov (1987) *Phys. Rev. D*, **35**, 4014.

Kirsten, T. (1984) *Inst. Phys. Conf. Ser.*, **71**, 251.

Kirsten, T. (1986) (1986) in *'86 Massive Neutrinos in Astrophysics and Particle Physics, Proceedings of the Sixth Moriond Workshop*, edited by O. Fackler and J. Tran Thanh Van (Gif-sur-Yvette: Editions Frontiéres) p. 119.

Kocharov, G.E. (1972) Ioffe Physico – Technical Institute Report No. 298 (Leningrad).

Kolb, E.W., M.S. Turner, and T.P. Walker (1986) *Phys. Lett.*, **175B**, 478.

Kolb, E.W., A. Stebbins, and M. Turner (1987) *Phys. Rev. D*, **35**, 3598.

Konopinski, E.J. (1966) *Theory of Beta Radioactivity* (Oxford: Clarendon) p. 399.

Kopylov, A.V. and V.N. Garvin (1984) *Pis'ma Astron. Zh.*, **10**, 154.

Kotov, V.A. (1986) *Solar Phys.*, **100**, 53.

Krastev, P.I. and S.T. Petcov (1988) *Phys. Lett.*, **207B**, 64.

Krauss, A., H.W. Becker, H.P. Trautvetter, and C. Rolfs (1987) *Nucl. Phys.*, **A467**, 273.

Krauss, L.M. (1987) *Nature*, **329**, 689.

Krauss, L.M., K. Freese, D.N. Spergel, and W.H. Press (1985) *Ap. J.*, **299**, 1001.

Krauss, L.M., S.L. Glashow, and D.N. Schramm (1984) *Nature*, **310**, 191.

Krauss, L.M. and S. Tremaine (1988) *Phys. Rev. Lett.*, **60**, 176.

Krauss, L.M. and F. Wilczek (1985) *Phys. Rev. Lett.*, **55**, 122.

Krauss, L.M. and F. Wilczek (1985) *Prog. Theor. Phys.*, **78**, 1428.

Krofcheck, D., *et al.* (1985) *Phys. Rev. Lett.*, **55**, 1051.

Krofcheck, D., *et al.* (1987) *Phys. Lett.*, **B189**, 299.

Krook, M. and T.T. Wu (1976) *Phys. Rev. Lett.*, **36**, 1107.

Kuchowicz, B. (1976) *Rep. Prog. Phys.*, **39**, 291.

Kuhn, J.R., K.G. Libbrecht, and R.H. Dicke (1986) *Nature*, **319**, 128.

Kume, H. *et al.* (1983) *N.I.M.*, **205**, 443.

Kuo, T.K., and J. Pantaleone (1986) *Phys. Rev. Lett.*, **57**, 1805.

Kuz'min, V.A. (1966) *Sov. Phys. - JETP*, **22**, 1050.

Kuz'min, V.A. and G.T. Zatsepin (1966) in *Proceedings of the Sixteenth International Cosmic Ray Conference*, Kyoto, Japan (Tokyo: University of Tokyo) p. 1023.

Kuzminov, V.V., A.A. Pomansky, and V.L. Chihladze (1988) *Nucl. Inst. Meth.*, **A271**, 257.

Kyuldjev, A.V. (1984) *Nucl. Phys.*, **B243**, 387.

Lamb, D.Q., F. Melia, and T.J. Loredo (1988) in *Supernova 1987A in the Large Magellanic Cloud, Proceedings of the Fourth George Mason Astronomy Workshop*, Fairfax, Virginia, edited by M. Kafatos and A. Michalitsianos (Cambridge: Cambridge University Press) p. 204.

Lanczos, K. (1964) *J. Soc. Ind. Appl. Math.*, **B1**, 86.

Landau, L.D. (1932) *Phys. Z. U.S.S.R.*, **1**, 426.

Langacker, P., *et al.* (1983) *Phys. Rev. D*, **27**, 1228.

Langacker, P. (1986) in *Weak and Electromagnetic Interactions in Nuclei, Proceedings of the International Symposium*, Heidelberg, Germany, edited by H.V. Klapdor (Berlin: Springer-Verlag) p. 879.

Langacker, P., S.T. Petcov, G. Steigman, and S. Toshev (1987) *Nucl. Phys.*, **B282**, 589.

Lanou, R.E., H.J. Maris, and G.M. Seidel (1987) *Phys. Rev. Lett.*, **58**, 2498.

Lanzerotti, L.J. and R.S. Raghavan (1981) *Nature*, **293**, 122.

Lattimer, J.M. and J. Cooperstein (1988) *Phys. Rev. Lett.*, **61**, 23.

Lattimer, J.M. and A. Yahil (1988) in *Origin and Distribution of the Elements*, edited by G.J. Mathews (Singapore: World Scientific Press) p. 444.

Leckrone, D.S. (1971) *Astron. Astrophys.*, **11**, 387.

Leibacher, J.W., R.W. Noyes, J. Toomre, and R.K. Ulrich (1985) *Sci. Am.*, **253**, 48.

Leibacher, J.W., and R.F. Stein (1971) *Astrophys. Lett.*, **7**, 191.

Leighton, R.B., R.W. Noyes, and G.W. Simon (1962) *Ap. J.*, **135**, 474.

Lewis, R.R. (1980) *Phys. Rev. D*, **21**, 663.

Lim, C.-S. and W.J. Marciano (1988) *Phys. Rev. D*, **37**, 1368.

Lin, P.B. and H.S. Hudson (1976) *Solar Phys.*, **50**, 153.

Lingenfelter, R.E., R. Ramaty, R.J. Murphy, and B. Kozlovsky, (1985) in *Solar Neutrinos and Neutrino Astronomy*, edited by M.L. Cherry, W.A. Fowler, and K. Lande (New York: American Institute of Physics) Conference Proceedings No. 126, p. 121.

Littleton, J.E., H.M. Van Horn, and H.L. Helfer (1972) *Ap. J.*, **173**, 677.

Loeb, A. (1989) *Phys. Rev. D* (to be published).

Longo, M.J. (1988) *Phys. Rev. Lett.*, **60**, 173.

Lowry, M.M., R.T. Kouzes, F. Loeser, A.B. McDonald, and R.A. Naumann (1987) *Phys. Rev. C*, **35**, 1950.

Lui, J. (1987) *Phys. Rev. D*, **35**, 3447.

Luyten, W.J. (1968) *Mon. Not. R. Astron. Soc.*, **139**, 221.

MacDonald, W.M., M.N. Rosenbluth, and W. Chuck (1957) *Phys. Rev.*, **107**, 305.

Mahler, H.J., P.J. Doe, and H.H. Chen (1983) *IEEE Trans. Nucl. Sci.*, **30**, 86.

Maki, Z., M. Nakagawa, and S. Sakata (1962) *Prog. Theor. Phys.*, **28**, 870.

Mannheim, P.D. (1988) *Phys. Rev. D*, **37**, 1935.

Marciano, W.J. and Z. Parsa (1986) *Ann. Rev. Nucl. Part. Sci.*, **36**, 171.

Martin, M.J. and P.H. Blickart-Toft (1970) *Nucl. Data Tables A*, **8**, 1.

Martoff, C.J. (1987) *Science*, **237**, 507.

Marx, G. and I. Lux (1970) *Acta. Phys. Acad. Sci. Hung.*, **28**, 63.

Matthews, J. (1988) in *Supernova 1987A in the Large Magellanic Cloud, Proceedings of the Fourth George Mason Astronomy Workshop*, Fairfax, Virginia, edited by M. Kafatos and A. Michalitsianos (Cambridge: Cambridge University Press) p. 151.

Maxwell, J.C. (1890) in *The Scientific Papers of James Clark Maxwell*, edited by W.D. Niven, (New York: Dover), Vol. 2, p. 37.

Mayle, R., J.R. Wilson, and D.N. Schramm (1987) *Ap. J.*, **318**, 288.

Mazurek, T.J. (1974) *Nature*, **252**, 287.

McCuskey, S.W. (1966) *Vistas Astr.*, **7**, 141.

McNaught, R.M. (1987) IAU Circular No. 4316.

Messiah, A. (1986) (1986) in *'86 Massive Neutrinos in Astrophysics and Particle Physics, Proceedings of the Sixth Moriond Workshop*, edited by O. Fackler and J. Tran Thanh Van (Gif-sur-Yvette: Editions Frontiéres) p. 373.

Michaud, G. (1985) in *Solar Neutrinos and Neutrino Astronomy*, edited by M.L. Cherry, W.A. Fowler, and K. Lande (New York: American Institute of Physics) Conference Proceedings No. 126, p. 75.

Mikheyev, S.P. and A.Yu. Smirnov (1986a) *Sov. J. Nucl. Phys.*, **42**, 913.

Mikheyev, S.P. and A.Yu. Smirnov (1986b) *Sov. Phys. - JETP*, **64**, 4.

Mikheyev, S.P. and A.Yu. Smirnov (1986c) *Nuovo Cimento*, **9C**, 17.

Mikheyev, S.P. and A.Yu. Smirnov (1986d) in *Proceedings of Twelfth International Conference on Neutrino Physics and Astrophysics*, edited by T. Kitagaki and H. Yuta (Singapore: World Scientific) p. 177.

Mikheyev, S.P. and A.Yu. Smirnov (1988) *Sov. Phys. - JETP*, **65**, 230.

Moe, M.K. and F. Reines (1965) *Phys. Rev. B*, **140**, 992.

Morinaga, H. (1988) *Nucl. Inst. Meth.*, **A271**, 256.

Mössbauer, R. (1988) *Nucl. Inst. Meth.*, **A271**, 237.

Nakahata, M. (1988) PhD. Thesis, Faculty of Science, University of Tokyo, (preprint UT-ICEPP-88-01).

Nakahata, M. *et al.* (1986) *Phys. Rev.*, **55**, 3786.

Narayan, R. (1987) *Ap. J.*, **319**, 162.

Nauenberg, M. (1987) *Phys. Rev. D*, **36**, 1080.

Newman, M.J. (1986) in *Physics of the Sun*, edited by P.A. Sturrock, T.E. Holzer, D.M. Mihala, and R.K. Ulrich (Dordrecht: Reidel) Vol. III, p. 33.

Nicolaidis, A. (1988) *Phys. Lett.*, **B200**, 553.

Nolte, E. and M.K. Pavićević (editors) (1988) *Solar Neutrino Detection, Proceedings of the International Conference on Solar Neutrino Detection with ^{205}Tl and Related Topics*, Dubrovnik, Yugoslavia (Amsterdam: North Holland).

Nomoto, K. (1984) *Ap. J.*, **277**, 791.

Norman, E.B. and A.G. Seamster (1979) *Phys. Rev. Lett.*, **43**, 1226.

Notzold, D. (1987) *Phys. Lett.*, **B196**, 315.

Nozawa, S. (1987) private communication.

Nozawa, S., Y. Kohyama, T. Kaneta, and K. Kubodera (1986) *J. Phys. Soc. Jpn*, **55**, 2636.

Nussinov, S. (1976) *Phys. Lett.*, **B63**, 201.

Nussinov, S. (1987) *Phys. Rev. Lett.*, **59**, 2401.

Nussinov, S., *et al.* (1987) *Nature*, **329**, 134.

Okun, L.B. (1982) *Leptons and Quarks* (Amsterdam: North Holland) p. 130.

Okun, L.B. (1986) *Sov. J. Nucl. Phys.*, **44**, 546.

Oyama, Y., (1987) *Phys. Rev. Lett.*, **59**, 2604.

Oyama, Y., *et al.* (1986) *Phys. Rev. Lett.*, **56**, 991.

Oyama, Y., *et al.* (1987) *Phys. Rev. Lett.*, **59**, 2604.

Paczyński, B. (1984) *Ap. J.*, **284**, 670.

Pakvasa, S. and K. Tennakone (1972) *Phys. Rev.*, **28**, 1415.

Pallé, P.L., J.C. Perez, C. Regulo, T. Roca Cortes, G.R. Isaak, C.P. McLeod, and H.B. van der Raay (1987) *Astron. Astrophys.*, **170**, 114.

Parke, S.J. (1986a) *Phys. Rev. Lett.*, **57**, 1275.

Parke, S.J. (1986b) in *Proceedings Fourteenth SLAC Summer Institute on Particle Physics*, edited by E.C. Brennan (Stanford: SLAC) p. 349.

Parke, S.J. (1988) private communication.

Parke, S.J. and T.P. Walker (1986) *Phys. Rev. Lett.*, **57**, 2322.

Parker, E. (1974) *Astron. Space Sci.*, **31**, 261.

Parker, P.D. (1972) *Ap. J.*, **175**, 261.

Parker, P.D. (1986) in *Physics in the Sun*, edited by P.A. Sturrock, T.E. Holzer, D.M. Mihalas, and R.K. Ulrich (New York: Reidel) Vol I, p. 15.

Parker, P.D., J.N. Bahcall, and W.A. Fowler (1964) *Ap. J.*, **139**, 602.

Pavićević, M.K. (1988) *Nucl. Inst. Meth.*, **A271**, 287.

Peak, L.S and A.M. Bakich (1985) *Nucl. Inst. Meth.*, **A240**, 429.

Peimbert, M. and S. Torres-Peimbert (1977) *Mon. Not. R. Astron. Soc.*, **179**, 217.

Pekeris, C.L. (1939) *Ap. J.*, **88**, 189.

Petcov, S.T. (1987) *Phys. Lett.*, **B191**, 299.

Petcov, S.T. (1988) *Phys. Lett.*, **B200**, 373.

Petcov, S.T. and S. Toshev (1987) *Phys. Lett.*, **B187**, 120.

Pfeiffer, L., A.P. Mills, Jr., and E.A. Chandross (1978) in *Proceedings of Informal Conference on Status and Future of Solar Neutrino Research*, edited by G. Friedlander, (Upton: Brookhaven National Labortory) Report No. 50879, Vol. 2, p. 31.

Pietschmann, H. (1983) *Weak Interactions Formulae, Results and Derivations* (Vienna: Springer-Verlag).

Piran, T. and D.N. Spergel (1988) in *The Proceedings of the Moriond Astrophysics Meeting on Dark Matter*, edited by J. Audouze and J. Tran Thanh Van (in press).

Pizzochero, P. (1987) *Phys. Rev. D*, **36**, 2293.

Pollock, E.L. and B.J. Adler (1978) *Nature*, **275**, 41.

Pomansky, A.A. (1986) *Nucl. Inst. Meth.*, **B17**, 406.

Pomansky, A.A. (1988) *Nucl. Inst. Meth.*, **A271**, 254.

Pontecorvo, B. (1946) *Chalk River Report PD-205*.

Pontecorvo, B. (1958) *Sov. Phys. - JETP*, **6**, 429.

Pontecorvo, B. (1968) *Sov. Phys. - JETP*, **26**, 984.

Popper, D.M. (1980) *Ann. Rev. Astron. Astrophys.*, **18**, 115.

Poskanzer, A.M., R. McPherson, R.A. Esterlund, and P.L. Reeder (1966) *Phys. Rev.*, **152**, 995.

Prentice, A.J.R. (1973) *Mon. Not. R. Astron. Soc.*, **163**, 331.

Prentice, A.J.R. (1976) *Astron. Astrophys.*, **50**, 59.

Press, W.H. (1981) *Ap. J.*, **245**, 286.

Press, W.H. (1986) in *Physics of the Sun*, edited by P.A. Sturrock, T.E. Holzer, D.M. Mihala, and R.K. Ulrich (Dordrecht: Reidel) Vol. I, p. 77.

Press, W.H. and G.B. Rybicki (1981) *Ap. J.*, **248**, 751.

Press, W.H. and D.N. Spergel (1985) *Ap. J.*, **296**, 679.

Press, W.H. and D.N. Spergel (1988) in *Dark Matter in the Universe* edited by J.N. Bahcall, T. Piran, and S. Weinberg (Singapore: World Scientific) p. 206.

Pretzl, K., N. Schmitz, and L. Stodolsky (1987) (editors) *Low Temperature Detectors for Neutrinos and Dark Matter*, (Heidelberg: Springer-Verlag).

Raby, S. and G.B. West (1987) *Nucl. Phys.*, **B292**, 793.

Raby, S. and G.B. West (1988) *Phys. Lett.*, **B202**, 47.

Racah, G. (1932) *Nature*, **129**, 723.

Raffelt, G. and D. Seckel (1988) *Phys. Rev. Lett.*, **60**, 1793.

Raghavan, R.S. (1976) *Phys. Rev. Lett.*, **37**, 259.

Raghavan, R.S. (1978a) in *Proceedings of Informal Conference on Status and Future of Solar Neutrino Research*, edited by G. Friedlander, (Upton: Brookhaven National Labortory) Report No. 50879, Vol. 2, p. 1.

Raghavan, R.S. (1978b) in *Proceedings of Informal Conference on Status and Future of Solar Neutrino Research*, edited by G. Friedlander, (Upton: Brookhaven National Labortory) Report No. 50879, Vol. 2, p. 270.

Raghavan, R.S. (1979) *Liquid Argon Detectors: A new approach to observation of neutrinos from meson factories and collapsing stars*, Bell Laboratory Technical Memorandum TM #79-1131-31.

Raghavan, R.S. (1986) *Phys. Rev. D*, **34**, 2088.

Raghavan, R.S., X.-G. He, and S. Paksava (1988) *Phys. Rev. D*, **38**, 1317.

Raghavan, R.S. and S. Pakvasa (1988) *Phys. Rev. D*, **37**, 849.

Raghavan, R.S., S. Pakvasa, and B.A. Brown (1986) *Phys. Rev. Lett.*, **57**, 1801.

Raghavan, R.S., *et al.* (1988) *Design Concepts for Borex (Boron Solar Neutrino Experiment)*, AT&T Bell Laboratories (unpublished).

Rakavy, G. and G. Shaviv (1967) *Ap. J.*, **148**, 803.

Rapaport, J., *et al.* (1981) *Phys. Rev. Lett.*, **47**, 1518.

Rapaport, J., *et al.* (1985) *Phys. Rev. Lett.*, **54**, 2325.

Ratnatunga K.U. and S. van den Bergh (1989) *Ap. J.*, (to be published).

Raychaudhuri, P. (1984) *Solar Phys.*, **93**, 397.

Raychaudhuri, P. (1986) *Solar Phys.*, **106**, 421.

Reines, F. (1967) *Proc. Roy. Soc.*, **A301**, 159.

Reines, F. and W.R. Kropp (1964) *Phys. Rev. Lett.*, **12**, 457.

Rhodes, E.J., Jr., R.K. Ulrich, and G.W. Simon (1977) *Ap. J.*, **218**, 901.

Robertson, R.G.H. (1986, 1988) private communication

Rolfs, C. (1973) *Nucl. Phys.*, **A217**, 29.

Rolfs, C. and R.E. Azuma (1974) *Nucl. Phys.*, **A227**, 291.

Rolfs, C. and R.W. Kavanagh (1986) *Nucl. Phys.*, **A455**, 179.

Rolfs, C. and W. Rodney (1974) *Nucl. Phys.*, **A235**, 450.

Rolfs, C. and W. Rodney (1988) *Cauldrons in the Cosmos* (Chicago: University of Chicago Press).

Rolfs, C. and H.P. Trautvetter (1978) *Ann. Rev. Nucl. Part. Sci.* **28**, 115.

Rood, R.T. (1978) in *Proceedings of Informal Conference on Status and Future of Solar Neutrino Research*, edited by G. Friedlander, (Upton: Brookhaven National Labortory) Report No. 50879, Vol. 1, p. 175.

Rood, R.T. and R.K. Ulrich (1974) *Nature*, **252**, 366.

Rose, M.E. (1936) *Phys. Rev.*, **49**, 727.

Rose, M.E. (1957) in *Elementary Theory of Angular Momentum* (New York: Wiley).

Rose, M.E. (1961) *Relativistic Electron Theory* (New York: Wiley).

Rosen, S.P. and J.M. Gelb (1986) *Phys. Rev. D*, **34**, 969.

Rosen, S.P. and J.M. Gelb (1987) in *Proceedings of the Twenty-Third International Conference on High Energy Physics*, Berkeley, California, edited by. S. Loken (Singapore: World Scientific) p. 909.

Rosenbluth, M.N. and J.N. Bahcall (1973) *Ap. J.*, **184**, 9.

Ross, J.E. and L.H. Aller (1976) *Science*, **191**, 1223.

Ross, G.G. and E. Segré (1987) *Phys. Lett.*, **B197**, 45.

Rowley, J.K. (1978) in *Proceedings of Informal Conference on Status and Future of Solar Neutrino Research*, edited by G. Friedlander, (Upton: Brookhaven National Labortory) Report No. 50879, p. 265.

Rowley, J.K., B.T. Cleveland, and R. Davis, Jr., (1985) in *Solar Neutrinos and Neutrino Astronomy*, edited by M.L. Cherry, W.A. Fowler, and K. Lande (New York: American Institute of Physics) Conference Proceedings No. 126, p. 1.

Roxburgh, I.W. (1974) *Nature*, **247**, 220.

Roxburgh, I.W. (1985a) *Solar Phys.*, **100**, 21.

Roxburgh, I.W. (1985b) in *Solar Neutrinos and Neutrino Astronomy*, edited by M.L. Cherry, W.A. Fowler, and K. Lande (New York: American Institute of Physics) Conference Proceedings No. 126, p. 88.

Roy, A., *et al.* (1983) *Phys. Rev. D*, **28**, 770.

Rubbia, C. (1977) *The Liquid Argon Time Projection Chamber: A new concept for neutrino detectors*, EP Internal Report 77–8 CERN.

Rubbia, C., *et al.* (1985) *ICARUS: A Proposal for the Gran Sasso Laboratory*, INFN Publication INFN/AE-85-7.

Ruff, I. and J. Liszi, and K. Gombos (1985) *Ap. J.*, **289**, 409.

Ruff, I. and J. Liszi (1985) *Chem. Phys. Lett.*, **116**, 335.

Russell, H.N. (1919) *Publ. Astron. Soc. Pacific*, **31**, 129.

Sadoulet, B., J. Rich, M. Spiro, and D.O. Caldwell (1988) *Ap. J. Lett.*, **324**, L75.

Sakurai, K. (1979) *Nature*, **269**, 401.

Salam, A. (1968) in *Elementary Particle Theory*, edited by N. Svartholm (Stockholm: Almqvist and Wiksells) p. 367.

Salpeter, E.E. (1954) *Aust. J. Phys.*, **7**, 373.

Salpeter, E.E. (1957) *Rev. Mod. Phys.*, **29**, 244.

Salpeter, E.E. (1970) *Nature*, **225**, 165.

Sandage, A. (1982) *Ap. J.*, **252**, 553.

Sarantakos, S., A. Sirlin, and W. J. Marciano (1983) *Nucl. Phys.*, **B217**, 84.

Sato, K. (1975) *Prog. Theor. Phys.*, **53**, 595.

Sato, K. and H. Suzuki (1987a) *Phys. Rev. Lett.*, **58**, 2722.
Sato, K. and H. Suzuki (1987b) *Phys. Lett.*, **B196**, 267.
Sawyer, G.A., and J. A. Phillips (1953) LASL Report No. 1578.
Schaeffer, R., Y. Declais, and S. Jullian (1987) *Nature*, **330**, 6144.
Schatzman, E. (1951) *C. R.*, **232**, 1740.
Schatzman, E. (1969) *Astrophys. Lett.*, **3**, 139.
Schatzman, E. (1985) in *Solar Neutrinos and Neutrino Astronomy*, edited by M.L. Cherry, W.A. Fowler, and K. Lande (New York: American Institute of Physics) Conference Proceedings No. 126, p. 69.
Schatzman, E. and A. Maeder (1981) *Astron. Astrophys.*, **96**, 1.
Schramm, D.N. (1987) *Comm. Nucl. Part. Phys.*, **17**, 239.
Schramm, D.N., R. Mayle, and J.R. Wilson (1986) *Nuovo Cimento*, **9C**, 493.
Schröder, U., *et al.* (1987) *Nucl. Phys.*, **A467**, 240.
Schwarzschild, M. (1958) *Structure and Evolution of the Stars* (Princeton: Princeton University Press).
Schwarzschild, M. and R. Härm (1965) *Ap. J.*, **142**, 855.
Schwarzschild, M. and R. Härm (1973) *Ap. J.*, **184**, 5.
Scott, R.D. (1976) *Nature*, **264**, 729.
Sears, R.L. (1964) *Ap. J.*, **140**, 477.
Severny, A.B. (1984) *Nature*, **307**, 247.
Sextro, R.G., R.A. Gough, and J. Cerny (1974) *Nucl. Phys.*, **A234**, 130.
Shapiro, S.L. and S.A. Teukolsky (1983) *Black Holes, White Dwarfs, and Neutron Stars* (New York: Wiley).
Shaviv, G. and J.N. Bahcall (1969) *Ap. J.*, **155**, 135.
Shaviv, G. and G. Beaudet (1968) *Astrophys. Lett.*, **2**, 17.
Shaviv, G. and E.E. Salpeter (1968) *Phys. Rev. Lett.*, **21**, 1602.
Shaviv, G. and E.E. Salpeter (1971) *Ap. J.*, **165**, 171.
Shelton, I. (1987) IAU Circular No. 4316.
Shirley, D.A. (1964) *Rev. Mod. Phys.*, **36**, 339.
Sienkiewicz, R., B. Paczyński, and S.J. Ratcliff (1988) *Ap. J.*, **326**, 392.
Sienkiewicz, R., J.N. Bahcall, and B. Paczyński, (1989) *Ap. J.* to be published .
Sinclair, D., *et al.* (1986) *Nuovo Cimento*, **C9**, 308.
Sotirovsky, P (1988) *Nucl. Inst. Meth.*, **A271**, 238.
Spergel, D.N. and J.N. Bahcall (1987) *Phys. Lett.*, **B200**, 366.
Spergel, D.N., T. Piran, A. Loeb, J. Goodman, and J.N. Bahcall (1987) *Science*, **237**, 1471.
Spergel, D.N. and W.H. Press (1985) *Ap. J.*, **294**, 663.
Spruit, H.C. (1987) in *The Internal Solar Angular Velocity*, edited by B.R. Durney and S. Sofia (Dordrecht: Reidel) p. 185.
Stacey, F.D. (1985) in *Physics of the Earth* (New York: Wiley) p. 17.
Stefanov, M.A. (1988) *JETP Lett.*, **47**, 1.

Steigman, G., C.L. Sarazin, H. Quintana, and J. Faulkner (1978) *A.J.*, **83**, 1050.

Steinberg, R.I. (1976) *Bull. Am. Astron. Soc.*, **21**, 528.

Steinberg, R.I., K. Kwiatkowski, W. Meanhaut, and N.S. Wall (1975) *Phys. Rev. D*, **12**, 2582.

Stodolsky, L. (1988) *Phys. Lett.*, **B201**, 353.

Stothers, R. and D. Ezer (1973) *A.J.*, **13**, 45.

Subramanian, A. (1983) *Current Science*, **52**, 342.

Sur, B. and R.N. Boyd (1985) *Phys. Rev. Lett.*, **54**, 485.

Suslov, Y.P. (1968) *Proceedings of the Conference on Electron Capture and Higher Order Processes in Nuclear Decays*, Debrecen, Hungary, edited by D. Berenyi (Budapest: Etvos Lorand Physical Society) p. 21.

Sutherland, P., *et al.* (1976) *Phys. Rev. D*, **13**, 2700.

Suzuki, A. (1986) in *Proceedings of Twelfth International Conference on Neutrino Physics and Astrophysics*, Sendai, Japan, edited by T. Kitagaki and H. Yuta (Singapore: World Scientific) p. 306.

Suzuki, A. (1987) in *Proceedings of the Workshop on Elementary-Particle Picture of the Universe*, edited by M. Yoshimura, Y. Totsuka, and K. Nakamura (KEK: National Laboratory for High Energy Physics) p. 136.

Taddeucci, T. N., *et al.* (1982) *Phys. Rev. C*, **25**, 1094.

Taddeucci, N.T., *et al.* (1987) *Nucl. Phys.*, **A469**, 125.

Takahara, M. and K. Sato (1987) *Mod. Phys. Lett.*, **A2**, 293.

Takahashi, K. and R.N. Boyd (1988) *Ap. J.*, **327**, 1009.

Takita, M., *et al.* (1986) *Phys. Rev. D*, **34**, 902.

Tammann, G. (1982) in *Supernovae: A Survey of Current Research*, edited by M.J. Rees and R.J. Stoneham (Dordrecht: Reidel) p. 371.

Tassoul, M. (1980) *Ap. J. Supp.*, **43**, 469.

Tassoul, M. and J.L. Tassoul (1984a) *Ap. J.*, **279**, 384.

Tassoul, M. and J.L. Tassoul (1984b) *Ap. J.*, **286**, 350.

Tayler, R.J. (1970) in *The Stars: Their Structure and Evolution*, Wykeham Science Series, No. 10 (London: Wykeham Publications).

Taylor, J.H. and R.N. Manchester (1977) *Ap. J.*, **215**, 885.

Taylor, J.H. and J.M. Weisberg (1982) *Ap. J.*, **253**, 908.

Tegnér, P.E. and Chr. Bargholtz (1983) *Ap. J.*, **272**, 311.

Toomre, J. (1986) in *Solar Seismology from Space*, edited by R.K. Ulrich, J.W. Harvey, E.J. Rhodes and J. Toomre, (Pasadena: JPL-PUB-84-84).

Toshev, S. (1987a) *Phys. Lett.*, **B196**, 170.

Toshev, S. (1987b) *Phys. Lett.*, **B198**, 551.

Totsuka, Y. (1987) in *Proceedings of Seventh Workshop on Grand Unification, ICOBAN'86*, Toyama, Japan, edited by J. Arafune (Singapore: World Scientific) p. 118.

Totsuka, Y. (1988) (private communication on behalf of the Kamiokande II collaboration).

Tubbs, D.L. and D.N. Schramm (1975) *Ap. J.*, **201**, 467.

Turck-Chièz, S., S. Cahen, M. Cassè, and C. Doom (1988) *Ap. J.*, **355**, 415.

Ulrich, R.K. (1969) *Ap. J.*, **158**, 427.

Ulrich, R.K. (1970) *Ap. J.*, **162**, 993.

Ulrich, R.K. (1971) *Ap. J.*, **168**, 57.

Ulrich, R.K. (1982) *Ap. J.*, **258**, 404.

Ulrich, R.K. (1986) *Ap. J. Lett.*, **306**, L37.

Ulrich, R.K. and E.J. Rhodes, Jr. (1977) *Ap. J.*, **218**, 521.

Ulrich, R.K. and E.J. Rhodes, Jr. (1983) *Ap. J.*, **265**, 551.

Ulrich, R.K., E.J. Rhodes, Jr., S. Tomczyk, P.J. Dumont, and W.M. Brunish (1983) in *Science Underground*, edited by M.M. Nieto, W.C. Haxton, C.M. Hoffman, E.W. Kolb, V.D. Sandberg, and J.W. Toevs (New York: American Institute of Physics) Conference Proceedings No. 96, p. 66.

Vaidya, S.C., *et al.* (1983) *Phys. Rev. D*, **27**, 486.

Vandakurov, Yu.V. (1967) *Astron. Zh.*, **44**, 786.

VandenBerg, D.A. (1983) *Ap. J. Supp.*, **51**, 29.

VandenBerg, D.A. (1985) *Ap. J. Supp.*, **58**, 711.

VandenBerg, D.A. and R.A. Bell (1985) *Ap. J. Supp.*, **58**, 561.

van den Bergh, S., R.D. McClure, and R. Evans (1987) *Ap. J.*, **323**, 44.

Veretyonkin, E.P., V.W. Gavrin, and E.A. Yanovich (1985) *Sov. Atom. Energy*, **58**, 82.

Voloshin, M.B., M.I. Vysotskii, and L.B. Okun (1986a) *Sov. J. Nucl. Phys.*, **44**, 440.

Voloshin, M.B., M.I. Vysotskii, and L.B. Okun (1986b) *Sov. Phys. - JETP*, **64**, 446.

Voloshin, M.B. and M.I. Vysotskii (1986) *Sov. J. Nucl. Phys.*, **44**, 544.

Volvodsky, A.V., V.L. Dadykin, and O.G. Ryazhskaya (1970) *Prib. Tekh. Eksp.*, **1**, 85.

Walker, T.P., E.W. Kolb, and M.S. Turner (1985) Fermi Laboratory Conf. 85/99-A.

Warburton, E.K. (1986) *Phys. Rev. C*, **33**, 303.

Wasserburg, G.J., D.A. Papanastassiou, and T. Lee (1980) in *Early Solar System Processes and the Present Solar System*, Corso Soc. Italiana di Fisica, Bologna.

Wasserburg, G.J., F. Tera, D.A. Papanastassiou, and J.C. Huneke (1977) *Earth Planet. Scie. Lett.*, **35**, 294.

Weinberg, S. (1967) *Phys. Rev. Lett.*, **19**, 1264.

Weinberg, S. (1987) *Int. J. Mod. Phys. A*, 2301.

Weizsaker, C.F. (1937) *Physik Z*, **38**, 176.

Weizsaker, C.F. (1938) *Physik Z*, **39**, 663.

Wielen, R. (1974) *Highlights Astr.*, **3**, 395.

Werntz, C. and J.G. Brennan (1973) *Phys. Rev. C*, **8**, 1545.

Wheeler, J.C. and A.G.W. Cameron (1975) *Ap. J.*, **196**, 601.

Wheeler, J.C. and J.A. Wheeler (1983) in *Science Underground*, edited by M.M. Nieto, W.C. Haxton, C.M. Hoffman, E.W. Kolb, V.D. Sandberg, and J.W. Toevs (New York: American Institute of Physics) Conference Proceedings No. 96, p. 214.

Wilkinson, D.H. (1982) *Nucl. Phys.*, **A377**, 474.

Wilkinson, D.H. and D.E. Alburger (1971) *Phys. Rev. Lett.*, **26**, 1127.

Willson, R.C. C.H. Duncan, and J. Geist (1980) *Science*, **207**, 177.

Willson, R.C., S. Gulkis, M. Janssen, H.S. Hudson, and G.A. Chapman (1981) *Science*, **211**, 700.

Willson, R.C. and J.R. Hickey (1977) in *The Solar Output and Its Variation*, edited by O.R. White (Boulder: Colorado Associated Universities) p. 112.

Willson, L.A., G.H. Bowen, and C. Struck-Marcell (1987) *Comments on Astrophysics*, **XII**, 17.

Wilson, J.R., *et al.* (1986) *Ann. New York Acad. Sci.*, **470**, 267.

Wolfendale, A.W., E.C.M. Young, and R. Davis, Jr. (1972) *Nature*, **238**, 1301.

Wolfenstein, L. (1978) *Phys. Rev. D*, **17**, 2369.

Wolfenstein, L. (1979) *Phys. Rev. D*, **20**, 2634.

Wolfenstein, L. (1986) in *Proceedings of the Twelfth International Conference on Neutrino Physics and Astrophysics*, Sendai, Japan, edited by T. Kitagaki and H. Yuta (Singapore: World Scientific) p. 1.

Wolfsberg, K. *et al.* (1985) in *Solar Neutrinos and Neutrino Astronomy*, edited by M.L. Cherry, W.A. Fowler, and K. Lande (New York: American Institute of Physics) Conference Proceedings No. 126, p. 196.

Woosley, S.E. (1988) *Ap. J.*, **330**, 218.

Woosley, S.E. and T.A. Weaver (1986) *Ann. Rev. Astron. Astrophys.*, **24**, 205.

Woosley, S.E., J.R. Wilson, and R. Mayle (1986) *Ap. J.*, **309**, 19.

Wyatt, A. (1984) *Physica B*, **126**, 392.

Yanagida, T. (1978) *Prog. Theor. Phys.*, **B315**, 66.

Zatsepin, G.T. (1968) *JETP Lett.*, **8**, 205.

Zatsepin, G.T. (1983) in *Proceedings of the Eighth International Workshop on Weak Interactions and Neutrinos*, Javea, Spain, edited by A. Morales (Singapore: World Scientific).

Zatsepin, G.T., A.V. Kopylov, and E.K. Shirokova (1981) *Sov. J. Nucl. Phys.*, **33**, 200.

Zeldovich, Y.B. and Y.B. Guseinov (1965) *JETP Lett.*, **1**, 109.

Zener, C. (1932) *Proc. Roy. Soc. A* **137**, 696.

Zhang, W., *et al.* (1988) *Phys. Rev. Lett.*, **61**, 385.

Zyskind, J.L. and P.D. Parker (1979) *Nucl. Phys.*, **A320**, 404.

Index

The entries in boldface type
contain the primary discussion of
the indicated topics.

A

absorption cross sections for
  neutrinos, **195–214**
  allowed, 195
  angular distribution, 197
  antineutrino cross section, 431
  at individual energies, 209
  atomic effects, 199
  bound state capture, 200
  computation, 201–207
  for solar sources, 207–209
  forbidden effects, 212–214
  from muon decay, 212–214
  $ft_{\frac{1}{2}}$-value, 197–198, 202–205
  general formulae, 195–200
  isotopic analogue states, 205
  (p,n) experiments, 206–207
  uncertainties, 178–182, 209–212
absorption reactions
  measure individual energies, 25,
    401
^{37}Ar
  see also ^{37}Cl experiment

^{71}Ga experiments, 346
^{115}In experiment, 415
  neutrino calibration line, 146
  neutrinos, 146
^{40}Ar experiment, 24–26, 397,
  408–412
  angular distribution of recoil
    electrons, 409
  background effects, 409
  "beam-off", 410
  characteristic gamma rays, 410
  characteristics, 397
  cross section uncertainties, 211
  detection reaction, 409
  events from neutrino–electron
    scattering, 398
  hep neutrinos, 412
  methane replacement, 410
  minimum rate, 412
  MSW effect, 411
  prediction by standard model,
    411
  scattering versus absorption, 411
  smoking gun, 410
  solar neutrino spectrum, 409
  stellar collapse, 457–459
  threshold energy, 409
  time-projection chamber, 409

555